ANNUAL REVIEW OF ECOLOGY AND SYSTEMATICS

EDITORIAL COMMITTEE (1973)

ANNUAL REVIEW OF ECOLOGY AND SYSTEMATICS

RICHARD F. JOHNSTON, *Editor*
University of Kansas

PETER W. FRANK, *Associate Editor*
University of Oregon

CHARLES D. MICHENER, *Associate Editor*
University of Kansas

VOLUME 4

1973

ANNUAL REVIEWS INC.   4139 EL CAMINO WAY   PALO ALTO, CALIFORNIA 94306

ANNUAL REVIEWS INC.
Palo Alto, California, USA

International Standard Book Number: 0-8243-1404-2
Library of Congress Catalog Card Number: 71-135616

Assistant Editor   Kathleen A. Gardner
Indexers   Mary A. Glass
    Susan Tinker
Subject Indexer   Mary Sue Schnell

PRINTED AND BOUND IN THE UNITED STATES OF AMERICA

PREFACE

About two years ago we began inviting colleagues throughout the world to write critical reviews of the significant recent literature in their specific fields of interest. Results appear, in part, in this volume of *ARES.* The intervening time has brought manuscripts of impressive scope, the publication of Volumes 2 and 3 of *ARES,* and plans for Volumes 5 and 6. Because we always have two volumes organized or in production at any given time, we doubtless think that *ARES* offers a more comprehensive view of ecology and systematics than does our average reader; but the view is in fact broad and if it transcends the scope of any one person's interests, that is as it should be.

The proper study of man is ecology, as one pedant insists, and we would suggest only that the danger in this attitude is that there are no definable limits to the subject. But, as our constant readers know, we have resisted the temptations of subsuming the whole world and content ourselves with ecology and systematics as these words are ordinarily meant.

Herewith in Volume 4 you will find assessments of ecosystems, of community structure, of ecologic metastructure, of reproduction and of aggression as ecologic variables, and of aspects of the genetic structure of populations; also of the morphology of protists, plants, and animals and what it may mean to the way in which man classifies them. There are discussions of energy transfer and of the whole range of its effects running the gamut from effects on individual trophic behavior to effects on systems thermodynamics.

We are as usual pleased and flattered to have had the cooperation and assistance of our authors. We appreciate also the help and advice of Robert R. Schutz, formerly of Annual Reviews Inc., in planning this volume, and the multiform help and guidance of Assistant Editors Barbara Murphy and Kathleen Gardner at various stages in the gestation of Volume 4.

THE EDITORS AND EDITORIAL COMMITTEE

CONTENTS

RESILIENCE AND STABILITY OF ECOLOGICAL SYSTEMS

❖ 4050

C. S. Holling

Institute of Resource Ecology, University of British Columbia, Vancouver, Canada

INTRODUCTION

Individuals die, populations disappear, and species become extinct. That is one view of the world. But another view of the world concentrates not so much on presence or absence as upon the numbers of organisms and the degree of constancy of their numbers. These are two very different ways of viewing the behavior of systems and the usefulness of the view depends very much on the properties of the system concerned. If we are examining a particular device designed by the engineer to perform specific tasks under a rather narrow range of predictable external conditions, we are likely to be more concerned with consistent nonvariable performance in which slight departures from the performance goal are immediately counteracted. A quantitative view of the behavior of the system is, therefore, essential. With attention focused upon achieving constancy, the critical events seem to be the amplitude and frequency of oscillations. But if we are dealing with a system profoundly affected by changes external to it, and continually confronted by the unexpected, the constancy of its behavior becomes less important than the persistence of the relationships. Attention shifts, therefore, to the qualitative and to questions of existence or not.

Our traditions of analysis in theoretical and empirical ecology have been largely inherited from developments in classical physics and its applied variants. Inevitably, there has been a tendency to emphasize the quantitative rather than the qualitative, for it is important in this tradition to know not just that a quantity is larger than another quantity, but precisely how much larger. It is similarly important, if a quantity fluctuates, to know its amplitude and period of fluctuation. But this orientation may simply reflect an analytic approach developed in one area because it was useful and then transferred to another where it may not be.

Our traditional view of natural systems, therefore, might well be less a meaningful reality than a perceptual convenience. There can in some years be more owls and fewer mice and in others, the reverse. Fish populations wax and wane as a natural condition, and insect populations can range over extremes that only logarithmic

1

transformations can easily illustrate. Moreover, over distinct areas, during long or short periods of time, species can completely disappear and then reappear. Different and useful insight might be obtained, therefore, by viewing the behavior of ecological systems in terms of the probability of extinction of their elements, and by shifting emphasis from the equilibrium states to the conditions for persistence.

An equilibrium centered view is essentially static and provides little insight into the transient behavior of systems that are not near the equilibrium. Natural, undisturbed systems are likely to be continually in a transient state; they will be equally so under the influence of man. As man's numbers and economic demands increase, his use of resources shifts equilibrium states and moves populations away from equilibria. The present concerns for pollution and endangered species are specific signals that the well-being of the world is not adequately described by concentrating on equilibria and conditions near them. Moreover, strategies based upon these two different views of the world might well be antagonistic. It is at least conceivable that the effective and responsible effort to provide a maximum sustained yield from a fish population or a nonfluctuating supply of water from a watershed (both equilibrium-centered views) might paradoxically increase the chance for extinctions.

The purpose of this review is to explore both ecological theory and the behavior of natural systems to see if different perspectives of their behavior can yield different insights useful for both theory and practice.

Some Theory

Let us first consider the behavior of two interacting populations: a predator and its prey, a herbivore and its resource, or two competitors. If the interrelations are at all regulated we might expect a disturbance of one or both populations in a constant environment to be followed by fluctuations that gradually decrease in amplitude. They might be represented as in Figure 1, where the fluctuations of each population over time are shown as the sides of a box. In this example the two populations in some sense are regulating each other, but the lags in the response generate a series of oscillations whose amplitude gradually reduces to a constant and sustained value for each population. But if we are also concerned with persistence we would like to know not just how the populations behave from one particular pair of starting values, but from all possible pairs since there might well be combinations of starting populations for which ultimately the fate of one or other of the populations is extinction. It becomes very difficult on time plots to show the full variety of responses possible, and it proves convenient to plot a trajectory in a phase plane. This is shown by the end of the box in Figure 1 where the two axes represent the density of the two populations.

The trajectory shown on that plane represents the sequential change of the two populations at constant time intervals. Each point represents the unique density of each population at a particular point in time and the arrows indicate the direction of change over time. If oscillations are damped, as in the case shown, then the trajectory is represented as a closed spiral that eventually reaches a stable equilibrium.

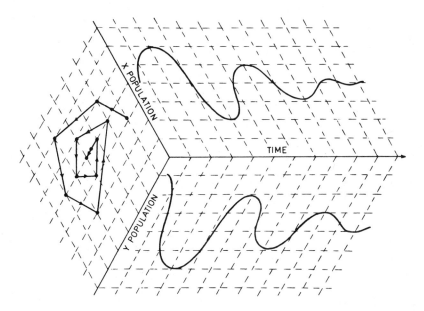

Figure 1 Derivation of a phase plane showing the changes in numbers of two populations over time.

We can imagine a number of different forms for trajectories in the phase plane (Figure 2). Figure 2a shows an open spiral which would represent situations where fluctuations gradually increase in amplitude. The small arrows are added to suggest that this condition holds no matter what combination of populations initiates the trajectory. In Figure 2b the trajectories are closed and given any starting point eventually return to that point. It is particularly significant that each starting point generates a unique cycle and there is no tendency for points to converge to a single cycle or point. This can be termed "neutral stability" and it is the kind of stability achieved by an imaginary frictionless pendulum.

Figure 2c represents a stable system similar to that of Figure 1, in which all possible trajectories in the phase plane spiral into an equilibrium. These three examples are relatively simple and, however relevant for classical stability analysis, may well be theoretical curiosities in ecology. Figures 2d–2f add some complexities. In a sense Figure 2d represents a combination of a and c, with a region in the center of the phase plane within which all possible trajectories spiral inwards to equilibrium. Those outside this region spiral outwards and lead eventually to extinction of one or the other populations. This is an example of local stability in contrast to the global stability of Figure 2c. I designate the region within which stability occurs as the domain of attraction, and the line that contains this domain as the boundary of the attraction domain.

The trajectories in Figure 2e behave in just the opposite way. There is an internal region within which the trajectories spiral out to a stable limit cycle and beyond

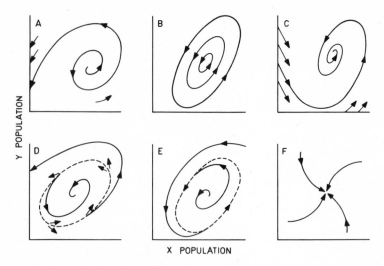

Figure 2 Examples of possible behaviors of systems in a phase plane; (a) unstable equilibrium, (b) neutrally stable cycles, (c) stable equilibrium, (d) domain of attraction, (e) stable limit cycle, (f) stable node.

which they spiral inwards to it. Finally, a stable node is shown in Figure 2f in which there are no oscillations and the trajectories approach the node monotonically. These six figures could be combined in an almost infinite variety of ways to produce several domains of attraction within which there could be a stable equilibrium, a stable limit cycle, a stable node, or even neutrally stable orbits. Although I have presumed a constant world throughout, in the presence of random fluctuations of parameters or of driving variables (Walters 39), any one trajectory could wander with only its general form approaching the shape of the trajectory shown. These added complications are explored later when we consider real systems. For the moment, however, let us review theoretical treatments in the light of the possibilities suggested in Figure 2.

The present status of ecological stability theory is very well summarized in a number of analyses of classical models, particularly May's (23–25) insightful analyses of the Lotka-Volterra model and its expansions, the graphical stability analyses of Rosenzweig (33, 34), and the methodological review of Lewontin (20).

May (24) reviews the large class of coupled differential equations expressing the rate of change of two populations as continuous functions of both. The behavior of these models results from the interplay between (*a*) stabilizing negative feedback or density-dependent responses to resources and predation, and (*b*) the destabilizing effects produced by the way individual predators attack and predator numbers respond to prey density [termed the functional and numerical responses, as in Holling (11)]. Various forms have been given to these terms; the familiar Lotka-Volterra model includes the simplest and least realistic, in which death of prey is caused only by predation, predation is a linear function of the product of prey and predator populations, and growth of the predator population is linearly proportional to the same product. This model generates neutral stability as in Figure 2b, but the assumptions are very unrealistic since very few components are included, there are no explicit lags or spatial elements, and thresholds, limits, and nonlinearities are missing.

These features have all been shown to be essential properties of the predation process (Holling 12, 13) and the effect of adding some of them has been analyzed by May (24). He points out that traditional ways of analyzing the stability properties of models using analytical or graphical means (Rosenzweig & MacArthur 33, Rosenzweig 34, 35) concentrate about the immediate neighborhood of the equilibrium. By doing this, linear techniques of analysis can be applied that are analytically tractable. Such analyses show that with certain defined sets of parameters stable equilibrium points or nodes exist (such as Figure 2c), while for other sets they do not, and in such cases the system is, by default, presumed to be unstable, as in Figure 2a. May (24), however, invokes a little-used theorem of Kolmogorov (Minorksy 26) to show that all these models have either a stable equilibrium point or a stable limit cycle (as in Figure 2e). Hence he concludes that the conditions presumed by linear analysis are unstable, and in fact must lead to stable limit cycles. In every instance, however, the models are globally rather than locally stable, limiting their behavior to that shown in either Figures 2c or 2e.

There is another tradition of models that recognizes the basically discontinuous features of ecological systems and incorporates explicit lags. Nicholson and Bailey initiated this tradition when they developed a model using the output of attacks and survivals within one generation as the input for the next (29). The introduction of this explicit lag generates oscillations that increase in amplitude until one or other of the species becomes extinct (Figure 2a). Their assumptions are as unrealistically simple as Lotka's and Volterra's; the instability results because the number of attacking predators at any moment is so much a consequence of events in the previous generation that there are "too many" when prey are declining and "too few" when prey are increasing. If a lag is introduced into the Lotka-Volterra formulation (Wangersky & Cunningham 40) the same instability results.

The sense one gains, then, of the behavior of the traditional models is that they are either globally unstable or globally stable, that neutral stability is very unlikely, and that when the models are stable a limit cycle is a likely consequence.

Many, but not all, of the simplifying assumptions have been relaxed in simulation models, and there is one example (Holling & Ewing 14) that joins the two traditions initiated by Lotka-Volterra and Nicholson and Bailey and, further, includes more realism in the operation of the stabilizing and destabilizing forces. These modifications are described in more detail later; the important features accounting for the difference in behavior result from the introduction of explicit lags, a functional response of predators that rises monotonically to a plateau, a nonrandom (or contagious) attack by predators, and a minimum prey density below which reproduction does not occur. With these changes a very different pattern emerges that conforms most closely to Figure 2d. That is, there exists a domain of attraction within which there is a stable equilibrium; beyond that domain the prey population becomes extinct. Unlike the Nicholson and Bailey model, the stability becomes possible, although in a limited region, because of contagious attack. [Contagious attack implies that for one reason or another some prey have a greater probability of being attacked than others, a condition that is common in nature (Griffiths & Holling 9).] The influence of contagious attack becomes significant whenever predators become abundant in relation to the prey, for then the susceptible prey receive the burden of attention, allowing more prey to escape than would be expected by random contact. This "inefficiency" of the predator allows the system to counteract the destabilizing effects of the lag.

If this were the only difference the system would be globally stable, much as Figure 2c. The inability of the prey to reproduce at low densities, however, allows some of the trajectories to cut this reproduction threshold, and the prey become extinct. This introduces a lower prey density boundary to the attraction domain and, at the same time, a higher prey density boundary above which the amplitudes of the oscillations inevitably carry the population below the reproduction threshold. The other modifications in the model, some of which have been touched on above, alter this picture in degree only. The essential point is that a more realistic representation of the behavior of interacting populations indicates the existence of at least one domain of attraction. It is quite possible, within this domain, to imagine stable equilibrium points, stable nodes, or stable limit cycles. Whatever the detailed configuration, the existence of discrete domains of attraction immediately suggests important consequences for the persistence of the system and the probability of its extinction.

Such models, however complex, are still so simple that they should not be viewed in a definitive and quantitative way. They are more powerfully used as a starting point to organize and guide understanding. It becomes valuable, therefore, to ask what the models leave out and whether such omissions make isolated domains of attraction more or less likely.

Theoretical models generally have not done well in simultaneously incorporating realistic behavior of the processes involved, randomness, spatial heterogeneity, and an adequate number of dimensions or state variables. This situation is changing very rapidly as theory and empirical studies develop a closer technical partnership. In what follows I refer to real world examples to determine how the four elements that tend to be left out might further affect the behavior of ecological systems.

SOME REAL WORLD EXAMPLES

Self-Contained Ecosystems

In the broadest sense, the closest approximation we could make of a real world example that did not grossly depart from the assumptions of the theoretical models would be a self-contained system that was fairly homogenous and in which climatic fluctuations were reasonably small. If such systems could be discovered they would reveal how the more realistic interaction of real world processes could modify the patterns of systems behavior described above. Very close approximations to any of these conditions are not likely to be found, but if any exist, they are apt to be fresh water aquatic ones. Fresh water lakes are reasonably contained systems, at least within their watersheds; the fish show considerable mobility throughout, and the properties of the water buffer the more extreme effects of climate. Moreover, there have been enough documented man-made disturbances to liken them to perturbed systems in which either the parameter values or the levels of the constituent populations are changed. In a crude way, then, the lake studies can be likened to a partial exploration of a phase space of the sorts shown in Figure 2. Two major classes of disturbances have occurred: first, the impact of nutrient enrichment from man's domestic and industrial wastes, and second, changes in fish populations by harvesting.

The paleolimnologists have been remarkably successful in tracing the impact of man's activities on lake systems over surprisingly long periods. For example, Hutchinson (17) has reconstructed the series of events occurring in a small crater lake in Italy from the last glacial period in the Alps (2000 to 1800 BC) to the present. Between the beginning of the record and Roman times the lake had established a trophic equilibrium with a low level of productivity which persisted in spite of dramatic changes in surroundings from *Artemesia* steppe, through grassland, to fir and mixed oak forest. Then suddenly the whole aquatic system altered. This alteration towards eutrophication seems to have been initiated by the construction of the Via Cassia about 171 BC, which caused a subtle change in the hydrographic regime. The whole sequence of environmental changes can be viewed as changes in parameters or driving variables, and the long persistence in the face of these major changes suggests that natural systems have a high capacity to absorb change without dramatically altering. But this resilient character has its limits, and when the limits are passed, as by the construction of the Roman highway, the system rapidly changes to another condition.

More recently the activities of man have accelerated and limnologists have recorded some of the responses to these changes. The most dramatic change consists of blooms of algae in surface waters, an extraordinary growth triggered, in most instances, by nutrient additions from agricultural and domestic sources.

While such instances of nutrient addition provide some of the few examples available of perturbation effects in nature, there are no controls and the perturbations are exceedingly difficult to document. Nevertheless, the qualitative pattern seems consistent, particularly in those lakes (Edmundson 4, Hasler 10) to which sewage has been added for a time and then diverted elsewhere. This pulse of disturbance characteristically triggers periodic algal blooms, low oxygen conditions, the sudden disappearance of some plankton species, and appearance of others. As only one example, the nutrient changes in Lake Michigan (Beeton 2) have been accompanied by the replacement of the cladoceran *Bosmina coregoni* by *B. Longirostris, Diaptomus oregonensis* has become an important copepod species, and a brackish water copepod *Eurytemora affinis* is a new addition to the zooplankton.

In Lake Erie, which has been particularly affected because of its shallowness and intensity of use, the mayfly *Hexagenia,* which originally dominated the benthic community, has been almost totally replaced by oligochetes. There have been blooms of the diatom *Melosira binderana,* which had never been reported from the

United States until 1961 but now comprises as much as 99% of the total phytoplankton around certain islands. In those cases where sewage has been subsequently diverted there is a gradual return to less extreme conditions, the slowness of the return related to the accumulation of nutrients in sediments.

The overall pattern emerging from these examples is the sudden appearance or disappearance of populations, a wide amplitude of fluctuations, and the establishment of new domains of attraction.

The history of the Great Lakes provides not only some particularly good information on responses to man-made enrichment, but also on responses of fish populations to fishing pressure. The eutrophication experience touched on above can be viewed as an example of systems changes in driving variables and parameters, whereas the fishing example is more an experiment in changing state variables. The fisheries of the Great Lakes have always selectively concentrated on abundant species that are in high demand. Prior to 1930, before eutrophication complicated the story, the lake sturgeon in all the Great Lakes, the lake herring in Lake Erie, and the lake whitefish in Lake Huron were intensively fished (Smith 37). In each case the pattern was similar: a period of intense exploitation during which there was a prolonged high level harvest, followed by a sudden and precipitous drop in populations. Most significantly, even though fishing pressure was then relaxed, none of these populations showed any sign of returning to their previous levels of abundance. This is not unexpected for sturgeon because of their slow growth and late maturity, but it is unexpected for herring and whitefish. The maintenance of these low populations in recent times might be attributed to the increasingly unfavorable chemical or biological environment, but in the case of the herring, at least, the declines took place in the early 1920s before the major deterioration in environment occurred. It is as if the population had been shifted by fishing pressure from a domain with a high equilibrium to one with a lower one. This is clearly not a condition of neutral stability as suggested in Figure 2b since once the populations were lowered to a certain point the decline continued even though fishing pressure was relaxed. It can be better interpreted as a variant of Figure 2d where populations have been moved from one domain of attraction to another.

Since 1940 there has been a series of similar catastrophic changes in the Great Lakes that has led to major changes in the fish stocks. Beeton (2) provides graphs summarizing the catch statistics in the lakes for many species since 1900. Lake trout, whitefish, herring, walleye, sauger, and blue pike have experienced precipitous declines of populations to very low values in all of the lakes. The changes generally conform to the same pattern. After sustained but fluctuating levels of harvest the catch dropped dramatically in a span of a very few years, covering a range of from one to four orders of magnitude. In a number of examples particularly high catches were obtained just before the drop. Although catch statistics inevitably exaggerate the step-like character of the pattern, populations must have generally behaved in the way described.

The explanations for these changes have been explored in part, and involve various combinations of intense fishing pressure, changes in the physical and chemical environment, and the appearance of a foreign predator (the sea lamprey) and foreign competitors (the alewife and carp). For our purpose the specific cause is of less interest than the inferences that can be drawn concerning the resilience of these systems and their stability behavior. The events in Lake Michigan provide a typical example of the pattern in other lakes (Smith 37). The catch of lake trout was high, but fluctuated at around six million pounds annually from 1898 to 1940. For four years catches increased noticeably and then suddenly collapsed to near extinction by the 1950s due to a complete failure of natural reproduction. Lake herring and whitefish followed a similar pattern (Beeton 2: Figure 7). Smith (37) argues that the trigger for the lake trout collapse was the appearance of the sea lamprey that had spread through the Great Lakes after the construction of the Welland Canal. Although lamprey populations were extremely small at the time of the collapse, Smith argues that even a small mortality, added to a commercial harvest that was probably at the maximum for sustained yield, was sufficient to cause the collapse. Moreover, Ricker (31) has shown that fishing pressure shifts the age structure of fish populations towards younger ages. He demonstrates that a point can come where only slight increases in mortality can trigger a collapse of the kind noted for lake trout. In addition, the lake trout was coupled in a network of competitive and predatory interconnections with other species, and pressures on these might have contributed as well.

Whatever the specific causes, it is clear that the precondition for the collapse was set by the harvesting of fish, even though during a long period there were no obvious signs of problems. The fishing activity, however, progressively reduced the resilience of the system so that when the inevitable unexpected event occurred, the populations collapsed. If it had not been the lamprey, it would have been something else: a change in climate as part of the normal pattern of fluctuation, a change in the chemical or physical environment, or a change in competitors or predators. These examples again suggest distinct domains of attraction in which the populations forced close to the boundary of the domain can then flip over it.

The above examples are not isolated ones. In 1939 an experimental fishery was started in Lake Windermere to improve stocks of salmonids by reducing the abundance of perch (a competitor) and pike (a predator). Perch populations were particularly affected by trapping and the populations fell drastically in the first three years. Most significantly, although no perch have been removed from the North Basin since 1947, populations have still not shown any tendency to return to their previous level (Le Cren et al 19).

The same patterns have even been suggested for terrestrial systems. Many of the arid cattle grazing lands of the western United States have gradually become invaded and dominated by shrubs and trees like mesquite and cholla. In some instances grazing and the reduced incidence of fire through fire prevention programs allowed invasion and establishment of shrubs and trees at the expense of grass. Nevertheless, Glendening (8) has demonstrated, from data collected in a 17-year experiment in which intensity of grazing was manipulated, that once the trees have gained sufficient size and density to completely utilize or materially reduce the moisture supply, elimination of grazing will not result in the grassland reestablishing itself. In short, there is a level of the state variable "trees" that, once achieved, moves the system from one domain of attraction to another. Return to the original domain can only be made by an explicit reduction of the trees and shrubs.

These examples point to one or more distinct domains of attraction in which the important point is not so much how stable they are within the domain, but how likely it is for the system to move from one domain into another and so persist in a changed configuration.

This sampling of examples is inevitably biased. There are few cases well documented over a long period of time, and certainly some systems that have been greatly disturbed have fully recovered their original state once the disturbance was removed. But the recovery in most instances is in open systems in which reinvasion is the key ingredient. These cases are discussed below in connection with the effects of spatial heterogeneity. For the moment I conclude that distinct domains of attraction are not uncommon within closed systems. If such is the case, then further confirmation should be found from empirical evidence of the way processes which link organisms operate, for it is these processes that are the cause of the behavior observed.

Process Analysis

One way to represent the combined effects of processes like fecundity, predation, and competition is by using Ricker's (30) reproduction curves. These simply represent the population in one generation as a function of the population in the previous generation, and examples are shown in Figures 3a, c, and e. In the simplest form, and the one most used in practical fisheries management (Figure 3a), the reproduction curve is dome-shaped. When it crosses a line with slope 1 (the straight line in the figures) an equilibrium condition is possible, for at such cross-overs the popula-

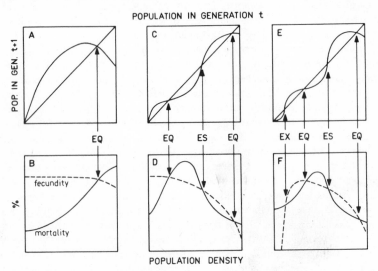

Figure 3 Examples of various reproduction curves (a, c, and e) and their derivation from the contributions of fecundity and mortality (b, d, and f).

tion in one generation will produce the same number in the next. It is extremely difficult to detect the precise form of such curves in nature, however; variability is high, typically data are only available for parts of any one curve, and the treatment really only applies to situations where there are no lags. It is possible to deduce various forms of reproduction curves, however, by disaggregating the contributions of fecundity and mortality. The three lower graphs in Figure 3b, 3d, and 3f represent this disaggregation of their counterpart reproduction curves. The simplest types of reproduction curve (Figure 3a) can arise from a mortality that regularly increases with density and either a constant fecundity or a declining one. With fecundity expressed as the percentage mortality necessary to just balance reproduction, the cross-over point of the curves represents the equilibrium condition. But we know that the effects of density on fecundity and mortality can be very much more complicated.

Mortality from predation, for example, has been shown to take a number of classic forms (Holling 11, 13). The individual attack by predators as a function of prey density (the functional response to prey density) can increase with a linear rise to a plateau (type 1), a concave or negatively accelerated rise to a plateau (type 2), or an S-shaped rise to a plateau (type 3). The resulting contribution to mortality from these responses can therefore show ranges of prey density in which there is direct density dependence (negative feedback from the positively accelerated portions of the type 3 response), density independence (the straight line rise of type 1), and inverse dependence (the positive feedback from the negatively accelerated and plateau portions of the curves). There are, in addition, various numerical responses generated by changes in the number of predators as the density of their prey increases. Even for those predators whose populations respond by increasing, there often will be a limit to the increase set by other conditions in the environment. When populations are increasing they tend to augment the negative feedback features (although with a delay), but when populations are constant, despite increasing prey density, the percent mortality will inevitably decline since individual attack eventually saturates at complete satiation (the plateaux of all three functional responses). In Figures 3d and 3f the mortality curves shown summarize a common type. The rising or direct density-dependent limb of the curve is induced by increasing predator populations and by the reduced intensity of attack at low densities, shown by the initial positively accelerated portion of the S-shaped type 3 response. Such a condition is common for predators with alternate prey, both vertebrates (Holling 14) and at least some invertebrates (Steele 38). The declining inverse density-dependent limb is induced by satiation of the predator and a numerical response that has been reduced or stopped.

Fecundity curves that decline regularly over a very wide range of increasing population densities (as in Figure 3d) are common and have been referred to as *Drosophila*-type curves (Fujita 6). This decline in fecundity is caused by increased competition for oviposition sites, interference with mating, and increased sterility. The interaction between a dome-shaped mortality curve and a monotonically decreasing fecundity curve can generate equilibrium conditions (Figure 3d). Two stable equilibria are possible, but between these two is a transient equilibrium designated as the escape threshold (ES in Figure 3). Effects of random changes on populations or parameters could readily shift densities from around the lower equilibrium to above this escape threshold, and in these circumstances populations would inevitably increase to the higher equilibrium.

The fecundity curves are likely to be more complex, however, since it seems inevitable that at some very low densities fecundity will decline because of difficulties in finding mates and the reduced effect of a variety of social facilitation behaviors. We might even logically conclude that for many species there is a minimum density below which fecundity is zero. A fecundity curve of this Allee-type (Fujita 6) has been empirically demonstrated for a number of insects (Watt 42) and is shown in Figure 3f. Its interaction with the dome-shaped mortality curve can add another transient equilibrium, the extinction threshold (EX in Figure 3f). With this addition there is a lower density such that if populations slip below it they will proceed inexorably to extinction. The extinction threshold is particularly likely since it has been shown mathematically that each of the three functional response curves will intersect with the ordinate of percent predation at a value above zero (Holling 13).

Empirical evidence, therefore, suggests that realistic forms to fecundity and mortality curves will generate sinuous reproduction curves like those in Figures 3c and 3e with the possibility of a number of equilibrium states, some transient and some stable. These are precisely the conditions that will generate domains of attraction, with each domain separated from others by the extinction and escape thresholds. This analysis of process hence adds support to the field observations discussed earlier.

The behavior of systems in phase space cannot be completely understood by the graphical representations presented above. These graphs are appropriate only when effects are immediate; in the face of the lags that generate cyclic behavior the reproduction curve should really produce two values for the population in generation $t + 1$ for each value of the population in generation t. The graphical treatment of Rosenzweig & MacArthur (33) to a degree can accommodate these lags and cyclic behavior. In their treatment they divide phase planes of the kind shown in Figure 2 into various regions of increasing and decreasing x and y populations. The regions are separated by two lines, one representing the collection of points at which the prey population does not change in density ($dx/dt = 0$, the prey isocline) and one in which the predator population does not so change ($dy/dt = 0$, the predator isocline). They deduce that the prey isocline will be dome-shaped for much the same reason as described for the fecundity curves of Figure 3f. The predator isocline, in the simplest condition, is presumed to be vertical, assuming that only one fixed level of prey is necessary to just maintain the predator population at a zero instantaneous rate of change.

Intersection of the two isoclines indicates a point where both populations are at equilibrium. Using traditional linear stability analysis one can infer whether these equilibrium states are stable (Figure 2c) or not (Figure 2a). Considerable importance is attached to whether the predator isocline intersects the rising or falling portion of the prey isocline. As mentioned earlier these techniques are only appropriate near equilibrium (May 24), and the presumed unstable conditions in fact generate stable limit cycles (Figure 2e). Moreover, it is unlikely that the predator isocline is a vertical one in the real world, since competition between predators at high predator densities would so interfere with the attack process that a larger number of prey would be required for stable predator populations. It is precisely this condition that was demonstrated by Griffiths & Holling (9) when they showed that a large number of species of parasites distribute their attacks contagiously. The result is a "squabbling predator behavior" (Rosenzweig 34, 35) that decreases the efficiency of predation at high predator/prey ratios. This converts an unstable system (Figure 2a) to a stable one (Figure 2c); it is likely that stability is the rule, rather than the exception, irrespective of where the two isoclines cross.

The empirical evidence described above shows that realistic fecundity and mortality (particularly predation) processes will generate forms that the theorists might tend to identify as special subsets of more general conditions. But it is just these special subsets that separate the real world from all possible ones, and these more realistic forms will modify the general conclusions of simpler theory. The ascending limb of the Allee-type fecundity curve will establish, through interaction with mortality, a minimum density below which prey will become extinct. This can at the same time establish an upper prey density above which prey will become extinct because the amplitude of prey fluctuations will eventually carry the population over the extinction threshold, as shown in the outer trajectory of Figure 2d. These conditions alone are sufficient to establish a domain of attraction, although the boundaries of this domain need not be closed. Within the domain the contagious attack by predators can produce a stable equilibrium or a stable node. Other behaviors of the mortality agents, however, could result in stable limit cycles.

More realistic forms of functional response change this pattern in degree only. For example, a negatively accelerated type of functional response would tend to make the domain of attraction somewhat smaller, and an S-shaped one larger. Limitations in the predator's numerical response and thresholds for reproduction of predators, similar to those for prey, could further change the form of the domain. Moreover, the behaviors that produce the sinuous reproduction curves of Figures 3c and 3e can add additional domains. The essential point, however, is that these systems are not globally stable but can have distinct domains of attraction. So long as the populations remain within one domain they have a consistent and regular form of behavior. If populations pass a boundary to the domain by chance or through intervention of man, then the behavior suddenly changes in much the way suggested from the field examples discussed earlier.

The Random World

To this point, I have argued as if the world were completely deterministic. In fact, the behavior of ecological systems is profoundly affected by random events. It is important, therefore, to add another level of realism at this point to determine how the above arguments may be modified. Again, it is applied ecology that tends to supply the best information from field studies since it is only in such situations that data have been collected in a sufficiently intensive and extensive manner. As one example, for 28 years there has been a major and intensive study of the spruce budworm and its interaction with the spruce-fir forests of eastern Canada (Morris

27). There have been six outbreaks of the spruce budworm since the early 1700s (Baskerville 1) and between these outbreaks the budworm has been an exceedingly rare species. When the outbreaks occur there is major destruction of balsam fir in all the mature forests, leaving only the less susceptible spruce, the nonsusceptible white birch, and a dense regeneration of fir and spruce. The more immature stands suffer less damage and more fir survives. Between outbreaks the young balsam grow, together with spruce and birch, to form dense stands in which the spruce and birch, in particular, suffer from crowding. This process evolves to produce stands of mature and overmature trees with fir a predominant feature.

This is a necessary, but not sufficient, condition for the appearance of an outbreak; outbreaks occur only when there is also a sequence of unusually dry years (Wellington 43). Until this sequence occurs, it is argued (Morris 27) that various natural enemies with limited numerical responses maintain the budworm populations around a low equilibrium. If a sequence of dry years occurs when there are mature stand of fir, the budworm populations rapidly increase and escape the control by predators and parasites. Their continued increase eventually causes enough tree mortality to force a collapse of the populations and the reinstatement of control around the lower equilibrium. The reproduction curves therefore would be similar to those in Figures 3c or 3e.

In brief, between outbreaks the fir tends to be favored in its competition with spruce and birch, whereas during an outbreak spruce and birch are favored because they are less susceptible to budworm attack. This interplay with the budworm thus maintains the spruce and birch which otherwise would be excluded through competition. The fir persists because of its regenerative powers and the interplay of forest growth rates and climatic conditions that determine the timing of budworm outbreaks.

This behavior could be viewed as a stable limit cycle with large amplitude, but it can be more accurately represented by a distinct domain of attraction determined by the interaction between budworm and its associated natural enemies, which is periodically exceeded through the chance consequence of climatic conditions. If we view the budworm only in relation to its associated predators and parasites we might argue that it is highly unstable in the sense that populations fluctuate widely. But these very fluctuations are essential features that maintain persistence of the budworm, together with its natural enemies and its host and associated trees. By so fluctuating, successive generations of forests are replaced, assuring a continued food supply for future generations of budworm and the persistence of the system.

Until now I have avoided formal identification of different kinds of behavior of ecological systems. The more realistic situations like budworm, however, make it necessary to begin to give more formal definition to their behavior. It is useful to distinguish two kinds of behavior. One can be termed stability, which represents the ability of a system to return to an equilibrium state after a temporary disturbance; the more rapidly it returns and the less it fluctuates, the more stable it would be. But there is another property, termed resilience, that is a measure of the persistence of systems and of their ability to absorb change and disturbance and still maintain the same relationships between populations or state variables. In this sense, the budworm forest community is highly unstable and it is because of this instability that it has an enormous resilience. I return to this view frequently throughout the remainder of this paper.

The influence of random events on systems with domains of attraction is found in aquatic systems as well. For example, pink salmon populations can become stabilized for several years at very different levels, the new levels being reached by sudden steps rather than by gradual transition (Neave 28). The explanation is very much the same as that proposed for the budworm, involving an interrelation between negative and positive feedback mortality of the kinds described in Figures 3d and 3f, and random effects unrelated to density. The same pattern has been described by Larkin (18) in his simulation model of the Adams River sockeye salmon. This particular run of salmon has been characterized by a regular four-year periodicity since 1922, with one large or dominant year, one small or subdominant, and two years with very small populations. The same explanation as described above has been proposed with the added reality of a lag. Essentially, during the dominant year limited numerical responses produce an inverse density-dependent response as in the descending limb of the mortality curves of Figure 3d and 3f. The abundance of the prey in that year is nevertheless sufficient to establish populations of predators that have a major impact on the three succeeding low years. Buffering of predation by the smolts of the dominant year accounts for the larger size of the subdominant. These effects have been simulated (Larkin 18), and when random influences are imposed in order to simulate climatic variations the system has a distinct probability of flipping into another stable configuration that is actually reproduced in nature by sockeye salmon runs in other rivers. When subdominant escapement reaches a critical level there is about an equal chance that it may become the same size as the dominant one or shrivel to a very small size.

Random events, of course, are not exclusively climatic. The impact of fires on terrestrial ecosystems is particularly illuminating (Cooper 3) and the periodic appearance of fires has played a decisive role in the persistence of grasslands as well as certain forest communities. As an example, the random perturbation caused by fires in Wisconsin forests (Loucks 21) has resulted in a sequence of transient changes that move forest communities from one domain of attraction to another. The apparent instability of this forest community is best viewed not as an unstable condition alone, but as one that produces a highly resilient system capable of repeating itself and persisting over time until a disturbance restarts the sequence.

In summary, these examples of the influence of random events upon natural systems further confirm the existence of domains of attraction. Most importantly they suggest that instability, in the sense of large fluctuations, may introduce a resilience and a capacity to persist. It points out the very different view of the world that can be obtained if we concentrate on the boundaries to the domain of attraction rather than on equilibrium states. Although the equilibrium-centered view is analytically more tractable, it does not always provide a realistic understanding of the systems' behavior. Moreover, if this perspective is used as the exclusive guide to the management activities of man, exactly the reverse behavior and result can be produced than is expected.

The Spatial Mosaic

To this point, I have proceeded in a series of steps to gradually add more and more reality. I started with self-contained closed systems, proceeded to a more detailed explanation of how ecological processes operate, and then considered the influence of random events, which introduced heterogeneity over time.

The final step is now to recognize that the natural world is not very homogeneous over space, as well, but consists of a mosaic of spatial elements with distinct biological, physical, and chemical characteristics that are linked by mechanisms of biological and physical transport. The role of spatial heterogeneity has not been well explored in ecology because of the enormous logistic difficulties. Its importance, however, was revealed in a classic experiment that involved the interaction between a predatory mite, its phytophagous mite prey, and the prey's food source (Huffaker et al 15). Briefly, in the relatively small enclosures used, when there was unimpeded movement throughout the experimental universe, the system was unstable and oscillations increased in amplitude. When barriers were introduced to impede dispersal between parts of the universe, however, the interaction persisted. Thus populations in one small locale that suffer chance extinctions could be reestablished by invasion from other populations having high numbers—a conclusion that is confirmed by Roff's mathematical analysis of spatial heterogeneity (32).

There is one study that has been largely neglected that is, in a sense, a much more realistic example of the effects of both temporal and spatial heterogeneity of a population in nature (Wellington 44, 45). There is a peninsula on Vancouver Island in which the topography and climate combine to make a mosaic of favorable locales for the tent caterpillar. From year to year the size of these locales enlarges or contracts depending on climate; Wellington was able to use the easily observed changes in cloud patterns in any year to define these areas. The tent caterpillar, to add a further element of realism, has identifiable behavioral types that are determined not by genetics but by the nutritional history of the parents. These types represent a range from sluggish to very active, and the proportion of types affects the shape of the easily visible web the tent caterpillars spin. By combining these defined differences of behavior with observations on changing numbers, shape of webs, and changing cloud patterns, an elegant story of systems behavior emerges. In a favorable year locales that previously could not support tent caterpillars now can, and populations are established through invasion by the vigorous dispersers from other locales. In these new areas they tend to produce another generation with a high proportion of vigorous behavioral types. Because of their high dispersal behavior and the small area of the locale in relation to its periphery, they then tend to leave in greater numbers than they arrive. The result is a gradual increase in the proportion of more sluggish types to the point where the local population collapses. But, although its fluctuations are considerable, even under the most unfavorable conditions there are always enclaves suitable for the insect. It is an example of a population with high fluctuations that can take advantage of transient periods of favorable conditions and that has, because of this variability, a high degree of resilience and capacity to persist.

A further embellishment has been added in a study of natural insect populations by Gilbert & Hughes (7). They combined an insightful field study of the interaction between aphids and their parasites with a simulation model, concentrating upon a specific locale and the events within it under different conditions of immigration from other locales. Again the important focus was upon persistence rather than degree of fluctuation. They found that specific features of the parasite-host interaction allowed the parasite to make full use of its aphid resources just short of driving the host to extinction. It is particularly intriguing that the parasite and its host were introduced into Australia from Europe and in the short period that the parasite has been present in Australia there have been dramatic changes in its developmental rate and fecundity. The other major difference between conditions in Europe and Australia is that the immigration rate of the host in England is considerably higher than in Australia. If the immigration rate in Australia increased to the English level, then, according to the model the parasite should increase its fecundity from the Australian level to the English to make the most of its opportunity short of extinction. This study provides, therefore, a remarkable example of a parasite and its host evolving together to permit persistence, and further confirms the importance of systems resilience as distinct from systems stability.

SYNTHESIS

Some Definitions

Traditionally, discussion and analyses of stability have essentially equated stability to systems behavior. In ecology, at least, this has caused confusion since, in mathematical analyses, stability has tended to assume definitions that relate to conditions very near equilibrium points. This is a simple convenience dictated by the enormous analytical difficulties of treating the behavior of nonlinear systems at some distance from equilibrium. On the other hand, more general treatments have touched on questions of persistence and the probability of extinction, defining these measures as aspects of stability as well. To avoid this confusion I propose that the behavior of ecological systems could well be defined by two distinct properties: resilience and stability.

Resilience determines the persistence of relationships within a system and is a measure of the ability of these systems to absorb changes of state variables, driving variables, and parameters, and still persist. In this definition resilience is the property of the system and persistence or probability of extinction is the result. Stability, on the other hand, is the ability of a system to return to an equilibrium state after a temporary disturbance. The more rapidly it returns, and with the least fluctuation, the more stable it is. In this definition stability is the property of the system and the degree of fluctuation around specific states the result.

Resilience versus Stability

With these definitions in mind a system can be very resilient and still fluctuate greatly, i.e. have low stability. I have touched above on examples like the spruce budworm forest community in which the very fact of low stability seems to introduce high resilience. Nor are such cases isolated ones, as Watt (41) has shown in his analysis of thirty years of data collected for every major forest insect throughout Canada by the Insect Survey program of the Canada Department of the Environment. This statistical analysis shows that in those areas subjected to extreme climatic conditions the populations fluctuate widely but have a high capability of absorbing periodic extremes of fluctuation. They are, therefore, unstable using the restricted definition above, but highly resilient. In more benign, less variable climatic regions the populations are much less able to absorb chance climatic extremes even though the populations tend to be more constant. These situations show a high degree of stability and a lower resilience. The balance between resilience and stability is clearly a product of the evolutionary history of these systems in the face of the range of random fluctuations they have experienced.

In Slobodkin's terms (36) evolution is like a game, but a distinctive one in which the only payoff is to stay in the game. Therefore, a major strategy selected is not one maximizing either efficiency or a particular reward, but one which allows persistence by maintaining flexibility above all else. A population responds to any environmental change by the initiation of a series of physiological, behavioral, ecological, and genetic changes that restore its ability to respond to subsequent unpredictable environmental changes. Variability over space and time results in variability in numbers, and with this variability the population can simultaneously retain genetic and behavioral types that can maintain their existence in low populations together with others that can capitalize on chance opportunities for dramatic increase. The more homogeneous the environment in space and time, the more likely is the system to have low fluctuations and low resilience. It is not surprising, therefore, that the commerical fishery systems of the Great Lakes have provided a vivid example of the sensitivity of ecological systems to disruption by man, for they represent climatically buffered, fairly homogeneous and self-contained systems with relatively low variability and hence high stability and low resilience. Moreover, the goal of producing a maximum sustained yield may result in a more stable system of reduced resilience.

Nor is it surprising that however readily fish stocks in lakes can be driven to extinction, it has been extremely difficult to do the same to insect pests of man's crops. Pest systems are highly variable in space and time; as open systems they are much affected by dispersal and therefore have a high resilience. Similarly, some Arctic ecosystems thought of as fragile may be highly resilient, although unstable. Certainly this is not true for some subsystems in the Arctic, such as Arctic frozen soil, self-contained Arctic lakes, and cohesive social populations like caribou, but these might be exceptions to a general rule.

The notion of an interplay between resilience and stabilty might also resolve the conflicting views of the role of diversity and stability of ecological communities. Elton (5) and MacArthur (22) have argued cogently from empirical and theoretical points of view that stability is roughly proportional to the number of links between species in a trophic web. In essence, if there are a variety of trophic links the same flow of energy and nutrients will be maintained through alternate links when a species becomes rare. However, May's (23) recent mathematical analyses of models of a large number of interacting populations shows that this relation between increased diversity and stability is not a mathematical truism. He shows that randomly assembled complex systems are in general less stable, and never more stable, than less complex ones. He points out that ecological systems are likely to have evolved to a very small subset of all possible sets and that MacArthur's conclusions, therefore, might still apply in the real world. The definition of stability used, however, is the equilibrium-centered one. What May has shown is that complex systems might fluctuate more than less complex ones. But if there is more than one domain of attraction, then the increased variability could simply move the system from one domain to another. Also, the more species there are, the more equilibria there may be and, although numbers may thereby fluctuate considerably, the overall persistence might be enhanced. It would be useful to explore the possibility that instability in numbers can result in more diversity of species and in spatial patchiness, and hence in increased resilience.

Measurement

If there is a worthwhile distinction between resilience and stability it is important that both be measurable. In a theoretical world such measurements could be developed from the behavior of model systems in phase space. Just as it was useful to disaggregate the reproduction curves into their constituent components of mortality and fecundity, so it is useful to disaggregate the information in a phase plane. There are two components that are important: one that concerns the cyclic behavior and its frequency and amplitude, and one that concerns the configuration of forces caused by the positive and negative feedback relations.

To separate the two we need to imagine first the appearance of a phase space in which there are no such forces operating. This would produce a referent trajectory containing only the cyclic properties of the system. If the forces were operating, departure from this referent trajectory would be a measure of the intensity of the forces. The referent trajectories that would seem to be most useful would be the neutrally stable orbits of Figure 2b, for we can arbitrarily imagine these trajectories as moving on a flat plane. At least for more realistic models parameter values can be discovered that do generate neutrally stable orbits. In the complex predator-prey model of Holling (14), if a range of parameters is chosen to explore the effects of different degrees of contagion of attack, the interaction is unstable when attack is random and stable when it is contagious. We have recently shown that there is a critical level of contagion between these extremes that generates neutrally stable orbits. These orbits, then, have a certain frequency and amplitude and the departure of more realistic trajectories from these referent ones should allow the computation of the vector of forces. If these were integrated a potential field would be represented with peaks and valleys. If the whole potential field were a shallow bowl the system would be globally stable and all trajectories would spiral to the bottom of the bowl, the equilibrium point. But if, at a minimum, there were a lower extinction threshold for prey then, in effect, the bowl would have a slice taken out of one side, as suggested in Figure 4. Trajectories that initiated far up on the side of the bowl would have amplitude that would carry the trajectory over the slice cut out of it. Only those

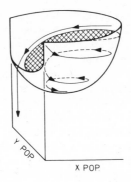

Y POP

X POP

Figure 4 Diagramatic representation showing the feedback forces as a potential field upon which trajectories move. The shaded portion is the domain of attraction.

trajectories that just avoided the lowest point of the gap formed by the slice would spiral in to the bowl's bottom. If we termed the bowl the basin of attraction (Lewontin 20) then the domain of attraction would be determined by both the cyclic behavior and the configuration of forces. It would be confined to a smaller portion of the bottom of the bowl, and one edge would touch the bottom portion of the slice taken out of the basin.

This approach, then, suggests ways to measure relative amounts of resilience and stability. There are two resilience measures: Since resilience is concerned with probabilities of extinction, firstly, the overall area of the domain of attraction will in part determine whether chance shifts in state variables will move trajectories outside the domain. Secondly, the height of the lowest point of the basin of attraction (e.g. the bottom of the slice described above) above equilibrium will be a measure of how much the forces have to be changed before all trajectories move to extinction of one or more of the state variables.

The measures of stability would be designed in just the opposite way from those that measure resilience. They would be centered on the equilibrium rather than on the boundary of the domain, and could be represented by a frequency distribution of the slopes of the potential field and by the velocity of the neutral orbits around the equilibrium.

But such measures require an immeanse amount of knowledge of a system and it is unlikely that we will often have all that is necessary. Hughes & Gilbert (16), however, have suggested a promising approach to measuring probabilities of extinction and hence of resilience. They were able to show in a stochastic model that the distribution of surviving population sizes at any given time does not differ significantly from a negative binomial. This of course is just a description, but it does provide a way to estimate the very small probability of zero, i.e. of extinction, from the observed mean and variance. The configuration of the potential field and the cyclic behavior will determine the number and form of the domains of attraction, and these will in turn affect the parameter values of the negative binomial or of any other distribution function that seems appropriate. Changes in the zero class of the distribution, that is, in the probability of extinction, will be caused by these parameter values, which can then be viewed as the relative measures of resilience. It will be important to explore this technique first with a number of theoretical models so that the appropriate distributions and their behavior can be identified. It will then be quite feasible, in the field, to sample populations in defined areas, apply the appropriate distribution, and use the parameter values as measures of the degree of resilience.

APPLICATION

The resilience and stability viewpoints of the behavior of ecological systems can yield very different approaches to the management of resources. The stability view emphasizes the equilibrium, the maintenance of a predictable world, and the harvesting of nature's excess production with as little fluctuation as possible. The resilience view emphasizes domains of attraction and the need for persistence. But extinction is not purely a random event; it results from the interaction of random events with those deterministic forces that define the shape, size, and characteristics of the domain of attraction. The very approach, therefore, that assures a stable maximum sustained yield of a renewable resource might so change these deterministic conditions that the resilience is lost or reduced so that a chance and rare event that previously could be absorbed can trigger a sudden dramatic change and loss of structural integrity of the system.

A management approach based on resilience, on the other hand, would emphasize the need to keep options open, the need to view events in a regional rather than a local context, and the need to emphasize heterogeneity. Flowing from this would be not the presumption of sufficient knowledge, but the recognition of our ignorance; not the assumption that future events are expected, but that they will be unexpected. The resilience framework can accommodate this shift of perspective, for it does not require a precise capacity to predict the future, but only a qualitative capacity to devise systems that can absorb and accommodate future events in whatever unexpected form they may take.

Literature Cited

1. Baskerville, G. L. 1971. *The Fir-Spruce-Birch Forest and the Budworm.* Forestry Service, Canada Dept. Environ., Fredericton, N. B. Unpublished
2. Beeton, A. D. 1969. Changes in the environment and biota of the Great Lakes. *Eutrophication: Causes, Consequences, Correctives.* Washington DC: Nat. Acad. Sci.
3. Cooper, C. F. 1961. The ecology of fire. *Sci. Am.* 204:150–6, 158, 160
4. Edmondson, W. T. 1961. Changes in Lake Washington following increase in nutrient income. *Verh. Int. Ver. Limnol.* 14:167–75
5. Elton, C. S. 1958. *The Ecology of Invasions by Animals and Plants.* London: Methuen
6. Fujita, H. 1954. An interpretation of the changes in type of the population density effect upon the oviposition rate. *Ecology* 35:253–7
7. Gilbert, N., Hughes, R. D. 1971. A model of an aphid population—three adventures. *J. Anim. Ecol.* 40:525–34

8. Glendening, G. 1952. Some quantitative data on the increase of mesquite and cactus on a desert grassland range in southern Arizona. *Ecology* 33:319–28

9. Griffiths, K. J., Holling, C. S. 1969. A competition submodel for parasites and predators. *Can. Entomol.* 101:785–818

10. Hasler, A. D. 1947. Eutrophication of lakes by domestic sewage. *Ecology* 28:383–95

11. Holling, C. S. 1961. Principles of insect predation. *Ann. Rev. Entomol.* 6:163–82

12. Holling, C. S. 1966. The functional response of invertebrate predators to prey density. *Mem. Entomol. Soc. Can.* 48:1–86

13. Holling, C. S. 1965. The functional response of predators to prey density and its role in mimicry and population regulations. *Mem. Entomol. Soc. Can.* 45:1–60

14. Holling, C. S., Ewing, S. 1971. Blind man's buff: exploring the response space generated by realistic ecological simulation models. *Proc. Int. Symp. Statist. Ecol.* New Haven, Conn.: Yale Univ. Press 2:207–29

15. Huffaker, C. D., Shea, K. P., Herman, S. S. 1963. Experimental studies on predation. Complex dispersion and levels of food in an acarine predator-prey interaction. *Hilgardia* 34:305–30

16. Hughes, R. D., Gilbert, N. 1968. A model of an aphid population—a general statement. *J. Anim. Ecol.* 40:525–34

17. Hutchinson, G. E. 1970. Ianula: an account of the history and development of the Lago di Monterosi, Latium, Italy. *Trans. Am. Phil. Soc.* 60:1–178

18. Larkin, P. A. 1971. Simulation studies of the Adams River Sockeye Salmon *(Oncarhynchus nerka). J. Fish. Res. Bd. Can.* 28:1493–1502

19. Le Cren, E. D., Kipling, C., McCormack, J. C. 1972. Windermere: effects of exploitation and eutrophication on the salmonid community. *J. Fish. Res. Bd. Can.* 29:819–32

20. Lewontin, R. C. 1969. The meaning of stability. *Diversity and Stability of Ecological Systems, Brookhaven Symp. Biol.* 22:13–24

21. Loucks, O. L. 1970. Evolution of diversity, efficiency and community stability. *Am. Zool.* 10:17–25

22. MacArthur, R. 1955. Fluctuations of animal populations and a measure of community stability. *Ecology* 36:533–6

23. May, R. M. 1971. Stability in multi-species community models. *Math. Biosci.* 12:59–79

24. May, R. M. 1972. Limit cycles in predator-prey communities. *Science* 177:900–2

25. May, R. M. 1972. Will a large complex system be stable? *Nature* 238:413–14

26. Minorsky, N. 1962. *Nonlinear Oscillations.* Princeton, N.J.: Van Nostrand

27. Morris, R. F. 1963. The dynamics of epidemic spruce budworm populations. *Mem. Entomol. Soc. Can.* 31:1–332

28. Neave, F. 1953. Principles affecting the size of pink and chum salmon populations in British Columbia. *J. Fish. Res. Bd. Can.* 9:450–91

29. Nicholson, A. J., Bailey, V. A. 1935. The balance of animal populations—Part I. *Proc. Zool. Soc. London* 1935:551–98

30. Ricker, W. E. 1954. Stock and recruitment. *J. Fish. Res. Bd. Can.* 11:559–623

31. Ricker, W. E. 1963. Big effects from small causes: two examples from fish population dynamics. *J. Fish. Res. Bd. Can.* 20:257–84

32. Roff, D. A. 1973. Spatial heterogeneity and the persistence of populations. *J. Theor. Pop. Biol.* In press

33. Rosenzweig, M. L., MacArthur, R. H. 1963. Graphical representation and stability condition of predator-prey interactions. *Am. Natur.* 97:209–23

34. Rosenzweig, M. L. 1971. Paradox of enrichment: destabilization of exploitation ecosystems in ecological time. *Science* 171:385–7

35. Rosenzweig, M. L. 1972. Stability of enriched aquatic ecosystems. *Science* 175:564–5

36. Slobodkin, L. B. 1964. The strategy of evolution. *Am. Sci.* 52:342–57

37. Smith, S. H. 1968. Species succession and fishery exploitation in the Great Lakes. *J. Fish. Res. Bd. Can.* 25:667–93

38. Steele, J. H. 1971. Factors controlling marine ecosystems. *The Changing Chemistry of the Oceans,* ed. D. Dryssen, D. Jaquer, 209–21. Nobel Symposium 20, New York: Wiley

39. Walters, C. J. 1971. Systems ecology: the systems approach and mathematical models in ecology. *Fundamentals of Ecology,* ed. E. P. Odum. Philadelphia: Saunders. 3rd ed.

40. Wangersky, P. J., Cunningham, W. J. 1957. Time lag in prey-predator population models. *Ecology* 38:136–9

41. Watt, K. E. F. 1968. A computer approach to analysis of data on weather, population fluctuations, and disease. *Biometeorology, 1967 Biology Colloquium,* ed. W. P. Lowry. Corvallis, Oregon: Oregon State Univ. Press

42. Watt, K. E. F. 1960. The effect of population density on fecundity in insects. *Can. Entomol.* 92:674–95
43. Wellington, W. G. 1952. Air mass climatology of Ontario north of Lake Huron and Lake Superior before outbreaks of the spruce budworm and the forest tree caterpillar. *Can. J. Zool.* 30: 114–27
44. Wellington, W. G. 1964. Qualitative changes in populations in unstable environments. *Can. Entomol.* 96:436–51
45. Wellington, W. G. 1965. The use of cloud patterns to outline areas with different climates during population studies. *Can. Entomol.* 97:617–31

DESERT ECOSYSTEMS: ENVIRONMENT AND PRODUCERS

❖ 4051

Imanuel Noy-Meir

Department of Botany, Hebrew University, Jerusalem, Israel

INTRODUCTION

The purpose of this review is to examine present knowledge on structure and function of the ecosystems of deserts or arid lands, terms used synonymously here. It focuses attention on features distinctly characteristic of deserts, i.e. common to all or most of them but not to most other ecosystems. It explores the implications of these characteristics for systems analysis and simulation modelling of arid ecosystems, and reviews recent efforts in these directions. The evidence includes results from the rapidly ramifying recent studies in desert ecology, in particular those under the International Biological Program (IBP). Though subjects in desert ecology are being reviewed separately fairly frequently (11, 44, UNESCO Arid Zone Research series), reviews with an integrated approach, such as that by Ross (91) on arid Australian ecosystems, are still rare.

The desert ecosystem is first considered as a whole system, with a sketching out of its dominant diagnostics and some of their deducible consequences. These are then examined in detail with reference to the components of climate, soil, and plants, which bring in more factual evidence and some complicating effects. Sections on consumers (including man) and decomposers, and on feedbacks and modelling in arid ecosystems will be included in the next volume of this series.

Definitions and Characteristics of Arid Ecosystems

The classification used here is generally consistent with the terms and maps of Meigs as used by McGinnies et al (63):

Extreme arid (E)—less than 60–100 mm mean annual precipitation;
Arid (A)—from 60–100 mm to 150–250 mm;
Semiarid (S)—from 150–250 mm to 250–500 mm.

The higher limits refer to areas with high evaporativity in the growing season (e.g. subtropical summer rainfall regions). The limit between A and E corresponds roughly to the limit between diffuse natural vegetation and vegetation contracted to favorable sites only (67). The limit between S and A is roughly the drier limit

25

of diffuse dryland farming; the limit between semiarid and nonarid zones is where such farming becomes a reasonably reliable operation. In this review ecosystems of all three zones are considered, but with emphasis on the typical A zone. Life in extreme deserts is scarce and little known, while semiarid ecosystems often have some features of grasslands or woodlands.

There are three main obvious attributes of these arid ecosystems, one almost by definition, two others by correlation with the first: (a) precipitation is so low that water is the dominant controlling factor for biological processes; (b) precipitation is highly variable through the year and occurs in infrequent and discrete events; (c) variation in precipitation has a large random (unpredictable) component.

Let us now ignore the exceptions and define desert ecosystems as "water-controlled ecosystems with infrequent, discrete, and largely unpredictable water inputs."

What are the implications of this definition for the behavior of the system, in particular the patterns and dynamics of energy flow in it and the adaptive strategies of its organisms? What are the implications for our attempts to understand this behavior and to represent it by models (conceptual, graphical, and mathematical)?

Water-Controlled Ecosystems

Attribute a means that the rates of energy flows to and within the ecosystem are controlled by levels of available water, directly or indirectly. Energy flow in a radiation- or temperature-controlled ecosystem may be well represented and understood by a classical diagram (Figure 1a), in which energy transfer is controlled by energy levels in the donor and recipient components. Most important, the flow of energy into the ecosystem, photosynthesis, is controlled by the level of radiant and/or heat energy available to the plants. But Figure 1a would be a meaningless model for a desert ecosystem if it did not represent the levels of available water which (rather than energy levels) determine the rate of energy inflow. These levels could be introduced as external factors, but it would be more meaningful to draw up a water flow model alongside the energy flow model, utilizing the fact that water moves in the system through essentially the same compartments and paths as energy and carbon (Figure 1b). The most important link between the two is the fact that the water status of the plant, through the stomatal control mechanism, influences the rates of both photosynthesis (A, energy and CO_2 inflow) and transpiration (T, water outflow). Changes in plant water content are usually small compared to the transpiration flow, so that the latter is almost equal to water uptake from the soil. Hence both A and T are in effect controlled by available soil moisture. They are also dependent similarly on other factors influencing the stomata (light, temperature, air humidity) and on the amount of vegetation. Thus the water-controlled nature of arid ecosystems is essentially due to the tight coupling of energy inflow with water outflow, or indeed with water throughflow in the soil-plant-atmosphere path. Or the vegetation in an arid system may be regarded as a converter of a water inflow to an energy inflow. The critical factors for production are those determining the water inflow and the efficiency of the conversion.

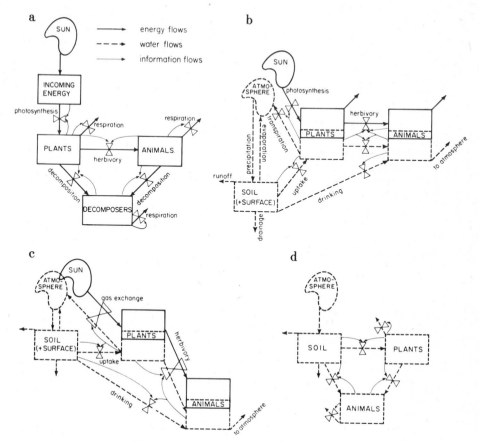

Figure 1 Compartment models of desert ecosystems: (*a*) Energy flow model; (*b*) Energy and water flow models combined (decomposers not shown); (*c*) Same, simplified; (*d*) Water flow model alone.

There may be other couplings. Herbivory and carnivory usually involve transfer of both food (energy, C) and water from prey to consumer by the same process. In arid ecosystems the rate of food consumption may often be controlled by the availability of water, in the food or as surface drinking water, and by the water (and heat) balance of the animal (62). Thus secondary as well as primary energy flows are coupled to, and often dominated by, corresponding water flows (Figure 1c). Indeed, one is tempted to drop the energy model altogether and regard the water flow model as a self-sufficient representation of life processes in a desert ecosystem (Figure 1d). Most organisms are fairly homeohydric, so that the amount of water in any particular biological compartment is a good measure of the amount of living material in it; in poikilohydric organisms (seeds, microorganisms) water content is closely related to biological activity. Such a model would be structurally similar to the energy flow model of an energy-controlled system (Figure 1a). Both have three trophic levels, flows between which are in general controlled by levels in donor and recipient compartments. The most significant difference is that in the water flow system the first trophic level, soil water, has no positive feedback control over its inflow comparable to the feedback from growing plants to photosynthesis.

This similarity highlights an important property of water as a limiting factor in an ecosystem; like energy, but unlike most nutrients, water is not recycled in the system but cascades through it (if we define an ecosystem locally rather than on a global scale). The amount of water recycled from plants and animals back to soil is negligible, and relatively little evaporated or transpired water is recycled locally (e.g. as dew). Most of it is lost from the local ecosystem by convection, to be precipitated in a distant ecosystem. Water is essentially a noncyclable, periodically exhaustible resource, replenished only by new input.

Thus even if our interest is in the trophic energy balance of the ecosystem or any of its subsystems, in an arid ecosystem study of this would be meaningless without considering its water balance as well. If we define a local ecosystem, including vegetation, animals, the root layer of the soil, and the canopy layer of the atmosphere, then the balance for any period is

$$P = R + \Delta S + D + E + \Delta V + T + \Delta A + L \qquad\qquad 1.$$

where P = precipitation, R = runoff/runon (all horizontal flows across the boundaries), ΔS = change in soil (and surface) storage, D = drainage (vertical flow beyond the root layer), E = evaporation (from soil surface), ΔV = change in vegetation storage, T = transpiration, ΔA = change in animals storage, and L = evaporative losses from animals.

ΔV, ΔA, and L are usually negligible compared to the rest, and ΔS is small for periods of one or several years. The component driving the energy flow to the biotic subsystem is the amount of transpired water

$$T = P - R - D - E \qquad\qquad 2.$$

Precipitation is the input or "driving variable." It is not controlled by factors within the local ecosystem, but its partition between the biologically active T and the "losses" R, D, E, and the partition of T in space and time and between organisms, to a large extent are controlled by such factors. The nature of the input is discussed first, then the factors affecting its partition.

DESERT CLIMATE: RAINFALL, THE MASTER INPUT

Systems with Discontinuous Input

While temperature, radiation, and nutrient input to ecosystems vary fairly continuously over the year, precipitation usually comes in discontinuous packages. In arid regions there are only 10–50 rainy days a year, occuring in 3–15 rain events or clusters of rainy days, of which probably no more than 5–6 (sometimes only one) are sufficiently large to affect biotic parts of the system.

Thus the input driving the system comes in "pulses" of very short duration relative to the periods of zero input between them. The response of the system, or any of its parts, to a single input pulse may itself be a pulse (Figure 2a). After a long

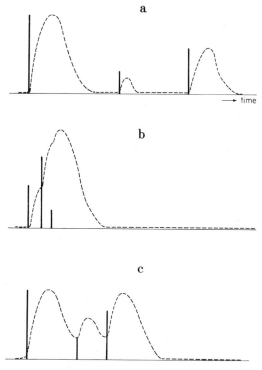

Figure 2 System response to input pulses: (*a*) widely separated, (*b*) clustered, (*c*) with intermediate spacing.

dry period the ecosystem (or most of it) is in some inactive steady state or "zero state." An effective rain event activates biological processes (in particular production and reproduction), and biomass of plants and animals builds up. These processes exhaust the ration of available water supplied by the rain. After a usually short growth period water becomes limiting and both processes and biomass decrease again to a steady state (which may or may not be equal to the previous one). The response of the system to a sequence of rain events depends on the time interval between events relative to the "relaxation time" of the system in response to the individual events. If the former is much larger (e.g. in hot deserts with aseasonal rain) the response will be a series of simple pulses (Figure 2a). If it is much smaller (rain events clustered, markedly seasonal rainfall, slow response, e.g. in cool, winter-rainfall deserts) the effects of the input pulses will accumulate to produce a single larger response pulse (Figure 2b); the total rain of the season may then be considered a single input pulse. In intermediate situations there will be a composite response of distinct but partially cumulative pulses (Figure 2c). The concept and techniques of impulse response used in engineering systems analysis may be applicable at least in the first two cases.

Since many physical and biological processes in deserts occur in fairly discrete pulses, and many responses are of a "trigger" type, Bridges et al (10) have proposed construction of simulation models of deserts in terms of discrete events and qualitative states rather than continuous processes and variables. This was applied, for instance, to models predicting the "phenological states" of plant types from weather conditions in the current and previous seasons.

Westoby (114) and Bridges et al (10) have also questioned the usefulness in deserts of the "level-regulating-flows" paradigm (Figure 3a) which is the now classical representation of each compartment in an ecosystem model. Instead, they suggest the "pulse and reserve" paradigm (Figure 3b). A trigger (e.g. a rain event) sets off a pulse of production (e.g. of annuals). Much of this pulse is lost rapidly by mortality or consumption but some is diverted back into a reserve (e.g. seeds). The reserve compartment loses only slowly during the no-growth period and from it the next pulse is initiated. The authors noted that: (a) flexible transition between an inactive resistant and an active (susceptible) state is highly adaptive in an intermittently favorable environment, (b) the prevalence of this pattern among desert organisms explains the long-term stability of the system despite its extreme short-term variabil-

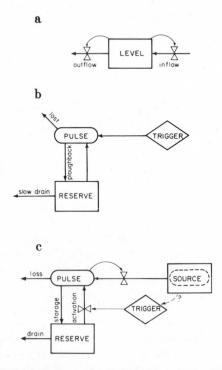

Figure 3 Graphical representations of a biological component in an ecosystem model. (a) The "level-controlling-flows" paradigm. (b) The "pulse-and-reserve" paradigm (from 10, 114). (c) Modified pulse-and-reserve module.

ity, (c) this stability will be endangered only by mechanisms causing overexploita-tion of the reserves or consistent prevention of backflow to reserves (which may be different from the mechanisms causing short-term variations), and (d) organisms at higher trophic levels must adapt either by adopting a pulse-reserve pattern them-selves (e.g. insects), by utilizing reserves of other organisms (e.g. seed-eaters), or by flexible feeding habits, using whatever pulse or reserve is available at any time.

Pulse and reserve modules of various organisms may be combined into a causal (information flow) diagram model of a desert ecosystem (114). Adaptation of Figure 3b to an energy/matter flow model requires some modification (Figure 3c): (a) an inflow to the pulse from a source (a pulse or a reserve of another component) and (b) the effect of both the trigger and the reserve level on the reserve-to-pulse initiation flow.

Systems with Stochastic Input

The master input to arid systems is not only discontinuous but also stochastic. The variation in timing and magnitude of precipitation events has a large random component. This creates special problems for climatologists trying to describe desert climate (inadequacy of averages), for hydrologists and ecologists trying to simulate it in models (inadequacy of deterministic input), and for organisms trying to live in it (optimization in an unpredictable environment).

It is useful to consider rainfall variation at several time-scales, attempting at each to separate some persistent components (pattern) from the random ones.

YEARS   The increase in between-year variability with decreasing mean rainfall is well documented in all arid zones. This variation seems to be mostly random. Stories about cycles of good and drought years (with a half-period of 3, 5, or 7 years) are common in the folklore of arid zone people (and in some publications) but have rarely, if ever, been demonstrated statistically. Throughout Australia, Maher (65) found no such persistences and showed runs of wet and dry years to be only random, with a binomial distribution. McDonald (61) found no year-to-year autocorrelation in Arizona. He did detect fluctuations within a period of about 50 years [as found also in the Negev (94)] but these accounted for only 10% of the variation.

MONTHS   There is more persistence in monthly variations. In some arid and semi-arid zones there are consistently timed rainy seasons in winter (mediterranean), summer (monsoonal), or spring and fall. In some, rainfall is aseasonal or erratic, i.e. randomly distributed throughout the year (parts of Australia and Sahara), and there are many intermediates between seasonal and erratic. Winter rain is generally the more reliable in areas where it is greater than or about equal to summer rain (e.g. in southeastern Australia), but the opposite is true in some of the mixed-season areas (61) and of course in summer rainfall areas.

DAYS   There is a definite tendency of rainy days to occur in sequential runs, which may be expressed by a first-order Markov chain, with different probabilities of rain after rainy and dry days. Fitzpatrick & Krishnan (30) in central Australia found persistence of wet and dry pentads (5-day periods) with a good fit to the Markov model.

WITHIN-DAYS   The distribution of rain intensity at a scale of hours and minutes is important for accurate runoff prediction and is expressed by hydrologists as intensity-duration functions. In arid stations these seem to be parallel with (though lower than) those for humid stations (94).

Meteorologists have long recognized the need to supplement mean annual and monthly rainfall data in arid zones by tables of probabilities of different amounts of yearly, seasonal, and monthly rain, and of weekly and daily rain in different times of the year. Of the various functions which have been fitted to these distributions, the "incomplete gamma" (3, 87) and a Poisson-geometric distribution (31) have been most successful.

While stochastic input has been used in hydrological models (26), ecologists often simply use sequences of actual rainfall data or permutations of these. However, in the first simulation model of an arid ecosystem, Goodall (35, 36) did use stochastic input with a first-order Markov model for raindays and an empirical probability distribution of rain quantity per rainday. Repeated randomized runs then provided a distribution and thus confidence limits for outputs. The development of realistic stochastic models for rainfall at all time scales, and the estimation of their parameters for sites, will become increasingly important as ecosystem modelling emerges from a validation stage to a predictive and applicative one.

The main problem that randomness poses to desert organisms is in the adjustment of their responses to environmental signals so as to optimize growth and survival. In particular, what signals should be used to trigger the activation flow from reserve to an active pulse (e.g. germination, shoot-growth in perennials, breeding in animals) and how much of the reserves should be used? Secondly, when and how fast should the storage flow start? (e.g. seed setting, translocation to reserves). In a predictably seasonal environment any of a number of correlated signals (the first shower, temperature, photoperiod) may serve as indicator of the start of a growing season in which reproductive success is almost certain; there are also several reliable indicators of its approaching end. Organisms in such environments are likely to respond to any one of these simple signals and can commit all or most of their reserves to each seasonal pulse.

When timing and magnitude of rainfall are uncertain, full response to a simple signal (e.g. a light rain) may be premature and may severely decrease rather than increase the reproductive potential (particularly for short-lived organisms). In a study of optimal strategy in random environments with special reference to germination of annuals (19, 20), two main conclusions were drawn: (a) The optimal germination fraction (in general, reserve commitment fraction) decreases as the probability of an unsuccessful outcome increases; thus in highly uncertain environments the optimal strategy is one of cautious opportunism. In such environments, longevity of reserve forms is also of high adaptive value. (b) Growth is optimized by maximizing correlation between the external signals for activation and a successful outcome. In an uncertain environment, with low correlations between the various signals and between signals and their outcomes (e.g. a desert), this requires organisms to process a larger amount of environmental information to regulate their responses (and to pay the cost of this processing and of missed opportunities due to cautious response).

Indeed, seed longevity and seed heterogeneity, which allows delayed or differential germination, are common in desert plants (52, and papers reviewed there). Many of them also have complex germination regulation mechanisms which attune their response rather finely to a precise combination of environmental factors or to a sequence of events (27, 38, 52, 74).

Spatial Variation in Rainfall

Imposed on the temporal variation, and interacting with it, is the spatial variation of rainfall, persistent and random, at all scales. Persistent differences occur not only at the regional scale but also at scales of 0.1–10 km. In particular, the orographic increase of rainfall with altitude (4) and the effects of direction and speed of wind, degree of slope, and rain angle on differences in rainfall between windward and leeward slopes should be important in hilly arid regions (25). In a 1 km² watershed in an extreme arid area Sharon (95) has reported an inverse orographic effect (valleys consistently receiving 40% more rain than ridges), probably related to local wind patterns.

Random spatial variation may be expressed by the lack of correlation in daily, monthly, or yearly rainfalls between two stations. The steepness at which this correlation decreases with distance depends on the size of rain systems and is an indicator of the spottiness of rainfall. It is greater for summer (thunderstorm) than for winter (cyclonic) rain (61), and seems to increase from humid to semiarid and to arid regions (96). For stations 50–200 km apart correlations are often very low even for seasonal or annual totals (61, 94). Daily rainfall is often localized at a much smaller scale. In several areas a considerable proportion of rainfall was found to come as thunderstorm "cells" of 3–8 km diameter, randomly distributed in space (31, 96) and discharging rain on a patch or a strip of land.

This frequently high spatial variation, both persistent and random, has obvious implications for interpolation of rainfall records and for input to hydrological models (26). It can hardly be ignored in ecological modelling in arid zones.

Spatial variation in rainfall (in addition to runoff redistribution and edaphic diversity) is one of the causes of patchiness in desert environments, affecting both species diversity and the adaptive behavior of organisms. It offers highly mobile organisms some compensation for the hazards of high and unpredictable temporal variability. Low spatial correlations mean that at a time of drought in one locality there is still a fair probability of favorable conditions in some other part of the region of which mobile organisms can take advantage. Opportunistic migration, "following the rains," is indeed known for some birds (64) and large mammals (e.g. 75) in arid zones and may be obligatory for their survival. The inclusion of such nomadic populations in ecosystem models requires modelling at a regional rather than a local scale.

Other Climatic Factors

Radiation as such is often assumed not to be a limiting factor in deserts. This may require some caution in view of stomatal behavior of plants which often restricts photosynthesis to periods of low evaporativity (e.g. early morning), when radiation is also low.

Evaporativity (E_o = potential or free-water evaporation), correlated with radiation, temperature, wind, and air humidity deficit, is much higher than precipitation in arid climates in most periods. Being the "evaporative demand" on evaporation and transpiration, evaporativity has a significant effect on the water balance and biological processes tied to it. This is evident in vegetation differences between north- and south-facing slopes, and in many "drought-evading" behavioral adaptations of plants and animals which utilize the marked difference in E_o between day and night.

Temperature often influences plant and animal activities in deserts to an extent which requires modification of the earlier approximation that "deserts are water-controlled ecosystems," though temperature effects are usually in close interaction with the water factor. Rainfall seasonality in relation to temperature has a strong modifying effect on plant growth dynamics. When rain occurs in a warm season (low-latitude or summer-rainfall deserts, e.g. Northern Australia, Sahara, Sahel), both soil moisture and temperature are simultaneously optimal and an almost immediate and very rapid growth pulse follows (e.g. 92). Production is unlikely to be significantly affected by too high temperatures as long as sufficient moisture is available. Where rain or snow fall in a cold season (high-latitude or altitude winter-rainfall deserts, e.g. parts of Central Asia, Great Basin), root and shoot growth are almost completely inhibited by low temperatures until spring, even though moisture is available. Since evaporation losses are also low in winter one may assume as a first approximation that the cold season precipitation is stored until the growing season starts. However, the eventual utilization and production from this water may sometimes be reduced by after-effects of an extremely cold winter (13).

In arid zones where rain falls in a cool winter (mediterranean-type climates), growth is slowed but not fully inhibited by winter temperatures, often after being initiated in autumn. Hence in these deserts spring production is greatly enhanced by autumn rains.

The effects of temperature on growth may be partly compensated for by adaptation of plants to prevailing temperatures. Species from warm-season rainfall zones have higher optimal temperatures for photosynthesis, and in some species temperature acclimation occurs during the growing season (14, 103). In arid zones with two rainfall seasons, different sets of species germinate after summer rain and after winter rain, due to different temperature requirements for germination (71, 112).

THE SOIL: STORE AND REGULATOR IN THE WATER FLOW SYSTEM

A discussion of the role of the soil in arid ecosystems is inseparable from a discussion of the ecosystem water balance and its dynamics. The edaphic factors which are often so prominent in arid zones operate almost always by modification of the water regime. The soil acts as: (*a*) a temporary store for the precipitation input, allowing its use by organisms; (*b*) a regulator controlling the partition of this input between the major outflows: runoff, drainage, evaporation, and uptake transpiration, and of the latter (biologically active) flow between different organisms.

Some aspects of these flows relevant in deserts are considered in order of increasing time lag after a rain event.

Infiltration, Runoff, and Horizontal Redistribution

Most of the water input (rain + runon) at any point either infiltrates the soil or runs off the surface within minutes to hours. Interception by plants causes only minor evaporative losses in arid zones (due to low cover), but may, in conjunction with stemflow, create marked patterns of soil wetting under and around shrubs and trees (86, 99). Surface storage for more than a few hours occurs in deserts only in low sites receiving runon, with low-permeability soils.

Detailed mathematical models of the infiltration/runoff partition at a point, with a resolution time of minutes, have been based on generalized flow equations (40, 41) or on explicit functions of time (15, 99). An approximation is provided by empirical functions relating daily runoff to daily rainfall (26, 106) and expressing the increase in runoff proportion with rain intensity for an area with given surface properties. Both types of functions have been used in models predicting runoff from whole watersheds in arid zones (15, 26, and others from the Tucson group).

Runoff from sandy and stony surfaces is usually lower than from clayey and silty ones, particularly if the latter are crust-forming (28). Cover of dead and living vegetation usually increases infiltration in arid zones (106) by reducing rain impact and probably some physical or chemical modifications of the surface (59).

Modelling and measurement of infiltration/runoff are mostly done either by soil physicists for uniform areas up to 1 m², or by hydrologists interested in the water output from the main channel of a large (1–100 km²) heterogeneous catchment. However, much of the ecological significance of these processes is at scales between 1 m² and 100 km², especially as they concern horizontal redistribution of water *within* the catchment. Even in a rain event producing no channel flow, runoff from some areas (sources) may become runon to others (sinks) and infiltrate there. The infiltration input at any point may be much lower or higher than precipitation, depending on position in the landscape, surface properties, and vegetation. The ensuing spatial variation in soil moisture has significant effects on diversity and production in arid zones. These are widespread, but easiest to demonstrate in areas with regular microtopographical patterns, e.g. the mulga grove-intergroves (98, 84), gilgai plains (16), or furrowed fields. In extreme arid zones it is this redistribution which enables any vegetation to survive in the sink areas at all (51, 67, 119). Runoff models with emphasis on within-catchment redistribution are badly needed.

Vertical Redistribution, Storage Capacity, and Drainage

The movement of infiltrated water down the soil profile can be accurately described by a generalized flow equations model taking account of the relations between water content, water potential (ψ), and conductivity in each layer (e.g. 40, 41). In most soils a more or less sharp and stable "wetting front" on which the concept of field capacity is based, is discernible for some time. Though it ignores the slow and often significant diffusion across the front (28), the wetting front is a useful abstraction for comparing vertical distribution in different soils from a single parameter: the storage capacity C_w, the difference in volumetric water content between soil at field capacity and "dry" soil. This parameter is 3–6% for sands, 7–15% for loams and silts, 15–25% for clays, and decreases linearly with stone content. The "depth of wetting" by a given rain P is P/C_w, hence proportionally larger for sandy and stony than for fine soils.

Therefore, in coarse-textured soils more water is generally lost by drainage (deep percolation) beyond the root zone. However, it is characteristic of the water balance in arid zones that the depth of wetting by prevalent rains is normally not greater than maximal rooting depth. Hence all soil moisture is evaporated or transpired, and layers beyond that depth are permanently dry (111, 43). Substantial drainage and groundwater recharge flows occur mostly in runon areas and channels and in unvegetated deep-wetting soils (e.g. dunes; 85). In most other sites in arid regions the term D in the water balance is zero or negligible.

Impermeable layers (e.g. of clay, marl) in the profile modify the vertical distribution by causing accumulation of water above them. However, in hard rocks and calcified horizons there often are enough cracks to allow (or even facilitate) the passage of water and roots, thus seemingly shallow soils on such substrates in arid zones may in fact be deep soils ecologically.

Evaporation and the Inverse Texture Effect

Evaporation from the soil may be simulated by the same flow model used for infiltration and redistribution (41). When the surface is wet evaporation is close to the demand E_o, but as the top layers dry out desiccation of deeper layers slows down progressively. The upper 5 or 10 cm are mostly dry within 5–25 days in arid climates and plants have little chance to extract water from this layer. It takes many weeks for evaporation to dry out the 10–30 cm layer, so that roots can effectively compete with it there. For many months there is little direct evaporative loss from beyond 30 cm (34, 85, 86). Thus the total evaporation loss E is proportional to the storage capacity of the top 20 or 30 cm; it will be considerably lower from sandy, gravelly, and rocky soils than from fine soils (2). Also, E will be higher if a given rainfall is distributed over more events. In general, a higher proportion of rain will evaporate in summer (higher E_o) than in winter; therefore summer rains are considered "less effective" than winter rains.

Vegetation cover reduces radiation and wind speed at the soil surface and hence reduces evaporation. Once established, a leafy plant thus controls to some extent the loss of its own resource, as well as creating a favorable microenvironment for other plants and animals. The proportion of the area thus affected is small in arid ecosystems, but may be ecologically significant.

Movement of vapor along soil temperature gradients can be important in dry soils under high daily radiation (90). The main effect is a nocturnal upward flow of moisture to the surface (66) causing subterranean dew. A similar seasonal flow should occur in autumn, but possible biological effects have not yet been demonstrated.

Condensation of atmospheric moisture as dew is common in coastal deserts, but much less common inland. Whether its absorption by higher plants in the hours before re-evaporating contributes significantly to their water balance is controversial (e.g. 66, 28). It certainly allows a morning pulse of photosynthesis in lichens and algae (54) and is utilized also by animals. According to Walter (111), fog is a useful supplement to soil moisture for plants in the Namib desert, and in parts of the South American coastal desert vegetation depends almost entirely on "combing" moisture out of the fog.

The fact that in arid zones evaporation from upper layers, rather than drainage from deeper ones, causes the largest loss of soil moisture, is the main cause of the "inverse texture effect." In humid climates sandy and rocky soils are considered dry (due to low C_w) and carry relatively poor vegetation. In arid and semiarid climates throughout the world they usually support taller and denser perennial vegetation than do finer soils (e.g. 5, 97, 111). The same vegetation may occur at lower rainfall on coarse soils than it does on fine ones (101, 77). The balance point between the advantage of coarser texture (less E) and its disadvantage (more D) occurs somewhere between 300 and 500 mm rainfall. Thus the inverse texture effect is really diagnostic of arid and semiarid ecosystems as defined above. Other factors contributing to the inverse texture effect are the lower runoff from coarse soils (105) and the reduced evaporation from a stony surface.

Transpiration and Water Uptake By Roots

The total amount of water taken up and transpired by plants (T) depends mostly on what remains after the unavoidable losses by runoff, drainage, and surface evaporation, and also on the rate of uptake from the 5–40 cm layer, in competition with evaporation. The rate of water flow through the soil-plant-atmosphere path depends on the difference in ψ between soil and atmosphere and on the resistances between them, among which the stomatal resistance, r_s, is most frequently dominant. Plants are able to regulate r_s (and hence the flow) rather tightly in response to changes in water demand (E_o) and supply (ψ in soil); this mechanism is essential for optimization of water use in arid conditions. Increased soil resistance due to drying around roots seems to be usually compensated for by root growth (H. P. Van Keulen, personal communication; 86).

Whether transpiration (and growth) is a linear or a step function of soil water content has long been controversial. It is now clear that the shape of the curve depends on E_o and root density (21, 99), or actually on the effectiveness of the plant in adjusting water supply to demand. Over a wide range T may be linear with soil water potential (69). While in crop plants zero transpiration or wilting point is reached at −10 to −30 bars, an ability to use soil moisture down to −100 (or −150) bars seems to be common for arid zone perennials, xerohalophytes, sclerophyllous shrubs, and grasses (69, 99, 115, 47). Local differences in root density and uptake are yet another source of horizontal and vertical patterns in soil moisture dynamics about which little is known and which could be important in arid communities (86).

Water Balance Models

A number of dynamic models of the overall ecosystem water balance have recently been developed. WATBAL (60) calculates weekly changes in soil moisture storage (S) assuming $(E + T)/E_o$ to be a step function of S; it was tested with moderate success on an arid grassland (115). A model developed by Specht (102) originally for perennial evergreen vegetation computes a monthly balance, assuming a linear relationship and the regulation of transpiration by the vegetation for year-long water availability. Model NEGEV (93) included a water balance subroutine which considered daily runoff, evapotranspiration (linear with S), and vertical redistribution between layers, based on the field capacity assumption. All these models are consciously simplistic, in using crude empirical approximations for most processes, in not separating evaporation from transpiration (processes which differ in controlling factors and in ecological significance), and in not considering root growth to different layers (important in annual communities). T and E were separated in models by Ross & Lendon (92) and Tadmor & Van Keulen (unpublished); the latter also modelled root growth. Basic models of soil water flow (e.g. 41) could be applied to ecosystem water balance once an uptake-transpiration term was included (76).

There are at present few data on comprehensive water balance dynamics of natural arid and semiarid ecosystems to validate even the simpler models. Separation of evaporation from water uptake, and of the vertical and horizontal components of the latter, presents some technical difficulties but is essential for a real understanding of desert systems. Some detailed studies have been done in central Australia (98, 115); a number of fairly comprehensive ecosystem water balance studies are now under way in arid Australia, Israel, and the USA.

Salinity and its Effect on Water Relations

Many fine and some medium textured soils in deserts have a horizon where soluble salts (mostly NaCl, some sulfates) accumulate. The origin of the salt (apart from hydromorphic saline depressions not considered here) may sometimes be the parent material; in deserts within 50–150 km of a coast there is a large input of air- and rain-borne salt (118). Accumulation is due to the lack of deep percolation and leaching in deserts; the depth of maximum accumulation varies between 10 and 100 cm, depending on the normal depth of wetting [(thus on rainfall, infiltration, and texture (105, 117)]. The water stored in this horizon is often a significant proportion of the total moisture store. Its water potential is lower due to the addition of an osmotic component to the matric potential. Therefore, this water can be used by plants only insofar as they reduce their internal ψ, or couple water uptake to salt uptake ("salt exclusion" mechanisms may be useful in wet saline soils, but not in dry ones). To prevent salt from accumulating indefinitely in the active tissues there must be a salt outflow by excretion or by accumulation in special organs. Xerohalophytic plants growing on dry saline soils indeed have high internal osmotic potentials and efficient mechanisms for salt uptake, transport, and secretion (e.g. 37). The salinity factor has so far been neglected in models of arid ecosystems, which for many deserts is hardly justifiable.

DESERT VEGETATION: CONVERTING WATER TO ENERGY

The Distribution of Soil Water in Space, Time, and Water Potential: An Opportunity for Niche Diversification

Soil water in deserts is far from being a single homogeneous resource; it is highly diversified in several dimensions. The water stores in different soil layers differ widely in the frequency at which they are filled, in the rate at which they are emptied by evaporation, and in the types of energy investment needed to gain access to them (Figure 4). Plants with different time strategies, root systems, and other special mechanisms have adapted to utilize each of them. The dominant type in each site will be the one which has the largest competitive advantage in the utilization of the largest store there. Usually there will be enough water in other stores to allow types with specialized niches, overlapping partly or not at all with that of the dominant, to coexist in the same site.

The surface layer (0–2 cm) water is too transient a pulse to be used by vascular plants. However, in many arid zones it is utilized by algae and lichens, which become photosynthetically active upon wetting. The lifetime of the 10–30 cm store

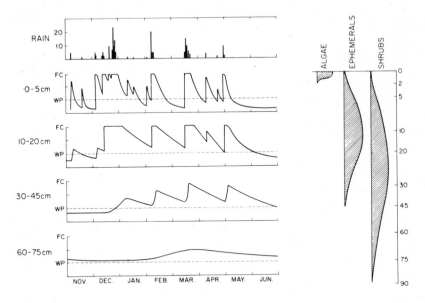

Figure 4 Left: Seasonal dynamics of soil moisture in different layers (schematized from data from desert shrubland on loessial plain; WP, FC – apparent "wilting point" and "field capacity" moistures). *Right:* Vertical distribution of activity of plant types (scale distorted to fit left part).

is just long enough for fast-growing ephemerals (annuals and herbaceous perennials) to take up most of it before it evaporates and to complete a significant production pulse and reproductive cycle. They utilize also part of the 2–10 cm store, which is important for germination and establishment, but they can hardly depend on it alone for reproduction unless there is a sequence of rain events. Many ephemerals, once established, can use water also from the 30–60 cm layer, and in certain soils and in the absence of shrubs, even down to 120 cm. However, in these layers (and deeper ones if they receive moisture) shrubs have the advantage of maintaining a deep perennial root system. These stores, only slightly affected by evaporation, are a more stable reserve resource which can be used at a slow and well-regulated rate to maintain perennial structures, and possibly some level of photosynthesis, during long dry periods. However, while trees and shrubs specialize in using the deeper stores for drought survival, many of them have not given up the production pulse from the 10–30 cm layer. They usually have an extensive and fairly dense horizontal root system there (in addition to deeper vertical roots), augmented in wet periods by deciduous rootlets. There is strong competition for the water in this layer between direct evaporation, ephemerals, and shrubs (particularly shrub seedlings) and between different species of each type. The long-term composition of the vegetation at each site in the desert is mainly determined by the terms of this competition.

The relative advantages of different types and species depend on the site characteristics which affect the vertical distribution of available moisture (or moisture-period probabilities), in particular, rainfall distribution and soil texture. Frequent very light rains (0.5–4 mm) will benefit only surface cryptogams. A number of light rains of 10–20 mm at intervals of a few weeks, which just maintain available moisture in the top 30 cm, will be highly advantageous for ephemerals. They also will enhance shrub growth, but shrub survival depends on the replenishment of the deeper reserve at least once every 1–2 years by 40–100 mm falling in a short period (or a period of low evapotranspiration).

The effect of texture on the moisture profile is no less important. In clayey, silty, or loamy soils with high water capacity most desert rains do not penetrate beyond 30 cm, thus favoring plants with shallow roots and rapid growth pulses, particularly ephemerals. In sandy, gravelly, or rocky soils, where capacity is lower, less water will be stored in this zone and much will percolate deeper, shifting the advantage to deep-rooted perennials (111, 77).

Texture also affects the form of relationship between water content and water potential. In sand it becomes very steep below −15 bars so that relatively little additional water becomes available between −15 and −50 (or −150) bars. In loams, and even more so in clays, the curve is such that the amount of this marginal water is far from negligible compared to the normal available water. In these soils there is an advantage to plants which can reduce their internal ψ (by osmotic potential or negative turgor) and take up water from soil well beyond −15 bars.

Soil salinity increases the relative proportion of marginal water and thus enhances the effect of fine texture in favoring plants able to extract potential water at low ψ, particularly by salt throughflow. Tolerance to high concentrations of specific ions is also required (9, 68). These requirements are met by a specialized group of xerohalophytes, both annuals (mostly summer-growing Chenopodiaceae) and shrubs (Chenopodiaceae, Zygophyllaceae, Tamaricaceae).

Another specialization is induced by soil heterogeneity, in particular, stoniness. In a rocky or stony soil, moisture is very unevenly distributed in the profile, being concentrated in soil pockets and fissures, at soil-rock interfaces, and under stones (28). This favors plants with flexible "exploring" roots, capable of penetrating cracks, following tortuous paths, and expanding whenever a favorable pocket is found. Some desert perennials are able to develop such roots, while others are apparently restricted to homogeneous soils. Still others seem to be specifically adapted to the moisture regime of a certain combination of soil layers (e.g. sand upon loam, loess upon chalk).

The importance of the space-time stratification of soil moisture and of the corresponding adaptive stratification of root systems and cycle durations of different plants has been recognized by desert ecologists (49, 111, 77, 28). This multidimensional partitioning of the most important resource allows the coexistence of a number of plant species and types in every site, which in turn affects animal diversity. It also allows a more complete and efficient utilization of this resource for primary production. While the general principles are well understood, the interactions are often complex and many phenomena not fully explained. The application of simulation modelling to this central problem in the desert system would be useful.

Problems of Water Use Efficiency

Primary production (A) in arid ecosystems depends on the part of water input used by plants (T) and on the efficiency (A/T) of its use for energy and CO_2 fixation. In an analysis of crop yields and water use in arid (high radiation and evaporativity) conditions, de Wit (116) found a good fit to a linear relation $A = gT/E_o$ or $A/T = g/E_o$ (see also 104); i.e. efficiency is inversely related to average evaporativity in the growing season, where g is a species constant more or less independent of climate and moisture. The theoretical explanation was that when radiation (and E_o) is high, photosynthesis is saturated and invariant, while transpiration continues to increase. Another interpretation is that in these conditions the two processes are perfectly coupled, i.e. their rates depend on the same resistances to gas diffusion in the stomatal pathway (mostly r_s itself). Any regulation of r_s by external and internal factors will not affect the ratio A/T, which will be equal to the ratio of the demands for photosynthesis (g, expressing the photosynthetic capacity) and transpiration (E_o). Only the introduction of substantial additional resistances, which affect A and T differently, will cause decoupling and deviation from this relationship.

The main prediction of the de Wit equation (lower yield per unit water transpired at higher E_o) has not been directly tested in natural arid communities. The often observed "lower effectiveness" of rain in summer and in low latitude deserts is consistent with it (but may also be caused entirely by higher evaporation losses). Decreases in the A/T ratio during summer and at midday have occasionally, but not consistently, been found in gas exchange measurements (14, 53).

One may expect desert plants to have adaptations which increase the efficiency of water use for either seed production or survival-maintenance, by increasing g, reducing E_o, or decoupling photosynthesis from transpiration. Very common are "behavioral" adaptations in which the plant uses stomatal control on A and T so that both activities occur mostly (or exclusively) in seasons, days, or hours of low E_o. Decreases in rates of gas exchange during summer and at midday have been observed in several desert perennials (14, 29, 47). The reduction of leaf surface in dry periods is another mechanism of adaptation (80). In many desert plants E_o of periods when most A and T occur is probably much lower than the average E_o. The major limitation is that photosynthesis cannot occur at night when E_o is lowest.

This is overcome in succulents, particularly desert cacti, by "adaptive decoupling" using the crassulacean acid metabolism (107, 108, 83). The capacity to fix and store large quantities of CO_2 in the dark allows these plants a separation in time of the light-requiring processes of photosynthesis from CO_2 uptake. The latter occurs at night, when E_o and T are very low, while in dry periods gas exchange during the day is negligible. Thus high overall A/T ratios may be achieved.

Behavioral restriction of transpiration to periods of low E_o will result in slow, but efficient growth. This will be advantageous for plants with a water reserve to which they exclusively have access (e.g. in deeper layers, under rocks). However, for plants in a competitive situation (e.g. shallow-rooted plants), where the water supply is exhausted also by direct evaporation or by other species, selection will favor unrestricted, rapid, though inefficient, transpiration (20). Most desert ephemerals have apparently adopted the latter strategy, while many shrubs have specialized in slow and regulated use of exclusive water stores (102). This also requires an ability to survive for long periods at low levels of activity.

High photosynthetic capacity (g), as a possible mechanism for higher water use efficiency (at given E_o), is characteristic of plants with the C_4 photosynthetic pathway, many of which occur in arid zones (82). Though higher A/T have been found in C_4 plants in gas exchange measurements over short periods (100, 14), annual production (absolute or per unit water) in semiarid communities dominated by them does not seem to be much higher than in those dominated by C_3 plants (J. K. Marshall, personal communication; 14). Possibly the lower intrinsic efficiency (g) of the latter is compensated for by stricter behavioral restriction of activity to low E_o. The adaptive value of higher g in C_4 plants may not be so much in higher productivity per se, but in the ability to produce at all in periods and latitudes with high E_o where they have the advantages of high light saturation and flexible temperature acclimation (14).

Adaptive decoupling of photosynthesis from transpiration by an increased mesophyll resistance to water flow may be important in some desert shrubs in which photosynthesis levels are maintained while transpiration decreases drastically after midday (53). On the other hand, desert plants usually avoid any decoupling which reduces efficiency; e.g. cuticular transpiration is usually very low (79, 100).

Ecological Types and Pulse-Reserve Strategies in Desert Plants

In attempting to apply the pulse-reserve paradigm (10, 114) to desert plants one may ask what is the reserve by which each species survives dry periods, and how does each species regulate the transfer between the reserve and the active pulse. This question is related to the distinction of the ecological types of desert plants (28, 51, 111, 119). The types differ in the degree to which their activity is geared to rainfall pulses and in their kinds of reserves: whether water or energy, internal or external, above or below ground.

POIKILOHYDRICS   These plants maintain all structures under extreme changes in hydration; their activity pulses are perfectly in phase with moisture pulses. Transition between active and reserve forms involves only rapid and reversible biochemical changes upon wetting and drying. This group includes algae and lichen, which are very common on soil and stones in deserts (32, 54), and a few ferns and higher plants (111).

EPHEMERALS   These are plants which during dry periods maintain no photosynthetically active parts but only a special resistant reserve form, from which a new pulse is initiated and which is in turn replenished from it. The transfer between the two may involve a certain lag in the response to rainfall, but the growth pulse is still closely tied to transients of available moisture in the upper (5–30 cm) soil layer. These species are often referred to as "drought evaders" and are a majority in the flora of most deserts. Two subtypes are:

(a) *Annuals*   The only reserve of annuals is seeds, which store energy and nutrients but not water and hence depend on external water for reactivation. Both activation (germination) and storage (seed setting) are irreversible once triggered. Reserve biomass is small compared to peak active biomass, the latter building up mostly from its own production.

(b) *Perennial ephemeroids*   (geophytes and hemicryptophytes)   Storage organs of these plants (bulbs, rhizomes) often contain water reserves, as well as a reserve of carbohydrates and protein which may be comparable in magnitude to peak vegetative biomass. Hence active biomass can be built up rapidly upon the first signal of a rainy season, and flowering can be independent of rainfall. Reactivation and storage flows may be more flexible and reversible than in annuals. Strategies in this group vary in the degree and timing of reserve commitment and in the timing of the storage flow (e.g. throughout the growth period or only towards its end).

DROUGHT PERSISTENTS   This group includes all perennials which maintain some photosynthesis throughout dry periods; they must have reserves of both water and energy to account for inevitable transpiration and respiration losses of the active tissues.

Sizeable energy and nutrient reserves in woody parts above and below ground are an adaptation of trees and shrubs generally, but those in arid zones require also an external or internal water reserve. The many special adaptations of these true xerophytes have been the subject of many morphological, physiological, and ecological studies (e.g. reviewed in 79). They vary in the level of activity maintained in drought and hence in their dependence on reserves.

(a) *Fluctuating persistents*   Many desert shrubs reduce photosynthetically active biomass and transpiring surface considerably during dry seasons by shedding most leaves or stems or by replacing them with smaller, denser leaves with lower gas exchange rates (80, 28, 51, 23, 24). The root system is also reduced by shedding rootlets in the dry upper layer. Thus, while a residual activity is maintained, water and energy losses are low and relatively small water reserves (in soil and perhaps in wood) are sufficient.

(b) *Stationary persistents*   In other desert shrubs and trees the green biomass is maintained at a nearly constant level throughout the year. These require a larger and more reliable water reserve, as well as tight control of cuticular and stomatal transpiration, high A/T efficiencies (e.g. by decoupling), and tolerance of low internal water potential and of high internal temperatures resulting from suppression of transpiration. All or most of these are characteristics of the two main types of true drought-enduring plants.

Evergreen shrubs may be sclerophyllic *(Larrea, Acacia)*, leafless with green stems ("retamoid"), or xerohalophytes with semisucculent leaves (many chenopods). Many of them have been shown to take up water from soil beyond −100 bars, and/or to have specialized root systems. These attributes enable them to utilize slowly and efficiently external water reserves in deep, saline, or rocky layers, that are not available to other plants.

Succulents, typically the cacti of American deserts, often have fairly shallow root systems and thus little external water reserves. Instead they accumulate a large internal water reserve, from which the green cortex (and some activity) is maintained during droughts.

There are indications that in both evergreens and succulents, even though the green biomass fluctuates only little, most net production occurs during the short favorable periods (78, 83); thus most production must be translocated to reserves. In these and other desert perennials ephemerals and fluctuating persistents, translocation, storage, and mobilization of reserve materials must be important. Little is known about these processes and how they are regulated; even the form and location of the main energy reserves are often not fully identified. The expected seasonal changes in reserves were detected in some plants (22, 48) but not in others (78, 103).

The distinctions between active and reserve biomass and between the patterns of production and translocation in different types should therefore be important elements in models of arid ecosystems. They are of consequence in the modelling not only of primary production but also of herbivory and its effect on long-term production potential (109). An interesting aspect is the analysis of the different pulse-reserve strategies as optimized strategies in different environments.

Primary Productivity and Biomass in Arid Lands

An examination of data from various arid regions shows that the average annual net above-ground primary production varies between 30 and 200 g/m² in the arid zone and between 100 and 600 g/m² in the semiarid zone (e.g. 17, 81, 73, 88, 12, 113, 55, 8). Estimates of below-ground production are scarce, but are given in some of the Russian papers. It seems that total production may be 100–400 g/m² for arid, 250–1000 g/m² for semiarid communities.

These data, as well as some which directly relate productivity to precipitation (56, 111), suggest that a fair proportion of variation in productivity (Y) in arid ecosystems could be accounted for by a linear regression on precipitation:

$$Y = b(P - a) \qquad (Y = 0 \text{ if } P < a)$$

where a may be interpreted as the total of "ineffective precipitation" or water losses (evaporation and runoff) and b as the average water use efficiency of the community. The "zero-yield intercept" a is between 25 and 75 mm/year. The efficiency b is between 0.5 and 2 (mg dry matter/g water) [or (g/m²)/mm] for above-ground production, from which one may estimate it to be 1–6 mg/g for total production. This is considerably lower than the A/T values of 5–50 mg/g measured for individual desert plants over periods of hours by gas exchange measurements (14, 29, 53, 100), but similar in magnitude to efficiencies reported for irrigated crops in arid climates over a growing period (68, 104). Apparently adaptations of desert plants for more efficient water use just compensate for energy losses due to the irregularity of the water supply.

The accumulation of standing live plant biomass and the turnover rate (productivity/biomass) in deserts depend on the dominant type. In ephemeral communities there is 100% turnover of shoots (and of root biomass in annuals) during the growth period of 2–5 months. The numbers for net annual production are equal to peak biomass (mean biomass has little meaning). In fluctuating perennials annual foliage production may be 50–95% of peak foliage biomass (81), but when stems and roots are included, production is probably only 20–40% of biomass, which in communities where this type is dominant amounts to 150–600 g/m² above ground or 400–2500 g/m² total (12, 73, 88, 113). These turnover rates are higher than in forest or tundra (88). In arid and semiarid communities of stationary drought-persistent trees, shrubs, and cacti, annual production may be only 10–20% of a standing biomass of 300–1000 g/m² above ground (e.g. 17), i.e. possibly 600–4000 g/m² including roots.

The distribution of biomass between roots and shoots also differs greatly between life forms. The root/shoot ratio of crop plants and trees is often observed to increase in dry conditions (probably a mechanism for adjustment of water supply to demand). However, not all desert plants have high root/shoot ratios. In many desert winter annuals it is apparently not much higher than in nondesert annuals (0.2–0.5). In perennials ratios of below-/above-ground biomass should be interpreted carefully, as the fractions include not only active roots and shoots but also reserve organs (wood, rhizomes); the ratio may often reflect mostly the distribution of the latter. For perennial grasses and forbs in arid and semiarid regions, ratios between 1 and 20 have been reported (89). For shrubs it is usually between 1 and 3, but in the cold deserts of Central Asia values of 6–12 are common (e.g. 58). On the other hand, in some shrubs in semiarid Australia (45, 12) ratios as low as 0.2–0.3 have been found. It seems that a high below-/above-ground ratio is not a characteristic of desert vegetation generally; it may be more closely related to certain life forms or to temperature regimes than to aridity.

Reproduction and Population Dynamics in Desert Plants

A major problem of desert plants is how to regulate time and intensity of flowering, seed-setting, dispersal, and germination so as to maximize successful reproduction in an environment where periods suitable for each of these processes are short and uncertain. As discussed above, seed longevity and a cautiously opportunistic strategy, with seed diversification and maximum utilization of environmental signals, are expected and usually found in desert annuals (18, 19, 38, 52, 74). Other adaptations of this group are the ability to produce some seeds (at least one per plant) even when conditions allow only limited vegetative growth, high seed yields in favorable conditions, and dispersal mechanisms which allow both continued occupation of safe microhabitats and search for new ones (27, 28, 52).

Desert perennials differ in that successful reproduction may occur less frequently, without endangering the population (depending on life expectation). Seed longevity is less critical and the seeds of some desert shrubs are viable for only a few months. Germination of some perennials follows a strategy of trial-and-error almost every year; in others it depends on climatic sequences which occur only once in several years (39, 28, 70), and which are presumably correlated with high probability of successful establishment. In some arid shrub and tree communities in Australia and Israel germination seems to occur only after death or disturbance of the mature population.

There are few detailed studies of plant population dynamics in arid zones. Summarizing 40 years of observation at Koonamore on Australian shrub communities recovering from grazing, Hall, Specht & Eardley (39) found irregular fluctuations superimposed on both the increasing trend and the eventual steady state. These fluctuations, with periods of 2–5 years, seemed to be mostly in response to sequences of dry and wet years. In some populations turnover was fairly rapid, while in others there was no change in 40 years!

These features make the definition of succession and climax in desert communities problematic. Kassas (50) distinguishes successional changes in desert vegetation from seasonal and accidental (due to random climatic fluctuations) changes. Thus the climax is defined to include these irregular fluctuations in composition. Succession (i.e. long-term trends) in deserts is usually allogenic in response to geomorphological processes (50, 5); autogenic succession occurs in dune stabilization and in the formation of mounds around shrubs. Changes in human and stock pressure may induce drastic successional or degradational trends in arid communities, as documented for instance in North America (42).

Competition and Other Interactions Between Plant Populations

Two opposite views on the importance of competition in arid communities are possible (33):(a) the harsh environment controls density so that competition rarely has a chance to occur; (b) there is intense competition for the scarce limiting resource—water. Assumption a may be true where the environment is not only arid but also extremely unstable, i.e. the frequency of catastrophes is high in relation to population growth rate (extreme deserts, erodible surfaces). In many arid communities, periods between disasters appear to be long enough for densities to build up to levels where intense competition for water does occur. In mature arid shrub communities root systems may occupy most of the area even where canopy is only 3–5%. Evidence for within-species competition is the regular spatial pattern sometimes observed in desert shrub populations (e.g. 110); in other cases evidence may have been obscured by habitat microheterogeneity. Mortality due to competition for water has been indicated in desert annuals populations (7).

Competitive inhibition of shrub seedlings by mature shrubs, to a distance 5 times the canopy radius, has been demonstrated by Friedman (33). Many phenomena in the distribution of species and communities in arid and semiarid zones can be explained only by assuming strong between-species competition for water (e.g. 111). The yield of forage grasses and forbs in semiarid rangelands is inversely related to density of woody perennials (e.g. 6).

Some desert shrubs produce allelopathic substances that inhibit germination and growth of other species, but the significance of this in the field cannot always be proved (72). Salinization of the soil surface by salt accumulating and excreting halophytes, with consequent inhibition of nonhalophytes, is apparently common (16, 57).

Positive effects of shrubs and trees on other plants, as expressed in spatial association, are also often observed in deserts (1, 72). The microenvironmental modifications involved are partly atmospheric (reduction of radiation, temperature, wind, and evaporativity) and partly edaphic (increased organic and nutrient contents, accumulation of windblown sand and silt). Other mechanisms are concentrations of windblown seeds and protection from grazing.

Acknowledgments

Much of the work on the review was done during a summer term in which I had the privilege of staying at the US Desert Biome Center in Logan, Utah. I am grateful in particular to David Goodall, John Hanks, Ron Kickert, and Mark Westoby for their comments on parts of the draft, and to many others of the Biome who agreed to citation of their Research Memoranda or helped in obtaining references. I also wish to thank Michael Evenari and my colleagues in Jerusalem, Patricia Paylore who made available a large bibliography through the information system of the Office of Arid Lands Studies at the University of Arizona, and all those arid zone ecologists from different countries who sent me reprints or preprints.

This work was partly supported by NSF Grant GB 15886 to the US/IBP Desert Biome Program.

Literature Cited

1. Agnew, A. D. Q., Haines, R. W. 1960. Studies on the plant ecology of the Jazira of Central Iraq. *Bull. Coll. Sci. Univ. Baghdad* 5:41–60
2. Alizai, H. V., Hulbert, L. C. 1969. Effects of soil texture on evaporative loss and available water in semi-arid climates. *Soil Sci.* 110:328–32
3. Barger, G. L., Thom, H. C. S. 1949. Evaluation of drought hazard. *Agron. J.* 41:519–26
4. Battan, L. J., Green, C. R. 1971. Summer rainfall over the Santa Catalina Mountains. *Univ. Ariz. Inst. Atmos. Phys., Tech. Rep. 22*
5. Beadle, N. C. W. 1948. *The Vegetation and Pastures of New South Wales.* Dept. Conservation, NSW, Sydney: Govt. Printer. 281 pp.
6. Beale, I. F. 1973. Tree density effects on yields of herbage and tree components in southwest Queensland mulga scrub. *Trop. Grassl.* 7:In press
7. Beatley, J. C. 1967. Survival of winter annuals in the northern Mojave Desert. *Ecology* 48:745–50
8. Beatley, J. C. 1969. Biomass of desert winter annual populations in southern Nevada. *Oikos* 20:261–73
9. Boukhris, M., Lossaint, P. 1972. Specifité biogéochemique des plantes gypsophiles de Tunisie. *Oecol. Plant.* 7:45–68
10. Bridges, K. W., Wilcott, C., Westoby, M., Kickert, R., Wilkin, D. 1972. Nature: a guide to ecosystem modelling. *Ecosystem Modelling Symposium, AIBS Meeting, Minneapolis*
11. Brown, G. W., Ed. 1968, 1972. *Desert Biology,* Vols. I, II. New York: Academic
12. Burrows, W. H. 1972. Productivity of an arid zone shrub *(Eremophila gilesii)* community in southwest Queensland. *Aust. J. Bot.* 20:317–30
13. Burygin, V. A., Markova, L. E. 1972. Winter weather conditions and fodder production of ephemeral pastures in Uzbekistan. See Ref. 89, 114–116
14. Caldwell, M. M., Moore, R. T., White, R. S., De Puit, E. J. 1972. Gas exchange of Great Basin shrubs. *Desert Biome Res. Memo. #20*
15. Chapman, T. G. 1970. Optimization of a rainfall-runoff model for an arid zone catchment. *Int. Assoc. Sci. Hydrol. Publ.* 96:127–144
16. Charley, J. L. 1959. *Soil salinity-vegetation patterns in western New South Wales and their modification by over-* grazing. PhD thesis. Univ. New England, Armidale
17. Chew, R. M., Chew, A. E. 1965. The primary productivity of a desert shrub *(Larrea tridentata)* community. *Ecol. Monogr.* 35:355–75
18. Cohen, D. 1966. Optimizing reproduction in a randomly varying environment. *J. Theor. Biol.* 12:119–29
19. Cohen, D. 1967. Optimizing reproduction in a randomly varying environment when a correlation may exist between the conditions at the time a choice has to be made and the subsequent outcome. *J. Theor. Biol.* 16:1–14
20. Cohen, D. 1970. Expected efficiency of water utilization in plants under different competition and selection regimes. *Isr. J. Bot.* 19:50–54
21. Cowan, I. R. 1965. Transport of water in the soil-plant-atmosphere system. *J. Appl. Ecol.* 2:221–38
22. Coyne, P. I., Cook, C. W. 1970. Seasonal carbohydrate reserve cycles in eight desert range species. *J. Range Manage.* 23:438–44
23. Cunningham, G. L., Strain, B. R. 1969. Ecological significance of seasonal leaf variability in a desert shrub. *Ecology* 50:400–8
24. DePuit, E. J., Caldwell, M. M. 1973. Seasonal pattern of net photosynthesis of *Artemisia tridentata* in northern Utah. In press
25. Desert Biome. 1972. HILRAIN. *Modelling Report Series No. 13*
26. Duckstein, L., Fogel, M. M., Kisiel, C. K. 1972. A stochastic model of runoff producing rainfall for summer type storms. *Water Resour. Res.* 8, No. 3
27. Evenari, M. 1963. Zur Keimungsökologie zweier Wüstenpflanzen. *Mitt. Flor.-Soziol. Arbeitsgemeinsch. Stolzenau/Weser* 10:70–81
28. Evenari, M., Shanan, L., Tadmor, N. H. 1971. *The Negev; The Challenge of a Desert.* Cambridge, Mass.: Harvard Univ. Press. 345 pp.
29. Evenari, M., Schulze, E.-D., Lange, O. L. 1972. Eco-physiological investigations in the Negev desert. III. The diurnal course of carbon dioxide exchange and transpiration and its balance in regard to primary production. See Ref. 89, 66–70
30. Fitzpatrick, E. A., Krishnan, A. I. 1967. A first-order Markov model for assessing rainfall discontinuity in Central Australia. *Arch. Meteorol. Geophys. Bioklimatol. B* 15:242–59

31. Fogel, M. M., Duckstein, L. 1969. Point rainfall frequencies in convective storms. *Water Resour. Res.* 5: 1229–37

32. Friedman, E. I., Galun, M. 1972. Desert algae, lichens and fungi. In *Desert Biology*, Vol. II, ed. G. W. Brown. New York: Academic

33. Friedman, J. 1971. The effect of competition by adult *Zygophyllum dumosum* Boiss. on seedlings of *Artemisia herba-alba* Asso in the Negev Desert of Israel. *J. Ecol.* 59:775–82

34. Gardner, H. R., Hanks, R. J. 1966. Effect of sample size and environmental conditions on evaporation of water from soil. *U.S. Dept. Agr. Conserv. Res. Rep. No. 9*

35. Goodall, D. W. 1967. Computer simulation of changes in vegetation subject to grazing. *J. Indian Bot. Soc.* 46: 356–62

36. Goodall, D. W. 1970. Simulating the grazing situation. In *Concepts and Models of Biomathematics: Simulation Techniques and Methods*, ed. F. Heinmets, 211–36. New York: Dekker

37. Greenway, H., Osmond, C. B. 1970. Ion relations, growth and metabolism of *Atriplex* at high external electrolyte concentrations. *The Biology of Atriplex*, ed. R. Jones, 49–56. Canberra: CSIRO

38. Gutterman, Y., Witztum, A., Evenari, M. 1967. Seed dispersal and germination in *Blepharis persica*. *Isr. J. Bot.* 16:213–34

39. Hall, E. A., Specht, R. L., Eardley, C. M. 1964. Regeneration of the vegetation on Koonamore vegetation reserve, 1926–1962. *Aust. J. Bot.* 12:205–64

40. Hanks, R. J., Bowers, S. A. 1962. Numerical solution of the moisture flow equation for infiltration into layered soils. *Soil Sci. Soc. Am. Proc.* 26: 530–34

41. Hanks, R. J., Klute, A., Bresler, E. 1969. A numeric method for estimating infiltration, redistribution, drainage and evaporation of water from soil. *Water Resour. Res.* 5:1064–69

42. Hastings, J. R., Turner, R. M. 1965. *The Changing Mile: an Ecological Study of Vegetation Change with Time in the Lower Mile of an Arid and Semi-Arid Region.* Tucson: Univ. Arizona. 317 pp.

43. Hillel, D., Tadmor, N. H. 1962. Water regime and vegetation in the Central Negev highlands of Israel. *Ecology* 43: 33–41

44. Hills, E. S., Ed. 1966. *Arid Lands.* London: Methuen. 461 pp.

45. Jones, R., Hodgkinson, K. C. 1970. Root growth of rangeland chenopods: morphology and production of *Atriplex nummularia* and *Atriplex vesicaria.* See Ref. 37, 77–85

46. Jurinak, J. J., Griffin, R. A. 1972. Factors affecting the movement and distribution of anions in desert soils. *Desert Biome Res. Memo.* #38

47. Kappen, L., Lange, O. L., Schulze, E.-D., Evenari, M., Buschbom, V. 1972. Extreme water stress and photosynthetic activity of the desert plant *Artemisia herba-alba* Asso. *Oecologia* 10:177–82

48. Karimov, H. H., Cherner, R. I. 1972. Winter vegetation and summer dormancy of plants in the arid zone of Tajikistan. See Ref. 89, 117–18

49. Kassas, M. 1952. Habitat and plant communities in Egyptian desert. I. Introduction. *J. Ecol.* 40:342–51

50. Kassas, M. 1966. Plant life in deserts. See Ref. 44, 145–80

51. Kassas, M., Imam, M. 1959. Habitat and plant communities in the Egyptian desert. IV. The gravel desert. *J. Ecol.* 47:289–310

52. Koller, D. 1972. Environmental control of seed germination. In *Seed Biology*, ed. T. T. Kozlowski, 1–101. New York: Academic

53. Lange, O. L., Schulze, E.-D. 1972. Ecophysiological investigations in the Negev desert. I. The relationship between transpiration and net photosynthesis measured with a mobile field laboratory. See Ref. 89, 57–62

54. Lange, O. L., Schulze, E-D., Koch, W. 1970. Experimentell-ökologische Untersuchungen an Flechten der Negev-Wüste. II. CO_2-Gaswechsel und Wasserhaushalt von Krust-und Blattflechten am natürlichen Standort während der sommerlichen Trockenperiode. *Flora* 159:525–8

55. LeHouerou, H.-N. 1972. An assessment of the primary and secondary production of the arid grazing systems of North Africa. See Ref. 89, 168–72

56. Lieth, H. 1972. Modelling the primary productivity of the world. *UNESCO, Nature and Resources.* In press

57. Litav, M. 1957. The influence of *Tamarix aphylla* on soil composition in the Northern Negev of Israel. *Bull. Res. Counc. Isr. D* 6:38–45

58. Litvinova, N. P. 1972. Productivity of high mountain deserts (Pamirs). See Ref. 89, 143–8

59. Lyford, F. P., Qashu, H. K. 1969. Infiltration rates as affected by desert vegetation. *Water Resour. Res.* 5:1373–76

60. McAlpine, J. R. 1970. Estimating pasture growth periods and droughts from simple water balance models. *Proc. Int. Grassland Congr., 11th* 484–88

61. McDonald, J. E. 1956. Variability of precipitation in an arid region: a survey of characteristics for Arizona. *Univ. Ariz. Inst. Atmos. Phys., Tech. Rep. 1*

62. Macfarlane, W. V., Howard, B. 1970. Water in the physiological ecology of ruminants. In *The Physiology of Digestion and Metabolism in the Ruminant*, ed. A. T. Philipson, 362–74. Aberdeen: Oriel

63. McGinnies, W. G., Goldman, B. J., Paylore, P., Eds. 1968. *Deserts of the World.* Tucson: Univ. Arizona. 188 pp.

64. Maclean, G. L. 1970. The biology of the larks (Alavidae) of the Kalahari sandveld. *Zool. Afr.* 5:7–39

65. Maher, J. V. 1967. Drought assessment by statistical analysis of rainfall. *Aust. N.Z. Assoc. Advan. Sci. Symp. Drought, Melbourne*, p. 57–71

66. Migahid, A. A. 1961. The drought resistance of Egyptian desert plants. *Arid Zone Res.* 16:213–33

67. Monod, Th. 1954. Modes "contracté" et "diffus" de la végétation saharienne. *Proc. Symp. Biol. Hot Cold Deserts*, 35–44

68. Moore, R. T., Breckle, S. W., Caldwell, M. M. 1972. Mineral ion composition and osmotic relationships of *Atriplex confertifolia* and *Eurotia lanata. Oecologia* 11:67–78

69. Moore, R. T., White, R. S., Caldwell, M. M. 1972. Transpiration of *Atriplex confertifolia* and *Eurotia lanata* in relation to soil, plant and atmospheric moisture stresses. *Can. J. Bot.* 50: 2411–18

70. Morello, J. 1951. Multiplicación de arbustos en las mesetas araucanas del valle de Santa Maria. *Bol. Soc. Argent. Bot.* 3:207–17

71. Mott, J. J. 1972. Germination studies on some annual species from an arid region of Western Australia. *J. Ecol.* 60:293–304

72. Muller, C. H. 1953. The association of desert annuals with shrubs. *Am. J. Bot.* 40:53–60

73. Nechayeva, N. T. 1970. *Vegetation of Central Karakum and its Productivity.* Ashkhabad: Ylym. 171 pp. (in Russian)

74. Negbi, M., Evenari, M. 1961. The means of survival of some desert summer annuals. *Arid Zone Res.* 16: 249–59

75. Newsome, A. E. 1965. The abundance of red kangaroos, *Megaleia rufa*, in Central Australia. *Aust. J. Zool.* 13: 269–87

76. Nimah, M. N. 1972. *Model for estimating soil water flow, water content, evapotranspiration and root extraction.* PhD thesis. Utah State Univ., Logan

77. Noy-Meir, I. 1970. *Component analysis of semi-arid vegetation in Southeastern Australia.* PhD thesis. Aust. Nat. Univ., Canberra

78. Oechel, W. C., Strain, B. R., Odening, W. R. 1972. Tissue water potential, photosynthesis, ^{14}C-labeled photosynthate and growth in the desert shrub *Larrea divaricata* Cav. *Ecol. Monogr.* 42:127–41

79. Oppenheimer, H. R. 1960. Adaptation to drought: xerophytism. *Arid Zone Res.* 15:105–38

80. Orshan, G. 1963. Seasonal dimorphism of desert and Mediterranean chamaephytes and its significance as a factor in their water economy. In *The Water Relations of Plants*, ed. A. J. Rutter, F. H. Whitehead, 206–222. London: Blackwell

81. Orshan, G., Diskin, S. 1968. Seasonal changes in productivity under desert conditions. In *Functioning of Terrestrial Ecosystems, Proc. UNESCO Symp. Copenhagen*, 191–201

82. Osmond, C. B. 1970. Carbon metabolism in *Atriplex* leaves. See Ref. 37, 17–21

83. Patten, D. T. 1972. Productivity and water stress in cacti. *Desert Biome Res. Memo.* #17

84. Perry, R. A. 1970. The effect on grass and browse production of various treatments on a mulga community in central Australia. *Proc. Int. Grassland Congr., 11th* 63–66

85. Prill, R. C. 1968. Movement of moisture in the unsaturated zone in a dune area, southwestern Kansas. *US Geol. Surv. Prof. Paper* 600D

86. Qashu, H. K., Evans, D. D., Wheeler, M. L. 1972. Soil factors influencing water uptake by plants under desert conditions. *Desert Biome Res. Memo.* #37

87. Ramirez, L. E. 1971. *Development of a procedure for determining spatial and time variation of precipitation in Venezuela.* PhD thesis. Utah State Univ., Logan; also, *Utah State Univ. Water Res. Lab. Publ.* PRWG 69–3

88. Rodin, L. E., Bazilevich, N. I., Miroshnichenko, Y. M. 1972. Productivity and biogeochemistry of *Artemisietea* in the Mediterranean area. See Ref. 89, 193–8

89. Rodin, L. E., Ed. 1972. *Ecophysiological Foundation of Ecosystems Produc-*

tivity in Arid Zones. Leningrad: Nauka. 232 pp.

90. Rose, C. W. 1968. Evaporation from bare soil under high radiation conditions. *Trans. Int. Congr. Soil Sci., 9th* 1:57–66

91. Ross, M. A. 1969. An integrated approach to the ecology of arid Australia. *Proc. Ecol. Soc. Aust.* 4:67–81

92. Ross, M. A., Lendon, C. 1973. Productivity of *Eragrostis eriopoda* in a mulga community. *Trop. Grassl.* 7:In press

93. Seligman, N. G., Tadmor, N. H., Noy-Meir, I., Dovrat, A. 1971. An exercise in simulation of a semi-arid Mediterranean grassland. *Bull. Rech. Agron. Gembloux, Semaire des problèmes méditerranéens,* 138–43

94. Shanan, L., Evenari, M., Tadmor, N. H. 1967. Rainfall patterns in the Central Negev desert. *Isr. Explor. J.* 17:163–84

95. Sharon, D. 1970. Areal patterns of rainfall in a small watershed. *Int. Assoc. Sci. Hydrol. Publ.* 96:3–11

96. Sharon, D. 1972. The spottiness of rainfall in a desert area. *J. Hydrol.* 17: 161–75

97. Shreve, F. 1942. The desert vegetation of North America. *Bot. Rev.* 8:195–246

98. Slatyer, R. O. 1961. Methodology of a water balance study conducted on a desert woodland (*Acacia aneura*) community. *Arid Zone Res.* 16:15–26

99. Slatyer, R. O. 1967. *Plant-Water Relationships.* New York: Academic. 366 pp.

100. Slatyer, R. O. 1970. Carbon dioxide and water vapour exchange in *Atriplex* leaves. See Ref. 37, 23–29

101. Smith, J. 1949. Distribution of tree species in the Sudan in relation to rainfall and soil texture. *Sudan Min. Agr. Bull.* No. 4

102. Specht, R. L. 1972. Water use by plant communities in the arid zone of Australia. See Ref. 89, 48–52

103. Strain, B. R. 1969. Seasonal adaptations in photosynthetis and respiration in four desert shrubs in situ. *Ecology* 50: 511–13

104. Tadmor, N. H., Evenari, M., Shanan, L. 1972. Primary production of pasture plants as function of water use. See Ref. 89, 151–7

105. Tadmor, N. H., Orshan, G., Rawitz, E. 1962. Habitat analysis in the Negev desert of Israel. *Bull. Res. Counc. Isr. D* 11:148–73

106. Tadmor, N. H., Shanan, L. 1969. Runoff inducement in an arid region by removal of vegetation. *Soil Sci. Soc. Am. Proc.* 33:790–4

107. Ting, I. P., Dugger, W. M. 1968. Nonautotrophic carbon dioxide metabolism in cacti. *Bot. Gaz.* 129:9–15

108. Ting, I. P., Johnson, H. B., Szarek, S. R. 1972. Net CO_2 fixation in Crassulacean Acid Metabolism plants. In *Net Carbon Dioxide Assimilation in Higher Plants,* ed. C. C. Black, 26–53. Cotton

109. Trlica, M. J., Cook, C. W. 1971. Defoliation effects on the carbohydrate reserves of desert range species. *J. Range Manage.* 24:418–25

110. Waisel, Y. 1971. Patterns of distribution of some xerophytic species in the Negev, Israel. *Isr. J. Bot.* 20:101–10

111. Walter, H. 1964. *Die Vegetation der Erde,* Vol. I. Jena, Germany: G. Fischer

112. Went, F. W. 1949. Ecology of desert plants. II. The effect of rain and temperature on germination and growth. *Ecology* 30:1–13

113. West, N. 1972. Biomass and nutrient dynamics of some major cold desert shrubs. *Desert Biome Res. Memo.* #15

114. Westoby, M. 1972. *Problem-Oriented Modelling: a Conceptual Framework.* Presented at Desert Biome Information Meeting, Tempe, Arizona

115. Winkworth, R. E. 1970. The soil water regime of an arid grassland (*Eragrostis eriopoda* Benth.) community in Central Australia. *Agr. Meteorol.* 7:387–99

116. Wit, C. T. de. 1958. *Transpiration and Crop Yields.* Wageningen, Netherlands: Inst. Biol. Scheikundig Onderzoek Landbouwgewasser, Med. 59

117. Yaalon, D. H. 1964. Downward movement and distribution of anions in soil profiles with limited wetting. *Experimental Pedology,* 157–164. London: Butterworths

118. Yaalon, D. H. 1964. Airborne salts as an active agent in pedogenetic processes. *Proc. Int. Congr. Soil Sci., 8th* 5:997–1000. Bucharest: Acad. Soc. Repub. Romania

119. Zohary, M. 1962. *Plant Life of Palestine: Israel and Jordan.* New York: Ronald. 262 pp.

THE STRUCTURE OF LIZARD COMMUNITIES

❖ 4052

Eric R. Pianka

Department of Zoology, University of Texas, Austin, Texas

Strictly speaking, a community is composed of all the organisms that live together in a particular habitat. Community structure concerns all the various ways in which the members of such a community relate to and interact with one another, as well as community-level properties that emerge from these interactions, such as trophic structure, energy flow, species diversity, relative abundance, and community stability. In practice, ecologists are usually unable to study entire communities, but instead interest is often focused on some convenient and tractable subset (usually taxonomic) of a particular community or series of communities. Thus one reads about plant communities, fish communities, bird communities, and so on. My topic here is the structure of lizard communities in this somewhat loose sense of the word (perhaps assemblage would be a more accurate description); my emphasis is on the niche relationships among such sympatric sets of lizard species, especially as they affect the numbers of species that coexist within lizard communities (species density).

So defined, the simplest (and perhaps least interesting) lizard communities would be those that contain but a single species, as, for instance, northern populations of *Eumeces fasciatus*. At the other extreme, probably the most complex lizard communities are those of the Australian sandridge deserts where as many as 40 different species occur in sympatry (20). Usually species densities of sympatric lizards vary from about 4 or 5 species to perhaps as many as 20. Lizard communities in arid regions are generally richer in species than those in wetter areas; therefore, because almost all ecological studies of entire saurofaunas have been in deserts (18, 20, 25), this paper emphasizes the structure of desert lizard communities. As such, I review mostly my own work. Other studies on lizard communities in nondesert habitats are, however, cited where appropriate.

Historical factors such as degree of isolation and available biotic stocks (particularly the species pools of potential competitors and predators) have profoundly shaped lizard communities. Thus one reason the Australian deserts support such very rich lizard communities may be that competition with, and perhaps predation pressures from, snakes, birds, and mammals are reduced on that continent (20).

Climate is also a major determinant of lizard species densities. The effects of various other historical factors, such as the Pleistocene glaciations, on lizard communities are very difficult to assess but may be considerable.

One of the strongest tools available to ecologists is the comparison of ecological systems which are historically independent but otherwise similar. Observations on pairs of such systems allow one to determine the degree of similarity in evolutionary outcome. Moreover, under certain circumstances such natural experiments may even allow some measure of control over such historical variables as the Pleistocene glaciations. For example, faunas of independently evolved study areas with similar climates and vegetative structure should differ primarily in the effects of history upon them.

This paper consists of two major sections. In the first, "Patterns Within Communities," I briefly review fundamental aspects of community structure and lizard niches to establish a frame of reference and to lay the groundwork for the remainder of the paper. Next I discuss ways of quantifying these niche relationships. In the second section, "Comparisons Between Communities," I use these methods to examine and compare three independently evolved desert-lizard systems in some detail; this section is not a review of the literature but a quantitative summary of much of my own research over the last ten years.

PATTERNS WITHIN COMMUNITIES

The number of species coexisting within communities can differ in four distinct ways: (a) More diverse communities can contain a greater variety of available resources, and/or (b) their component species may, on the average, use a smaller range of these available resources (the former corresponds roughly to "more niches," "a larger total niche space," or "more niche dimensions," and the latter to "smaller niches"). (c) Two communities with identical ranges of resources and average utilization patterns per species can also differ in species density with changes in the average degree of overlap in the use of available resources; thus greater overlap implies that more species exploit each resource (this situation can be described as "smaller exclusive niches" or "greater niche overlap"). (d) Finally, some communities may not contain the full range of species they could conceivably support and species density might then vary with the extent to which available resources are actually exploited by as many different species as possible (that is, with the degree of saturation with species or with the number of so-called empty niches). MacArthur (11) summarized all but the fourth of the above factors with a simple equation for the number of species in a community N

$$N = \frac{R}{\bar{U}}\left(1 + C\,\frac{\bar{O}}{\bar{H}}\right) \qquad \text{1.}$$

where R is the total range of available resources actually exploited by all species, $\bar{U}$ the average niche breadth or the range of resources used by an average species, C a measure of the potential number of neighbors in niche space, increasing more or less geometrically with the number of niche dimensions (below), and $\overline{O}/\overline{H}$ the relative amount of niche overlap between an average pair of species. MacArthur improved Equation 1 to handle situations in which resources are not distributed uniformly

$$D_s = \frac{D_r}{\overline{D}_u} \left(1 + \overline{C}\,\overline{a} \right) \qquad\qquad 2.$$

where D_s is the diversity of species in the community, D_r is the overall diversity of the resources exploited by all species, $\overline{D}_u$ is the mean diversity of utilization or the niche breadth of an average species, $\overline{C}$ measures the average number of potential niche neighbors as before, and $\overline{a}$ is a measure of the average amount of niche overlap (MacArthur called this the mean competition coefficient). I return to Equation 2 below after considering various aspects of the niche relationships of lizards and how they can be quantified. Results presented here, however, depend in no way upon the validity of MacArthur's equation.

Niche Dimensions

Animals partition environmental resources in three basic ways: temporally, spatially, and trophically; that is, species differ in times of activity, the places they exploit, and/or the foods they eat. Such differences in activities separate niches, reduce competition, and presumably allow the coexistence of a variety of species (8, 11). Among lizards these three fundamental niche dimensions are often fairly distinct and more or less independent of each other, although they sometimes interact; for example, the mode of foraging can influence all three niche dimensions. For convenience I first treat each major niche dimension separately (below) and then briefly examine ways in which they interact. Rather than refer to "the trophic and temporal dimensions of the niche," etc, I use verbal shorthand and speak of the food niche, time niche, etc.

All else being equal (number of species, niche breadths, niche overlaps, etc), a greater number of effective niche dimensions results in fewer immediate actual neighbors in niche space; moreover, pairs of potential competitors with high overlap along one niche dimension may often overlap relatively little or not at all along another niche dimension, presumably reducing or eliminating competition between them.

TIME NICHE   To the extent that being active at different times leads to exploitation of different resources, such as prey species, temporal separation of activities may reduce competition between lizard species. Perhaps the most conspicuous temporal separation of activities is the dichotomy of diurnal and nocturnal lizards, which are entirely nonoverlapping in the time dimension. However, more subtle temporal differences in daily and seasonal patterns of activity are widespread among lizards, both within and between species. In the North American Sonoran desert, for example, *Uta stansburiana* emerge early in the day and comprise the vast majority of the lizards encountered during the cool morning hours (Table 1). Later, small *Cnemido-*

Table 1 Statistics on time of activity of four species of lizards in the Sonoran desert, expressed as time[a] since sunrise, during the period when temperatures are rising. All means are significantly different (t - tests, $P < .01$).

Species	$\overline{X}$[b]	S.E.[c]	s[d]	N[e]	95% Confidence Limits of Means
Uta stansburiana	3.67	0.06	1.39	470	3.55–3.79
Cnemidophorus tigris	4.11	0.05	1.33	669	4.01–4.21
Callisaurus draconoides	4.60	0.09	1.32	204	4.42–4.78
Dipsosaurus dorsalis	5.83	0.27	1.71	40	5.29–6.37

[a]in hundredths of an hour   [d]standard deviation
[b]arithmetic mean   [e]sample size (number of lizards)
[c]standard error of the mean

phorus tigris appear, while still later larger *C. tigris* emerge. As air and substrate temperatures rise with the daily march of temperature other species such as *Callisaurus draconoides* and *Dipsosaurus dorsalis* become active (Table 1). Similar patterns of gradual sequential replacement of species during the day occur in Australian skinks of the genus *Ctenotus* (21) and in lacertid lizards in the Kalahari desert of southern Africa (25). Daily patterns of activity also change seasonally with later emergence during cooler winter months than in warm summer ones (4, 13, 21, 23, 27, 30, 31, 46). Species with bimodal daily activity patterns during warm months (early and late in the day) often have a unimodal activity period during cooler months (13, 21, 30, 31, 46). Such seasonal changes in the time of activity presumably allow a lizard to encounter a similar thermal environment and microclimate over a period of time when the macroclimate is changing. Standardizing times of activities to "time since sunrise" (diurnal species) or "time since sunset" (nocturnal species) corrects for such seasonal shifts in time of activity and greatly facilitates comparison among species (Table 1) as well as comparisons between communities (below). Body temperatures of active individuals often reflect the time of activity reasonably well (21), although body temperature can be strongly affected by microhabitat(s) as well (4, 13, 14, 21, 26, 30, 32, 44). Thus species that emerge earlier in the day frequently have lower active body temperatures than those that emerge later; indeed, body temperature can sometimes be used as an indicator of time (21) or thermal (36, 41, 43) niche. The anatomy and size of a lizard's eyes are another useful indicator of its time niche; large eyes and elliptical pupils almost invariably indicate nocturnal activity (48).

PLACE NICHE   The use of space varies widely among lizard species. A few are entirely subterranean (fossorial), many others are completely terrestrial, while still others are almost exclusively arboreal. Various degrees of semifossorial and semiarboreal activity also occur. Microhabitat differences among species are often pronounced even within these groups. Thus some terrestrial species forage primarily in the open spaces between plants, whereas others forage mainly under or within plants, the plants sometimes having a particular life form. Similar subtle differences in the use of various parts of the vegetation also occur among arboreal lizard species, especially *Anolis* (35, 36, 39, 41, 43). Some lizard species are strongly restricted to a rock-dwelling (saxicolous) existence. In addition to such microhabitat specificity, various species have specialized in their habitat requirements. Thus different sets of species of Australian desert lizards are restricted to sandridge, sandplain, and shrubby habitats respectively (21, 28). As defined here the place niche is more inclusive than Rand's (35) structural niche, as it includes both habitat and microhabitat preferences. Exactly where in the environmental mosaic a lizard forages, as well as its mode of foraging in that space, is perhaps its most important ecological attribute.

Lizards that exploit space in different ways have evolved a variety of morphological adaptations for the use of space (21, 30, 33, 37); such anatomical traits are often accurate indicators of their place niche. Thus fossorial species typically have either very reduced appendages or none at all. Diurnal arboreal lizards are usually long-tailed and slender. Terrestrial species that forage in the open between shrubs and/or grass clumps generally have long hind legs relative to their size, while those that forage closer to cover or within dense clumps of grass usually have proportionately shorter hind legs (21, 30, 33). Lamellar structure often reveals arboreal or terrestrial activity as well as the texture of the substrate exploited (1). Moreover, terrestrial geckos have proportionately larger eyes than arboreal ones (33, 48).

FOOD NICHE   Most lizards are insectivorous and fairly opportunistic feeders, taking without any obvious preference whatever arthropods they encounter within a broad range of types and sizes. Smaller species or individuals, however, do tend to eat smaller prey than larger species or individuals (6, 21, 33, 38, 39, 43); also, differences in foraging techniques (below) and place and time niches often result in exposure to a different spectrum of prey species. Rather few lizard species have evolved severe dietary restrictions; among these are the ant specialists *Phrynosoma* and *Moloch* (17, 31, 32), termite specialists such as *Rhynchoedura* and *Typhlosaurus* (7, 33), various herbivorous lizards which include *Ctenosaurus, Dipsosaurus, Sauromalus,* and *Uromastix,* and secondary carnivores such as *Crotaphytus, Heloderma, Lialis,* and *Varanus* which prey primarily upon the eggs and young of vertebrates and the adults of smaller species (17, 19, 22, 24). All the above foods are at least temporarily very abundant making food specialization advantageous (12). Just as lamellar structure and hind leg proportions reflect the place niche of a lizard, head proportions, jaw length, and dentition frequently prove to be useful indicators of the food niche (6, 21, 38), especially of the sizes and kinds of prey eaten.

Another, somewhat more behavioral, aspect of a lizard's food niche concerns the way in which it hunts for prey. Two extreme types of foragers have been recognized (17, 40, 42): a lizard may either actively search out prey (widely foraging strategy) or wait passively until a moving prey item offers itself and then ambush the prey (sit-and-wait strategy). Normally the success of the sit-and-wait method requires a fairly high prey density, high prey mobility, and/or a low energy demand by the predator (40, 42). The effectiveness of the widely foraging tactic also depends on the density and mobility of prey and the predator's energy needs, but in this case the distribution of prey in space and the searching abilities of the predator may take on considerable importance (40, 42). Clearly, this dichotomy is artificial and these two tactics actually represent pure forms of a variety of possible foraging strategies. However, the dichotomy has substantial practical value because the actual foraging techniques used by lizards are often strongly polarized. Thus most teids and skinks and many varanid and lacertid lizards are very active and widely foraging, typically on the move continually; in contrast, almost all iguanids, agamids, and geckos are relatively sedentary sit-and-wait foragers. These differences in the mode of foraging presumably influence the types of prey encountered, thus affecting the composition of a lizard's diet.

INTERACTIONS BETWEEN THE TIME, PLACE, AND FOOD NICHES   Place niches and food niches of lizards change in time, both during the day and with the seasons. In the early morning, when ambient air and substrate temperatures are relatively low, lizards typically locate themselves in the warmer microhabitats of the environmental mosaic, such as depressions in the open sun or the sunny side of a rock, slope, sandridge, or tree trunk. Often an animal orients its body at right angles to the sun's beams, thereby maximizing heat gained from the sun. Later in the day as environmental temperatures rise the same lizards usually spend most of their time in the cooler patches in the environmental mosaic, such as shady spots underneath shrubs or trees (4, 26, 27, 41). Finally, as the surface gets still hotter many lizards retreat into cool burrows; certain species, such as *Amphibolurus inermis,* climb up off the ground into cooler air and face into the sun, minimizing their heat load due to solar irradiation (4, 26). Thus time of activity strongly affects a lizard's place niche and its habitat and microhabitat requirements may dictate periods when the animal can be active.

Similarly, the composition of the diet of many lizards changes as the relative abundances of different types of prey fluctuate with the seasons (and probably within a day). Nocturnal lizards clearly encounter a different spectrum of potential prey items than diurnal lizards, and those that forage in different places usually encounter different prey. The mode of foraging or the way in which a lizard uses space can influence both its place and food niches; thus widely foraging species typically have broader place niches than sit-and-wait species, while the latter type of foragers often tend to have broader food niches than the former. Recall that pairs of lizard species with high overlap along one niche dimension, say microhabitat, may have low overlap along another niche dimension such as foods eaten, effectively reducing interspecific competition between them.

Niche Breadth and Niche Overlap

In addition to the differences in times of activity and use of space and foods noted above, lizard species differ in the spans of time over which they are active as well as the ranges of spatial and trophic resources they exploit. As outlined above, such differences in niche breadth may have a considerable impact upon the structure and diversity of lizard communities. Following MacArthur (11), niche breadth along any single dimension is here quantified using Simpson's index of diversity

$$B = 1/\sum_{i}^{n} p_i^2$$

3.

where p_i represents the proportion of the ith time period (or microhabitat or food type) actually used; B varies from unity to n depending upon the p_i values. Niche breadths based on a different number of p_i categories can be compared after standardizing them by dividing by n. Overall niche breadth along several niche dimensions can be estimated either as the product or the geometric mean of the breadths along each component dimension (recall that the lower bound on B is one) or by the arithmetic mean of the latter breadths.

Niche overlap also varies among lizard species and between communities. Overlap along any single niche dimension can be quantified in a wide variety of ways (2, 5, 10, 21, 34, 39, 47). Here I use still another measure of overlap, based upon Levins' (10) formula for α

$$a_{jk} = \frac{\sum_{i}^{n} p_{ij} p_{ik}}{\sum_{i}^{n} p_{ij}^2} \qquad a_{kj} = \frac{\sum_{i}^{n} p_{ij} p_{jk}}{\sum_{i}^{n} p_{ik}^2}$$

4.

where p_{ij} and p_{ik} are the proportions of the ith resource used by the jth and the kth species respectively. The above equations have been used to estimate the so-called competition coefficients (10, 11, 47), and give different α values for each partner in a niche overlap pair provided that niche breadths (the inverse of the denominators in Equation 4) differ. Here I use the following multiplicative measure of overlap

$$O_{jk} = O_{kj} = \frac{\sum_{i}^{n} p_{ij} p_{ik}}{\sqrt{\sum_{i}^{n} p_{ij}^2 \sum_{i}^{n} p_{ik}^2}}$$

5.

where the p_{ij} and p_{ik} are defined as before (I am indebted to Selden Stewart for suggesting this equation). Equation 5 is symmetric and gives a single overlap value for each niche overlap pair; it can never generate values less than zero or greater than one [Equation 4, however, does give one α value (of a pair) that is greater than unity provided niche breadth and overlap are high]. Overall niche overlap along several niche dimensions can be estimated by the product of the overlaps along each component dimension (10, 21), although this procedure may either overestimate or underestimate overall overlap (H. S. Horn, personal communication; R. M. May, unpublished). Thus if niches are completely separated along any single niche dimension both niche overlap along that dimension and overall niche overlap are zero.

COMPARISONS BETWEEN COMMUNITIES

During the last decade I have studied in some detail three independently derived and evolved, but otherwise basically comparable, sets of desert lizard communities at similar latitudes in western North America, southern Africa, and Western Australia. Here I use data from these studies to quantify and compare various parameters of lizard niches. Although lizards were studied on 32 different study areas (below) I lump data from various study areas within each continental desert-lizard system here for brevity and clarity (a more detailed area by area analysis will be undertaken elsewhere). A few allopatric species pairs are thus treated as though they are sympatric, but the vast majority of the species considered are sympatric on one or more study areas.

The number of sympatric lizard species on 14 North American desert study areas varies from 4 to 11, with either 4 or 5 sympatric species in the northernmost Great Basin desert, 6–8 species in the more southern Mojave and Colorado deserts, and 9–11 species in the still more southerly Sonoran desert (16–18). (The analysis to follow includes only 10 southern North American desert study areas.) Ten study areas in the Kalahari desert of southern Africa support 12–18 sympatric species of lizards (25). In the Western Australian desert 18–40 species of lizards occur together in sympatry on eight different study sites (20, 21, 33). In addition to such censuses of lizard species densities, I gathered supporting data on the physiography, climate, vegetation, and faunas of each of the 32 desert study areas (15–18, 20, 21, 25, 28–31).

The actual diversity of lizards observed on all sites within each desert-lizard system, estimated using the relative abundances of the various species in my collections (below) as p_i's in Equation 3, are: North America = 3.0 (28% of the maximum possible diversity of 11), Kalahari = 12.5 (60% of the maximum possible diversity of 21), and Australia = 19.0 (32% of the maximum possible diversity of 59). (These are crude approximations of the actual lizard diversities, both because real relative abundances doubtless differ somewhat from the relative abundances in my samples and because not all species actually occur in sympatry.)

Time of activity and microhabitat were recorded for most active lizards encountered. Table 2 lists the average numbers of species in five basic time and/or place niches in each desert system (see also below). Wherever possible, lizards were collected; these specimens[1] allowed analysis of stomach contents. Twenty basic prey categories, corresponding roughly to various orders of arthropods, were distinguished. Both the numbers and volume of prey items in each category were recorded for every stomach.

I used these data on time of activity, microhabitat usage, and stomach contents for the following analyses of the time, place, and food niches of desert lizards. The numbers of lizards active at different times were grouped by species into 22 hourly categories expressed in time since sunrise for diurnal species (14 categories) and time since sunset for nocturnal ones (limitations on human endurance allowed only 8

[1]Some 5000 North American lizards, over 6000 Kalahari lizards, and nearly 4000 Australian ones, all of which are now lodged in the Los Angeles County Museum.

Table 2   Average numbers of species of lizards in five basic niche categories on study areas in the three desert systems. The percentage of the average total number of species in each system is also given.

Niche Category	North America[a]		Kalahari[b]		Australia[c]	
	$\overline{X}$	%	$\overline{X}$	%	$\overline{X}$	%
diurnal terrestrial	5.7	69	6.3	43	14.4	51
diurnal arboreal	1.2	14	1.9	13	2.6	9
nocturnal terrestrial	1.4	17	3.5	24	7.6	27
nocturnal arboreal	0.0	0	1.6	11	2.6	9
fossorial	0.0	0	1.4	10	1.1	4
totals	8.3	100	14.8	101	28.3	100

[a]10 different southern study areas

[b]10 study areas

[c]8 study areas

nocturnal hourly categories); these 22 time categories were used as p_i's in the above equations. Fifteen basic microhabitat categories were recognized and used as p_i's. Time and place niche breadths and overlaps were calculated for desert lizards in these three independently evolved systems of lizard communities using Equations 3 and 5 and the above data on the numbers of lizards active at different times and in different microhabitats. The overall span or diversity of time of activity of all the lizards in each continental desert system (D_r in Equation 2), as well as the microhabitats used by them, were estimated using Equation 3 and the proportions of each time period or microhabitat type as computed from grand totals summed over all lizard species. Stomach content data (prey items by volume[2]) allowed similar calculations of food niche breadths and overlaps, as well as the average and overall diversity of foods eaten by all lizards, $\overline{D}_u$ and D_r, in each of the above deserts. Mean niche breadths of all the species in a given community ($\overline{D}_u$ in Equation 2) were also calculated for the time and place niches. Average niche overlap along each niche dimension in any particular community was calculated as the arithmetic mean of all interspecific overlaps (calculated from Equation 5); products of these values were also computed to estimate overall niche overlap.

Diversity of Resources Used by Lizards

The overall diversity of times of activity of all lizards (D_r for the time niche) in each desert-lizard system was computed using Equation 3 and the proportional representation of the 22 hourly time categories among all species (recall that these categories are expressed in hours since sunrise or sunset and that they therefore correct

[2]Prey items in the same 20 categories by numbers of items, rather than their volumetric importance, and prey in 34 size categories (irrespective of type) were also examined, but are not considered further here because there is very little niche separation in either of these two aspects of the food niche.

somewhat for seasonal shifts in activity patterns). Overall diversity of time of activity thus computed is quite low in North America (5.9 or only 27% of the maximum possible value of 22) and nearly twice as large in the Kalahari and Australia (11.6 and 11.7 respectively, or about 53% of the possible maximum). A major factor contributing to the greater diversity of time of activity in the Kalahari and Australia is the increased numbers of nocturnal lizards in the southern hemisphere (Table 2), although the diversity of time of activity of diurnal lizards is also somewhat higher in these two deserts than in North America. Lizards are active year around in the Kalahari and Australia and they were sampled over the entire year, while the seasonal period of activity is shorter in North America and lizards were sampled only over a six-month period. Whatever the reason(s) for this difference between the desert systems, the more diverse communities of the Kalahari and Australia certainly exhibit much greater temporal variation in their times of activity on both a daily and a seasonal basis than the less diverse North American lizard community.

Overall microhabitat diversity, computed using Equation 3 and the 15 basic microhabitat categories as exploited by all the lizards in each system, represents D_r for the place niche; again, it is very low in North America (3.3 or only 22% of the maximal value of 15), where the vast majority of lizards were first sighted in the open sun, and considerably higher in the Kalahari (8.8 or 59% of maximum) and Australia (8.2 or 55% of maximum). These differences in the diversity of microhabitats actually used by lizards are due partly to an increased incidence of arboreal and subterranean lizards in the two deserts of the southern hemisphere (Table 2), although more animals are also first sighted in the shade of various types of plants (Table 3). Nocturnality is much more prevalent in the Kalahari and Australia (Table 2) and contributes to the increased use of shade in these lizard communities (nocturnal lizards were arbitrarily assigned to shade categories in Table 3, although this somewhat confounds place and time niches).

Somewhat surprisingly, the overall diversity of foods eaten by all the lizards[3] in a community, or D_r for the food niche, is lowest in the Kalahari (4.4 or 22% of the maximal value of 20), intermediate in Australia (7.4 or 37% maximum), and highest in the least diverse lizard communities of North America (8.7 or 44% of maximum). The low diversity of foods eaten by Kalahari lizards stems from the preponderance of termites in the diets of these lizards (Table 4). Examination of Table 4 shows that the proportions of various prey categories actually eaten by lizards differ markedly among the desert systems. For example, although termites are a major food item in all three deserts, their fraction of the total prey eaten by all lizards is considerably higher in the Kalahari (41.3%) than in either of the other deserts (16.5 and 15.9%). Prominent prey in the Australian desert are vertebrates (24.8%), especially lizards, and ants (16.4%). By volume, beetles constitute 18.5% of the food eaten by North American desert lizards, 16.3% of that eaten by Kalahari lizards, but only 7.3% of the Australian desert lizard diet.

[3]Computed using Equation 3 and the proportion of the total volume of food in each of 20 prey categories in the stomachs of all the lizards collected in a series of communities from each desert-lizard system.

Table 3 Microhabitats actually used by all lizards in three different desert systems. Nocturnal lizards assigned to shade categories. Numbers (N) and percentages (%).

Microhabitat Category	North America		Kalahari		Australia	
	N	%	N	%	N	%
Subterranean	0	0.0	579	12.1	17	0.5
Terrestrial						
open sun	1335	45.3	890	18.6	596	19.0
grass sun	92	3.1	155	3.2	314	10.0
bush sun	883	30.0	547	11.4	192	6.2
tree sun	103	3.5	126	2.6	31	1.0
other sun	95	3.2	6	0.1	14	0.4
open shade	49	1.7	546	11.4	547	17.4
grass shade	2	0.1	274	5.7	525	16.6
bush shade	165	5.6	765	15.9	221	6.9
tree shade	30	1.0	179	3.7	81	2.6
other shade	72	2.4	18	0.4	43	1.3
Arboreal						
low sun	12	0.4	125	2.6	56	1.5
low shade	6	0.2	109	2.3	224	7.0
high sun	50	1.8	200	4.2	91	2.0
high shade	51	1.8	276	5.8	250	7.7
TOTALS	2945	100.1	4795	100.0	3202	100.1

Table 4 Major prey items in the stomachs of all lizards in three different desert systems by volume in cubic centimeters.

Prey Category	North America		Kalahari		Australia	
	volume	percentage	volume	percentage	volume	percentage
spiders	50	1.6	36	3.1	54	3.4
scorpions	23	0.7	33	2.9	22	1.4
ants	307	9.7	155	13.6	261	16.4
locustidae	364	11.5	70	6.1	138	8.7
blattidae	100	3.2	4	0.4	37	2.3
beetles	587	18.5	187	16.3	117	7.3
termites	525	16.5	473	41.3	253	15.9
homoptera-hemiptera	31	1.0	15	1.3	30	1.9
lepidoptera	68	2.1	16	1.4	9	0.5
all larvae	384	12.1	41	3.6	80	5.0
Miscellaneous arthropods	225	7.0	76	6.6	107	6.7
vertebrates	246	7.8	26	2.3	395	24.8
plants	262	8.3	13	1.2	89	5.6
TOTALS	3172	100.1	1145	100.0	1592	99.9

To give each niche dimension equal weight the above estimates of D_r were standardized by dividing by the number of p_i categories and multiplying by 100, thus expressing the diversity of use of resources as a percentage of the maximal possible resource diversity along a given niche dimension. The overall diversity of resources used by all lizards in all three niche dimensions was then computed as the product of the above three standardized D_r values divided by 1000. So estimated, overall diversity of resources used is lowest in North America (25.9), intermediate in the Kalahari (68.9), and highest in Australia (107.5); moreover, these estimates of the size of the lizard niche space are directly proportional to observed lizard diversities in the various deserts (above).

Differences in Niche Breadth

Niche breadths for the food, place, and time niches, as well as their products (overall niche breadth) were calculated for 91 species of desert lizards in 10 families on the three continents. Frequency distributions and averages of all the species in each desert-lizard system are shown for each niche dimension in Figure 1; these mean niche breadths represent the average diversity of utilization of each niche dimension, or $\bar{D}_u$ in Equation 2, by the lizards in a given system. In all three deserts average time niche breadths are very similar, though their frequency distributions differ

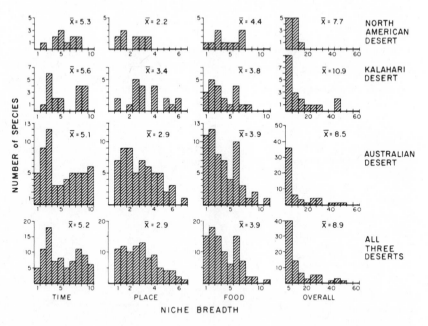

Figure 1 Frequency distributions of niche breadths of 91 species of desert lizards along three major niche dimensions in three deserts. Overall niche breadths, computed as the products of the standardized breadths along each component dimension, weight each niche dimension equally. See text for discussion.

(Figure 1). The frequency distribution of time niche breadth of North American lizards is fairly continuous, but these distributions are distinctly bimodal in the Kalahari and Australia where most nocturnal species have relatively narrow time niches while diurnal ones generally have comparatively broader time niches. (The narrow time niches of nocturnal lizards are probably an artifact due to the shorter nighttime sampling period; however, this bias is similar in all three deserts and should not generate differences between the desert systems.) Place niche breadths are more evenly distributed than time niche breadths, although the distributions are skewed with more narrow place niches than broad ones (Figure 1); place niches are smallest in North America ($\overline{x} = 2.2$, or 15% of maximal value), intermediate in Australia ($\overline{x} = 2.9$, or 19% of maximum), and broadest in the Kalahari ($\overline{x} = 3.4$, or 23% of maximum). In all three deserts food niche breadths appear to be distinctly bimodal, suggesting a natural dichotomy of food specialists versus food generalists (Figure 1). Average food niche breadth is fairly similar in all three deserts and is largest in North America.

Because species with broad niches along one dimension often, though by no means always,[4] have narrow niches along another dimension, overall niche breadths are strongly skewed with the majority of species having rather narrow overall niches (Figure 1). Nevertheless, a few species in the Kalahari and Australia with broader than average niches along all three niche dimensions have extremely broad overall niches (Figure 1). Average overall niche breadth is smallest in North America (7.7), intermediate in Australia (8.5), and largest in the Kalahari (10.9). However, overall niche breadths, as well as average overall niche breadths, do not differ strikingly between the desert systems; indeed, if anything, overall niches tend to be slightly larger in the more diverse communities, rather than smaller as might have been anticipated.

Niche Dimensionality

Any given niche dimension's potential to separate niches, and thus its potential effectiveness in reducing interspecific competition, should be roughly proportional to the ratio of the overall diversity of use of that niche dimension divided by the diversity of utilization by an average species, or $D_r/\overline{D}_u$. Table 5 summarizes much of the above discussion and lists the ratios of $D_r/\overline{D}_u$ for each major niche dimension in the three desert-lizard systems. Estimates for each niche dimension are also multiplied to give overall estimates (products of the standardized estimates for each component dimension). Thus measured, the dimension with the greatest apparent potential to separate niches in North America is food, which, by the same criteria, is a comparatively negligible niche dimension in the Kalahari; conversely, by these standards place and time niches seem to have a much greater potential to separate niches of Kalahari lizards than North American ones (Table 5). All three niche dimensions, especially place and time, appear to have the potential to separate niches of Australian lizards. The products of the $D_r/\overline{D}_u$ ratios for all three dimen-

---

[4]Product moment correlation coefficients among niche breadths along various dimensions range from –0.38 to 0.40 and are generally weak and seldom statistically significant.

Table 5 Estimates of various niche parameters (see text and Table 6).

Desert and Niche Dimension	D_r	$\overline{D}_u$	$D_r/\overline{D}_u$	$\overline{C}$	Mean Overlap (all pairs)	Mean Overlap (nonzero pairs)
North America						
time	25.4	24.2	1.05	3.0	0.58	0.86
place	22.0	14.6	1.51	3.0	0.34	0.55
food	43.7	22.0	1.98	1.2	0.46	0.49
overall	25.9	7.7	3.34	-9.5	0.09	0.23
Kalahari						
time	52.7	25.4	2.07	11.7	0.43	0.78
place	58.9	22.8	2.58	13.3	0.29	0.38
food	22.2	18.8	1.18	14.9	0.64	0.64
overall	68.9	10.9	6.34	12.2	0.08	0.27
Australia						
time	53.3	23.1	2.31	22.9	0.32	0.54
place	54.8	19.1	2.87	19.2	0.29	0.35
food	36.8	19.3	1.90	28.4	0.32	0.36
overall	107.5	8.5	12.62	17.3	0.03	0.13

sions (Table 5), which should be proportional to the overall potential for niche separation, increase from North America (3.3) to the Kalahari (6.3) to Australia (12.6), as might be expected. Hence, as measured by $D_r/\overline{D}_u$, the potential for niche partitioning seems to be greater in more diverse lizard communities; moreover, this potential is directly proportional to actual lizard diversities observed.

Differences in Niche Overlap

Figure 2 shows the frequency distributions of niche overlap values for all interspecific pairs along each niche dimension in the three desert systems (calculated using Equation 5). Estimates of overall overlap, computed as the products of the overlap along the three niche dimensions, are shown at the right of the figure. Although there are some striking differences and trends in overlap patterns,[5] among both niche dimensions and deserts, overall overlaps are uniformly low in all three deserts (Figure 2 and Tables 5 and 6). The vast majority of interspecific pairs overlap very little or not at all when all three dimensions are considered. This is demonstrated by low overall overlap values and by the size of the "zero" classes of overall overlap in the various deserts (Tables 5 and 6). Table 5 gives averages both for all overlap pairs and for only those pairs which overlap somewhat (that is, all pairs other than those with zero overlap) for each niche dimension and for overall overlap estimates. Provided average niche breadth ($\overline{D}_u$) remains relatively constant, the number of possible nonoverlapping pairs increases markedly as overall niche space (D_r) increases. Hence the average niche overlap of pairs with some overlap is of interest as it should reflect the limiting similarity and/or maximal tolerable overlap

[5]For instance, distributions of time niche overlap are distinctly bimodal in all three deserts (particularly North America and the Kalahari), reflecting the nonoverlapping times of activity of nocturnal and diurnal species.

Table 6 Summary of overall niche overlap patterns (see text and Table 5).

Desert System	Total Number of Overlap Pairs	Number of Zero Overlap Pairs	Zero Overlap Pairs as % of Total	Number of Nocturnal-Diurnal (ND) Pairs	ND Pairs as % of Zero Overlap Pairs	Number of Non-ND Pairs with Zero Overlap	Non-ND Pairs with Zero Overlap as % of Zero Overlap Pairs	Non-ND Pairs with Zero Overlap as % of all Overlap Pairs
North America	55	37	67%	18	49%	19	51%	35%
Kalahari	171	101	59%	78	77%	23	23%	13%
Australia	1596	1255	78%	680	54%	575	46%	36%

in each desert system. Although a substantial number of nonoverlapping pairs are nocturnal-diurnal species pairs, many non-nocturnal–diurnal pairs also do not overlap (Table 6). The proportion of such zero overlap pairs is distinctly lower in the Kalahari desert, where only 23% of the non-nocturnal–diurnal pairs do not overlap, than in North America and Australia (51 and 46% respectively). Furthermore, the average overlap among all nonzero overlap pairs tends to be somewhat greater in the Kalahari and North America than in Australia, suggesting that maximal tolerable niche overlap is lower in the latter desert (Table 5).

Although niche overlap values are far from normally distributed (Figure 2), arithmetic means [especially of the nonzero overlap values (Table 5)] do reflect differences between the various niche dimensions and deserts. Average overlap in microhabitat is low and generally similar in all three deserts, while average overlaps in the time and food niches are considerably more variable (Figure 2 and Table 5). Average time niche overlap is high in North America, while both average food and time niche overlaps are high in the Kalahari. In Australia, average niche overlap values are low along all three niche dimensions (Table 5). As a result, overall overlap is distinctly lower in Australia than in the other two desert systems. Thus overall niche overlap seems to vary inversely with lizard species diversity.

Numbers of Neighbors in Niche Space

By far the most difficult parameter to estimate in Equation 2 is the number of neighbors in niche space C (indeed, MacArthur did not indicate how one might

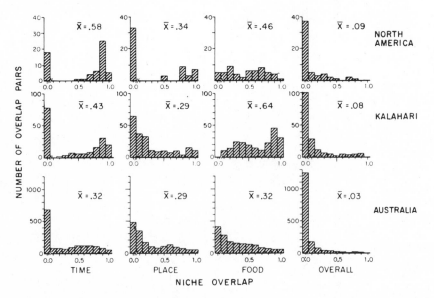

Figure 2 Frequency distributions of niche overlap values of desert lizards along three major niche dimensions in three deserts. See text for discussion.

attempt to estimate C). This quantity cannot be estimated satisfactorily from my data in an independent way; however, $\overline{C}$ can be calculated by simply rearranging Equation 2 to solve for $\overline{C}$

$$\overline{C} = \frac{\left\{ (\overline{D}_u / D_r) D_s - 1 \right\}}{\overline{O}}$$

6.

Values of $\overline{C}$ estimated by substituting various estimates of other parameters (above) into Equation 6 are listed in Table 5. These values appear to be reasonable for any single niche dimension. However, the estimate of the number of neighbors in overall niche space (all three niche dimensions) is actually negative for North America. Estimates of the number of neighbors in overall niche space are much higher and more reasonable in the Kalahari and Australia (Table 5).

As indicated earlier, communities can differ in species diversity with differences in the extent to which they contain as many different species as they can support. The negative estimate of the number of neighbors in overall niche space in North America suggests that lizard diversity in these deserts may actually be lower than it could potentially be, or that these deserts may not be truly saturated with species. Further, the complete absence of any fossorial lizards or any which are both nocturnal and arboreal in North America (Table 2) suggests that these niches either (a) do not exist, (b) are unoccupied, or (c) are occupied but by another kind of animal (see next section). (Indeed, I would be quite surprised if a successful climbing gecko such as the Australian *Gehyra variegata* were unable to invade the North American desert without a simultaneous extinction of another nocturnal animal.)

Reciprocal Relations With Other Taxa

The ecological roles of lizards and various other taxa, especially birds and mammals, are strongly interdependent (9). Thus lizards may capitalize on variability of primary production, and this might be a factor contributing to their relative success over birds in desert regions (18, 20, 25). There are proportionately more species of ground-dwelling insectivorous birds in the Kalahari than there are in Australia (29), suggesting that competition between birds and lizards may be keener in southern Africa than it is in Australia. Figure 3 plots the number of bird species against the number of lizard species on 27 study areas representative of each desert system. As the total number of species increases, the numbers of bird species increase faster than lizard species in North America and the Kalahari, whereas in Australia lizards increase faster than birds. This figure suggests a sharp upper bound on the number of sympatric lizard species in North America and the Kalahari, but no such limit in Australia. Exactly the reverse seems to be true of birds in the three continental desert systems; that is, a distinct upper limit on bird species diversity appears to exist in Australia, but not in either North America or the Kalahari. The reasons for such differences between the three desert systems are elusive and must remain conjectural (9). There are very few migratory bird species in Australia, whereas a number of migratory birds periodically exploit the North American and Kalahari deserts; competitive pressures from these migrants must have their effects upon the lizard

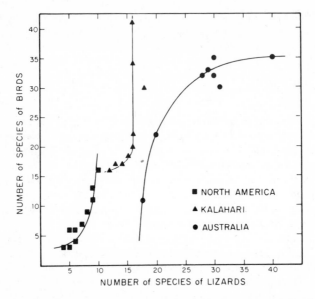

Figure 3 Number of species of birds plotted against the number of lizard species on various study areas within three desert systems. See text.

communities in the latter two desert systems. The higher incidence of arboreal, fossorial, and nocturnal lizard species in the Kalahari and Australia, as compared with North America (Table 2), are probably related to fundamental differences in the niches occupied by other members of these communities such as arthropods, snakes, birds, and mammals (20, 25). These differences in the composition and structure of the various communities presumably have a historical basis. Thus southern Africa has an exceptionally rich termite fauna, which in turn may have allowed the evolution of termite-specialized subterranean *Typhlosaurus* species (7). The prevalence of nocturnality among Kalahari and Australian lizards may arise from variations among systems in either or both of the following: (*a*) differences in the diversity of available nocturnal resources, such as nocturnal insects, or (*b*) differences in the numbers and/or densities of insectivorous and carnivorous nocturnal birds and mammals. The mammalian fauna of the Australian desert is conspicuously impoverished, and the snake fauna less so; in this desert system varanid and pygopodid lizards are ecological equivalents of carnivorous mammals and snakes, respectively, in North America and the Kalahari (20, 25). Such usurpation of the ecological roles of other taxa in the other deserts has expanded the diversity of resources exploited by Australian desert lizards (20).

Within-Habitat and Between-Habitat Diversity

Overall species diversities in an area (as opposed to point diversities) can differ in a way that is included neither in Equation 2 nor in the above analysis of niche relationships. Thus only the so-called "within-habitat" component of diversity (11, 25) was considered above (indeed, for brevity and clarity, data from various different study areas within each desert-lizard system were lumped for the above analyses). The other way in which communities can differ in species diversity is through differences in species composition from area to area or habitat to habitat within a study area (no study area is perfectly homogeneous); such horizontal turnover in species composition represents the so-called "between-habitat" component of diversity (11). To estimate the amount of between-habitat diversity in each of the above desert-lizard systems I calculated coefficients of community similarity[6] for every pair of lizard communities within each continental desert system (25). Community similarity values are high and rather uniform in the North American desert ($\bar{x} =$ 0.67, S. E. $= 0.019$, $s = 0.153$, $N = 66$) and the Kalahari desert ($\bar{x} = 0.67$, S. E. $= 0.015$, $s = 0.127$, $N = 66$), indicating little difference between study areas in species composition (i.e. a low between-habitat component of diversity). However, community similarity values are significantly lower (t-tests, $P < 0.01$) in the Australian desert ($\bar{x} = 0.49$, S. E. $= 0.027$, $s = 0.144$, $N = 28$), demonstrating that this component of diversity is greater in that desert system. Habitat specificity is much more pronounced in Australian desert lizards than it is in North American or Kalahari desert lizards (20, 25, 28). For example, although both the Kalahari and the Australian deserts are characterized by long stabilized sandridges, only a single species [*Typhlosaurus gariepensis* (7)] is specialized to Kalahari sandridges whereas ten lizard species are sandridge specialists in Australia (20, 28).

TAXONOMIC COMPONENTS OF LIZARD SPECIES DENSITY

Because closely related species are often ecologically similar and therefore in strong competition when they occur together, Elton (3) suggested that competitive exclusion should occur more frequently between pairs of congeneric species than between more distantly related pairs of species. Moreover, he reasoned that if this argument is valid fewer pairs of congeneric species should occur within natural communities than in a random sample of species and genera from a broader geographic area which includes several to many different communities. Frequent cases of abutting allopatry (parapatry) of congeners seem to support this argument. Elton examined the numbers of congeneric species in portions of many different natural communities and found evidence for such a paucity of congeners, even in spite of the bias towards an increased number of congeneric pairs due to the possibility of inclusion of two or more communities (and thus abutting allopatric congeneric pairs) in his samples. Although his numerical analysis has since been shown to be incorrect (49), his argument is still reasonable and worthy of consideration. Using a corrected statistical approach, Williams (49) failed to find fewer congeners than expected in a variety of natural communities (indeed, he found more than expected in many). Terborgh

[6]Community similarity (*CS*) is simply X/N, where X is the number of species common to two communities and N is the total number of different species occurring in either; thus *CS* equals one when two communities are identical, and zero when they share no species.

& Weske (45) also used this corrected method to calculate the expected numbers of congeneric species pairs in Peruvian bird communities, and found that these communities were not impoverished with congeneric pairs, thus refuting any increased incidence of competitive exclusion among congeners in this particular avifauna. Similar analyses of the saurofaunas of the Kalahari and the Australian deserts are summarized in Figures 4 and 5. Again, the observed numbers of congeneric pairs are not conspicuously or consistently lower than expected.

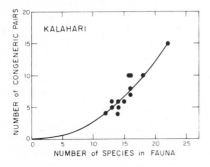

Figure 4 Dots represent the actual numbers of pairs of congeneric species of lizards observed on ten study areas in the Kalahari desert. Curve is the expected number of such pairs in a random subsample of the entire fauna.

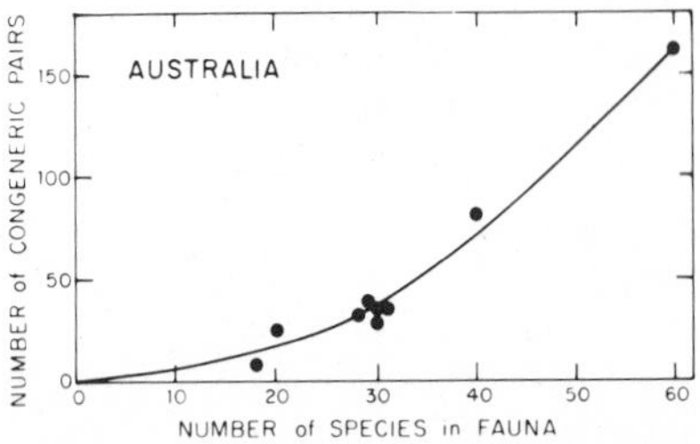

Figure 5 Dots are the actual numbers of pairs of congeneric species of lizards observed on eight Australian desert study areas. Curve represents the number of such pairs expected in a random subsample of the entire fauna.

CONCLUDING REMARKS

Interpretation of the structure of desert lizard communities has become steadily more difficult as the amount of information increases. Early in these studies, I expected to find much more pronounced similarities between these independently evolved, but otherwise basically similar ecological systems. Although a few crude ecological equivalents can be found among the different desert-lizard systems (26, 27, 30, 32), the ecologies of most species are quite disparate and unique. As seen above, the diversity of resources actually used by lizards along various niche dimensions, as well as the amount of niche overlap along them, differs markedly among the desert systems; moreover, the relative importance of various niche dimensions in separating niches varies. Thus food is a major dimension separating the niches of North American lizards, whereas in the Kalahari food niche separation is slight and differences in the place and time niches are considerable. All three niche dimensions are important in separating the niches of Australian desert lizards. Overall niche overlap is least in the most diverse lizard communities of Australia. Differences in diversity between the three continental systems stem from differences in the overall diversities of resources used by lizards or the size of the lizard niche space, as well as from differences in overall niche overlap, but are not due to conspicuous differences in overall niche breadths. Factors underlying these observed differences in diversity of utilized resources and niche overlap are poorly understood at present, but probably involve some of the following: (*a*) the degree to which any given system is truly saturated with species, (*b*) differences in the available range of resources among deserts that stem from historical factors, such as diversification of termites, reciprocal relations with other taxa, and the usurpation of their ecological roles, (*c*) differences between desert systems in the extent of spatial heterogeneity and habitat complexity which alter the degree of habitat specificity and the between-habitat component of diversity, and (*d*) other factors, such as possible differences in climatic stability and predictability, which might affect tolerable niche overlap.

ACKNOWLEDGMENTS

My research has benefited from contacts with numerous persons and parties, too many to enumerate here. I am grateful to Henry Horn, Chris Smith, Robert Colwell, Raymond Huey, William Parker, Larry Gilbert, and my wife Helen for reading this manuscript, and to Glennis Kaufman for much help in data analysis. Finally, the project would have been impossible without the financial assistance provided by the National Institutes of Health and the National Science Foundation.

Literature Cited

1. Collette, B. B. 1961. Correlations between ecology and morphology in anoline lizards from Havana, Cuba and southern Florida. *Bull. Mus. Comp. Zool. Harvard Univ.* 125:137–62
2. Colwell, R. K., Futuyma, D. J. 1971. On the measurement of niche breadth and overlap. *Ecology* 52:567–76
3. Elton, C. S. 1946. Competition and the structure of ecological communities. *J. Anim. Ecol.* 15:54–68
4. Heatwole, H. 1970. Thermal ecology of the desert dragon, *Amphibolurus inermis. Ecol. Monogr.* 40:425–57
5. Horn, H. S. 1966. Measurement of overlap in comparative ecological studies. *Am. Natur.* 100:419–24
6. Hotton, N. 1955. A survey of adaptive relationships of dentition to diet in the North American Iguanidae. *Am. Midl. Natur.* 53:88–114
7. Huey, R. B., Pianka, E. R., Egan, M. E., Coons, L. W. 1974. Ecological shifts in sympatry: Kalahari fossorial lizards (*Typhlosaurus*). *Ecology* 55:In press
8. Hutchinson, G. E. 1957. Concluding remarks. *Cold Spring Harbor Symp. Quant. Biol.* 22:415–27
9. Lein, M. R. 1972. A trophic comparison of avifaunas. *Syst. Zool.* 21:135–50
10. Levins, R. 1968. *Evolution in Changing Environments.* Princeton: Princeton Univ. Press. 120 pp.
11. MacArthur, R. H. 1972. *Geographical Ecology: Patterns in the Distribution of Species.* New York: Harper and Row. 269 pp.
12. MacArthur, R. H., Pianka, E. R. 1966. On optimal use of a patchy environment. *Am. Natur.* 100:603–09
13. Mayhew, W. 1968. Biology of desert amphibians and reptiles. *Desert Biology,* ed. G. W. Brown, 195–356. New York: Academic
14. Parker, W. S., Pianka, E. R. 1973. Notes on the ecology of the iguanid lizard, *Sceloporus magister. Herpetologica* 29: 143–52
15. Parker, W. S., Pianka, E. R. Comparative ecology of populations of the lizard *Uta stansburiana.* Unpublished
16. Pianka, E. R. 1965. *Species diversity and ecology of flatland desert lizards in western North America.* PhD thesis. Univ. Wash., Seattle. 212 pp.
17. Pianka, E. R. 1966. Convexity, desert lizards, and spatial heterogeneity. *Ecology* 47:1055–59
18. Pianka, E. R. 1967. On lizard species diversity: North American flatland deserts. *Ecology* 48:333–51
19. Pianka, E. R. 1968. Notes on the biology of *Varanus eremius. West. Aust. Natur.* 11:39–44
20. Pianka, E. R. 1969. Habitat specificity, speciation, and species density in Aus-

tralian desert lizards. *Ecology* 50:498–502

21. Pianka, E. R. 1969. Sympatry of desert lizards *(Ctenotus)* in western Australia. *Ecology* 50:1012–30

22. Pianka, E. R. 1969. Notes on the biology of *Varanus caudolineatus* and *Varanus gilleni. West. Aust. Natur.* 11:76–82

23. Pianka, E. R. 1970. Comparative autecology of the lizard *Cnemidophorus tigris* in different parts of its geographic range. *Ecology* 51:703–20

24. Pianka, E. R. 1970. Notes on the biology of *Varanus gouldi flavirufus. West. Aust. Natur.* 11:141–44

25. Pianka, E. R. 1971. Lizard species density in the Kalahari desert. *Ecology* 52:1024–29

26. Pianka, E. R. 1971. Comparative ecology of two lizards. *Copeia* 1971:129–38

27. Pianka, E. R. 1971. Ecology of the agamid lizard *Amphibolurus isolepis* in Western Australia. *Copeia* 1971:527–36

28. Pianka, E. R. 1972. Zoogeography and speciation of Australian desert lizards: an ecological perspective. *Copeia* 1972:127–45

29. Pianka, E. R., Huey, R. B. 1971. Bird species density in the Kalahari and the Australian deserts. *Koedoe* 14:123–30

30. Pianka, E. R., Parker, W. S. 1972. Ecology of the iguanid lizard *Callisaurus draconoides. Copeia* 1972:493–508

31. Pianka, E. R., Parker, W. S. Ecology of the Desert Horned Lizard, *Phrynosoma platyrhinos.* Unpublished

32. Pianka, E. R., Pianka, H. 1970. The ecology of *Moloch horridus* (Lacertilia: Agamidae) in Western Australia. *Copeia* 1970:90–103

33. Pianka, E. R., Pianka, H. Comparative ecology of twelve species of nocturnal lizards (Gekkonidae) in the Western Australian desert. Unpublished

34. Pielou, E. C. 1972. Niche width and niche overlap: a method for measuring them. *Ecology* 53:687–92

35. Rand, A. S. 1964. Ecological distribution in anoline lizards of Puerto Rico. *Ecology* 45:745–52

36. Rand, A. S., Humphrey, S. S. 1968. Interspecific competititon in the tropical rain forest: ecological distribution among lizards at Belem, Para. *Proc. US Nat. Mus.* 125:1–17

37. Sage, R. D. 1973. Convergence of the lizard faunas of the chaparral habitats in central Chile and California. *The Convergence in Structure of Ecosystems in Mediterranean Climates,* ed. H. Mooney. New York: Springer-Verlag

38. Schoener, T. W. 1967. The ecological significance of sexual dimorphism in size in the lizard *Anolis conspersus. Science* 155:474–77

39. Schoener, T. W. 1968. The *Anolis* lizards of Bimini: resource partitioning in a complex fauna. *Ecology* 49:704–26

40. Schoener, T. W. 1969. Models of optimal size for solitary predators. *Am. Natur.* 103:277–313

41. Schoener, T. W. 1970. Nonsynchronous spatial overlap of lizards in patchy habitats. *Ecology* 51:408–18

42. Schoener, T. W. 1971. The theory of foraging strategies. *Ann. Rev. Ecol. Syst.* 2:369–404

43. Schoener, T. W., Gorman, G. C. 1968. Some niche differences in three lesser antillean lizards of the genus *Anolis. Ecology* 49:819–30

44. Soulé, M. 1968. Body temperatures of quiescent *Sator grandaevus* in nature. *Copeia* 1968:622–23

45. Terborgh, J., Weske, J. S. 1969. Colonization of secondary habitats by Peruvian birds. *Ecology* 50:765–82

46. Tinkle, D. W. 1967. The life and demography of the side-blotched lizard, *Uta stansburiana. Misc. Publ. Mus. Zool., Univ. Mich.* No. 132:1–182

47. Vandermeer, J. H. 1972. Niche theory. *Ann. Rev. Ecol. Syst.* 3:107–32

48. Werner, Y. L. 1969. Eye size in geckos of various ecological types (Reptilia: Gekkonidae and Sphaerodactylidae). *Isr. J. Zool.* 18:291–316

49. Williams, C. B. 1964. *Patterns in the Balance of Nature and Related Problems in Quantitative Ecology.* New York: Academic. 324 pp.

Reprinted from
ANNUAL REVIEW OF ECOLOGY AND SYSTEMATICS
Volume 4, 1973

GENETIC VARIATION AMONG VERTEBRATE SPECIES

❖4053

Robert K. Selander and Walter E. Johnson[1]

Department of Zoology, University of Texas, Austin, Texas

INTRODUCTION

The early mutationists believed that single mutations could produce species, but by 1940 studies of hybrids had demonstrated that species differences involve numerous genic and chromosomal mutations, individually of small effect (18). Similarly, the "classical" model of genetic population structure, which assumed widespread homozygosity of "wild-type" alleles, was refuted by research on viability and phenotypic characters, indicating the existence of much genic polymorphism in populations (76). In 1966 the concept of a high level of polymorphism as the normal condition in populations was convincingly established by direct measurements in *Drosophila* (42) and humans (25).

The acquisition by evolutionary geneticists of electrophoretic and other techniques for demonstrating allelic variation at loci encoding polypeptides permits a new and extensive examination of variation in populations and of genetic differences between races and species. This review of allozymic variation in vertebrates deals with 1. genic variation in populations; 2. local genetic heterogeneity; 3. geographic variation in allele frequencies; 4. genetic similarity among populations and species; and 5. systematic applications of molecular population data.

GENIC VARIATION IN POPULATIONS

Estimates of genic heterozygosity *(H)* are available for 24 species of vertebrates with continental distributions (Table 1). Extrapolation from small sets of loci to entire genomes is risky because of our ignorance of the composition of genomes and the proportion of total DNA involved in structural genes (54). Yet the general consistency of estimates suggests that they index some basic parameter of genic variation (see also 74). Values for tetrapods are fairly consistent at 5–6%, but those for invertebrates are higher, averaging 15% (67). The evidence indicates that most

---

[1]Present address: Department of Biology, Western Michigan University, Kalamazoo, Michigan.

populations are polymorphic at thousands of loci and that each individual has a unique protein complement.

The vertebrate-invertebrate difference in heterozygosity is not consistent with theories relating variability to species number or gene flow (33), but is predicted by Richard Levins' theory of adaptive strategies in relation to environmental grain (67).

In vertebrates inhabiting islands and caves where populations are small or subject to periodic bottlenecking, severely reduced genic variability has been detected. For example, populations of the old-field mouse *(Peromyscus polionotus)* on barrier islands and peninsulas off the Gulf coast of Florida are half as variable as those on the adjacent mainland (Figure 1) (70).

Heterozygosity in *P. polionotus* varies geographically on the mainland, increasing from 5% in South Carolina and Georgia to 9% in peninsular Florida. Whether this cline is related to aspects of ecological amplitude or "niche width" is problematical. Because the species has only recently occupied the northern part of its range, the reduced variability there may reflect a range-front phenomenon in small pioneering populations. Following the method used by Lewontin (40) in analyzing human population diversity, we can express proportions of the total genetic diversity in *P. polionotus* in terms of Shannon's information coefficient. A sample from a 20-acre area in peninsular Florida may contain 84% of the total genetic "information" in the species, whereas as little as 14% is represented on Santa Rosa Island.

Table 1   Estimates of genic heterozygosity in vertebrate populations[a]

| | | | Heterozygosity (*H*) | |
Organism	Number of species	Number of loci	Mean	Range
Fish (Tetra)	1	17	.112	–
Salamander	1	22	.049	–
Lizards	4	15–29	.058	.05–.07
Sparrow	1	15	.059	–
Rodents	15	17–41	.056	.01–.09
Seal	1	19	.030	–
Man (European)	1	70	.067	–
Mean			.0580	

[a]From Selander & Kaufman (67), with additional data from Webster (80) and Nevo & Shaw (55)

In cave populations of the fish *Astyanax mexicanus* (2) and in some species of *Anolis* lizards on South Bimini Island (81), there is an absence or severe reduction of variability. Similarly, near or complete fixation at all loci occurs in populations of *Peromyscus eremicus* on small islands in the Gulf of California (Avise et al, in preparation). A marked reduction in numbers of the elephant seal *(Mirounga angustirostris)* in the last century appears to have caused a loss of genic variability (7). Genetic drift has been invoked to account for reduced variability in these insular or quasi-insular situations, but it is possible that the effect results in part from density-dependent selection (12, 13) or adaptation to narrow ranges of environmental heterogeneity. In any event these cases demonstrate that high levels of variability are not essential to the existence of natural populations.

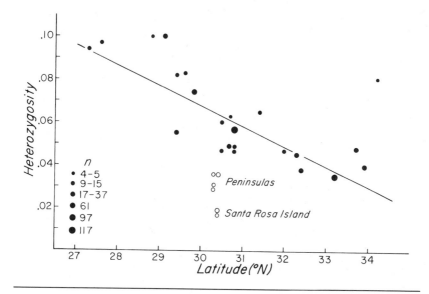

Figure 1 Weighted linear regression of H on latitude (decimal scale) of locality in mainland samples (dots) of *Peromyscus polionotus*. $H = .344-.0092\ L$; $F_{(1,\ 21)} = 48.8^{***}$. Data from Selander et al (70).

Shown in Table 2 are the results of surveys of allozymic variation in the rodents *Thomomys* [2 species], *Dipodomys* [3], *Sigmodon* [2], *Peromyscus* [4], and *Mus* [3 semispecies]; a passerine bird, *Zonotrichia* [1]; the lizards *Sceloporus* [3], *Anolis* [4], and *Uta* [1]; and a fish, *Astyanax* [1] (see 67 for references). The data are from studies attempting to assess levels of overall genic variability, in which both monomorphic and polymorphic loci were scored. Hence the results estimate the probability that a particular protein will be polymorphic in local populations of a given vertebrate species. There is a wide range of variation among proteins, with some, such as PGM-1, polymorphic in most species, and others, such as G6PD, rarely variable. Esterases are highly variable, and muscle and other "general," nonenzymatic proteins are conservative. Proteins that are frequently polymorphic tend to have high heterozygosity values, but five proteins (ME, PGI, ADH, IPO, and ALB) are exceptional in that, while they occur in polymorphic state with only average frequency, they are highly heterozygous in those species in which they vary.

For species of *Drosophila,* allozymic variability is lower in "glucose-metabolizing enzymes" (Group I) than in other enzymes (Group II) (4, 37). The survey of vertebrates summarized in Table 2 fails to demonstrate such a relationship, but the

Table 2 Allozymic variability in proteins in populations of vertebrates

Protein[a]	Number of species surveyed	Species polymorphic Number	Species polymorphic %	Mean heterozygosity (H) Total species	Mean heterozygosity (H) Polymorphic species
Group I					
Super. NAD-MDH	23	4	17	.0066	.0380
Mito. NAD-MDH	24	4	17	.0119	.0712
Super. ME	11	3	27	.0553	.2027
6PGD	23	17	74	.0840	.1136
G6PD	12	1	8	.0028	.0330
aGPD	23	15	65	.0676	.1036
Super. IDH	21	12	57	.0719	.1257
Mito. IDH	18	4	22	.0031	.0140
LDH-1	24	12	50	.0469	.0938
LDH-2	24	10	42	.0127	.0305
PGI	21	12	57	.0410	.1550
PGM-1	24	19	79	.1072	.1354
PGM-2 or -3	15	8	53	.1280	.2400
Mean (n = 13)			43.7	.04915	.10435
Group II					
ADH	16	7	44	.0908	.2074
SDH	6	0	0	.0000	.0000
Super. GOT	21	12	57	.0475	.0832
Mito. GOT	17	3	18	.0018	.0100
IPO[b]	18	4	22	.0454	.2045
Esterases[c]	16 (4.25/sp.)	1.88	44	.1341	.3041
Mean (n = 6)			30.8	.05327	.13487
Group III					
ALB	23	7	30	.0610	.2006
TRF	18	12	67	.1033	.1550
HB (2 loci)[d]	17	7	21	.0605	.1470
General proteins[e]	24 (3.17/sp.)	0.25	8	.0054	.0682
Mean (n = 5)			29.4	.05816	.14356
Grand mean (n = 24)			37.50	.051933	.119813

[a]*Group I:* Glucose-metabolizing enzymes; *Group II:* Other enzymes; *Group III:* Nonenzymatic proteins.

[b]Homology across species uncertain for indophenol oxidase.

[c]68 esterases, or a mean of 4.25 loci per species; 30 loci polymorphic. Values are means for all loci.

[d]Observed values for polymorphism and heterozygosity in hemoglobin halved to obtain mean for a and β loci.

[e]76 "general proteins," or a mean of 3.17 loci per species; 6 loci polymorphic. Values are means for all loci.

sample of Group II enzymes is perhaps too small for an adequate test. Cohen et al (14) have reported unusually low variability in human glycolytic enzymes.

The data in Table 2 provide a basis for predicting the likelihood that races or congeneric species will differ at a given locus. In *Dipodomys,* MDH-1, a conservative protein, is monomorphic for the same allozyme in ten species and weakly polymorphic in one, whereas TRF, a highly variable protein, is polymorphic in eight species and shows little sharing of allozymes among species (28).

LOCAL GENETIC HETEROGENEITY

Few attempts have been made to study deme size and organization or to analyze genetic heterogeneity at the level of the local population, with the consequence that we have little understanding of the effects of geographic and ecologic factors and of social behavior and other aspects of "culture" on the genetic structure of species (48). However, extensive data on blood-group and serum proteins have been accumulated by human geneticists. Of special interest are investigations dealing with relatively unaccultured, tribal populations of American Indians (48, 85), Australian aborigines (34), and African Pygmies (10). These studies attempt to understand the genetic structure of human populations as they existed in the precivilization period.

There is considerable genetic differentiation between villages in tribes of South American Indians (50). In terms of the distance function of Cavalli-Sforza & Edwards (9), the mean pair-wise genetic distance between seven Makiritare villages is 0.356, and that between seven Yanomama villages is 0.330 (79). Since the mean pair-wise distance between Indian tribes of Central and South America is 0.385 (49), it is apparent that intratribal variation is almost as great as intertribal. The complex of factors responsible for intratribal differentiation includes the founder effect, genetic drift, cleavage of villages along familial lines, nonrandom patterns of intervillage migration, and differential fertility in a polygynous mating system (11).

For vertebrates other than man, the most extensive analysis of genetic microdifferentiation deals with the house mouse *(Mus musculus).* From theoretical considerations developed in studies of *t*-alleles, Lewontin & Dunn (41) predicted that populations are subdivided into tribes (family groups) with an effective breeding size below that at which random fluctuations and inbreeding are important in determining gene frequencies. This interpretation has been confirmed by experimental studies and field investigations (1, 66). Populations become divided into territorial tribes composed of a dominant male and several females and subordinate males. Intertribal migration is rare, and the genetically effective size of the units probably is less than ten. Given the social system of the house mouse, genetic heterogeneity among tribes is inevitable, even in large, continuously distributed populations, and even if strong heterotic or other forms of balancing selection are involved in maintenance of the polymorphisms. Populations inhabiting different barns on the same farm often are heterogeneous in allele frequencies, apparently because they are established by a few founders that rapidly multiply to saturate the available space, thereby reducing and eventually eliminating interbarn migration. In large barns spatial clustering of similar genotypes among adult mice has been demonstrated, providing direct evidence of population subdivision (66).

The long-term evolutionary effects of subdivision in house mice may be minor because the average "life span" of the tribal units is short, and long-distance dispersal of young occurs while the parental populations remain statically subdivided. For these reasons we disagree with the view (1) that the isolation of demes necessarily eliminates gene flow as a factor in the maintenance of species integrity.

Levin et al (39) have suggested that genetic drift does not adequately account for the low observed frequencies of lethal t-alleles in populations of *Mus*. Their argument is based on Petras' studies (59) of the *Es-2* locus suggesting that the breeding-unit sizes and migration rates of populations exceed those in which drift could have an important effect. Yet estimates of deme size based on the "Wahlund effect" are unreliable (66). This is especially true for the *Es-2* locus because the heterozygote *Es-2ª/Es-2ᵇ* cannot be consistently distinguished from the homozygote *Es-2ᵇ/Es-2ᵇ*, the *Es-2ª* allele having a "null" phenotype. Work with inbred strains (15) has shown that, because the Darwinian fitness of subordinate males is very low, the effective population size is sufficiently small for drift to affect the frequency of t-alleles.

Research on the house mouse confirms Wright's contention that random drift due to sampling error in small populations is an important factor governing gene frequencies. The view that drift rarely occurs in natural populations (22) is unwarranted. For an individual or small population, stochastic processes may play the dominant role in determining genotype or gene frequency.

GEOGRAPHIC VARIATION

In some widely distributed vertebrate species, one allele predominates at each polymorphic locus in all parts of the range, and allele frequencies are remarkably uniform regionally. This pattern is also characteristic of *Drosophila* (3) and many other invertebrates. At the other extreme are species such as *Peromyscus polionotus* that are geographically variable at all polymorphic loci (70). Vertebrates are on the average more variable geographically than are species of *Drosophila*, a circumstance perhaps reflecting lesser degrees of gene flow among regional populations.

Four major patterns of geographic variation were found in 17 polymorphic loci recorded in *P. polionotus*. 1. At the *Pgi-1* locus, there is a widespread polymorphism involving two alleles. *Pgi-1ᶜ* predominates in all mainland samples and is fixed on the barrier islands off the coast of the Florida Panhandle. 2. At the *Trf-1* locus, most mainland populations are again polymorphic, but a different minor allele occurs in each of the three major parts of the range, and one of the western insular populations has a unique allele in moderate frequency. 3. The *Alb-1* locus is strongly variable geographically, with *Alb-1ª* nearly fixed in peninsular Florida and *Alb-1ᵇ* at a similar high frequency in the Florida Panhandle. 4. The *Ldh-1* and *Pgm-3* loci exemplify a pattern in which mainland populations are monomorphic and western insular populations are polymorphic, having an allele not elsewhere represented.

Patterns 2 and 3 suggest that *P. polionotus* was once divided into three regional populations, one in peninsular Florida, another in Georgia, and a third in western Florida, although the continental range of the species is presently continuous. Circumstantial evidence of the action of selection in maintaining the variation is provided by pattern 4. In view of the isolation of the insular populations from one another and the strong trend for fixation (presumably due to drift) of the predominant mainland allele in these small populations, it seems unlikely that a mutant would have persisted on five or six islands unless it were maintained by some form of balancing selection.

In the cricket frog *Acris crepitans,* no variation was found in several plasma esterases, liver LDH, GOT, GDH, ME, ALB, and hemoglobin peptide 4 (17). However, the geographic distribution of variant transferrins, hemoglobins, and liver esterases suggests that the species has differentiated into separate population groups occurring in the Great Plains, Louisiana Delta, and Appalachian regions. But, the pattern of variation in allele frequencies at the heart LDH locus is not concordant with that of the other variable loci (61).

The interpretation of geographic variation in allele frequencies in adaptive terms is unlikely to be convincing when the analysis is limited to correlation with environmental factors (46, 69). Temperature, precipitation, and other variables often are so strongly correlated with latitude, longitude, or altitude that it is difficult to relate patterns of variation to environmental factors, independent of geographic effects. Other problems of interpretation also arise. For example, a low level of geographic variation, reflecting a recent expansion of range from a small refugium, could be interpreted as evidence of uniformity in the selective regime. It has been maintained (33) that geographic uniformity results from the flow of selectively neutral or nearly neutral alleles, but for widely distributed species with large populations this thesis is unsatisfactory (3).

Only when differences in functional properties of polymorphic proteins can be related to the distribution of relevant environmental variables is an adaptive interpretation convincing. For a two-allele *Ldh* locus in the minnow *Pimephales promelas,* Merritt (46) demonstrated a functional basis for the maintenance of the polymorphism and for a north-south cline in allele frequencies by determining that the LDHs of the homozygote of the allele that is common in the north have lower substrate affinities for pyruvate at 25°C than do LDHs of the other genotypes. In other work with esterases of fishes, Koehn (35; see also 36) was able to explain variation in allele and zygotic frequencies by relating genotypic variation in optimal temperature for enzyme activity to geographic patterns of temperature variation. Whether most protein polymorphisms will yield to this type of analysis is questionable, but the biochemical properties of proteins must be studied if we are to understand the meaning of polymorphic variation.

GENETIC SIMILARITY AMONG POPULATIONS AND SPECIES

Many measures have been developed to express genetic similarity or distance among populations (73, 82). In most of our work we have employed a coefficient that measures the average geometric distance *(D)* between allele frequency vectors on a scale from 0 to 1 (60). Genetic similarity *(S)* is defined as 1-*D*.

Estimates of genetic similarity between conspecific populations and congeneric species are presented in Tables 3 and 4. Table 4 also includes some comparable estimates of similarity for species of *Drosophila*. A wide range of values is expected in species comparisons, since differences reflect both the amount of genetic modification accompanying speciation and changes accumulating thereafter. Comparisons of similarity values for species of different groups of organisms cannot be very meaningful because (*a*) taxonomic convention in defining specific and generic limits varies from group to group, (*b*) the term "sibling species" merely denotes an arbitrarily defined degree of morphological similarity that varies from one group to another, and (*c*) the average age of species varies from genus to genus and among higher groups of organisms. Yet there is a pattern.

Conspecific Populations

Coefficients of similarity for continental populations are generally in the .90s (Table 3) and tend to decrease with distance between populations. As shown for *Peromyscus polionotus,* physical barriers to gene flow also decrease similarity. Insular populations and those on the closely adjacent mainland are as divergent ($\overline{S}$ = .92) as those from different continental regions. The most divergent populations are those strongly isolated from one another on the western and eastern barrier islands of Florida, with $\overline{S}$ = .84, a value similar to those obtained for some semispecies pairs (Table 4). Similarly, unusually low values are recorded for cave populations of

Table 3 Genetic similarity between conspecific populations

| Genus and species | Number of loci | Rogers' coefficient of genetic similarity (S) | | Reference |
		Mean	Range	
Rodents				
Dipodomys (11 species)	18	.97	.92–1.00	28
Sigmodon hispidus	23	.98	.98–1.00	29
Mus musculus	41	.95	.93–.98	68
Peromyscus polionotus	32	.95	.82–.99	70
Peromyscus floridanus	39	.97	.96–.98	71
Thomomys bottae	27	.93	.90–.96	58
Lizards				
Uta stansburiana	19	.89	.77–.98	45
Anolis (2 species)	25	.75	.69–.82	81
Fish				
Astyanax mexicanus				2
6 surface populations	17	.96	.94–.98	
3 cave populations	17	.83	.79–.92	
cave vs surface	17	.82	.77–.90	

Table 4 Genetic similarity between congeneric species

Relationship and genus	Number of species	Number of loci	Rogers' coefficient of genetic similarity (S)		Reference
			Mean	Range	
Semispecies					
Mus	2	41	.79	–	68
Sceloporus	2	20	.73	–	24
Sigmodon	2	23	.76	–	29
Thomomys	2	27	.84	–	58
Sibling species					
Peromyscus	2	22	.81	–	72
Dipodomys	11	18	.61	.31–.89	28
Drosophila	18	18	.50	.23–.86	26
Drosophila	6	24	.50	.30–.77	47
Nonsibling species					
Anolis	4	25	.21	.16–.29	81
Peromyscus	2	32	.32	–	71
Drosophila	27	18	.18	.08–.29	26

Astyanax mexicanus, insular and continental subspecies of *Anolis carolinensis,* and insular populations of *A. sagrei* (Table 3).

Semispecies

Semispecies are pairs or groups of allopatric populations that hybridize but for various reasons maintain the integrity of their respective gene pools (6, 44). Because semispecies are in some sense intermediate between races and full species, the average degree of genic difference between them may approximate the amount of modification of gene pools normally involved in the process of geographic speciation. An intensively studied pair of semispecies are the mice *Mus musculus musculus* and *M. m. domesticus,* which freely hybridize in a zone across the Jutland Peninsula of Denmark and south through Germany. An assessment of genic similarity between the parental forms yielded a value of $\bar{S} = .79$ (Table 4). Most of the transition in gene frequencies and morphological characters occurs over a distance of 20 km (Figure 2). Patterns of introgression vary among loci, but whether selection directly affects the loci or the segments of chromosomes that they mark is unknown. Apparently alleles (or chromosomal segments) that are relatively compatible with both gene pools introgress farther than do those having large disruptive effects on these complexes.

The center of the hybrid zone lies in an area of transition in mean annual rainfall, and the east-west asymmetry in width is paralleled by a more gradual transition in precipitation in the west than in the east. However, the cline in genetic character is much steeper than that in precipitation or other environmental factors. At several loci minor (mutant) alleles occur more frequently in the hybrid zone than elsewhere,

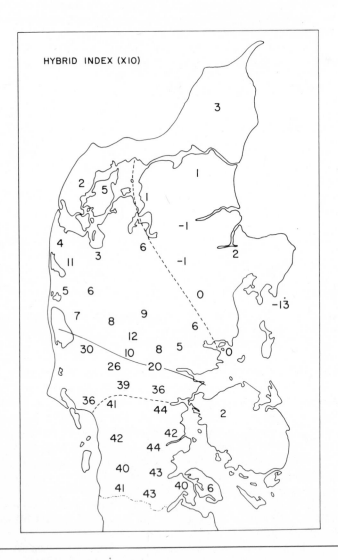

Figure 2 Geographic variation in hybrid index scores based on six polymorphic loci in *Mus musculus* in Jutland. Dashed lines indicate limits of detectable introgression into *M. m. musculus* (northern form) and *M. m. domesticus* (southern form). The center of the zone of hybridization is indicated by the continuous line. From Hunt & Selander (27).

an association suggesting that introgression causes a partial breakdown in the co-adaptations of gene pools, modifying selective constraints on the incorporation of mutant alleles (27).

Chromosomal and/or genic imbalance has also been implicated in the failure of semispecies pairs of lizards of the *Sceloporus grammicus* complex (24) and of pocket gophers (*Thomomys bottae* and *T. umbrinus*) (58) to exchange genes, despite extensive hybridization in zones of contact.

Species

Degrees of morphological and genic similarity between species are correlated. In *Peromyscus* estimates of genic similarity range from $\bar{S} = .81$ for the siblings *P. gossypinus* and *P. leucopus* to $\bar{S} = .32$ for the morphologically very dissimilar species *P. polionotus* and *P. floridanus* (Table 4). A similar range was found for species of *Dipodomys*. Nonsibling species are less similar than siblings, with coefficients generally less than .30 (Table 4).

An intergeneric comparison of several species of *Dipodomys* and *Perognathus hispidus* (both in the family Heteromyidae) yielded a value of $\bar{S} = .16$ (.07 to .19). If heteromyids are representative of rodents, the allozymic technique will be of limited value in systematics at levels much above the genus, beyond which it is likely that alleles will not be shared at any loci (28).

Other Estimates of Genetic Relationship

Nei (51, 52; see also 57) has developed a statistical method for estimating the number of net codon differences per locus from gene frequency data: Genetic distance, $D = -\log_e I$, where I is the normalized identity of genes between populations X and Y, and the unit of D is average codon difference per locus. This approach involves several simplifying assumptions, but has the practical advantage of permitting comparisons of intra- and interpopulational differences. As noted by Nei (52), the relationship between normalized identity of genes and distance can be precisely specified only if we assume that the alleles are unaffected by selection. In comparisons of conspecific populations there is little difference between estimates of distance derived by Nei's formula and minimum estimates of codon difference based on the obvious consideration that there must be at least one between any pair of different alleles.

For semispecies of *Mus,* Nei's (52) estimates of the number of codon differences (detected by electrophoresis) are 0.03 or less per locus between local populations and 0.13 to 0.20 per locus for populations of the two forms. Nei & Roychoudhury (53) have estimated for Caucasian, Negro, and Japanese populations of man that the net interpopulational codon difference for protein loci studied electrophoretically is 0.01–0.03 per locus, whereas the intrapopulational difference is 0.09–0.14 per locus. Thus the genetic differences between populations are small relative to intrapopulational differences (see also 40). Indeed, Nei & Roychoudhury (53) conclude that human racial group differences are no greater than those between local populations of the house mouse or between geographic populations of *Drosophila pseudoobscura.* However, a source of error in these and other estimates of genetic distance is that the component of genetic variance contributed by sampling error is neither eliminated nor standardized. Because the allele frequency data for Danish house mice were based on samples of only 15 to 20 individuals, the data may not provide a sound base for comparison with human variation.

In an analysis of allele frequencies at five loci in three nominal species of macaques (*Macaca*), Weiss & Goodman (83) obtained an index of genetic distance only slightly larger than that derived from a comparison (based on six loci) between human populations of New Guinea and the combined Dravidian-Australian group. They conclude that "species of macaques appear to be no more different from each other than some races of man," but a more plausible hypothesis is that the macaques are conspecific (21).

SYSTEMATIC APPLICATIONS OF MOLECULAR POPULATION DATA

Allozymic information is potentially of great value in systematics, particularly at the specific and generic levels. The systematic worth of allozymic characters does not hinge on the question of the selective neutrality or nonneutrality of alleles. If in fact most amino acid substitutions underlying allelic variation are neutral or nearly so, so much the better for certain systematic objectives, since the probability of convergence will be minimal (86).

Several techniques have been employed to cluster populations on the basis of molecular genetic data, including minimum path and least squares methods of producing bifurcating "trees" (32). This approach has been applied to cattle breeds (31) and to human populations (10, 20, 78, 79). In an analysis of 18 neighboring villages in Bougainville, New Guinea, Friedlaender et al (23) demonstrated that anthropometric measures and blood group allele frequencies are related to linguistic, geographic, and migrational measures of "distance" and compared trees derived from anthropometrics and blood polymorphisms. The trees were topologically similar and approximated the known demographic and historical relationships of the village populations.

A genetic dendrogram for species of *Dipodomys* and *Perognathus*, constructed by agglomerative cluster analysis from a matrix of S coefficients (based on 18 loci) is in better accord with the karyotypic evidence of relationship than are morphologically based schemes (28). This type of dendrogram cannot be interpreted as a phylogeny, but is a useful visual summary of information on overall genetic resemblance contained in similarity or distance matrices.

Several immunological and other methods of studying the "comparative anatomy" of whole proteins or partial digests have been successfully applied to problems in vertebrate systematics (16, 19, 38), but the extensive literature on species differences in single proteins will not be reviewed here.

Seal's (65) study of carnivore hemoglobins yielded the remarkable finding that species of the superfamily Canoidea and the suborder Pinnipedia share a hemoglobin component of identical electrophoretic mobility. Electrophoretic and other indirect evidence of structural homogeneity led Seal to suggest that the primary structure of this component has persisted unchanged in lineages for 45 million years. Yet hemoglobin has been considered (56) to evolve at a constant rate, with a mean period of seven million years per effective amino acid substitution per chain in single lineages. Recently; Sarich's (63) microcomplement fixation studies of hemoglobins of canids and pinnipeds appear to have resolved this paradox by demonstrating underlying structural heterogeneity not detectable by electrophoresis.

CONCLUDING COMMENTS

The conclusion of geneticists of a generation past that species differ at hundreds or thousands of loci has been confirmed, and quantitative measures of genetic affinity have been applied to several vertebrate groups. The evidence now available suggests that species normally have major differences in allele frequencies at between a third and a half of the structural loci of the genome, but the range of variation is large. Studies of vertebrates and *Drosophila* have established that "slight morphological and/or ecological dissimilarity such as exists between sibling species cannot be taken as evidence of little genetic differentiation" (5). Contrary to the interpretation of Hubby & Throckmorton (26), the evidence from molecular genetics is not incompatible with the hypothesis that speciation normally involves major reorganizations of gene pools (18, 44). However, allozymic analyses have thus far not told us much about the extent of modification occurring in the actual process of speciation.

Arbitrarily defined standards of genetic similarity or distance should be avoided in judging systematic relationships. Because conspecific populations may differ genetically more than those of different species, "an approach that merely counts the number of gene differences is meaningless" (44).

If the average rate of accumulation of allelic differences at loci proves to be a function of absolute time or generation length (43, 64), it will be possible to estimate the age of species pairs by relating genetic distance to time. Nei (51) estimated that an average of 500,000 years was required to establish existing differences between pairs of sibling species of *Drosophila,* and Kimura & Ohta (33) suggested that the divergence of semispecies of the house mouse from a common ancestor occurred roughly two million years ago. The fossil record does not permit a testing of these estimates, but neither seems unreasonable. Crude estimates of times of divergence of several racial groups and subgroups of man, assuming constant evolutionary rates, are given by Cavalli-Sforza (8).

Finally, we call attention to the problem of the relative age of species in different vertebrate groups. Wallace et al (77) have reported that serological differences (measured by microcomplement fixation) in albumins between species in the frog genera *Rana* and *Hyla* are much greater than taxonomic considerations would suggest, in some cases exceeding those between mammals of different families or suborders. Adopting the position that amino acid substitutions accumulate in vertebrate lines at a constant rate (62, 84), these workers argue that contemporary amphibian species are much older and more distant genetically than are those of mammals. Yet from a morphological standpoint species of frogs do not on the average appear to be any more distantly "related" than congeneric species of other vertebrate groups. The "evolutionary clock" hypothesis and the alternate view that proteins evolve at variable rates (30, 75) clearly define an important area of investigation in evolutionary genetics.

ACKNOWLEDGMENTS

This review is based on a paper presented at the XVII International Congress of Zoology in Monaco, 1972. Our research has been supported by NIH grant GM-15769 and NSF grants GB-15664 and GB-37690. Johnson acknowledges the support of NIH training grant GM-00337.

Literature Cited

1. Anderson, P. K. 1970. Ecological structure and gene flow in small mammals. *Symp. Zool. Soc. London* 26:299–325
2. Avise, J. C., Selander, R. K. 1972. Evolutionary genetics of cave-dwelling fishes of the genus *Astyanax*. *Evolution* 26:1–19
3. Ayala, F. J. 1972. Darwinian *versus* non-Darwinian evolution in natural populations of *Drosophila*. *Proc. Sixth Berkeley Symp. Math. Stat. Prob.* 5:211–36
4. Ayala, F. J., Powell, J. R., Tracey, M. L. 1972. Enzyme variability in the *Drosophila willistoni* group. V. Genic variation in natural populations of *Drosophila equinoxialis*. *Genet. Res.* 20:19–42
5. Ayala, F. J., Mourão, C. A., Pérez-Salas, S., Richmond, R., Dobzhansky, T. 1970. Enzyme variability in the *Drosophila willistoni* group. I. Genetic differentiation among sibling species. *Proc. Nat. Acad. Sci. USA* 67:225–32
6. Bigelow, R. S. 1965. Hybrid zones and reproductive isolation. *Evolution* 19:449–58
7. Bonnell, M. L., Selander, R. K. 1973. Elephant seals: genetic consequences of near extinction. *Science.* In press
8. Cavalli-Sforza, L. L. 1969. Human diversity. *Proc. XII Int. Congr. Genet.* 3:405–17
9. Cavalli-Sforza, L. L., Edwards, A. W. F. 1967. Phylogenetic analysis models and estimation procedures. *Am. J. Hum. Genet.* 19:233–57
10. Cavalli-Sforza, L. L. et al 1969. Studies on African Pygmies. I. A pilot investigation of Babinga Pygmies in the Central African Republic (with an analysis of genetic distance). *Am. J. Hum. Genet.* 21:252–74
11. Chagnon, N. A., Neel, J. V., Weitkamp, L., Gershowitz, H., Ayres, M. 1970. The influence of cultural factors on the demography and pattern of gene flow from the Makiritare to the Yanomama Indians. *Am. J. Phys. Anthropol.* 32:339–49
12. Charlesworth, B., Giesel, J. T. 1972. Selection in populations with overlapping generations. IV. Fluctuations in gene frequency with density-dependent selection. *Am. Natur.* 106:402–11
13. Clarke, B. 1972. Density-dependent selection. *Am. Natur.* 106:1–13
14. Cohen, P. T. W., Omenn, G. S., Motulsky, A. G., Chen, S.-H., Giblett, E. R. 1973. Restricted variation in the glycolytic enzymes of human brain and erythrocytes. *Nature New Biol.* 241:229–33
15. DeFries, J. C., McClearn, G. E. 1972. Behavioral genetics and the fine structure of mouse populations: a study in microevolution. *Evol. Biol.* 5:279–91
16. Dessauer, H. C. 1969. Molecular data in animal systematics. In *Systematic Biology*, 325–57. Washington: Nat. Acad. Sci. (Publ. 1692)
17. Dessauer, H. C., Nevo, E. 1969. Geographic variation of blood and liver proteins in cricket frogs. *Biochem. Genet.* 3:171–88
18. Dobzhansky, T. 1970. *Genetics of the Evolutionary Process.* New York: Columbia Univ. Press
19. Feeney, R. E., Allison, R. G. 1969. *Evolutionary Biochemistry of Proteins.* New York: Wiley Interscience
20. Fitch, W. M., Neel, J. V. 1969. The phylogenetic relationships of some Indian tribes of Central and South America. *Am. J. Hum. Genet.* 21:384–97
21. Fooden, J. 1964. Rhesus and crab-eating macaques: intergradation in Thailand. *Science* 143:363–65
22. Ford, E. B. 1971. *Ecological Genetics.* London: Chapman & Hall. 3rd ed.
23. Friedlaender, J. S. et al 1971. Biological divergences in south-central Bougainville: an analysis of blood polymorphism gene frequencies and anthropometric measurements utilizing tree models, and a comparison of these variables with linguistic, geographic, and migrational "distances." *Am. J. Hum. Genet.* 23:253–70
24. Hall, W. P., Selander, R. K. 1973. Hybridization in karyotypically differentiated populations of the *Sceloporus grammicus* complex (Iguanidae). *Evolution* 27:In press

25. Harris, H. 1966. Enzyme polymorphisms in man. *Proc. Roy. Soc. London B* 164: 298–310
26. Hubby, J. L., Throckmorton, L. H. 1968. Protein differences in *Drosophila*. IV. A study of sibling species. *Am. Natur.* 102:193–205
27. Hunt, W. G., Selander, R. K. 1973. Biochemical genetics of hybridization in European house mice. *Heredity* 31:In press
28. Johnson, W. E., Selander, R. K. 1971. Protein variation and systematics in kangaroo rats (genus *Dipodomys*). *Syst. Zool.* 20:377–405
29. Johnson, W. E., Selander, R. K., Smith, M. H., Kim, Y. J. 1972. Biochemical genetics of sibling species of the cotton rat (*Sigmodon*). *Univ. Texas Publ.* 7213: 297–305
30. Jukes, T. H., Holmquist, R. 1972. Evolutionary clock: nonconstancy of rate in different species. *Science* 177:530–32
31. Kidd, K. K., Pirchner, F. 1971. Genetic relationships of Austrian cattle breeds. *Anim. Blood Groups Biochem. Genet.* 2:145–58
32. Kidd, K. K., Sgaramella-Zonta, L. A. 1971. Phylogenetic analysis: concepts and methods. *Am. J. Hum. Genet.* 23: 235–52
33. Kimura, M., Ohta, T. 1971. *Theoretical Aspects of Population Genetics.* Princeton: Princeton Univ. Press
34. Kirk, R. L., Sanghvi, L. D., Balakrishnan, V. 1972. A further study of genetic distance among Australian aborigines: nine tribes in the Northern Territory. *Humangenetik* 14:95–102
35. Koehn, R. K. 1969. Esterase heterogeneity: dynamics of a polymorphism. *Science* 163:943–44
36. Koehn, R. K., Perez, J. E., Merritt, R. B. 1971. Esterase enzyme function and genetical structure of populations of the freshwater fish, *Notropis stramineus*. *Am. Natur.* 105:51–69
37. Kojima, K., Gillespie, J., Tobari, Y. N. 1970. A profile of *Drosophila* species' enzymes assayed by electrophoresis. I. Number of alleles, heterozygosities, and linkage disequilibrium in glucose-metabolizing systems and other enzymes. *Biochem. Genet.* 4:627–37
38. Leone, C. A., Ed. 1964. *Taxonomic Biochemistry and Serology.* New York: Ronald
39. Levin, B. R., Petras, M. L., Rasmussen, D. I. 1969. The effect of migration on the maintenance of a lethal polymorphism in the house mouse. *Am. Natur.* 103: 647–61
40. Lewontin, R. C. 1972. The apportionment of human diversity. *Evol. Biol.* 6: 381–98
41. Lewontin, R. C., Dunn, L. C. 1960. The evolutionary dynamics of a polymorphism in the house mouse. *Genetics* 45: 705–22
42. Lewontin, R. C., Hubby, J. L. 1966. A molecular approach to the study of genic heterozygosity in natural populations. II. Amount of variation and degree of heterozygosity in natural populations of *Drosophila pseudoobscura*. *Genetics* 54:595–609
43. Lovejoy, C. O., Burstein, A. H., Heiple, K. G. 1972. Primate phylogeny and immunological distance. *Science* 176: 803–5
44. Mayr, E. 1970. *Populations, Species, and Evolution.* Cambridge: Harvard Univ. Press
45. McKinney, C. O., Selander, R. K., Johnson, W. E., Yang, S. Y. 1972. Genetic variation in the side-blotched lizard (*Uta stansburiana*). *Univ. Texas Publ.* 7213: 307–18
46. Merritt, R. B. 1972. Geographic distribution and enzymatic properties of lactate dehydrogenase allozymes in the fathead minnow, *Pimephales promelas*. *Am. Natur.* 106:173–84
47. Nair, P. S., Brncic, D., Kojima, K. 1971. Isozyme variations and evolutionary relationships in the *mesophragmatica* species group of *Drosophila*. *Univ. Texas Publ.* 7103:15–28
48. Neel, J. V. 1972. The genetic structure of a tribal population, the Yanomama Indians. I. Introduction. *Ann. Hum. Genet.* 35:255–59
49. Neel, J. V., Ward, R. H. 1970. Village and tribal genetic distances among American Indians and the possible implications for human evolution. *Proc. Nat. Acad. Sci. USA* 65:323–30
50. Neel, J. V., Ward, R. H. 1972. The genetic structure of a tribal population, the Yanomama Indians. VI. Analysis by *F*-statistics (including a comparison with the Makiritare and Xavante). *Genetics* 72:639–66
51. Nei, M. 1971. Interspecific gene differences and evolutionary time estimated from electrophoretic data on protein diversity. *Am. Natur.* 105:385–98
52. Nei, M. 1972. Genetic distance between populations. *Am. Natur.* 106:283–92
53. Nei, M., Roychoudhury, A. K. 1972. Gene differences between Caucasian, Negro, and Japanese populations. *Science* 177:434–35

54. Nei, M., Roychoudhury, A. K. 1973. Sampling variances of heterozygosity and genetic distance. Unpublished

55. Nevo, E., Shaw, C. R. 1972. Genetic variation in a subterranean mammal, *Spalax ehrenbergi. Biochem. Genet.* 7:235–41

56. Nolan, C., Margoliash, E. 1968. Comparative aspects of primary structures of proteins. *Ann. Rev. Biochem.* 3:727–90

57. Ohta, T. 1972. Evolutionary rate of cistrons and DNA divergence. *J. Mol. Evol.* 1:150–57

58. Patton, J. L., Selander, R. K., Smith, M. H. 1972. Genic variation in hybridizing populations of gophers (genus *Thomomys*). *Syst. Zool.* 21:263–70

59. Petras, M. L. 1967. Studies of natural populations of *Mus.* I. Biochemical polymorphisms and their bearing on breeding structure. *Evolution* 21:259–74

60. Rogers, J. S. 1972. Measures of genetic similarity and genetic distance. *Univ. Texas Publ.* 7213:145–53

61. Salthe, S. N., Nevo, E. 1969. Geographic variation of lactate dehydrogenase in the cricket frog, *Acris crepitans. Biochem. Genet.* 3:335–41

62. Sarich, V. M. 1970. Primate systematics with special reference to Old World monkeys. In *Old World Monkeys,* ed. J. R. Napier, P. H. Napier, 175–226. New York: Academic

63. Sarich, V. M. 1972. On the nonidentity of several carnivore hemoglobins. *Biochem. Genet.* 7:253–58

64. Sarich, V. M. 1972. Generation time and albumin evolution. *Biochem. Genet.* 7:205–12

65. Seal, U. S. 1969. Carnivora systematics: a study of hemoglobins. *Comp. Biochem. Physiol.* 31:799–811

66. Selander, R. K. 1970. Behavior and genetic variation in natural populations. *Am. Zool.* 10:53–66

67. Selander, R. K., Kaufman, D. W. 1973. Genic variability and strategies of adaptation in animals. *Proc. Nat. Acad. Sci. USA* 70:1875–77

68. Selander, R. K., Hunt, W. G., Yang, S. Y. 1969. Protein polymorphism and genic heterozygosity in two European subspecies of the house mouse. *Evolution* 23:379–90

69. Selander, R. K., Yang, S. Y., Hunt, W. G. 1969. Polymorphism in esterases of hemoglobin in wild populations of the house mouse *(Mus musculus). Univ. Texas Publ.* 6918:271–338

70. Selander, R. K., Smith, M. H., Yang, S. Y., Johnson, W. E., Gentry, J. B. 1971. Biochemical polymorphism and systematics in the genus *Peromyscus.* I. Variation in the old-field mouse *(Peromyscus polionotus.) Univ. Texas Publ.* 7103:49–90

71. Smith, M. H., Selander, R. K., Johnson, W. E. 1973. Biochemical polymorphism and systematics in the genus *Peromyscus.* III. Variation in the Florida deer mouse *(Peromyscus floridanus),* a Pleistocene relict. *J. Mammal.* 54:1–13

72. Smith, M. H., Selander, R. K., Johnson, W. E. 1973. Biochemical polymorphism and systematics in the genus *Peromyscus.* V. Variation in *Peromyscus gossypinus* and *P. leucopus.* In preparation

73. Sneath, P. H. A., Sokal, R. R. 1973. *Principles of Numerical Taxonomy.* San Francisco: Freeman. 2nd ed.

74. Soulé, M. E., Yang, S. Y., Weiler, M. G. W., Gorman, G. C. 1973. Island lizards: the genetic-phenetic variation correlation. *Nature* 242:191–93

75. Uzzell, T., Corbin, K. W. 1972. Evolutionary rates in cistrons specifying mammalian hemoglobin α- and β-chains: phenetic versus patristic measurements. *Am. Natur.* 106:555–73

76. Wallace, B. 1968. *Topics in Population Genetics.* New York: Norton

77. Wallace, D. G., Maxson, L. R., Wilson, A. C. 1971. Albumin evolution in frogs: a test of the evolutionary clock hypothesis. *Proc. Nat. Acad. Sci. USA* 68:3127–29

78. Ward, R. H. 1972. The genetic structure of a tribal population, the Yanomama Indians. V. Comparison of a series of genetic networks. *Am. J. Hum. Genet.* 36:21–43

79. Ward, R. H., Neel, J. V. 1970. Gene frequencies and microdifferentiation among the Makiritare Indians. IV. A comparison of a genetic network with ethnohistory and migration matrices; a new index of genetic isolation. *Am. J. Hum. Genet.* 22:538–61

80. Webster, T. P. 1973. Adaptive linkage disequilibrium between two esterase loci of a salamander. *Proc. Nat. Acad. Sci. USA* 70:1156–60

81. Webster, T. P., Selander, R. K., Yang, S. Y. 1972. Genetic variability and similarity in the *Anolis* lizards of Bimini. *Evolution* 26:523–35

82. Weiner, J. S., Huizinga, J., Eds. 1972. *The Assessment of Population Affinities in Man.* Oxford: Clarendon

83. Weiss, M. L., Goodman, M. 1971. Genetic structure and systematics of some macaques and man. In *Comparative Genetics in Monkeys, Apes and Man,* ed. A. B. Chiarelli, 129–51. London: Academic

84. Wilson, A. C., Sarich, V. M. 1969. A molecular time scale for human evolution. *Proc. Nat. Acad. Sci. USA* 63: 1088–93

85. Workman, P. L., Niswander, J. D. 1970. Population studies on Southwestern Indian tribes. II. Local genetic differentiation in the Papago. *Am. J. Hum. Genet.* 22:24–49

86. Zuckerkandl, E. 1963. Perspectives in molecular anthropology. In *Classification and Human Evolution,* ed. S. L. Washburn, 243–72. Chicago: Aldine

ENZYME POLYMORPHISM AND BIOSYSTEMATICS: THE HYPOTHESIS OF SELECTIVE NEUTRALITY

❖4054

George B. Johnson
Department of Biology, Washington University, St. Louis, Missouri

In the last few years there has been an explosion of information concerning electrophoretic variation at enzyme loci. These data are being increasingly employed in attempts to elucidate biosystematic and phylogenetic relationships. As the evolutionary role of these allozyme polymorphisms is not well understood, the assumptions inherent in such approaches warrant careful consideration. The following review addresses itself to an examination of the possible role of selection in maintaining enzyme polymorphisms in natural populations. Selected for discussion here are those papers which seem to me to bear importantly upon central issues; the literature citations are not intended to be comprehensive or complete.

EXPERIMENTAL ASPECTS OF ELECTROPHORETIC ANALYSIS

Before discussing either the patterns of allozyme variation which have been observed or their possible evolutionary significance, it is necessary to consider what has been examined: the classes of protein variants detectable by current methods, the organisms which have been examined for such variants, and the enzyme reactions which have been used as screens. It is necessary to state carefully the experimental question which allozyme surveys pose in order to evaluate possible limitations and bias in the results obtained.

Experimental approaches involving electrophoretic analysis have dealt with three related sorts of questions: (*a*) those concerning relative amounts of variation; (*b*) those concerning the genetic nature of polymorphic variation; (*c*) those employing comparisons of variant types. Work in each of these areas may entail important assumptions about the nature of the variation.

When assessing the levels of electrophoretic variation in a natural population, the assumption is generally made, implicitly or explicitly, that electrophoretically de-

93

tectable variants represent a constant fraction of the total variation. In simplest terms, the detectable variants are thought to represent unit alterations in charge of amino acid residues. The possibility that amino acid substitutions not involving a unit charge transition might affect protein tertiary structure so as to alter net charge has not often been considered. Nor has the very real possibility that the proportion of total polymorphism detectable electrophoretically might be different for different loci been addressed experimentally to this date. These problems could be approached directly by amino acid sequence analysis of proteins from single individuals of a natural population (20), or from pure lines developed from such individuals; alternatively, peptide fingerprinting might provide similar information more readily. In either case extensive protein purification presents an enormous technical hurdle. Difficult as this sort of assessment may be, however, it will be necessary before comparisons of levels of electrophoretic polymorphism reported at different loci may be interpreted with confidence.

It has become increasingly common in the literature of the past few years to assume that multiple electrophoretic bands constitute direct evidence of heterozygosity at a protein locus (6, 86, 150, 173–176, 179, 199). This is a particularly dangerous assumption. It has certainly proven true for many cases that have been experimentally examined for Mendelian inheritance (24, 55, 71, 72, 90, 114, 134, 139, 164). Various cases have been reported, however, in which similar patterns of variation arise from nongenetic causes. For many proteins, interactions with electrophoresis buffers will produce multiple bands (148, 169, 170). This phenomenon has been examined in detail (26–29). Electrophoretic heterogeneity has also been reported to result from molecular instability (2, 17, 127, 185), the binding to proteins of marker dyes (70), charged cellular cofactors such as NAD or NADP (74, 137), intracellular acids such as sialic acid (34), and various other factors (57, 98, 184). Thus failure to verify experimentally the Mendelian inheritance of observed variation may result in overestimation of levels of variability. For genetic interpretation of "heterozygote" gel banding patterns, independent genetic information is required. In its absence one is dealing with phenetic variation, not genetic polymorphism.

The use of enzyme polymorphism data in biosystematics involves the critical issue of homology assessment. Electrophoretic comparisons are really comparisons of rates of protein migration, and thus are sensitive to a wide range of experimental variables (35): temperature, voltage, buffer molarity and pH, gel uniformity, degree of gel sieving, etc. Because electrophoretic mobility is sensitive to experimental variation, a statement that two electrophoretic bands are homologous requires proper documentation. The strongest verification results from pooling the two putative homologous types and rerunning the mixture: heterogeneity is evidence against homology. In a large survey this approach is impractical. At a minimum in such surveys, however, statements of homology should be statistical statements. If two electrophoretic types are adjudged different, then a pooling of all electrophoretic observations should produce a distribution of mobilities which passes a statistical test of bimodality; similarly, homology should be documented by a statistical test of the homogeneity of observed electrophoretic mobility.

If the electrophoretic analysis is suitably standardized, a great deal more confidence may be placed in homology assessments. In the author's own work two internal markers are employed that travel in the same path through the gel that the sample traverses. Such internal standards operate as follows (78): for a given gel the distance moved by a standard substance may be taken as $kS_a t$, where k is an experimental constant determined by the extrinsic variables, S_a is an experimental constant determined by the intrinsic nature of the protein species, and t is time. The rate of movement relative to that of the front is then $kS_a t/kFt$, where kF is the rate at which the front moves. The ratio of the mobilities of two internal standards is thus $(kS_a t/kFt)/(kS_b t/kFt)$, or S_a/S_b. This ratio is independent of both the value of k and the position of the front. This dual-marker procedure provides rigorous standardization. The value of the ratio of the distances the markers move cannot vary from gel to gel unless k is not uniform from gel to gel; if extrinsic factors are essentially constant throughout individual runs, the ratio will not vary outside of a narrow range of error for runs performed several years apart. If they are not constant, such dual-marker standardization is a very sensitive estimator of extrinsic variation; slight changes in buffer, current, or temperature result in significantly different standard ratios (80).

When it is shown that variation of extrinsic factors has not significantly affected a given run, the mobility of an enzyme variant may be unambiguously characterized by the ratio of its movement to that of a standard marker. Once it has been established for a particular gel that k is regular, then any differences seen in the ratio of variant enzyme to standard must be due to a difference in the intrinsic nature of the variant.

Isoelectric focusing is becoming widely used as an alternative to electrophoresis (30, 41, 65, 73, 87, 118, 133, 187, 189). For enzymes which retain activity at their isoelectric point, it may provide a superior survey technique. In isoelectric focusing, an ampholine compound added to the gel establishes a pH gradient within it; the sample protein migrates along this gradient to the pH which corresponds to its isoelectric point. Because the net charge of the protein at its isoelectric point is zero, it does not migrate further, but remains at that equilibrium position. The equilibrium position depends only upon the protein's isoelectric point under experimental conditions, and is independent of the rate at which that position is approached (14, 108, 161, 182, 195). The method is thus potentially far less sensitive to experimental error than electrophoresis. Direct experimental determination of the pH at the equilibrium band position with a "spear" electrode or similar device yields a number which directly reflects the physical character of the protein; other workers investigating the same protein should observe the same isoelectric point. Homology assessment is thus direct, and straightforward statistical assessments of validity may be performed.

Much of the current literature reflects the sorts of assumptions and problems outlined in this section. The degree to which this may influence the significance of reported findings is difficult to determine. Until the experimental variance of data on enzyme polymorphisms is reported for individual characterizations, and genetic assignments documented, interpretation will remain difficult.

THE HYPOTHESIS OF SELECTIVE NEUTRALITY

The use of electrophoretic data to construct phylogenies or to determine genetic relationships depends critically upon the question of the evolutionary role of electrophoretic variants. Much of the usefulness of data on enzyme variation in biosystematics depends on the assumption that the electrophoretic variants are selectively equivalent, or nearly so. Thus an evaluation of the validity of these approaches requires an assessment of the hypothesis of selective neutrality.

The current controversy regarding selective neutrality has roots extending back several decades in the history of population genetics to the arguments of Fisher and Wright concerning the significance of genetic drift (51, 201, 202). This is still being argued at present in quite a different context. In the absence of experimental data on the levels of genic variation being maintained in natural populations, Kimura & Crow (92) suggested from theoretical considerations that maintaining such variation entails an evolutionary cost, or "genetic load," and that because of this the total amount of polymorphism in natural populations cannot be great; excessive genetic load would be expected to drive a population to extinction. [Subsequently, models involving truncation selection and gene interaction have been advanced which, if valid in nature, might act to reduce the cost (49, 130, 188, 189). There is, however, disagreement as to their applicability (131)]. With the advent of electrophoresis as a common tool for surveying genetic variation at enzyme loci [initially by Lewontin & Hubby in 1966 (71, 115)], it became apparent that the level of polymorphic variation at the enzyme loci of natural populations is quite high (11, 12, 19, 66, 68, 103, 113, 132, 140, 152, 156–158, 205)—far higher than could exist if the original genetic load concepts were correct. These concepts could be maintained in the face of this result most simply by assuming that no selectively important differences exist among the electrophoretic variants. Thus the hypothesis has been advanced (93, 95, 96), supported (5, 22, 39, 119, 178, 198), and contested (36, 64, 126, 194, etc) that the variant proteins contain only minor differences in tertiary structure which are sufficient to affect electrophoretic mobility but not to affect significantly the functioning of the enzyme. Because electrophoretically different proteins are seen as functionally identical, they are thought to affect the organism's fitness identically, the differences among them thus being neutral to the action of selection.

This concept of evolution at enzyme loci by the random fixation of selectively neutral or nearly neutral mutations was first advanced by Kimura in 1968 (91) to account for the apparent constancy of the rate of amino acid substitution in evolution. Comparison of amino acid differences between homologous proteins of different species, as he and others pointed out, might be explained by random drift. (A large literature has built up concerning molecular phylogeny, which we will consider later.) In the last several years Kimura & Ohta (93, 95, 141–143) and Maruyama (125) have strongly advanced the proposition that protein polymorphism and protein evolution should be regarded as different aspects of the same process. Most enzyme polymorphisms in natural populations are seen as reflecting random frequency drift.

The experimental evidence necessary to test Kimura's stochastic view of molecular evolution is only now becoming available. While some caution is necessary in evaluating this diverse body of information, it seems on the whole to offer evidence against Kimura's hypothesis.

Evidence Involving Specific Enzyme Loci

The earliest and most intuitively appealing arguments against selective neutrality arose from observations of very nonrandom biogeographic patterns in field data. It was pointed out by Prakash et al (157), and has been observed frequently since (18, 25, 105, 106, 163), that electrophoretic polymorphisms show a considerable uniformity of allele frequencies over a wide geographical range. If the subpopulations are effectively isolated from each other, then random processes cannot account for this identity. Kimura and others have argued that a relatively small amount of migration between subpopulations may produce a panmictic unit, and that such a genetically unified grouping might exhibit uniform allele frequencies even under selective neutrality (93, 94, 149). The force of this argument is difficult to evaluate in the absence of experimental data on the population dynamics of the systems studied. The appropriateness of the no-mutation migration model assumed by Kimura & Ohta has been criticized by Bulmer (23), who argues that models that also consider the effects of mutation lead to quite different predictions. Data on the temporal constancy of allele frequencies is not yet available, but offers a more promising test of neutrality.

With increasing frequency other biogeographical patterns of allozyme frequency are being reported. Among the simplest reported patterns are clines which correlate with latitude: in *Drosophila ananassae* of the Pacific area, the frequency of the major allele of the esterase *C* locus increases from north to south towards the equator, and then decreases again south of the equator (186); in *D. melanogaster* similar latitudinally correlated clines are seen along the US east coast for alcohol dehydrogenase (196) and a variety of other enzyme loci (76). Selander and co-workers have reported geographic patterns in the frequencies of a variety of alleles for the house mouse *Mus musculus* (172) and for the field mouse *Peromyscus polionotus* (176). Clinal variation has also been reported for marine organisms: allele frequencies of leucine aminopeptidase in the ectoproct *Schizoporella unicornis* parallel a gradient in water temperature (167). In the pierid butterfly *Colias meadii* apparent clines have been observed in heterozygosity at the α-glycerophosphate dehydrogenase locus within single genetic populations that correlate with altitude along a mountainside transect (80, 85); in this latter case differential migration may not be invoked to explain the observed cline in terms of selective neutrality, as the population dynamics of the population are well characterized. Allele frequency clines within single well-characterized populations offer the opportunity for particularly strong tests of neutrality.

The possibility that migration may produce observed clinal patterns may be similarly discounted when biogeographic patterns of variation are mirrored in separate noninterbreeding species. This is seen clearly in the study of the *D. willistoni* group undertaken by Ayala and co-workers (7, 12): biogeographic patterns in allele frequencies observed across South America for *D. willistoni* are very similar to those seen for the other species of the group. As the species do not interbreed in nature, their similar allele frequency patterns cannot be explained by small amounts of gene flow between them. Similar parallelism may be seen among the Hawaiian

Drosophila (162). However, the possibility that population sizes have "bottle-necked" in the recent past cannot be discounted in these cases, rendering the argument against neutrality less than compelling.

It can be reasonably argued that biogeographic, clinal, and other patterns of enzyme polymorphism may reflect linkage of neutral enzyme alleles to other loci which are being selected, but which are not themselves polymorphic (144). Such "associative overdominance" is so difficult to discount completely experimentally that it may be invoked to explain practically any conceivable pattern of allele frequencies at a locus.

Selection has been implicated by a different sort of evidence at the tetrazolium oxidase locus of the mussel *Modiolus* (102): significant differences were found in the proportion of heterozygotes characteristic of the various age classes examined. In young individuals heterozygotes were fewer than expected, but became more frequent with increasing age. This suggests differential larval mortality.

Other classes of evidence which argue against selective neutrality involve physiological or catalytic differences between allozymes. In many cases, allele frequencies were observed to correlate with environmental factors, particularly temperature. In *D. melanogaster* (15), the frequency of the esterase-6^F allele in laboratory populations was higher in populations raised at higher temperatures; in laboratory populations (59) of the same species, alcohol dehydrogenase allele frequencies were affected by levels of dietary ethanol. In the freshwater fish, *Catostomus clarkii* (99, 100) the allele frequencies of serum esterases sampled at locations along the Colorado River in Kansas seem correlated with ambient water temperature: the activity of the allozyme more frequent in southern populations increased as temperature increased from 0 to 37°C, while the activity of the allozyme more frequent in northern populations increased as temperature decreased. Temperature in this case thus might act to effect single gene heterosis. Differential response over a range of environmental temperatures has been reported for the lactate dehydrogenase allozymes of another freshwater fish, the minnow *Pimephales promelas* (129), where the substrate binding affinities responded differently to changes in temperature.

In the ant *Pogonomyrmex barbatus* (77) and in *Drosophila* (76, 104, 163), regression and principle component analysis of weather variables characteristic of sampled habitats have indicated that enzyme polymorphism significantly correlates with environment. The problems associated with such methods of correlation analysis have been reviewed (151, 180). It seems clear that correlation with selected environmental variables is only a first step in determining the identity of possible selective agents. Although little experimental data of this sort has been reported for plant populations, it would seem a very promising avenue of investigation, as plants might be expected to experience strong selective pressures directly relatable to microhabitat (21, 46, 75, 181).

One of the single most important experimental goals of research in enzyme polymorphism in the coming years will be to characterize the catalytic properties of allozyme variants. Allozymes of polymorphic enzymes alleles have been reported to differ in their biochemical properties. Differences were observed in substrate binding affinity (3, 56, 58, 101, 155, 193), in thermal effects upon that affinity (99,

129, 183), and in enzyme stability (13, 171). Harris (67) listed sixteen enzyme polymorphisms in man where quantitative differences between the common phenotypes have been found. It is difficult to believe that such catalytic differences between allozymes of a polymorphic locus have no functional effects; selective neutrality in these cases would be very surprising.

Conflicting results have been obtained from attempts to directly assess putative differences in fitness between polymorphic alleles at a locus. Typical of the negative findings is the extensive study by Yamazaki (203) of the relative fitness of alleles at the esterase-5 locus in *Drosophila*. He was able to detect no significant differences in fitness, and concluded that the allelic alternatives at this locus were either selectively equivalent or nearly so. Contrary results were obtained at a different locus in *Drosophila* by Ayala & Anderson (8). It might be argued that in the Yamazaki study the experimental environment was not exercising selection at this locus under the conditions of these experiments: if this esterase is involved in some aspect of dietary metabolism (such as the breakdown of plant secondary compounds), the selectively important variables in natural populations may not be present in cultures raised on laboratory media.

The possibility also exists that the pattern of environmental variation is of differential selective importance to the alleles; the Yamazaki experiments, conducted under uniform conditions, would not be responsive to such selection. Suggestions that this might indeed be the case have been reported by Long (117) and by Powell (153). Powell reported that the average heterozygosity of a collection of enzyme loci increased when experimental factors affecting diet and temperature were varied. What is required to clarify this issue is an experiment which considers a single locus of known function and a single environmental factor known to be selectively important to that locus. An example for *Drosophila* might be alcohol dehydrogenase, which is important in lipid metabolism (63) and for which K_m seems a function of temperature. The most straightforward experimental design would involve comparing populations raised in a constant environment, a predictably variable environment, and a more randomly variable environment; one may then ask whether or not relative fitness is a function of the predictability of the pattern of variation (characterized perhaps in terms of its information content, $n \log n$), as some theoretical considerations predict (16, 60, 62, 109–112, 168, 190).

General Arguments Concerning Selective Neutrality

It is often not clear when data concerning individual loci such as discussed above should be regarded as special cases and when they may be considered as illustrating a more general principle. Thus those individual cases where selection is implicated at an allozyme locus do not necessarily argue powerfully for the generality of selective significance. Many such cases would be required to support statements of generality convincingly. More general assessments of the hypothesis of selective neutrality are possible, however. Analyses of the statistical distribution of allele frequency data provide one such approach. Two such analyses are available; each reports quite different conclusions. One relates heterozygosity to allele frequencies, the other to numbers of alleles. The first analysis, presented by Yamazaki &

Maruyama (124, 204) and discussed by Crow (40), argued that polymorphic enzyme loci of natural populations do not exhibit the relationship which would theoretically be expected for heterosis or for strong directional selection. They concluded that strong selection is not involved. Vigorous objections have been raised to the theoretical approach employed (48). Nor does it seem appropriate to pool data obtained at all enzyme loci (different functional classes of enzyme may experience different modes of selection) in a bewildering variety of species (mice may experience quite a different selective regime than *Drosophila*). When analysis was restricted to *Drosophila* for purposes of comparability, and functional classes of enzyme reactions (discussed below) were treated separately, the result was not so clear-cut (Figure 1), and could be attributed as easily to a combination of balancing selection and detrimental alleles. Until the distribution expected under neutrality can be sharply delineated from that which might be expected to occur under reasonable assumptions of disruptive and heterotic selection, the approach seems inconclusive.

A different conclusion is reached when one considers the relationship between polymorphism and numbers of alleles at loci. In several natural populations, the "evenness" (equitability) of allele frequencies (e.g. 0.33, 0.33, 0.34 is a more "even" distribution of frequencies than is 0.90, 0.09, 0.01) has been observed to be greater for loci with higher numbers of alleles (81). The distribution expected for selective neutrality predicts the opposite, that a more skewed distribution of frequencies will be seen for loci of high allele number (84). This prediction depends upon the sampling theory developed by Ewens (47), and assumes equilibrium and panmixia in the analyzed populations. It is difficult to assess the validity of these assumptions, as the history and breeding structure of the populations are not known. If the assumptions are warranted, then the observed frequency distributions of loci with many alleles support the argument against selective neutrality. Again, one must be careful in pooling diverse sets of data. When such an analysis is restricted to *Drosophila,* the relationship described above seems general (Figure 2); when, however, data on insects, mammals, fish, etc, are all lumped together, a significant result is not obtained (97).

The Importance of Substrate Variability

Although the weight of the evidence discussed above is decidedly against the hypothesis of selective neutrality, the case as stated so far is convincing, rather than compelling. In the absence of knowledge of breeding structure, and without a means of excluding the possibility of associative overdominance, the hypothesis of neutrality seems still tenable, although precariously so. Another line of argument suggests that levels of enzyme polymorphism reflect physiological function, some functional classes being far more variable than others. No combination of linkage, migration, or breeding structure can render this relationship compatible with selective neutrality.

Two hypotheses have been advanced concerning the physiological role of enzyme polymorphism. The first of these, proposed in 1968 by Gillespie & Kojima (61), suggests that levels of enzyme polymorphism may reflect environmental variation in substrates; the second hypothesis, proposed by the author in 1971 (79), suggests that polymorphism occurs preferentially at regulatory reactions in metabolism.

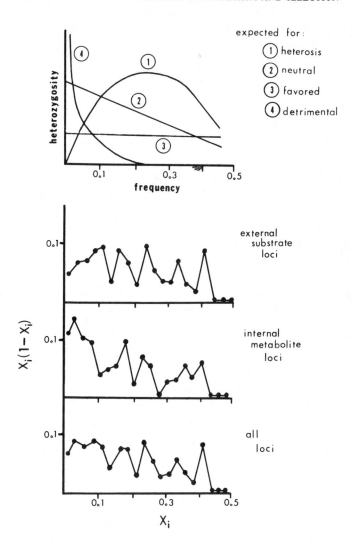

Figure 1 Reevaluation of Yamazaki & Maruyama (204). The theoretical expectations are as described by Crow (40); following his suggestion, the frequency of the most common allele is omitted from calculations. This, and the assumption that alternative homozygotes may differ selectively at heterotic loci, shifts the expected heterotic equilibrium to the left of 0.5. Data on all the *Drosophila* species of Table 1 are employed. Only when the two functional classes of enzyme loci (82) are pooled is a "neutral" curve obtained. It thus seems that the result of the analysis is dependent upon the identity of the loci being considered.

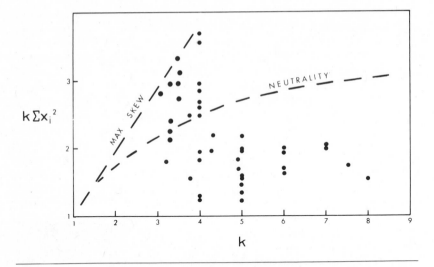

Figure 2 Skew in allele frequency distribution as an index of selection. The data employed are those of the thirteen *Drosophila* species cited in Table 1. The method of analysis is that of Johnson & Feldman (84): k = number of alleles at a frequency ≥ 0.01 in a sample of at least 100 genes; x_i = allele frequency; $k\Sigma x_i^2$ is a measure of the degree of skew in the allele frequency distribution. For loci with more than four alleles, the observed distributions are in every case far less skewed than the expectation under selective neutrality. This suggests that frequencies of rarer alleles are being maintained by selective foces at higher levels than would otherwise obtain.

Gillespie & Kojima (61) pointed out that in laboratory cultures of *Drosophila ananassae* far less heterozygosity was observed at the loci of enzymes involved in energy production than at other enzyme loci. Their later more detailed examination of natural populations (103) seemed to confirm their original observation. They went on to suggest that the greater variability in "non-glucose-metabolizing" enzymes might reflect greater variability in their substrates, as many of these substrates originate in the external environment.

In the several years since this dichotomy of enzyme types was proposed, there has been a massive expansion of electrophoretic investigation, resulting in a great deal of new experimental data. The number of flies collected from natural populations, examined electrophoretically, and reported upon now exceeds 30,000. Extensive work has also been reported on small vertebrates and on man. Analysis of this large body of data (38, 82, 147) strongly confirms the original hypothesis of Gillespie and Kojima. In Table 1, enzymes studied in 13 *Drosophila* species have been partitioned into two groups: (*a*) enzymes of broad specificity, many of which utilize substrates originating from the external environment; and (*b*) enzymes which utilize specific metabolically produced substrates. Clearly the latter group of enzyme loci is by far less variable. These differences are statistically significant (82).

The differing levels of polymorphic variability seen in Table 1 might be attributed to any of a variety of physiological interactions. Superior structural stability and differential substrate binding of heterologous (heterozygote) enzymes seem logical avenues of investigation.

The role of allele multiplicity in polymorphism may be quite different from that of heterozygosity as such. The simplest hypothesis, and the one which makes the most intuitive biochemical sense, is that polymorphisms exhibiting many alleles reflect qualitative, as opposed to quantitative, variability in available substrates.

Regulatory Enzymes and Metabolic Organization

The second major hypothesis involving the physiological role of enzyme polymorphism is that proposed by myself (79). I have suggested that "those enzymes which exert acute control over flow through (metabolic) pathways should be most individ-

Table 1   The importance of substrate source to enzyme polymorphism

Populations sampled	Enzymes with source of substrate internal		Enzymes with source of substrate external		Reference
	S^a	$\overline{k}^b$	S	$\overline{k}$	
D. willistoni (Guyana)	12	3.00	8	3.37	10
D. subobscura (several populations)	3	2.67	7	3.14	106
D. melanogaster (Raleigh)	8	1.62	5	2.00	103
D. equinoxialis (several populations)	12	2.33	8	3.25	11
D. pseudoobscura (Colorado)	3	2.67	5	3.40	157
D. obscura (Central Finland)	3	2.33	6	3.00	105
D. willistoni (Trinidad)	11	2.64	7	3.43	10
D. willistoni (Mirassol)	9	2.89	7	3.86	10
D. melanogaster (Japan)	12	1.33	5	1.80	103
D. pseudoobscura (California)	3	2.67	5	3.80	157
D. paulistorum (transitional)	4	2.25	5	3.40	158
D. bipectinata	11	2.27	6	3.67	205
D. parabipectinata	11	1.54	6	2.50	205
D. obscura (Lapland)	3	2.00	6	3.33	105
D. obscura (South Finland)	3	1.67	6	3.33	105
D. malerkotliana pallens	11	1.82	6	3.67	205
D. paulistorum (Andean-Brazilian)	2	1.00	9	2.44	158
D. athabasca (New Jersey)	9	1.33	5	3.40	103
D. simulans (Texas)	10	1.40	5	3.60	103
D. paulistorum (interior)	4	1.00	7	2.71	158
D. affinis (New Jersey)	7	1.43	5	4.00	103
D. paulistorum (Amazonian)	2	1.00	8	2.87	158

[a] S: number of loci samples with $n > 100$ genomes.

[b] $\overline{k}$: mean number of alleles observed at a frequency ≥ 0.01 in samples with $n > 100$ genomes. Classification of enzyme function is that of Johnson (82).

ually sensitive as sites of action of selective forces." The rationale of this argument is that selection must act ultimately upon the reproductive fitness of individuals. The contributions of particular metabolic sequences to that fitness must be considered in terms of overall pathway output rather than in terms of specific reactions. Changes at loci whose enzymes regulate flow through pathways would be expected to produce far greater alterations in fitness than changes affecting enzymes which do not regulate metabolic flow.

As a preliminary estimator of the regulatory character of polymorphic enzymes, I suggested that the equilibrium constant (K_{eq}) of the reaction is catalyzed, reasoning that thermodynamically irreversible reactions tend to be sites of metabolic regulation. This approach has been criticized by several workers (10, 177). Even in the absence of criticism, it has been clear that K_{eq} is at best only a crude estimator of regulatory involvement, factors other than physiological reversibility being equally important in regulation. The activity of the initial reaction of an anabolic pathway is often tightly controlled; similarly, branch points in metabolism are almost always regulated, as are multifunctional enzymes. The large array of enzyme loci which have been examined for polymorphism within the last two years includes many whose regulatory roles have been well characterized, permitting a more meaningful assessment of the hypothesis than was possible previously (83).

Table 2 presents a compilation of available data on levels of enzyme polymorphism. That the degree of heterozygosity at an enzyme locus correlates with the regulatory role of the enzyme seems to be a principle of broad validity. For *Drosophila,* the probability that polymorphism at regulatory and at nonregulatory loci is not significantly different is very small, $P < 0.01$. For small vertebrates, the probability is $P < 0.1$, and for man it is $P < 0.001$. This suggests that the physiological

Table 2 Enzyme polymorphism and metabolic function[a]

	HET		
	Drosophila (13 species)	Small vertebrates (22 species)	Humans
Variable Substrate Loci	0.24	0.22	0.18
Regulatory Reaction Loci	0.19	0.13	0.13
Nonregulatory Reaction Loci	0.08	0.07	0.005

[a]The analysis is that of Johnson (83). Data for *Drosophila* are those cited in Table 1; data for small vertebrates is that summarized by Selander & Johnson (177); data for humans is that reviewed by Harris (68). $\overline{\text{HET}}$ = mean heterozygosity.

role of polymorphisms at the loci of metabolic enzymes is related to metabolic regulation. Perhaps enzyme polymorphisms increase fitness by providing a more flexible response in a varying environment. The precise nature of the physiological advantage of enzyme multiplicity may be expected to differ for different classes of enzymes. Whatever the specific hypothesis advanced concerning the nature of the selective advantage, clear understanding will demand direct experimental examination of the kinetic character of the variants.

The dichotomy described in Table 2 argues strongly against the hypothesis that polymorphic alleles are selectively neutral. No such hypothesis can account for a correlation between enzyme polymorphism and enzyme function. Other general arguments against selective neutrality involve assumptions about the nature of the populations studied; this argument does not. The conclusions of these other arguments have relied primarily on data from those few loci with large numbers of alleles. The relationship of heterozygosity to metabolic function, by contrast, is equally apparent for loci with only two or three alleles; this suggests that polymorphism is influenced by selection generally, and that the multi-allele loci do not constitute a special case.

Modes of Selection

Few data exist which permit direct analysis of the mode of action of selection in maintaining polymorphism in natural populations. One possible method of analysis is to examine the distributions of allele frequencies; selection would be expected to affect allele frequency distributions. Moreover, normalizing selection will produce changes different from those resulting from disruptive selection. Comparison of the variances in allele frequency might provide a sensitive indicator of such differences between distributions. (Heterozygosity, by contrast, is an estimator of central tendency, and is not as sensitive to changes in the shapes of distributions.)

Over the last year the author has collaborated with M. Feldman on such an analysis. Following the suggestion of Cavalli-Sforza (31–33), we have characterized allele frequency distributions in terms of the normalized variance, $\sigma^2(pq)^{1/2}$. A similar approach has recently been employed by Lewontin & Krakauer (116). The mean frequency of each allele at a given locus examined in several populations will have associated with it a variance. If there are no selectively important differences between the alleles (e.g. the alleles respond identically to differences among the populations), then the frequency variances of the alleles at that locus must be the same. This would also be true if multiple alleles are being maintained by uniform heterotic selection. However, if disruptive selection is involved in determining the allele frequencies (different alleles responding to the array of populations differently), or if different heterotic combinations are not selectively equivalent, then the associated allele frequency variances will differ.

As is evident in Table 3, among South American *D. willistoni* (12) populations, many loci exhibit significant heterogeneity in allele frequency variances. This suggests the possible involvement of disruptive selection in maintaining polymorphism at these loci. Loci of enzymes that do not show significant heterogeneity in allele frequency variances are consistent with neutrality, or with heterotic selection (45,

Table 3   Heterogeneity of allele frequency variances[a]

	Locus	k	N	Significantly Heterogeneous?
Enzymes using	LAP-5	5	27	Yes
substrates of	EST-5	3	27	Yes
environmental	EST-7	5	24	No
origin	EST-4	4	16	No
	EST-6	3	12	Yes
	EST-2	4	8	No
	AdH	4	8	No
	OdH	3	8	Yes
	EST-3	3	4	No
Enzymes using	a-GPdH	3	8	No
internal metabolites	MdH-2	3	8	No
as substrates	KH-1	3	5	No
	HK-2	4	5	Yes
	Pgm	4	4	Yes
	Me	3	4	No
	IdH	3	4	Yes
Enzymes whose	ApH-1	4	23	No
natural substrates	AcpH-1	3	10	No
are unknown	To	4	7	No

[a]The analysis of significance employs data on *Drosophila willistoni* collected at a variety of South American locations (10). k = number of alleles at frequency ≥ 0.01; N = number of populations sampled for ≥ 75 genes. Variances are judged to be significantly heterogeneous at a level of $P < 0.05$. The assignment of AdH and OdH to metabolic category is arbitrary, as these enzymes may function in both respects (83).

50, 128, 132, 160, 200) operating at these loci. It is perhaps worth noting that among *Drosophila* the "external substrate" class of loci typically exhibit far greater numbers of alleles than do the "internal substrate" loci (82), which is also consistent with differing modes of selection.

The finding of significant heterogeneity in allele frequency variances among the loci of *D. willistoni* constitutes direct evidence against the hypothesis of selective neutrality, at least insofar as it applies to these enzyme loci. It also serves to emphasize the importance of considering disruptive as well as heterotic modes of selection when analyzing adaptive relationships among populations of variable density (37, 165) occupying heterogeneous environments (4, 190).

ESTIMATES OF GENETIC SIMILARITY

Data on enzyme polymorphism are being widely used to estimate genetic similarity (107, 135, 136, 191, 192, 197). A variety of measures have been devised; several have been reviewed by Hedrick (69). These measures estimate in various ways the probability of selecting identical allozymes from two populations: the more the populations differ genetically, the lower this probability. The many problems associated with such measures of "genetic distance" have recently been reviewed (32).

Indices of genetic similarity offer promise of providing meaningful comparisons of the degree of genetic similarity of closely related species (159). As various of the allozyme loci may be subject to short-term selective pressures, significant levels of difference may occur within single species occupying diverse habitats. It is important, however, that the nature of the question being asked be kept clearly in mind. Comparisons of more distantly related groups may reflect both short-term adaptations to differing habitats and longer-term evolutionary divergence. For such groups uniquely occurring markers such as chromosomal inversions (which are usually monophyletic in origin) provide a more clear-cut approach to questions of phyletic or geographic origin (44, 154). For closely related groups, unique characteristic alleles may serve the same purpose (9).

When comparing different estimates of genetic similarity, it is important to consider the ecological sensitivity which these estimates may reflect. Thus the similarity of two populations might be characterized quite differently by two independent investigators, if one investigator chose to analyze an array of loci containing a greater proportion of internal, nonregulatory enzymes. It should not be assumed that estimates of genetic similarity based upon allozyme surveys are comparable.

Finally, it is worth emphasizing that the critical experimental data from which all such estimates of genetic similarity ultimately derive are statements of homology made by the investigator. The experimental errors involved in such statements of identity are generally not reported, and when differing bands exhibit similar mobilities this error in identification need not be small. This serves to point up the need for documenting homology statements with adequate statistical tests. In such a case a reader may determine objectively the degree of confidence associated with individual assessments of homology.

PHYLOGENY AND THE EVOLUTIONARY CLOCK

Amino acid sequence data have been obtained for proteins of a wide range of species (42, 138). For some proteins such as hemoglobin or cytochrome c, data exist on a sufficient array of species to permit construction of elaborate family trees (1, 53, 54, 88, 120–122) which seem similar to those based upon more traditional methods. Although not really supported by this data, the generalization has emerged in the literature that the rate of amino acid substitution of a given protein seems relatively constant over evolutionary time (43, 52, 89, 96, 146, 166), different proteins having different rates. This constancy is thought to support the hypothesis that many of the amino acid changes occurring during the evolution of a protein are selectively unimportant ones which have been fixed by random processes (93, 96, 145). Different proteins are seen as evolving at different rates because differing portions of the proteins are thought to be "nonessential" and thus tolerant of random changes in sequence (43, 96, 123). Under this argument it is not clear why the rate of amino acid substitution does not vary with generation time.

The use of "trees" in biological systematics and evolution involves a variety of assumptions (32). Primary among them is the assumption that the amount of divergence between two groups is proportional to the number of generations elapsed since their separation. Only in "random walk" situations is this assumption a safe one. If amino acid substitutions in the evolution of proteins occur primarily by genetic drift, then the number of substituted residues does indeed increase proportional to time. If, on the other hand, selective processes have been responsible for the amino acid substitutions, then the amount of divergence is proportional instead to the rate of change of the selective environment, and need not reflect the number of generations since separation.

The rate of evolution will be linear with time under the neutral hypothesis. Linearity is also compatible with selection. Under selection, if changes in selective values occur over short periods relative to the long periods of time reflected by "molecular phylogenies," then the observed rate of evolution will reflect the characteristic rate of change of selective values experienced at a particular locus. If that rate of change, the selective variability, remains approximately constant during the course of evolution, then a linear rate of amino acid substitution will be observed over long time periods.

Thus the argument that a stochastic evolutionary clock is implied by constancy of evolutionary rates is by no means compelling, as it is not the only hypothesis

Table 4 Physiological function and rate of evolution[a]

Protein	Unit Evolutionary Period [MY] (43)	Rate of Amino Acid Substitution (146)	Physiological & Functional Location
Fibrinopeptides	1.1	4.55	Circulates in blood
Proinsulin	1.9	2.63	Circulates in blood
Pancreatic Ribonuclease	2.1	2.38	Secreted into blood (intestine)
Lysozyme	5.3	0.94	Circulates in blood
Hemoglobin (a and β)	5.8	0.86	Red blood cells
Myoglobin	5.8	0.86	Red blood cells
Trypsinogen	10.8	0.46	Compartmentalized in lysozymes
Chymotrypsin	10.8	0.46	Compartmentalized in lysozymes
Insulin	15.8	0.32	Acts in cytoplasm
Glyceraldehyde-3-P dehydrogenase	18.4	0.27	Acts in cytoplasm
Cytochrome c	20.0	0.25	Acts in cytoplasm

[a]EUP values are those of Dickerson (43), some of which are preliminary.

consistent with available data. An equally coherent argument may be advanced involving selection. Differing rates of amino acid substitution in variable regions of a protein may reflect different levels of variability in selective environments. A protein functioning in a relatively constant physiological environment such as the interior of a eucaryotic cell might experience, over evolutionary time, far fewer major perturbations in its physiological environment than would a protein functioning in the bloodstream, where changes in physiological character may be more common (Table 4). This hypothesis is consistent with available information and with the rate of substitution being proportional to historical time, rather than the number of generations since separation.

Until such time as more convincing arguments are advanced to support the hypothesis of stochastic molecular evolution, molecular phylogenies constructed from comparisons of amino acid sequence data should be interpreted with care.

CONCLUSION

The rapidly increasing use in the past few years of electrophoretic techniques to survey levels of genetic polymorphism at enzyme loci has produced a great deal of information within a very short period of time. It is now clear that high levels of polymorphism are common in nature, and although the issue remains unsettled, it seems likely on the basis of current evidence that this polymorphism is maintained by selection. The evolutionary role of enzyme polymorphism may well be to provide metabolic flexibility in a changeable environment. Now that the broad outlines of the problem are emerging from this initial wave of investigation, more pointed questions may be addressed. Most important, the catalytic character of allozyme variants must be assessed before the details of their evolutionary roles can be understood. Also, the nature of the selective environments of the populations being sampled may be more pointedly characterized when the factors differentially affecting allozyme catalytic behavior are known. Undoubtedly allozyme data will see continued use in biosystematic and phylogenetic applications. In this they should be treated like any other group of taxonomic characters, with criteria for category assignments clearly stated and the possibility of selective interactions considered. In all of the above, careful consideration of enzyme metabolic function will be important.

Literature Cited

1. Air, G. M., Thompson, E. O. P., Richardson, B. J., Sharman, G. B. 1971. Amino-acid sequences of kangaroo myoglobin and haemoglobin and the date of marsupial-eutherian divergence. *Nature* 229:391–94
2. Akroyd, P. 1965. Acrylamide-gel electrophoresis of β-lactoglobulins stored in solutions of pH 8.7. *Nature* 208: 488–89
3. Albers, L. V., Dray, S., Knight, K. L. 1969. Allotypes and isozymes of rabbit

arylesterase. Allelic products with different enzymatic activities for the same substrate. *Biochemistry* 8: 4416–24
4. Antonovics, J. 1971. The effects of a heterogeneous environment on the genetics of natural populations. *Am. Sci.* 59:593–99
5. Arnheim, N., Taylor, C. E. 1969. Non-Darwinian evolution: consequences for neutral allelic variation. *Nature* 233: 900–2

6. Avise, J., Kitto, G. B. 1973. Phosphoglucose isomerase gene duplication in the bony fishes: An evolutionary history. In preparation

7. Ayala, F. J. 1972. Darwinian vs. non-Darwinian evolution in natural populations of *Drosophila. Proc. Berkeley Symp. Math. Stat. Prob., 6th,* ed. L. LeCam, J. Neyman, E. Scott, Vol. 5. Berkeley: Univ. Calif.

8. Ayala, F. J., Anderson, W. 1973. Evidence of natural selection in molecular evolution. *Nature New Biol.* 241: 274–76

9. Ayala, F. J., Powell, J. R. 1972. Allozymes as diagnostic characters of sibling species of *Drosophila. Proc. Nat. Acad. Sci. USA* 69:1094–96

10. Ayala, F. J., Powell, J. R. 1972. Enzyme variability in the *Drosophila willistoni* group. VI. Levels of polymorphism and the physiological function of enzymes. *Biochem. Genet.* 7:331–45

11. Ayala, F. J., Powell, J. R., Tracey, M.L. 1972. Enzyme variability in the *Drosophila willistoni* group. V. Genic variation in natural populations of *Drosophila equinoxialis. Genet. Res., Cambridge* 20:19–42

12. Ayala, F. J., Powell, J. R., Tracey, M. L., Mourao, C. A., Perez-Salas, S. 1972. Enzyme variability in the *Drosophila willistoni* group. IV. Genic variation in natural populations of *Drosophila willistoni. Genetics* 70:113–39

13. Babalola, O., Cancedda, R., Luzzatto, L. 1972. Genetic variants of glucose 6-phosphate dehydrogenase from human erythrocytes: Unique properties of the A⁻ variant isolated from "deficient" cells. *Proc. Nat. Acad. Sci. USA* 69: 946–50

14. Bates, L. S., Deyoe, C. W. 1972. Polyacrylamide gel electrofocusing and the ampholyte shift. *J. Chromatogr.* 73: 296

15. Beardmore, J. 1970. Ecological factors and the variability of gene-pools in *Drosophila.* In *Essays in Evolution and Genetics in Honor of Theodosius Dobzhansky,* ed. M. Hecht, W. Steere, 299–314. New York: Appleton

16. Beardmore, J., Levine, L. 1963. Fitness and environmental variation. I. A study of some polymorphic populations of *Drosophila pseudoobscura. Evolution* 17:121–29

17. Bennett, G. A., Shotwell, O. L., Hall, H. H. 1968. Hemolymph proteins of healthy and diseased larvae of the Japanese Beetle, *Popillia japonica. J. Invertebr. Pathol.* 11:112–18

18. Berger, E. M. 1971. A temporal survey of allelic variation in natural and laboratory populations of *Drosophila melanogaster. Genetics* 667:121–36

19. Berry, R. J., Murphy, H. M. 1970. The biochemical genetics of an island population of the house mouse. *Proc. Roy. Soc. B* 176:87–103

20. Boyer, S. H. 1972. Extraordinary incidence of electrophoretically silent genetic polymorphisms. *Nature* 239: 453–54

21. Bradshaw, A. 1972. Some of the evolutionary consequences of being a plant. *Evol. Biol.* 5:25–47

22. Bullini, L., Coluzzi, M. 1972. Natural selection and genetic drift in protein polymorphism. *Nature* 239:160–61

23. Bulmer, M. G. 1973. Geographical uniformity of protein polymorphisms. *Nature* 241:199–200

24. Burns, J. M., Johnson, F. M. 1967. Esterase polymorphism in natural populations of a sulfur butterfly, *Colias eurytheme. Science* 156:93–96

25. Burns, J. M., Johnson, F. M. 1971. Esterase polymorphism in the butterfly *Hemiargus isola:* Stability in a variable environment. *Proc. Nat. Acad. Sci. USA* 68:34–37

26. Cann, J. R. 1968. Recent advances in the theory and practice of electrophoresis. *Immunochemistry* 5:107–34

27. Cann, J. R., Goad, W. B. 1965. Theory of zone elctrophoresis of reversibly interacting systems. Two zones from a single macromolecule. *J. Biol. Chem.* 240:1162–64

28. Cann, J. R., Goad, W. B. 1968. Two or more electrophoretic zones from a single macromolecule. *Ann. NY Acad.* 151:638

29. Cann, J. R., Goad, W. B. 1968. The theory of transport of interacting systems of biological macromolecules. *Advan. Enzymol. Relat. Areas Mol. Biol.* 30:139

30. Catsimpoolas, N. 1971. Analytical scanning isoelectrofocusing: 3. Design and operation of an *in situ* scanning apparatus. *Anal. Biochem.* 44:411–26; 436–44

31. Cavalli-Sforza, L. L. 1966. Population structure and human evolution. *Proc. Roy. Soc. B* 164:362–79

32. Cavalli-Sforza, L. L. 1973. Some current problems of human population genetics. *Am. J. Hum. Genet.* 25:82–104

33. Cavalli-Sforza, L. L., Barrai, I., Edwards, A. W. 1964. Analysis of human evolution under random genetic drift. *Cold Spring Harbor Symp. Quant. Biol.* 29:9–20

34. Chen, S.-H., Sutton, H. E. 1967. Bovine transferrins: Sialic acid and the complex phenotype. *Genetics* 56:425–30
35. Chrambach, A., Rodbard, D. 1970. Polyacrylamide gel electrophoresis. *Science* 172:440–51
36. Clarke, B. 1970. Darwinian evolution of proteins. *Science* 168:1009–11
37. Clarke, B. 1972. Density-dependent selection. *Am. Natur.* 106:1–13
38. Cohen, P., Omenn, G. S., Motulsky, A. G., Chen, S. H., Giblett, E. R. 1973. Restricted variation in the glycolytic enzymes of human brain and erythrocytes. *Nature New Biol.* 241: 229–33
39. Corbin, K. W., Uzzell, T. 1970. Natural selection and mutation rates in mammals. *Am. Natur.* 104:37–53
40. Crow, J. 1972. The dilemma of nearly neutral mutations: How important are they for evolution and human welfare? *J. Hered.* 63:306–16
41. Csopak, H., Jonsson, M., Hallberg, B. 1972. Isoelectric fractionation of the *E. coli* alkaline phosphatase and resolution of the purified enzyme into isoenzymes. *Acta Chem. Scand.* 26:2412
42. Dayhoff, M. 1969. *Atlas of Protein Sequence*. Silver Spring, Md.: Nat. Biomed. Res. Found.
43. Dickerson, R. E. 1971. The structure of cytochrome C and the rates of molecular evolution. *J. Mol. Evol.* 1:26–45
44. Dobzhansky, T. 1972. Species of Drosophila. New excitement in an old field. *Science* 177:664–69
45. Efron, Y. 1973. Specific differences in maize alcohol dehydrogenase: possible explanation of heterosis at the molecular level. *Nature New Biol.* 241:41–42
46. Ehrlich, P. R., Raven, P. H. 1969. Differentiation of populations. *Science* 165:1228–32
47. Ewens, W. 1972. The sampling theory of selectively neutral alleles. *Theor. Pop. Biol.* 3:87–112
48. Ewens, W., Feldman, M. 1973. On the analysis of neutrality in protein polymorphism. In preparation
49. Felsenstein, J. 1971. On the biological significance of the cost of gene substitution. *Am. Natur.* 105:1–11
50. Fincham, J. R. 1971. Heterozygous advantage as a likely general basis for enzyme polymorphisms. *Heredity* 28:387–91
51. Fisher, R. A. 1929. *The Genetical Theory of Natural Selection*. Oxford: Clarendon, 2nd ed.; also, 1958. New York: Dover
52. Fitch, W. M. 1972. Does the fixation of neutral mutations form a significant part of observed evolution in proteins? *Brookhaven Symp. Biol.* 23:186
53. Fitch, W. M., Margoliash, E. 1967. Construction of phylogenetic trees. *Science* 155:279–84
54. Fitch, W. M., Margoliash, E. 1970. The usefulness of amino acid and nucleotide sequences in evolutionary studies. *Evol. Biol.* 4:67
55. Fox, D. J. 1971. The soluble citric acid cycle enzymes of *Drosophila melanogaster*. I. Genetics and ontogeny of NADP-linked isocitrate dehydrogenase. *Biochem. Genet.* 5:69–80
56. Frelinger, J. A. 1972. The maintenance of transferrin polymorphism in pigeons. *Proc. Nat. Acad. Sci. USA* 69:326–29
57. Gasser, D. L., Rowlands, D. T. Jr. 1972. Nongenetic determinants of human serum esterases. *Am. J. Pathol.* 67:501–10
58. Gibson, J. B. 1970. Enzyme flexibility in *Drosophila melanogaster*. *Nature* 227: 959–60
59. Gibson, J. B., Miklovich, R. 1971. Modes of variation in alcohol dehydrogenase in *Drosophila melanogaster*. *Experientia* 27:99–100
60. Gillespie, J. H. 1972. The effects of stochastic environments on allele frequencies in natural populations. *Theor. Pop. Biol.* 3:241–48
61. Gillespie, J. H., Kojima, K. 1968. The degree of polymorphisms in enzymes involved in energy production compared to that in nonspecific enzymes in two *Drosophila ananassae* populations. *Proc. Nat. Acad. Sci. USA* 61:582–85
62. Gooch, J., Schopf, T. 1972. Genetic variability in the deep sea: relation to environmental variability. *Evolution* 26:545–52
63. Greville, G. D., Tubbs, P. K. 1968. The catabolism of long chain fatty acids in mammalian tissues. *Essays Biochem.* 4:155–212
64. Haigh, J., Smith, J. M. 1972. Population size and protein variation in man. *Genet. Res., Cambridge* 19:73–89
65. Hanabusa, K., Kono, H., Matsuyama, S. 1972. Electrofocusing properties of pig phosphorylase isozymes. *J. Biochem.* 72:1261
66. Harris, H., 1971. Protein polymorphism in man. *Can. J. Genet. Cytol.* 13: 381–96
67. Harris, H. 1971. Annotation: polymorphism and protein evolution. The neutral mutation-random drift hypothesis. *J. Med. Genet.* 8:444–52

68. Harris, H., Hopkinson, D. 1972. Average heterozygosity per locus in man: an estimate based on the incidence of enzyme polymorphisms. *Ann. Hum. Genet.* 36:9–20
69. Hedrick, P. 1971. A new approach to measuring genetic similarity. *Evolution* 25:276–80
70. Hiebert, M., Gauldie, J., Hillcoat, B. L. 1972. Multiple enzyme forms from protein-bromphenol blue interaction during gel electrophoresis. *Anal. Biochem.* 46: 433–37
71. Hubby, J. L., Lewontin, R. C. 1966. A molecular approach to the study of genic heterozygosity in natural populations. I. The number of alleles at different loci in *Drosophila pseudoobscura*. *Genetics* 54:577–94
72. Hutton, J. J., Roderick, T. H. 1970. Linkage analyses using biochemical variants in mice. III. Linkage relationships of eleven biochemical markers. *Biochem. Genet.* 4:339–50
73. Illingworth, J. A. 1972. Anomalous behavior of yeast isocitrate dehydrogenase during isoelectric focusing. *Biochem. J.* 129:1125
74. Jacobson, K. B., Murphy, J. B., Knopp, J. A., Ortiz, J. R. 1972. Multiple forms of *Drosophila* alcohol dehydrogenase. III. Conversion of one form to another by nicotinamide adenine dinucleotide or acetone. *Arch. Biochem. Biophys.* 149: 22–35
75. Jain, S. K., Bradshaw, A. D. 1966. Evolutionary divergence among adjacent plant populations. I. The evidence and its theoretical analysis. *Heredity* 21: 407–41
76. Johnson, F., Schaffer, H. 1973. Genotype-environment relationships in populations of *D. melanogaster* from the Eastern U.S. *Biochem. Genet.* In press
77. Johnson, F. M., Schaffer, H. E., Gillaspy, J. E., Rockwood, E. 1969. Isozyme genotype-environment relationships in natural populations of the harvester ant, *Pogonomyrmex barbatus*, from Texas. *Biochem. Genet.* 3:429–50
78. Johnson, G. B. 1971. Analysis of enzyme variation in natural populations of the butterfly *Colias eurytheme*. *Proc. Nat. Acad. Sci. USA* 68:997–1001
79. Johnson, G. B. 1971. Metabolic implications of polymorphism as an adaptive strategy. *Nature* 232:347–48
80. Johnson, G. B. 1972. The selective significance of biochemical polymorphisms in *Colias* butterflies. PhD thesis. Stanford Univ., Stanford, Calif.
81. Johnson, G. B. 1972. Evidence that enzyme polymorphisms are not selectively neutral. *Nature New Biol.* 237:170–71
82. Johnson, G. B. 1973. The importance of substrate variability to enzyme polymorphism. *Nature.* In press
83. Johnson, G. B. 1973. Enzyme polymorphism and metabolism: Polymorphism among enzyme loci is related to metabolic function. In preparation
84. Johnson, G. B., Feldman, M. 1973. On the hypothesis that polymorphic enzyme alleles are selectively neutral. I. The evenness of allele frequency distribution. *Theor. Pop. Biol.* 4:In press
85. Johnson, G. B. 1973. Polymorphism for α-glycerophosphate dehydrogenase in *Colias* butterflies. I. Gradients in allele frequency corresponding to environmental gradients. In preparation
86. Johnson, W. E., Selander, R. K. 1971. Protein variation and systematics in kargaroo rats (genus *Dipodomys*). *Syst. Zool.* 20:377–405
87. Jeppsson, J. O., Berglund, S. 1972. Thin-layer isoelectric focusing for haemoglobin screening and its application of haemoglobin malmo. *Clin. Chim. Acta* 40:153
88. Jukes, T. H., Holmquist, R. 1972. Estimation of evolutionary changes in certain homologous polypeptide chains. *J. Mol. Biol.* 64:163–79
89. Jukes, T. H., Holmquist, R. 1972. Evolutionary clock: nonconstancy of rate in different species. *Science* 177:530–32
90. Kahler, A. L., Allard, R. W. 1970. The genetics of isozyme variants in barley I Esterases. *Crop Sci.* 10:444–48
91. Kimura, M. 1968. Evolutionary rate at the molecular level. *Nature* 217:624–62
92. Kimura, M., Crow, J. 1964. The number of alleles that can be maintained in a finite population. *Genetics* 49:725–38
93. Kimura, M., Ohta, T. 1971. Protein polymorphism as a phase of molecular evolution. *Nature* 229:467–69
94. Kimura, M., Maruyama, T. 1971. Pattern of neutral polymorphism in a geographically structured population. *Genet. Res., Cambridge* 18:125–31
95. Kimura, M., Ohta, T. 1972. Mutation and evolution at the molecular level. *Fogarty International Center Conference (NIH), October 1972*
96. King, J. L., Jukes, T. H. 1969. Non-Darwinian evolution. *Science* 164: 788–98
97. Kirby, G., Halliday, R. 1973. Another view of neutral alleles in natural populations. *Nature* 243:463–464

98. Kitto, G. B., Wassarman, P. M., Kaplan, N. O. 1966. Enzymatically active conformers of mitochondrial malate dehydrogenase. *Proc. Nat. Acad. Sci. USA* 56:578–85

99. Koehn, R. K. 1969. Esterase heterogeneity: dynamics of a polymorphism. *Science* 163:943, 944

100. Koehn, R. K., Rasmussen, D. I. 1967. Polymorphic and monomorphic serum esterase heterogeneity in catostomid fish populations. *Biochem. Genet.* 1: 131–44

101. Koehn, R. K., Perez, J. E., Merritt, R. B. 1971. Esterase enzyme function and genetical structure of populations of the freshwater fish, *Notropis stramineus*. *Am. Natur.* 105:51–69

102. Koehn, R., Turano, F., Mitton, J. 1972. Population genetics of marine pelecypods. II. Genetic zonation on *Modiolus demissus*. *Evolution* 26:In press

103. Kojima, K., Gillespie, J. H., Tobari, Y. N. 1970. A profile of *Drosophila* species' enzymes assayed by electrophoresis. I. Number of alleles, heterozygosities, and linkage disequilibrium in glucose-metabolizing systems and some other enzymes. *Biochem. Genet.* 4: 627–37

104. Kojima, K. I., Smouse, P. E., Yang, S., Nair, P., Brncic, D. 1972. Isozyme frequency patterns in *Drosophila pavani* associated with geographical and seasonal variables. *Genetics* 72:721–31

105. Lakovaara, S., Saura, A. 1971. Genetic variation in natural populations of *Drosophila obscura*. *Genetics* 69: 377–84

106. Lakovaara, S., Saura, A. 1971. Genic variation in marginal populations of *Drosophila subobscura*. *Hereditas* 69: 77–82

107. Lakovaara, S., Saura, A., Falk, C. 1972. Genetic distance and evolutionary relationships in the *Drosophila obscura* group. *Evolution* 26:177–84

108. Laughton, E., Simmons, J., Chrambach, A. 1972. Instability of pH gradients in isoelectric focusing on polyacrylamide gel. *Anal. Biochem.* 49:109

109. Levene, H. 1953. Genetic equilibrium when more than one niche is available. *Am. Natur.* 87:331–33

110. Levins, R. 1962. Theory of fitness in a heterogeneous environment. I. The fitness set and adaptive function. *Am. Natur.* 96:361–78

111. Levins, R. 1968. *Evolution in changing environments*. Princeton: Princeton Univ.

112. Levins, R., MacArthur, R. 1966. Maintenance of genetic polymorphism in a heterogeneous environment. *Am. Natur.* 100:585–90

113. Lewis, W. H. P. 1971. Polymorphism of human enzyme proteins. *Nature* 230: 215–18

114. Lewis, W. H. P., Truslove, G. M. 1969. Electrophoretic heterogeneity of mouse erythrocyte peptidases. *Biochem. Genet.* 3:493–498

115. Lewontin, R. C., Hubby, J. L. 1966. A molecular approach to the study of genic heterozygosity in natural populations. II. Amount of variation and degree of heterozygosity in natural populations of *Drosophila pseudoobscura*. *Genetics* 54:595–609

116. Lewontin, R., Krakauer, J. 1973. Distribution of gene frequency as a test of the theory of the selective neutrality of polymorphisms. *Genetics*. In press

117. Long, T. 1970. Genetic effects of fluctuating temperature in populations of *Drosophila melanogaster*. *Genetics* 66: 401–6

118. Malthe-Soerenssen, D., Fonnum, F. 1972. Multiple forms of choline acetyltransferase in several species demonstrated by isoelectric focusing. *Biochem. J.* 127:229

119. Mao, S. H., Dessauer, H. C. 1971. Selectively neutral mutations, transferrins, and the evolution of natricine snakes. *Comp. Biochem. Physiol. A* 40:669–80

120. Margoliash, E. 1963. Primary structure and evolution of cytochrome C. *Proc. Nat. Acad. Sci. USA* 50:672–79

121. Margoliash, E., Smith, E. L. 1965. Structural and functional aspects of cytochrome C in relation to evolution. In *Evolving Genes and Proteins*, ed. V. Bryson, H. Vogel, 221–42. New York: Academic

122. Margoliash, E., Fitch, W. M. 1968. Evolutionary variability of cytochrome C primary structures. *Ann. NY Acad. Sci.* 151:359

123. Margoliash, E., Fitch, W. M., Dickerson, R. E. 1968. Molecular expression of evolutionary phenomena in the primary and tertiary structures of cytochrome C. *Brookhaven Symp. Biol.* 21:259–305

124. Maruyama, T. 1972. Some invariant properties of a geographically structured finite population: Distribution of heterozygotes under irreversible mutation. *Genet. Res., Cambridge* 20: 141–49

125. Maruyama, T. 1972. A note on the hypothesis: Protein polymorphism as a phase of molecular evolution. *J. Mol. Evol.* 1:368–70

126. Smith, J. M. 1972. Protein polymorphism. *Nature New Biol.* 237:31

127. McKenzie, H. A., Sawyer, W. H. 1966. Zone electrophoresis of β-lactoglobulins. *Nature* 212:161–63

128. Merrell, D. J. 1970. Limits on heterozygous advantage as an explanation of polymorphism. *J. Hered.* 60:180–182

129. Merritt, R. 1972. Geographic distribution and enzymatic properties of lactate dehydrogenase allozymes in the fathead minnow, *Pimephales promelas. Am. Natur.* 106:173–84

130. Milkman, R. D. 1967. Heterosis as a major cause of heterozygosity in nature. *Genetics* 55:493–95

131. Mukai, T., Schaffer, H., Cockerham, C. 1972. Genetic consequences of truncation selection at the phenotypic level in *Drosophila melanogaster. Genetics* 72:763–69

132. Murphy, H., Berry, R. 1971. Isoenzymes in the study of wild populations. *Biochem. J.* 119:8P–9P

133. Nakai, Y., Tsunewaki, K. 1971. Isozyme variations in *Aegilops* and *Triticum.* I. Esterase isozymes in *Aegilops* studied using the gel isoelectrofocusing method. *Jap. J. Genet.* 46:321–36

134. Narang, S., Kitzmiller, J. B. 1971. Esterase polymorphism in a natural population of Anopheles punctipennis. II. Analysis of the Est-C system. *Can. J. Genet. Cytol.* 13:771–76

135. Nei, M. 1971. Evolutionary time estimates from electrophoretic data on protein identity. *Am. Natur.* 105:385

136. Nei, M. 1972. Genetic distance between populations. *Am. Natur.* 106:283–92

137. Neithammer, D., Huennekens, F. 1971. Bound TPN as the determinant of polymorphism in methemoglobin reductase. *Biochem. Biophys. Res. Commun.* 45:345–50

138. Nolan, C., Margoliash, E. 1968. Comparative aspects of primary structure of proteins. *Ann. Rev. Biochem.* 37:727

139. O'Brien, S. J., MacIntyre, R. J. 1972. The α-glycerophosphate in *Drosophila melanogaster.* II. Genetic aspects. *Genetics* 71:127–38

140. Ohno, S., Stenius, C., Christian, L. C., Schipmann, G. 1969. *De novo* mutation-like events observed at the 6PGD locus of the Japanese quail, and the principle of polymorphism breeding more polymorphism. *Biochem. Genet.* 3:417–28

141. Ohta, T. 1972. Population size and rate of evolution. *J. Mol. Evol.* 1:305–14

142. Ohta, T., Kimura, M. 1971. Genetic load due to mutations with very small effects. *Jap. J. Genet.* 46:393–401

143. Ohta, T., Kimura, M. 1971. Functional organization of genetic material as a product of molecular evolution. *Nature* 233:118–19

144. Ohta, T., Kimura, M. 1971. Behavior of neutral mutants influenced by associated overdominant loci in finite populations. *Genetics* 69:247–60

145. Ohta, T., Kimura, M. 1971. Amino acid composition of proteins as a product of molecular evolution. *Science* 174:150–53

146. Ohta, T., Kimura, M. 1971. On the constancy of the evolutionary rate of cistrons. *J. Mol. Evol.* 1:18

147. Omenn, G. S., Cohen, P. T. W., Motulsky, A. G. 1971. Genetic variation in glycolytic enzymes in human brain (Abstr.). *Int. Congr. Hum. Genet., 4th, Excerpta Medica* 233:135

148. Parker, W. C., Bearn, A. G. 1963. Boric acid-induced heterogeneity of conalbumin by starch-gel electrophoresis. *Nature* 199:1184–1186.

149. Parsons, P. A., McKenzie, J. A. 1972. The ecological genetics of *Drosophila. Evol. Biol.* 5:87–132

150. Patton, J., Selander, R., Smith, M. 1972. Genic variation in hybridizing populations of gophers (genus *Thomomys*). *Syst. Zool.* 21:263–270

151. Perkins, J. M. 1972. The principal component analysis of genotype-environmental interactions and physical measures of the environment. *Heredity* 29:51

152. Petras, M. L., Reimer, J., Biddle, F., Martin, J. E., Linton, R. 1969. Studies of natural populations of *Mus.* V. A survey of nine loci for polymorphisms. *Can. J. Genet. Cytol.* 11:497–513

153. Powell, J. R. 1971. Genetic polymorphism in varied environments. *Science* 174:1035–36

154. Powell, J. R., Levene, H., Dobzhansky, T. 1972. Chromosomal polymorphism in *Drosophila pseudoobscura* used for diagnosis of geographic origin. *Evolution* 26:553–59

155. Powers, D. A. 1972. Hemoglobin adaptation for fast and slow water habitats in sympatric catostomid fishes. *Science* 177:360–62

156. Prakash, S. 1969. Genic variation in a natural population of *Drosophila persimilis. Proc. Nat. Acad. Sci. USA* 62:778–84

157. Prakash, S., Lewontin, R. C., Hubby, J. L. 1969. A molecular approach to the study of genic heterozygosity in natural populations. IV. Patterns of genic variation in central, marginal and isolated populations of *Drosophila pseudoobscura*. *Genetics* 61:841–58

158. Richmond, R. C. 1972. Enzyme variability in the *Drosophila willistoni* group. III. Amounts of variability in the superspecies, *D. paulistorum*. *Genetics* 70:87–112

159. Richmond, R. C. 1972. Genetic similarities and evolutionary relationships among the semispecies of *Drosophila paulistorum*. *Evolution* 26:536–44

160. Richmond, R. C., Powell, J. R. 1970. Evidence of heterosis associated with an enzyme locus in a natural population of *Drosophila*. *Proc. Nat. Acad. Sci. USA* 67:1264–67

161. Robinson, H. 1972. Comparison of different techniques for isoelectric focusing on polyacrylamide gel slabs using bacterial asparaginases. *Anal. Biochem.* 49:353

162. Rockwood, E., Kanapi, C., Wheeler, M., Stone, W. 1971. X. Allozyme changes during the evolution of Hawaiian *Drosophila*. *Studies in Genetics*, VI. *Texas Univ. Publ.* 7103:193–212

163. Rockwood-Sluss, E., Johnston, J., Heed, W. 1973. Allozyme-genotype-environment relationships. I. Variation in natural populations of *Drosophila pachea*. *Genetics* 73:135–46

164. Roderick, T. H., Ruddle, F. H., Chapman, V. M., Shows, T. 1971. Biochemical polymorphisms in feral and inbred mice *(Mus musculus)*. *Biochem. Genet.* 5:457–66

165. Roughgarden, J. 1971. Density-dependent natural selection. *Ecology* 52:453–68

166. Sarich, V. 1972. *Old World Monkeys*, ed. J. Napier, P. Napier. New York: Academic

167. Schopf, T., Gooch, J. 1971. Gene frequencies in a marine ectoproct: a cline in natural populations related to sea temperature. *Evolution* 25:286–89

168. Schopf, T., Gooch, J. 1971. A natural experiment using deep-sea invertebrates to test the hypothesis that genetic homozygosity is proportional to environmental stability. *Biol. Bull.* 141:401 (Abstr.)

169. Schmid, K., Polis, A. 1960. Electrophoretic behavior of normal human serum albumin at pH 4.0. II. Interpretation of electrophoretic patterns in terms

of a reversible protein interaction. *J. Biol. Chem.* 235:1321–25

170. Schmid, K., Polis, A., Hunziker, K., Fricke, R., Yayoshi, M. 1967. Partial characterization of the sialic acid-free forms of α_1-acid glycoprotein from human plasma. *Biochem. J.* 104:361

171. Schwartz, D., Laughner, W. J. 1969. A molecular basis for heterosis. *Science* 166:626–27

172. Selander, R. K., Yang, S. Y., Hunt, W. G. 1969. Polymorphism in esterases and hemoglobin in wild populations of the house mouse *(Mus musculus)*. *Studies in Genetics, V. Texas Univ. Publ.* 6918:271–338

173. Selander, R. K., Hunt, W. G., Yang, S. Y. 1969. Protein polymorphism and genetic heterozygosity in two European subspecies of the house mouse. *Evolution* 23:379–90

174. Selander, R. K., Yang, S. Y. 1969. Protein polymorphism and genic heterozygosity in a wild population of the house mouse *(Mus musculus)*. *Genetics* 63:653–67

175. Selander, R. K., Yang, S. Y., Lewontin, R. C., Johnson, W. E. 1970. Genetic variation in the horseshoe crab *(Limulus polyphemus)*, a phylogenetic "relic." *Evolution* 24:402–14

176. Selander, R. K., Smith, M., Yang, S. Y., Johnson, W. E., Gentry, J. 1971. Biochemical polymorphism and systematics in the genus *Peromyscus*. I. Variation in the old field mouse. *Texas Univ. Publ.* 7103:49

177. Selander, R. K., Johnson, W. E. 1973. *Proc. Int. Congr. Zool., 17th*. In press

178. Shaw, C. R. 1970. How many genes evolve? *Biochem. Genet.* 4:275–83

179. Sing, C. F., Brewer, G. J. 1971. Evidence for nonrandom multiplicity of gene products in 22 plant genera. *Biochem. Genet.* 5:243–51

179a. Smith, J. Maynard. See Ref. 126

180. Smouse, P. E., Kojima, K. I. 1972. Maximum likelihood analysis of population differences in allelic frequencies. *Genetics* 72:709–19

181. Snaydon, R., Davies, M. 1972. Rapid population differentiation in a mosaic environment. II. Morphological variation in *Anthoxanthum odoratum*. *Evolution* 26:390–405

182. Soderholm, J., Allestam, P., Wadstrom, T. 1972. A rapid method for isoelectric focusing in polyacrylamide gel. *Fed. Eur. Biochem. Soc. Lett.* 24:89

183. Somero, G. N., Hochachka, P. W. 1971. Biochemical adaptation to the environment. *Am. Zool.* 11:159–67

184. Spencer, E. M., King, Te P. 1971. Isoelectric heterogeneity of bovine plasma albumin. *J. Biol. Chem.* 246:201–8

185. Steele, M. W., Young, W. J., Childs, B. 1968. Glucose 6-phosphate dehydrogenase in *Drosophila melanogaster:* starch gel electrophoretic variation due to molecular instability. *Biochem. Genet.* 2:159–75

186. Stone, W. S., Kojima, K., Johnson, F. M. 1969. Enzyme polymorphisms in animal populations. *Jap. J. Genet.* 44, Suppl. 1:166–71

187. Summaria, L., Arzadon, L., Bernabe, P., Robbins, K. C. 1972. Studies on the isolation of the multiple molecular forms of human plasmin by isoelectric focusing methods. *J. Biol. Chem.* 247: 4691

188. Sved, J. A. 1968. Possible rates of gene substitution in evolution. *Am. Natur.* 102:283–92

189. Sved, J. A., Reed, T. E., Bodmer, W. F. 1967. The number of balanced polymorphisms that can be maintained in a natural population. *Genetics* 55:469–81

190. Thoday, J. 1972. Disruptive selection. *Proc. Roy. Soc. B* 182:109

191. Throckmorton, L. H. 1966. The use of biochemical characteristics for the study of problems of taxonomy and evolution in the genus *Drosophila. Studies in Genetics,* II. *Texas Univ. Publ.* 6205:415–87

192. Throckmorton, L. H. 1968. Biochemistry and taxonomy. *Ann. Rev. Entomol.* 13:99–114

193. Townson, H. 1972. Esterase polymorphism in *Aedes aegypti:* the genetics and K_m values of electrophoretically heterogeneous forms. *Ann. Trop. Med. Parasitol.* 66:255–66

194. Uzzell, T., Corbin, K. W. 1971. Fitting discrete probability distributions to evolutionary events. *Science* 172:1089–96

195. Vesterberg, O. 1972. Isoelectric focusing of proteins in polyacrylamide gels. *Biochim. Biophys. Acta* 257:11–19

196. Vigue, C., Johnson, F. 1973. Isozyme variability in species of the genus *Drosophila.* VI. Frequency-property-environment relationships of allelic alcohol dehydrogenases in *D. melanogaster. Biochem. Genet.* In press

197. Vrijenhoek, R. 1972. Genetic relationships of unisexual-hybrid fishes to their progenitors using lactate dehydrogenase isozymes as gene markers (*Poeciliopsis,* Poeciliidae). *Am. Natur.* 106: 754–66

198. Wallace, D. G., Maxson, L. R., Wilson, A. C. 1971. Albumin evolution in frogs: A test of the evolutionary clock hypothesis. *Proc. Nat. Acad. Sci. USA* 68: 3127–29

199. Webster, P., Selander, R. K., Yang, S. Y. 1972. Genetic variability and similarity in the *Anolis* lizards of Bimini. *Evolution* 26:523–35

200. Wills, C., Nichols, L. 1972. How genetic background masks single-gene heterosis in *Drosophila. Proc. Nat. Acad. Sci. USA* 69:323–25

201. Wright, S. 1931. Evolution in Mendelian populations. *Genetics* 16:97–159

202. Wright, S. 1970. Random drift and the shifting balance theory of evolution. In *Mathematical Topics in Population Biology,* ed. K. Kojima, 1–31. Berlin: Springer

203. Yamazaki, T. 1971. Measurement of fitness at the esterase-5 locus in *Drosophila pseudoobscura. Genetics* 67:579–603

204. Yamazaki, T., Maruyama, T. 1972. Evidence for the neutral hypothesis of protein polymorphism. *Science* 178:56–58

205. Yang, S., Wheeler, M., Bock, I. 1972. Isozyme variations and phylogenetic relationships in the *Drosophila bipectinata* species complex. *Studies in Genetics,* VII. *Texas Univ. Publ.* 7213: 213–27

THE ECOLOGY
OF AGGRESSIVE BEHAVIOR

❖ 4055

John A. King

Department of Zoology, Michigan State University, East Lansing, Michigan

INTRODUCTION

Aggression is a good English word without a suitable technical definition. Diplomats use the word "defense" when bombs are dropped in Viet Nam. Scientists substitute the words "agonistic behavior" in order to avoid hair-splitting decisions between defense and aggression. Agonistic behavior includes all behaviors associated with the contest or struggle between individuals. The initial attack, the retaliatory strike, the flight, the pursuit, the threat, and the retreat are parts of the agonistic encounter in which the roles of aggressor and defender are often reversed. Scientific precision enters when these units, like elements of a compound, are analyzed and measured. To the behaviorist, the units are postures, action patterns, vocalizations, and odors.

Species-specificity is omitted from the concept of agonistic behavior because the motor or action patterns in the predator-prey encounter or in interspecific competition are often the same as those patterns exhibited between conspecific rivals. The behavior is similar, but the function of the agonistic behavior changes from interspecific competition and predator-prey relationships to intraspecific sociality. Intraspecific aggression is the social equivalent of interspecific competition (101). Since the function of any behavior is not always apparent, the definition of agonistic behavior is often restricted to the subjects rather than the function of the encounter. The definition of agonistic behavior used here is any combative behavior involving a struggle or contest among individuals of the same species. Aggression refers to the initiation and attack phases of the agonistic encounter (see 47, p. 333). At the experimental level, operational definitions replace these general concepts.

The purpose of this review is to examine the concept of agonistic behavior as it pertains to ecology. First, a tentative verbal model of the role of agonistic behavior in ecology is presented as distilled from the relevant literature. Second, specific hypotheses derived from the model are stated (usually at the beginning of paragraphs). Third, recent tests of the hypotheses are presented and evaluated. Finally, recommendations for further tests are offered. The model and review initially deal with the mechanisms or causes of agonistic behavior and subsequently with the functions or consequences.

117

Comprehensive coverage of the literature on aggression can be found in *Psychological Abstracts, Biological Abstracts, Zoological Record,* or by any information retrieval service. A recent and reasonable treatment of aggression in men and animals appears in Johnson's book (50). Southwick (79) compiled a useful collection of historical and selected papers on animal aggression. Several volumes resulting from symposia on animal aggression have been published recently: Clemente & Lindsley (29), Garattini & Sigg (38), Eleftheriou & Scott (32), Esser (33). The older symposia of Carthy & Ebling (21) and Ulrich & King (92) contain some worthwhile contributions that have not been updated. Vernon (95) wrote a brief, psychologically oriented review of animal aggression. Numerous volumes on aggression and violence in man are available. Some, like those of Lorenz (63) and Ardrey (7, 8), are atavistically derived from animal studies. Nonhuman primate books usually contain sections on aggressive behavior. Other volumes or reviews on aggression will certainly appear by the time this review is published. In contrast to previously published reviews, this one will examine animal aggression primarily from an ecological and experimental perspective. The problems in this area are sufficiently profound and complex to arrest and intrigue the professional ecologist without his venturing into the realm of the social scientist and social philosopher.

A VERBAL MODEL

Agonistic behavior is an inherent capacity of most animals. Individuals and species vary in its expression and the circumstances or stimuli which bring it about. Internal changes of the organism caused by genes, maturation, or hormones can lower the threshold for the expression of agonistic behavior. The external environment, particularly alterations associated with the seasons, induces some of these internal changes. Other extroceptive stimuli, often those transmitted by conspecific individuals, can alter the physiological state of the animal sufficiently to modify its agonistic responses. Individuals, thus, variously primed by genetic endowments and physiological states become exposed to releasing situations. A highly primed individual may even seek stimuli that release agonistic responses. The types of stimuli eliciting agonistic behavior are diverse, depending on the species and prior associations with the stimuli. The initial noxious stimuli, often tactile, become associated with visual, auditory, or olfactory stimuli to produce retraction or retaliation.

Environmental changes not only bring about agonistic behavior, but they are also produced by it. The social environment is most readily altered by agonistic behavior. Losers in the agonistic encounter are injured, killed, driven away, or reproductively repressed. Winners gain social status, access to commodities, freedom of movement, and reproductive success. Frequent agonistic encounters affect the physiology of the population through the pituitary-adrenal-gonadal axis. These physiologically stressed individuals readily succumb to shortages of necessary resources, or they become exposed to predation and severe weather, or they fail to mature and reproduce. A population enduring the consequences of heightened agonistic behavior fluctuates in numbers, with surviving remnants contributing to a gene pool quite different from that of the original population. A species that once saturated the environment and then became decimated or locally exterminated can radically affect the abundance of species in the trophic levels above and below it. Such possible irreversible changes to the species and its environment become buffered by the evolution of restraints to overt aggressive behavior. Ritualized displays replace aggression, and maturity is delayed, reproduction curtailed, mortality reduced, dispersal insured, and dispersion enhanced.

The verbal model above suffices as well as a diagramatic model, with its blocks and arrows indicating inputs, outputs, feedbacks, and loops, until a formal mathematical statement can be proposed. In essence, the model outlines the causes or determinants of agonistic behavior and postulates its consequences in the ecosystem. Individual investigators will include fewer or more details and emphasize various aspects of the model other than those presented here. Few investigators, however, would attempt to eliminate agonistic behavior any more than reproductive or nutritional behavior from their analysis of the ecosystem. The advantage of a model is that it provides testable hypotheses which can now be examined. An attempt is made to state the hypotheses in a testable form. Only the evidence or references to the stated hypothesis are examined. The reader is invited to contribute and examine evidence familiar to him for and against each hypothesis.

DETERMINANTS OF AGONISTIC BEHAVIOR

Factors contributing to agonistic behavior occur within the organism and externally to it. An organism's genotype, hormonal balance, and neurology are internal characteristics which can alter its responsiveness to external stimuli. Conversely, external stimuli like photoperiods and social stimuli can affect endocrine secretions (12, 14, 35, 60, 64, 80, 99). Investigations of the effects of agonistic behavior on ecological parameters should include the manipulation of agonistic behavior with the recognition of constraints to these manipulations. For example, alterations of the genotype of the subjects may affect their agonistic behavior differently depending on the environmental conditions. Among inbred strains of mice, one strain was more aggressive than the other under low light intensity, but less aggressive under a bright light (54). The relative amounts of aggression exhibited by each strain can also vary with group size (93). Further treatment of these determinants can be found in the symposia volumes cited in the Introduction. Seasonal changes (100), availability of food (11, 30, 42, 77), crowding (3, 44), early experience (20, 52), and drugs (36) have also been implicated in the modification of agonistic behavior. Density has been implicated as probably contributing to aggression by ecologists more than other factors.

AGONISTIC BEHAVIOR INCREASES WITH POPULATION DENSITY   The chief proponents of this hypothesis have been Christian (23, 25, 27), Chitty (22), Wynne-Edwards (102), Calhoun (17), and their associates. The principal problem with this hypothesis is the concept of population density, which is purely descriptive. It explains nothing. Changes in population density do not cause or otherwise bring about increased agonistic behavior. As Vale et al (93) showed with different genetic mouse strains, one strain is aggressive and the other not aggressive at the same density. The frequency of social interactions, individual recognition, and social disorganization is often associated with density and explains increases in agonistic behavior better than a simple numerical description of the population (61). Indeed, density has not been shown to be a good predictor of agonistic behavior except at the ends of the density continuum. Nevertheless, density is relatively easy to measure and very easy to manipulate in experiments, so it has been used instead of more meaningful behavioral variables.

Does aggression increase with density in experimental populations? If density is manipulated, several alternatives should be considered in the design and analysis of the experiment: 1. whether the area should be held constant at the individual or population level, 2. how many populations should be used at each density, 3. what is the prior experience or the genetic relationship of the founding population, 4. how should the environment be structured and what is the disposition of food, water, nest sites, and refugia, 5. whether the agonistic behavior is measured for the entire population, the mean number of agonistic acts per individual, or the mean number per possible interaction, i.e. $N(N-1)/2$. The alternatives selected will place a limit on the conclusions of the experiment. A brief inspection of the last alternative could alter or reverse the conclusions. If the mean number of agonistic acts per possible interaction had been used in the experiment by Vale et al (93) instead of the number per individual, increased densities would have had little effect on the aggressive strains and actually reduced the level of aggression in the docile strains of mice. The relative merits of these alternatives will not be considered further because I do not recommend the manipulation of density as a meaningful independent variable. However, density, as a dependent variable, is useful. Thus when Vessey (96) altered the behavior of mice in growing populations by treating them with chloropromozine, his measures of increased density made a procedural advance, one which has not been pursued sufficiently.

The hypothesis that agonistic behavior increases with density has also been tested in the field. Since the physiological condition of field-caught animals has been examined in the laboratory, the same can be done for behavior. Sadlier (74) and Healey (45) brought deermice (*Peromyscus*) in from the field and examined their aggressive behavior. The sample sizes of these investigations were limited, but they introduced a procedure which has been expanded to mammoth size by C. Krebs and co-workers (51, 55, 56). He observed approximately 575 social encounters between male *Microtus pennsylvanicus* and 650 encounters between male *M. ochrogaster* (55). The interactions between females and species were not examined. Later Conley (31) made intersex and interspecies comparisons with *M. longicaudus* and *M. mexicanus*. These studies describe behavioral changes in oscillating populations over several generations, meaning genetic factors could be involved, but are not necessarily. These behavioral assays of wild caught animals have made a major procedural advance, but limitations of space prevent a thorough analysis.

Both Krebs and Conley have strengthened the correlation of agonistic behavior with population density. Their studies did not examine environmental or genetic mechanisms responsible for these changes, although Krebs did show independent correlations of transferrin alleles with density. The correlations of other demographic characteristics with density were ambiguous. In general, a reader of these studies is left uncertain about the causal relationships of agonistic behavior to population density. Part of the confusion may derive from the simultaneous examination of sympatric species of similar voles that also have large allopatric distributions. The strength of the comparison is reduced by the possibility of interspecific competition. The best test of the aggression-density hypothesis in the field at this stage of our research is probably the simplest: one species studied at a time in areas of allopatry.

Each population develops its own social norms, as Southwick (78) dramatically illustrated with confined populations of house mice. Thus it is possible that *Microtus* populations studied by Krebs and Conley were combinations of several social units, or just one special social group. Replicate populations or even independently analyzed sections of the population would have added certainty to their conclusions. The lack of experimental replication is particularly serious when one population serves as a model for other populations, as was the case in Krebs' study. Replications can be added without enormously expending effort if the size of the demes can be estimated and separately examined, or if the study plot is arbitrarily divided in separate sections, preferably noncontiguous. Further comments regarding the reliability and validity of these studies could be offered. However, the evidence is strong that some aspect of behavior changes in oscillating populations of voles. Density is only one correlated variable with these behavioral changes and need not be the causal mechanism.

ECOLOGICAL CONSEQUENCES OF AGONISTIC BEHAVIOR

The mechanisms responsible for agonistic behavior have less immediate relevance to ecology than have the consequences of such behavior. What ecological changes can be attributed to agonistic behavior? Although ecological parameters may ultimately include the form and content of the ecosystem by the flow of energy through various food webs, this review considers only those variables affecting the species exhibiting agonistic behavior. The role of a given species in the ecosystem is broader than that attributed directly to agonistic behavior. Perhaps the most critical variables a species contributes to an ecosystem are its numerical abundance and distribution. These dependent variables can also be measured quite satisfactorily and are frequently used by ecologists. Abundance and distribution are remote products of many intervening variables more directly related to behavior. Some of these intervening variables are recruitment, mortality, and movements. Thus the question shifts from how does agonistic behavior affect ecology, to how does it affect recruitment, mortality, and movement. Obviously these dependent variables can be further dissected into reproductive season, number and size of clutches or litters, perinatal, juvenile, and adult mortality, dispersal, home range, etc. It would be ideal to examine the effect of agonistic behavior on each of these measures, but we will have to look hard for such studies. Instead of these variables, the final products of population size and distribution have been measured, but these are also affected by factors other than behavior. Consequently, many studies not only measure the wrong dependent variable, but they also fail to manipulate the proper independent variable: agonistic behavior.

If one postulates that population size, birth rate, mortality rate, or dispersion patterns result from agonistic behavior, then the behavior must be systematically altered and the effect measured. It is not sufficient to measure the amount of agonistic behavior and some population parameter, although this is an acceptable first approximation of the effect. Correlated measures often obscure the critical variable affecting the population. The basic problem centers around the manipulation of agonistic behavior. Once ecologists recognize this problem, the mechanisms influencing agonistic behavior become relevant. We have reviewed recent developments in the manipulation of agonistic behavior through genetics, hormones, and eliciting stimuli. Now we consider how such alterations of agonistic behavior influence the population parameters of recruitment, mortality, and dispersion. Few experiments in this area are clear cut and, therefore, it will be necessary to resort to correlative investigations and even anecdotes.

Reproduction

Reproduction is the principle means of population recruitment. Reproduction is labile throughout the life cycle of most species, from the ripening of parental gonads, through the production and union of the gametes, the nurturing of the zygote, embryo, fetus, and neonate, to the final recruitment of the sexually mature individual into the population. Despite the evolution of sexual organs and elaborate sexual behavior, the loss or wastage of gametes severely restricts natality. Gametic loss can be prevented by retarding maturation of the gonads. Indeed, delayed maturation of the gonads is currently a popular explanation for curtailing population recruitment in mammals (25). Not only does deferment of maturity change the entire life table of a species productivity, but it also provides a surplus of individuals ready to enter the effective breeding population promptly when conditions are suitable. Can agonistic behavior delay growth or defer sexual maturity?

AGONISTIC BEHAVIOR RETARDS SEXUAL MATURITY   Most support for this hypothesis comes from laboratory populations of rodents, which exhibit high levels of agonistic behavior and also have an abundance of nonreproducing adults (see 25 for a review). The pituitary-adrenocortical-gonadal triad is often postulated as the means by which maturation is retarded because of the stress induced in dense populations. Agonistic behavior is not the only way to induce stress or activate the adrenal glands, nor is stress the only way to retard maturation. Regardless of the physiologic mechanisms responsible for retardation, the hypothesis is amenable to testing. One of the most direct tests is that of Healey (45), who examined juvenile growth and survival of deermice (*Peromyscus*) in the laboratory and field in response to different levels of aggression.

Healey (45) allowed four juvenile deermice to 1. invade laboratory enclosures either occupied by a pair of adults or 2. unoccupied, or 3. to invade an unoccupied area simultaneously with a pair of adults. The experiment was replicated twice.

Changes in mean weight for 8 juveniles in each condition over a period of two weeks were obtained. Each situation for the invasion was designed to alter adult aggression, with the enclosure already occupied by adults producing the most intense aggression. Juveniles invading the empty enclosure or invading with the adult pair experienced an increase in weight almost three times that of the juveniles invading the enclosure occupied by an adult pair. This weight gain difference was significant only at the 10% level, which indicates considerable variability among the juveniles. Two replications using the same adults are not sufficient for reliable conclusions, but the method has merit. In a similar experiment, aggressive levels were manipulated by sex of the resident, since females were less aggressive. Again the juveniles grew more and survived longer when invading enclosures with females than when invading enclosures occupied by males. Healey selected three aggressive males and three docile males in another experiment and put them into separate cages for two days. Then he introduced two juveniles into each cage and recorded their weight for seven days. The juveniles with the docile males grew significantly faster than those housed with the aggressive males. He next took his experiment to the field. All resident deermice were removed from two isolated woodlots. Then he released four aggressive males in the large woodlot and three aggressive males in the small woodlot. A week later 14 and 10 juveniles were released in the large and small lots, respectively. The procedure was repeated next by releasing only docile males into the woodlots. Several replications were made for both aggressive and docile males. Survival and growth rates of juveniles were significantly greater two weeks after release in woodlots inhabited by docile males than by aggressive males. Since some adults disappeared before the juveniles were released and since other adults changed from aggressive to docile and vice versa, the results are not conclusive. Nevertheless, these procedures advance the testing of the hypothesis that growth of juvenile deermice is retarded by the aggressive behavior of adults. Further challenges of the hypothesis are needed because the data on the retardation of maturation in confined *Peromyscus* populations are equivocal. The retardation found in freely growing populations (83) was not exhibited in populations consisting of four adult pairs and their offspring (or fostered offspring) less than 140 days of age (86). A report in this study also found no difference in maturation rates between populations with the adults removed and those with adults present. Agonistic behavior was not manipulated nor reported in these populations studies, but the results should prevent any premature conclusions regarding the deferment of maturation by adults in deermouse populations.

The hypothesis that maturation is retarded by adult aggression in voles *(Microtus)* has little support. A *Microtus* population studied for two years showed 100% of 50 adult males were wounded, often severely, during one year and about 33% of 29 males wounded during the following year when the population was less dense (26). In contrast to the evidence for aggression among the adult males, none of the sexually immature males were wounded either of the two years. Wounds were detected by examining the flesh side of the skin, which is a reliable, though final, measure of aggression. Christian concluded that aggressive biting did not occur until the males were sexually mature. The retardation of maturation he observed in this population does not appear to be attributable to aggression as measured by skin wounds. In a study by Getz (40), adult and juvenile voles captured together in multiple capture traps showed no injuries. The absence of injuries among voles confined together in small traps was attributed to a declining population, which is getting the cart before the horse. Although these observations are not appropriate tests of the hypothesis, they severely weaken it when applied to microtine populations. Both the techniques of examining inside of skins and multiple captures deserve further applications to situations requiring some measure of aggressive behavior in field populations.

The effect of aggression on the growth and sexual maturation of juveniles in a population may be species specific. The population dynamics of *Peromyscus* and *Microtus* differ in many respects (25) and conclusions derived from the study of one of these two most abundant North American rodents need not apply to the other. Our necessary and valid attempts to make generalizations across species, families, orders, and classes should not obscure reality, which may reside in the equally intriguing complexities of nature. The diversity and lability of behavior within as well as between species makes it particularly vulnerable to premature generalizations (43). Such generalizations cannot substitute for examinations of the detailed precision animals use to adjust behaviorally to their constantly changing environmental conditions.

AGONISTIC BEHAVIOR REDUCES REPRODUCTION IN MATURE INDIVIDUALS   In contrast to the preceeding hypothesis, this one includes individuals of reproductive age and maturity that fail to continue breeding or cease to breed because of increased levels of agonistic behavior. Since agonistic behavior could differentially affect each sex, the general hypothesis can be subdivided as follows: (*a*) Inter-male aggression reduces male fertility. (*b*) Inter-male aggression reduces female fertility. (*c*) Inter-sex aggression reduces female or male fertility. (*d*) Inter-female aggression reduces female fertility. (*e*) Inter-female aggression reduces male fertility. The appropriate test for each hypothesis requires controlled levels of aggression with replicates at each level and some valid measure of fertility, such as gonadal functioning or sexual performance. The size of the group can be an added variable, but group size is not part of the hypothesis. The question of whether or not the experimental conditions used to test these and other hypotheses occur in the field or under natural conditions has no relevance to the hypothesis. An entirely different and valid hypothesis is that the experimental conditions prevail in the field. Ecologists are often concerned with this hypothesis. The "field" or "natural conditions" defy definition and ultimately become as specific to time and place as the most rigidly controlled laboratory experiments. In fact, most natural conditions are so unique that generalizations from one time and place to another may be less valid than generalizing from laboratory experiment to the field.

INTER-MALE AGGRESSION REDUCES MALE FERTILITY   Aggression among males can result in some individuals mating and not others because the latter are killed, driven away, negatively conditioned to mating by the dominant male (psychological castration), or their gonadal activity is reduced. Extreme examples occur among harem or territorial species, in which one male excludes other males from breeding privileges and the excluded males show gonadal quiescence (59). In less extreme examples, the generality of this hypothesis might be questioned. In experimental populations of deermice, the arrested gonadal growth in males could be attributed to inter-male interactions (82) if males grouped after sexual maturity had smaller sexual organs than those males living alone with a female. No reduction in sexual organ size was noted in grouped males, although gonadal growth was inhibited in males grouped prior to sexual maturity (39). The grouping of males does not mean they will fight and no measure of aggression was reported in this carefully designed experiment, but increased adrenal size, which is often associated with agonistic behavior, was not correlated with size of sex organs. The positive correlation between testes weight and amount of wounding in a free-living *Microtus* population (26) also contests the generality of this hypothesis. Other variables like dominance, subordination, age, season, availability of females, and prior experience can also influence the validity of the hypothesis.

INTER-MALE AGGRESSION REDUCES FEMALE FERTILITY   This can result from females being stressed or caught in the "cross-fire" of aggressive males or from too few or too frequent matings by males involved in fighting. In laboratory rats, penile intromissions facilitate sperm transport through the cervix (1), but genital stimulation soon after the male partner ejaculates inhibits sperm transport (2). Inter-male aggression could possibly contribute to either of these circumstances (18), although specific studies have not been done. That fertility of a population might suffer as a consequence of inter-male aggression was apparently one reason Krebs (55) only examined aggression among males. Inter-male aggression could also serve as an index for all aggression in a population because Conley (31) found no sex differences. Adequate tests of this hypothesis have not been made, perhaps because of our captivation with the next hypothesis.

INTER-SEX AGGRESSION REDUCES FEMALE OR MALE FERTILITY   Courtship and threat displays or any sexual and aggressive behavior are so closely associated that it is often difficult to distinguish between them. Among some animals mating occurs only after a rather vicious battle between the sexual partners, i.e. weasels. In other species, courtship displays appear to reduce or eliminate aggression between the sexes before body contact can be established. If courtship displays are interrupted, aggressive activities are not subdued and mating fails to take place (89). Often the agonistic displays between mates insure that unrecognized intruders will be repulsed from the territory or nest site of the pair. Mating is prevented, of course, if either of the prospective mates repels the other by unrelenting displays of aggression. These descriptions which border on the anecdotal level cannot substitute for suitable tests of the hypothesis. Such tests would involve measures of fertility among pairs exhibiting different levels of aggression.

INTER-FEMALE AGGRESSION REDUCES FEMALE FERTILITY   In some territorial species, the female drives off invading females more actively than her male partner. This behavior can contribute to the increase of nonbreeding females without territories, such as the red grouse (98). Inter-female aggression can also produce nesting territories within the larger territory of polygynous males, such as the red-wing blackbird (69). In the absence of overt aggression the female pheromone of house mice inhibits the estrous cycle (99). Among species which establish dominance hierarchies, such as chickens, low ranking females produce fewer eggs (41). In one carefully designed study (68) with rabbits *(Oryctolagus)*, either the number of rabbits or their density was kept constant in enclosures of three different sizes. Several replications of each condition were made and the duration of the observation periods was proportional to the number of rabbits present in order to obtain comparable behavioral data. The amount of aggressive behavior exhibited by each sex in groups of two males and three females was the same, but increased with density when the space available was reduced. Female fertility declined, as indicated by reduced ovary size, number of ovulations, and corpora lutea, with increasing density and aggression. The reduction in female fertility with the increase of female aggression was confounded by the increase in male aggression. In contrast to increased density, female aggression increased about eight fold in a large area with a constant density, but more rabbits. Although male aggression only doubled, they were still more aggressive than females. The increased aggression among females was related to increased ovarian size but decreased number of ovulations and corpora lutea. These inconsistencies in physiological measures are less critical to population recruitment than the entire reproductive performance, which includes maternal care and survival of young. Reproductive success was correlated with the smallest number of rabbits, even when aggression levels were high. Other features of the experimental conditions, such as sexual behavior and size of nest burrows, affect reproduction in addition to the amount of aggression. This well-designed study illustrates the necessity of varying only one or two parameters at a time, while controlling others. Here control was exercised for density and numbers with behavior and reproduction used as dependent variables, obscuring the relationship between them. In conclusion, the hypothesis that inter-female aggression reduces fertility can be supported by bits of evidence collected from this and other experiments, but much of it is confounded and inferential.

INTER-FEMALE AGGRESSION REDUCES MALE FERTILITY   This hypothesis is stated merely to complete the symmetry. It has not been seriously proposed and is probably not worth testing. It may be relevant for those species with reversed sex roles.

Aggression can apparently reduce reproduction in some circumstances, but often the circumstances are poorly defined. Even the preceding attempt to define the sexes involved is not precise in respect to the number of individuals, sex ratios, season of year, ages, food available, or size of home range. Further consideration should be given to sex ratios (84) because the relative number of individuals of each sex can drastically affect both aggression and reproduction. Examination of age, season, sex ratio, etc may appear tedious, but they will approximate reality more than the catagorical statement that aggression reduces reproduction. Careful experimental investigation of this hypothesis may reveal that other behaviors associated with aggression are actually influencing reproduction. For example, differences in the aggressive behavior of male sticklebacks had less affect on the hatching success of their eggs than did the number of raiding males attracted by the father's defense activities (9). The more vigorously the father defended the eggs, the more vulnerable the eggs were to being eaten by raiding males. Perhaps raiders were attracted by the aggressive signals. The suggestion of intervening variables is far removed from testing them, but they alert one to the possibility of fallacious conclusions.

Mortality

AGGRESSION INCREASES MORTALITY   This inclusive hypothesis makes no distinction regarding periods in the life cycle when mortality is most influenced by aggression. Early in the life cycle mortality is almost equivalent to loss or wastage of the gametes, embryos, or infants. Later in the life cycle, measures of mortality may depend on the techniques used. For example, nesting mortality in nidicolous birds can be measured in natural populations quite accurately (48), but little is known regarding mortality in small rodents of comparable ages. The best estimates for nestling mortality in voles *(Microtus)* are derived from multiplying the number of pregnant females by mean litter size and then subtracting from this sum the number of juveniles captured in traps several weeks later (51). Infantile mortality calculated in this way is subject to considerable error. Consequently, the effect of aggression on infantile mortality in small rodents often comes from laboratory investigations.

AGGRESSION INCREASES INFANTILE MORTALITY   Infant animals are sometimes taken as prey by conspecific individuals and may even be consumed by one or both parents. Parental aggression can then influence infant mortality in two ways: 1. by protecting the infants and the integrity of the nest against conspecific invaders, even one of the parents, and 2. by attacking and killing their own offspring. A vigorous pursuit of an invader by the parent sometimes leaves the offspring unguarded and exposed to invasion by other individuals (9). On the other hand, frequent efforts to invade the nest site can break down the defenses of a guarding parent and expose the young to intruders (78). Both of these alternatives may develop at high densities because of the increased frequency of attempted nest invasions. Since the protection of the young and nest site depends upon the efforts of one or both parents, any increase in the attempted frequency of attacks can soon deplete the parental resources and fatally expose the young. This can occur without a change in aggression by either parents or invader. Indeed, Southwick (78) reported that young mice were merely trampled in their nests by intruders. Appropriate tests of the hypothesis would involve alterations of aggression in either the parents or potential invaders to see the effect of these alterations on infantile mortality. Despite the relevance of this hypothesis to population regulation, such manipulations have not been done. Any observed correlations between infant mortality and density are probably not the result of changes in aggressive behavior alone.

AGGRESSION INCREASES JUVENILE MORTALITY   Juveniles or subadults often suffer high mortality by agents other than the conspecific aggressor. They may be driven from their natal sites and exposed to predation or prevented from obtaining neces-sary resources by dominant adults. Probably few juveniles are directly killed by conspecific adults because they usually flee or disperse from encounters with adults (13). If any loss of the juveniles from a population is considered mortality (55), then the investigations by Healey (45) reported in the section on deferment of maturation substantially contribute to the testing of this hypothesis. Few other studies have directly attempted to manipulate the aggressiveness of the adults. This hypothesis is also considered under the topic of dispersal, since this age group is particularly prone to disperse.

AGGRESSION INCREASES ADULT MORTALITY   This classical hypothesis creates the image of a dramatic struggle between two powerful adult males engaged in fatal combat. Observers watching the combat rarely see death and more often find the combat replaced by threats, bluffs, displays, or avoidance. This substitution of overt, injurious combat by threat displays, vocalizations, and appeasement postures has been the subject of many scientific and popular writings (7, 63, 90, 102). The argument is usually that these substitutions are energy-conserving or life-preserving adaptations that have evolved through intraspecific competitions. Like many other evolutionary arguments about adaptations, the logic is sound and the evidence is weak. Although the loser may not suffer death in combat, its viability may be reduced later. For example, annual mortality in the red grouse of Scotland approxi-mates 65%, with 80% of the mortality attributed to subordinate birds without a territory (98). A loser that avoids death in combat is not assured of a long, fruitful life. If the primary function of aggression was to bring about increased mortality, one could argue that a swift, clean battle to death would be more efficient than bluffs and threats that drive an individual to a lingering, resource-consuming, inevitable doom. Obviously aggression should enhance viability and fertility of the most ag-gressive, dominant, or territorial individuals. This side of the coin makes more ecological and evolutionary sense and has more support than the thesis that aggres-sion is the mechanism for regulating populations by increasing mortality.

Dispersion

AGONISTIC BEHAVIOR AFFECTS DISPERSION   Spatial dispersion of animals can be random, even, or clumped. Random dispersion patterns are uncommon in most animals, but they provide a basis for comparison with other patterns (28). Clumps or aggregations of animals result from the attraction of individuals to favorable locations in physically and biotically heterogeneous environments, from restricted processes of dispersal, or from social interactions among individuals. Agonistic behavior can bring individuals together as well as disperse them. Fighting or threat displays can be reinforcing (87) and they can attract contestants and spectators alike. Such aggregations are usually temporary, like those of mice (6) patrolling territorial boundaries. More often than attraction to areas of aggression, the clump-ing is a sink for individuals forced into less habitable areas by socially dominant members (57). If agonistic behavior is necessary for dispersal, any reduction could bring about an aggregation of nondispersing individuals. In general, aggregations resulting from agonistic behavior are less likely than aggregations resulting from other social behavior, such as sexual behavior. Agonistic behavior tends towards even dispersion patterns.

AGONISTIC BEHAVIOR SPACES ANIMALS EVENLY   In contrast to clumped distributions, which can be attributed to nonsocial factors, even distributions require some type of social interaction. Therefore, evenly spaced patterns of animal distribution are caused by social interactions. The behavior need not be observed in order to draw this conclusion. This logical deduction has led to defining territories, spacing animals evenly, entirely in terms of space: a mutually exclusive area (71). The validity of this hypothesis has been examined at length in several excellent reviews (15, 16, 58). Since there is little doubt that agonistic behavior can space animals evenly, a more powerful statement of the hypothesis is that agonistic behavior is necessary.

AGONISTIC BEHAVIOR IS NECESSARY FOR EVEN DISPERSION   The value of this hypothesis lies in the ease of testing it by finding an even pattern of dispersion in the absence of agonistic behavior. The burden of proof immediately shifts to finding an agreeable definition of agonistic behavior. In the introduction, we accepted any behavior associated with a contest or struggle between individuals. The contest is apparent during threat displays, ambivalent oscillatory behavior across territorial borders, or even the songs or other vocalizations of birds. Contests become more nebulous, however, when a solitary individual deposits feces, urine, or a dermal gland secretion in the course of its wanderings. What does it mean when sometime later another individual may perform the same act at the same site? Is this territorial marking (85) or another form of agonistic behavior? Is there a contest when two individuals slowly alter their approach towards each other when sighted at a distance? These kinds of questions make it easier to define agonistic behavior in terms of spacing than to define spacing in terms of agonistic behavior (71). Indeed, the readily quantified dimensions of space and distance have been proposed as the basis for describing social interactions and organization (19, 33, 65).

THE TYPE OF AGONISTIC BEHAVIOR DETERMINES THE PATTERN OF DISPERSION
Tests of this hypothesis would require separate groups of the same species exhibiting different types of agonistic behavior; for example, visual displays versus vocal displays. Simultaneous comparisons are difficult, but sequential differences often occur. Territorial birds often vocalize during the formation of the territory, but after nesting the territories are maintained by visual displays or combats (34). The dispersion changes little during the breeding season but the size of the territory may be altered. Greater changes in spacing occur after the breeding when many territorial birds clump into flocks. Their vocal threat displays are replaced by direct attacks, chasings, or supplantings. Such sequential differences can easily confuse the cause with the effect because the flocking may have brought about changes in agonistic behavior. It is experimentally possible to alter simultaneously the types of agonistic

behavior displayed in different groups and measure the effect of these alterations on spacing.

THE AMOUNT OF AGONISTIC BEHAVIOR DETERMINES THE PATTERN OF DISPERSION
The pattern of dispersion is less likely to be altered by the amount of agonistic behavior than the size of the territory. Watson & Miller (97) found a positive correlation of 0.85 between aggression in male red grouse and territory size. On the other hand, in dense populations with small territories, agonistic encounters may be more frequent. These same relationships can sometimes be observed among birds roosting on a line. The most aggressive bird has the greatest length of line to itself (81). Manipulations of agonistic behavior in order to observe the effect upon spacing have just recently been undertaken by injecting birds with androgens on their breeding grounds (57, 91, 98).

Dispersal

AGONISTIC BEHAVIOR INCREASES DISPERSAL   In the absence of abundant evidence that aggression contributes to mortality, investigators have turned to this hypothesis because dispersal and mortality affect a local population similarly. Perhaps the subordinates are driven away instead of being killed. Furthermore, displaced individuals are usually easier to locate than dead ones. Either death or dispersal must account for surplus individuals produced by a well-established population. If surplus animals are driven out or somehow excluded from their natal site, they have three alternatives in addition to death. 1. They can colonize suitable, uninhabited areas. 2. They can establish themselves in marginal or suboptimal areas (Level 2 of Brown, 15). 3. They can become members of a floating population (Level 3), always ready to replace established residents. Few investigators question the existence of dispersal, the colonization of uninhabited areas, or the presence of transient individuals. The question is whether agonistic behavior is partly responsible for these events. An appealing, but speculative, article by Christian (24) proposes "that low-ranking, predominantly young individuals usually are forced to disperse" (p. 86). As an axiomatic correlary he also states that "a successful, dominant mammal will not leave its preferred habitat, and only animals forced to do so will migrate to other, marginal habitats." Of course, one can argue the opposite: that only the bold, dominant individuals explore and colonize new regions and the timid, subordinate individuals remain at home. Fortunately, these hypotheses are testable, but few tests have been done and we resort again to descriptive studies.

SUBORDINATE ANIMALS DISPERSE   Often cited tests of this hypothesis are those in which surplus individuals are released in an occupied area (10, 45, 66). It is assumed that the residents drive out the strangers if more residents than strangers remain after a brief period. The assumption may be correct, but the strangers have many attributes other than merely being subordinate. An appropriate test would require the simultaneous release of both dominant and subordinate individuals into an area equally familiar to both groups. Perhaps the closest approximation to this test is when young and adult deermice were simultaneously released into a woodlot (45). The relationship between dominant and subordinate deermice is often more complex than simple repulsion, as Hill (46) found when both were allowed to occupy an area twice the size of their previous range. The subordinate mouse moved into the nest box of the dominant mouse more often than it remained alone or was joined by the dominant mouse. This suggests that subordinant mice are attracted to dominant individuals rather than being repelled by them.

Another indirect procedure for testing the hypothesis that subordinate animals disperse involves the removal of individuals from an area and an examination of the invaders. Voles *(Microtus)* removed from the center of a large census tract tended (not significantly) to be males more more often than females, which suggested a "forced emigration of subordinate males in conjunction with the acceptance of many females by the dominant male" (94, p. 102). In another study by Myers & Krebs (67) all voles were removed continuously from two separate plots of approximately 0.7 ha each. One plot was isolated and the other was near three control plots, in which the voles were captured, marked, and released. Comparison of the voles from the control plots with those from the removal plots enabled the investigators to evaluate differences in age, sex, breeding condition, genotype, and behavior throughout two years of numerical fluctuations. The number of comparisons was large and some revealed significant differences, while others had no difference. In general, more males dispersed than females, but young (small) females dispersed more than any other age or sex group. Indeed, juvenile males were no more common emigrants than adult males. Dispersing voles had different genotypes than sedentary voles. In laboratory tests for behavioral differences among males, dispersing males were less exploratory (entered fewer maze alleys) and were somewhat less active than nondispersing males. Aggressive behavior differed between resident and dispersing *M. pennsylvanicus* males only during a brief period when the dispersing males were most aggressive. However, *M. ochrogaster* males were less aggressive when dispersing than the residents. These ambiguous results lead the authors to conclude that "aggressive behavior (tests) shows no clear differences between dispersing and resident males, although at peak population densities dispersing male *M. pennsylvanicus* tended to show more aggressive acts" (p. 76). These studies are not appropriate tests of either the hypothesis that subordinate animals disperse or that aggression increases dispersal, but they should prevent ready acceptance of both hypotheses.

The social conditions affecting dispersal can be systematically investigated as illustrated in a study by Savidge (75). Dispersal involves three distinct processes or steps: 1. departure from the natal site, 2. transversing a barrier, which is usually space, and 3. settling or colonizing a new area. Each of these three processes can be manipulated separately in order to determine their effect on dispersal. Social conditions can become tolerable at home if the barrier is severe or conditions in a new area are worse. If the barrier and the area of colonization are held constant, the home environment can be manipulated in respect to food, space, density, and social conditions. Savidge altered the social conditions of deermice *(Peromyscus maniculatus)* at the natal site by changing the parental combinations associated with their juvenile offspring. The rate of crossing an electrified barrier by the juveniles was not altered by the removal of (*a*) both parents, (*b*) father, (*c*) mother, or (*d*) neither parent. Juvenile crossings of the barrier did increase if the father also crossed or if the mother was aggressive. The sex of the juveniles did not affect the results. Since the aggressive character of the mother appeared to be the significant factor in the departure of both father and young, it is possible that the mothers differed genetically in this respect—a hypothesis currently under investigation.

Alternative hypotheses to aggression increasing dispersal are that aggression decreases dispersal (enhances group cohesion) or that aggression does not affect dispersal. Appropriate tests of these alternatives have not been undertaken, so a few arguments on their behalf may make them tenable. We have seen that threat displays and aggressive encounters can serve as reinforcers (88). The apparently positive correlation among gregarious animals and frequency of aggressive encounters could be interpreted to mean that aggression enhances group cohesion. Lorenz (62) once made an appealing argument that might be summarized by the catch phrase, "animals that fight together, keep together." Apparently many animals are more willing to engage in agonistic encounters than to separate themselves from the social group. These relationships between aggression and group cohesion could be greatly expanded, but do not warrant further discussion in the absence of experimental tests. The same could be said regarding dispersal in the absence of aggression. Since all animals disperse, but many species do not exhibit overt aggression, one might look to other factors than aggression for an explanation of dispersal. Frequently dispersal accompanies physiological or morphological changes during the life cycle of a species, i.e. wings in termites, gonadal maturation in mice. It has also been postulated that genetic differences separate the dispersants from the residents (49). The validity of these alternatives does not eliminate the possibility that in some species and under certain conditions aggression contributes to dispersal. The generality of this hypothesis, however, does require reservations.

Gene Pool

AGONISTIC BEHAVIOR ALTERS THE GENE POOL   Although logic compels the acceptance of this hypothesis, the experimental evidence is circumstantial. The basic requirement is that aggression differentially affects the fertility, mortality, or migration of various genotypes. We have examined the effect of aggression on each of these three variables without regard to the genotypes involved and we found the evidence equivocal for most variables. This lack of definitive conclusions may result in part from our concern with populations rather than individuals, which are the bearers of genetic material and are the units of selection. Individuals with genes closely correlated with a given behavior, e.g. aggression, will affect that behavior more in the next generation than genes unrelated to the behavior (53). If genes influence the exhibition of aggressive behavior, then the behavior will readily alter the frequency of those genes in subsequent generations as long as aggression is related to fitness. Genetic feedback from behavior is a short loop, which can lead to rapid behavioral evolution. Indeed, the proposed genetic changes in the multiannual oscillations of microtine populations rest upon this assumption of a rapid genetic feedback from behavior (22).

The gene pool can be altered by aggression in three ways, all of which depend on the genome of the aggressive parent. 1. If the aggression is genetically determined, then the descendents will become more aggressive. 2. If the aggression is phenotypic but has correlated genetic traits, those traits will become more numerous in the descendents. 3. If the aggression is phenotypic and the parents have a random sample of genes from the population, then only the immediate offspring will be affected by the aggression and not subsequent generations. These conditions imply a selective advantage to aggression. In a small population, even a random sample of genes carried by an aggressive parent will be affected by the stochastic processes common to genetic drift. Since aggression can have genetic determinants and since dominant individuals may leave more offspring (73), it should be relatively easy to test the hypothesis that the gene pool is altered by aggression. Most tests have revealed a restriction of gene flow through aggression, with resulting differences in gene frequency among several geographically close, but socially isolated demes (4, 72). In completely isolated populations, the initial advantage of aggressiveness in producing more descendents is quickly replaced by counter-selection which can lead to extinction of the population. Certainly, the genetic consequences of aggression, like other behaviors, have only begun to be investigated with the care given to restricted gene flow in rodent populations (5, 70).

The evolutionary consequences of aggression apparent from sexual dimorphism and the embellishment of horns, antlers, canine teeth, body size, and coloration need little comment. Some investigators will accept these characters as confirmation of the hypothesis that aggression alters the gene pool. Other investigators will be stimulated to study the processes that could be responsible for these aggressively related characters. Readings of the evolutionary record from extant life forms are open to many interpretations.

CONCLUSION

The purpose of this review has been to examine hypotheses derived from commonly held concepts that relate aggression to ecology. Aggression, or agonistic behavior in general, has been incorporated into theories of population regulation often without formal models and without critical evaluations of the evidence. Perhaps behavioral explanations of phenomena, like population regulation, are made when the explanations amenable to measurement and manipulation are unsatisfactory. Almost any type of behavior can provide an impenetrable refuge for an infirm theory. No attempt has been made here to offer a formal model for aggression, although one like those for feeding (76) and territoriality (37) is overdue. Consequently, the hypotheses examined were not formal derivations from a model. Other hypotheses might be substituted for those given here. However, many of these tentative hypotheses have not yet been sufficiently and appropriately tested. My search for evidence has not been comprehensive, but the examples selected mostly from rodent studies illustrate frequent problems in the collection and interpretation of the data. Hypotheses pertaining to the mechanisms or cause of aggression generally have stronger experimental support than those pertaining to the effects or functions of aggression. Apparently physiologists and psychologists have developed the experimental approach further than the ecologists. The impatience of ecologists for solutions to global problems has often led to a discrepancy between results and conclusions. Broad implications are often made from results, whose validity deserves careful inspection. These criticisms do not detract from the sound logical and experimental advances that have been made.

The problems associated with the relationship between aggression and populations require conceptual isolation, careful construction of hypotheses, and rigorous experimental challenges. Many hypotheses in this area are beyond the stage of describing possible relationships or merely demonstrating that aggression affects one variable or another. We can quantitatively postulate how much the variable will be influenced by aggression by using several levels of aggression and several replications at each level. The techniques of physiology and psychology can be used in manipulating aggression to prescribed levels. Conceptual isolation of the problem is often hampered by the fuzzy concept of natural or field conditions. This concept admits to so many uncontrolled variables that the investigator can escape the responsibility of adequately testing his postulated variable. Not only do experiments performed under "natural conditions" ignore the scientific premise that all variables are controlled or precisely manipulated, but they also defy the principle of repeatability. One cannot repeat "natural conditions" in time or space. Similar criticisms apply to those feeble attempts at recreating "natural conditions" in the laboratory, which usually lack validity, repeatability in other laboratories, or isolation and control of the critical variables. After a variable has been isolated, its influence measured, and its modification by other factors determined, the variable can be tested for its robustness by swamping it with all of the confounding variables in the field. This sequential program is offered for thought and not for action, because most ecologists are naturalists who are mentally or physically at work in the field, where they should carefully distinguish the trees from the forest.

Each investigator interprets his professional and social responsibilities differently. Each attempts to market his ideas and investigations to different consumers: colleagues, editors, granting agencies, theoreticians, scientists in other disciplines, scientific practitioners, the public. Material on aggression is a readily marketable commodity. Studies on aggression are mass produced and variously packaged for most consumers. Literature on aggression produced for large volume sales to the public often contains the "hard sell" with exaggerations, omissions of confounding details, and unwarranted generalizations. Literature produced for a few specialists may be trivial. For each of these approaches, one can only expect that each writer accomplish his selected task skillfully. The principal aim of this review has been to illustrate how this can be achieved by pointing out the need for strong hypotheses and solid tests of them. Perhaps the review has also cautioned scientific practitioners against premature applications to human welfare. Popular writers should find enough controversy revealed here to provide excitement for their messages. Aggression, like food and sex, will perennially be a topic for investigation, discussion, and entertainment.

ACKNOWLEDGMENTS

This review was written with the support of U.S.P.H.S. Research Grant EY-447 from the National Eye Institute and of Research Career Fellowship 5K3-HD-3081 from the National Institute of Child Health and Human Development. Lynwood G. Clemens critically read the manuscript.

Literature Cited

1. Adler, N. T. 1969. Effects of the male's copulatory behavior on successful pregnancy of the female rat. *J. Comp. Physiol. Psychol.* 69:613–22
2. Adler, N. T., Zoloth, S. R. 1970. Copulatory behavior can inhibit pregnancy in female rats. *Science* 168:1480–82
3. Alexander, B. K., Roth, E. M. 1971. The effects of acute crowding on aggressive behavior of Japanese monkeys. *Behaviour* 39:73–90
4. Anderson, P. K. 1964. Lethal alleles in *Mus musculus:* Local distribution and evidence for isolation of demes. *Science* 145:177–78
5. Anderson, P. K. 1970. Ecological structure and gene flow in small mammals. *Symp. Zool. Soc. London* 26:299–325
6. Anderson, P. K., Hill, J. L. 1965. *Mus musculus:* Experimental induction of territory formation. *Science* 148:1753–55
7. Ardrey, R. 1966. *The Territorial Imperative.* New York: Atheneum
8. Ardrey, R. 1970. *The Social Contract.* New York: Atheneum
9. Black, R. 1971. Hatching success in the three-spined stickleback (*Gasterosteus aculeatus*) in relation to changes in behavior during the parental phase. *Anim. Behav.* 19:532–41
10. Blair, W. F. 1940. A study of prairie deer-mouse populations in southern Michigan. *Am. Midl. Natur.* 24:273–305
11. Boice, R. 1970. Competitive feeding behaviour in captive *Terrapene c. carolina. Anim. Behav.* 18:703–10
12. Brockway, B. F. 1969. Roles of budgerigar vocalization in the integration of breeding behaviour. In *Bird Vocalization,* ed. R. A. Hinde, 131–58. Cambridge: Univ. Press
13. Bronson, F. H. 1964. Agonistic behaviour in woodchucks. *Anim. Behav.* 12:470–78
14. Bronson, F. H. 1971. Rodent pheromones. *Biol. Reprod.* 4:344–51
15. Brown, J. L. 1969. Territorial behavior and population regulation in birds. *Wilson Bull.* 81:293–329
16. Brown, J. L., Orians, G. H. 1970. Spacing patterns in mobile animals. *Ann. Rev. Ecol. Syst.* 1:239–62
17. Calhoun, J. B. 1952. The social aspects of population dynamics. *J. Mammal.* 33:139–59
18. Calhoun, J. B. 1962. *The Ecology and Sociology of the Norway Rat, Public Health Service Publ. No. 1008.* Bethesda: U.S. Dept. Health, Educ., Welfare
19. Calhoun, J. B. 1963. The social use of space. In *Physiological Mammalogy,* ed. W. V. Mayer, R. G. Van Gelder, 1:1–187. New York: Academic
20. Cairns, R. B., Nakelski, J. S. 1971. On fighting in mice: ontogenetic and experiential determinants. *J. Comp. Physiol. Psychol.* 74:354–64
21. Carthy, J. D., Ebling, F. J. 1964. *The Natural History of Aggression.* New York: Academic
22. Chitty, D. 1967. The natural selection of self-regulatory behaviour in animal populations. *Proc. Ecol. Soc. Aust.* 2:51–78
23. Christian, J. J. 1963. Endocrine adaptive mechanisms and the physiologic regulation of population growth. See Ref. 19, 1:189–353
24. Christian, J. J. 1970. Social subordination, population density, and mammalian evolution. *Science* 168:84–90
25. Christian, J. J. 1971. Population density and reproductive efficiency. *Biol. Reprod.* 4:248–94
26. Christian, J. J. 1971. Fighting, maturity, and population density in *Microtus pennsylvanicus. J. Mammal.* 52:556–67
27. Christian, J. J., Davis, D. E. 1964. Endocrines, behavior and populations. *Science* 146:1550–60
28. Clark, P. J., Evans, F. C. 1955. On some aspects of spatial pattern in biological populations. *Science* 121:397–98
29. Clemente, C. D., Lindsley, D. B. 1967. *Aggression and Defense.* Berkeley: Univ. Calif. Press
30. Colnaghi, G. 1971. Partitioning of a restricted food source in a territorial iguanid (*Anolis carolinensis*). *Psychon. Sci.* 23:59–60

31. Conley, W. H. 1971. *Behavor, demography, and competition in Microtus longicaudus and M. mexicanus.* PhD thesis. Texas Tech. Univ., Lubbock, Texas. 46 pp.
32. Eleftheriou, B. E., Scott, J. P. 1971. *The Physiology of Aggression and Defeat.* New York: Plenum
33. Esser, A. H., Ed. 1971. *Behavior and Environment.* New York: Plenum
34. Falls, J. B. 1969. Functions of territorial song in the white-throated sparrow. See Ref. 12, 207-32
35. Farner, D. S. 1964. The photoperiodic control of reproductive cycles in birds. *Am. Sci.* 52:137-56
36. Fox, K. A., Snyder, R. L. 1969. Effect of sustained low doses of diazepan on aggression and mortality in grouped male mice. *J. Comp. Physiol. Psychol.* 69:663-66
37. Fretwell, S. D., Lucas, H. L. 1969. On territorial behaviour and other factors influencing habitat distribution in birds. I. Theoretical development. *Acta Biotheor.* 19:16-36
38. Garattini, S., Sigg, E. B., Eds. 1969. *Aggressive Behaviour.* New York: Wiley
39. Gardner, R. H., Terman, C. R. 1970. The relationship between age of grouping and weight of selected organs of prairie deermice. *Res. Pop. Ecol.* 12:1-18
40. Getz, L. L. 1972. Social structure and aggressive behavior in a population of *Microtus pennsylvanicus. J. Mammal.* 53:310-17
41. Guhl, A. M. 1962. The behavior of chickens. In *The Behaviour of Domestic Animals,* ed. E. S. E. Hafez, 491-530. London: Bailliere, Tindall, Cox. 1st ed.
42. Hamby, W., Cahoon, D. D. 1971. The effect of water deprivation upon shock elicited aggression in the white rat. *Psychon. Sci.* 23:52
43. Harlow, H. F., Gluck, J. P., Suomi, S. J. 1972. Generalization of behavioral data between nonhuman and human animals. *Am. Psychol.* 27:709-16
44. Hazlett, B. A. 1968. Effects of crowding on the agonistic behavior of the hermit crab, *Pagurus bernhardus. Ecology* 49:573-75
45. Healey, M. C. 1967. Aggression and self-regulation of population size in deermice. *Ecology* 48:377-92
46. Hill, J. L. 1970. *Space utilization of Peromyscus: social and spatial factors.* PhD thesis. Mich. State Univ., East Lansing. 84 pp.

47. Hinde, R. A. 1970. *Animal Behaviour.* New York: McGraw-Hill
48. Holcomb, L. C. 1969. Breeding biology of the American goldfinch in Ohio. *Bird-banding* 40:26-43
49. Howard, W. E. 1965. Interaction of behavior, ecology, and genetics of introduced mammals. In *The Genetics of Colonizing Species,* ed. H. G. Baker, G. L. Stebbins, 461-80. New York: Academic
50. Johnson, R. N. 1972. *Aggression in Man and Animals.* Philadelphia: Saunders
51. Keller, B. L., Krebs, C. J. 1970. *Microtus* population biology: III Reproductive changes in fluctuating populations of *M. ochrogaster* and *M. pennsylvanicus* in southern Indiana, 1965-67. *Ecol. Monogr.* 40:263-94
52. King, J. A. 1957. Relationship between early social experience and adult aggressive behavior in inbred mice. *J. Genet. Psychol.* 90:151-66
53. King, J. A. 1967. Behavioral modification of the gene pool. In *Behavior-Genetic Analysis,* ed. J. Hirsch, 22-43. New York: McGraw-Hill
54. Klein, T. W., Howard, J., DeFries, J. C. 1970. Agonistic behavior in mice: Strain differences as a function of test illumination. *Psychon. Sci.* 19:177-78
55. Krebs, C. J. 1970. *Microtus* population biology: Behavioral changes associated with the population cycle in *M. ochrogaster* and *M. pennsylvanicus. Ecology* 51:34-52
56. Krebs, C. J., Keller, B. L., Tamarin, R. H. 1969. *Microtus* population biology: Demographic changes in fluctuating populations of *M. ochrogaster* and *M. pennsylvanicus. Ecology* 50:587-607
57. Krebs, J. R. 1971. Territory and breeding density in the great tit, *Parus major. Ecology* 52:2-22
58. Kummer, H. 1971. Spacing mechanisms in social behavior. In *Man and Beast,* ed. J. F. Eisenberg, 219-34. Washington: Smithsonian Inst. Press
59. Le Boeuf, B. J. 1972. Sexual behaviour in the northern elephant seal *Mirounga angustirostris. Behaviour* 41:1-26
60. Lehrman, D. S. 1959. Hormonal responses to external stimuli in birds. *Ibis* 101:478-96
61. Lloyd, J. A., Christian, J. J. 1967. Relationship of activity and aggression to density in two confined populations of house mice (*Mus musculus*). *J. Mammal.* 48:262-69

62. Lorenz, K. 1964. Ritualized fighting. See Ref. 21, 39–50
63. Lorenz, K. 1966. *On Aggression.* New York: Harcourt, Brace, World
64. Lott, D., Brody, P. N. 1966. Support of ovulation in the ring dove by auditory and visual stimuli. *J. Comp. Physiol. Psychol.* 62:311–13
65. McBride, G. 1964. A general theory of social organization and behaviour. *Univ. Queensl. Pap.* 1:73–110
66. Metzgar, L. H. 1971. Behavioral population regulation in the wood-mouse, *Peromyscus leucopus. Am. Midl. Natur.* 86:434–48
67. Myers, J. H., Krebs, C. J. 1971. Genetic, behavioral, and reproductive attributes of dispersing field voles *Microtus pennsylvanicus* and *Microtus ochrogaster. Ecol. Monogr.* 41:53–78
68. Myers, K., Hale, C. S., Mykytowycz, R., Hughes, R. L. 1971. The effects of varying density and space on sociality and health in animals. See Ref. 33
69. Orians, G. H. 1961. The ecology of blackbird (*Agelaius*) social systems. *Ecol. Monogr.* 31:285–312
70. Petras, M. L. 1967. Studies of natural populations of *Mus.* I. Biochemical polymorphisms and their bearing on breeding structure. *Evolution* 21:259–74
71. Pitelka, F. A. 1959. Numbers, breeding schedule, and territoriality in pectoral sandpipers of northern Alaska. *Condor* 61:233–64
72. Reimer, J. D., Petras, M. L. 1967. Breeding structure of the house mouse, *Mus musculus,* in a population cage. *J. Mammal.* 48:88–99
73. Robel, R. J. 1966. Booming territory size and mating success of the greater prairie chicken (*Tympanuchus cupido pinnatus*). *Anim. Behav.* 14:328–31
74. Sadleir, R. M. F. S. 1965. The relationship between agonistic behaviour and population changes in the deermouse, *Peromyscus maniculatus* (Wagner). *J. Anim. Ecol.* 34:331–52
75. Savidge, I. R. 1970. *Social factors contributing to the departure of Peromyscus maniculatus bairdi from their natal site.* PhD thesis. Mich. State Univ., East Lansing. 56 pp.
76. Schoener, T. W. 1971. Theory of feeding strategies. *Ann. Rev. Ecol. Syst.* 2:369–404
77. Smith, R. J. F. 1970. Effects of food availability on aggression and nest building in brook stickleback (*Culaea*

inconstans). *J. Fish. Res. Bd. Can.* 27:2350–55
78. Southwick, C. H. 1955. The population dynamics of confined house mice supplied with unlimited food. *Ecology* 36:212–25
79. Southwick, C. H. 1970. *Animal Aggression.* New York: Van Nostrand, Reinhold
80. Stark, B., Hazlett, B. A. 1972. Effects of olfactory experience on aggression in *Mus musculus* and *Peromyscus maniculatus. Behav. Biol.* 7:265–69
81. Stevenson, M. 1969. Agonistic behavior in the cowbird, *Molothrus ater. Am. Zool.* 9:571
82. Terman, C. R. 1965. A study of population growth and control exhibited in the laboratory by prairie deermice. *Ecology* 46:890–95
83. Terman, C. R. 1969. Weights of selected organs of deermice (*Peromyscus maniculatus bairdii*) from asymptotic laboratory populations. *J. Mammal.* 50:311–20
84. Terman, C. R., Sassaman, J. F. 1967. Sex ratio in deer mouse populations. *J. Mammal.* 48:589–97
85. Thiessen, D. D., Lindzey, G., Nyby, J. 1970. The effects of olfactory deprivation and hormones on territorial marking in the male Mongolian gerbil (*Meriones unguiculatus*). *Horm. Behav.* 1:315–25
86. Thomas, D., Terman, C. R. 1973. The effects of differential prenatal and postnatal social environments on sexual maturation of young prairie deermice. *Anim. Behav.* In press
87. Thompson, T. 1966. Operant and classically-conditioned aggressive behavior in Siamese fighting fish. *Am. Zool* 6:629–41
88. Thompson, T. 1969. Aggressive behaviour of Siamese fighting fish. See Ref. 38, 15–31
89. Tinbergen, N. 1951. *The Study of Instinct.* Oxford: Clarendon
90. Tinbergen, N. 1968. On war and peace in animals and man. *Science* 160:1411–18
91. Trobec, R. J., Oring, L. W. 1972. Effects of testosterone propionate implantation on lek behavior of sharp-tailed grouse. *Am. Midl. Natur.* 87:531–36
92. Ulrich, R. E., King, J. A. 1966. Recent findings in the experimental analysis of aggression. *Am. Zool.* 6:627–701
93. Vale, J. R., Vale, C. A., Harley, J. P. 1971. Interaction of genotype and population number with regard to

aggressive behavior, social grooming, and adrenal and gonadal weight in male mice. *Commun. Behav. Biol.* 6: 209–21

94. Van Vlick, D. B. 1968. Movements of *Microtus pennsylvanicus* in relation to depopulated areas. *J. Mammal.* 49:92–103

95. Vernon, W. M. 1969. Animal aggression: Review of research. *Genet. Psychol. Monogr.* 80:3–28

96. Vessey, S. H. 1967. Effects of chloropromazine on aggression in laboratory populations of wild house mice. *Ecology* 48:367–76

97. Watson, A., Miller, G. R. 1971. Territory size and aggression in a fluctuating red grouse population. *J. Anim. Ecol.* 40:367–83

98. Watson, A., Moss, R. 1971. Spacing as affected by territorial behavior, habitat and nutrition in red grouse (*Lagopus l. scoticus*). See Ref. 33, 92–111

99. Whitten, W. K. 1966. Pheromones and mammalian reproduction. In *Advances in Reproductive Physiology*, ed. A. McLaren, 1:155–77. New York: Academic

100. Wilson, A. R., Boelkins, R. C. 1970. Evidence for seasonal variation in aggressive behaviour by *Macaca mulatta. Anim. Behav.* 18:719–24

101. Wilson, E. O. 1971. Competitive and aggressive behavior. See Ref. 58, 183–217

102. Wynne-Edwards, V. C. 1962. *Animal Dispersion in Relation to Social Behaviour.* London: Oliver and Boyd

REPRODUCTIVE STRATEGIES OF MAMMALS

❖ 4056

Barbara J. Weir[1] and I. W. Rowlands[2]
Wellcome Institute of Comparative Physiology, Zoological Society of London,
Regent's Park, London, England

INTRODUCTION

The most important strategy in the life of any mammal is to beget offspring and thereby perpetuate the species. Although there is a considerable variety of reproductive patterns in nonmammalian vertebrates (Perry & Rowlands 63), the complexity of known mammalian reproduction (Asdell 5) is such as to confuse what may once have been the original pattern, if indeed there ever was just one. A polyphyletic origin of the mammals has frequently been postulated (Simpson 77, Dawson 20) from the Mesozoic synapsid reptiles; it is, therefore, remarkable that so many features of mammalian reproduction do coincide, rather than that there are differences from one species to the next regardless of generic, familial, or ordinal relationships. So far, exhaustive studies have been performed on not more than about 40 of the 4000 species of mammals (Walker 86) extant at the present time. It would be of immense interest to speculate on the possible variations that might be exhibited by the remaining 99% or even of any other 40 species chosen at random. The rodents represent about one third of the total number of mammalian species; the 10% of the 230 myomorph and sciuromorph genera that have been studied have displayed in their reproductive processes only differences of degree, not of kind, from each other and from other mammals (Mossman 53). And yet rodents of the smallest suborder, the Hystricomorpha (about 170 species), appear to be more diversified in their patterns of reproduction than any other group of related mammals.

With the exception of the Prototheria (Monotremata), which are oviparous, the mode of reproduction used by all mammals is the same, although in the evolution of internal fertilization and of viviparity certain major changes have taken place in the reproductive organs. There has been, for instance, a marked reduction in the size of the egg as the need no longer exists for the large volume of yolk present in the macrolecithal egg of oviparous species. The evolution of the Graafian follicle, char-

---

[1]Ford Foundation Research Fellow of the Zoological Society of London.
[2]Senior Research Fellow of the Zoological Society of London.

acteristic of all Metatheria (Marsupialia) and Eutheria (Placentalia), parallels the development of microlecithal eggs in these two subclasses of Mammalia.

Viviparity has been accompanied by the development of two transient structures of paramount importance, namely the corpus luteum and the allantoic placenta, in all Metatheria and Eutheria. The former structure, although present in some lower vertebrates, secretes progesterone to produce the progestational changes in the reproductive tract that are prerequisite for the occurrence of implantation and the formation of the decidua; the latter to provide the vascular connection needed for gaseous exchange and the passage of nutrients between the embryo and its mother until parturition. These two tissues have to be removed completely before reproduction can be repeated in any one individual; complex processes, not yet fully understood, are involved in the regression of luteal activity and in the expulsion of the feto-placental unit at parturition.

As far as is known, with the exception of the male marsupial mouse, *Antechinus stuartii,* which experiences only one cycle of spermatogenesis (Woolley 109), the processes leading to the reproduction of living young occur more than once in the life of every mammal. The testes are regulated for the production of very large numbers of spermatozoa and a fluid vehicle (seminal plasma) for the transfer of these gametes to the reproductive tract of the female at coitus. The maintenance of the accessory male organs of reproduction for the latter purpose forms the basis of the endocrinological function of the testis. The pattern of this activity is relatively simple and consequently is in marked contrast to the great differences observed in the reproductive mechanisms in the female mammal which form the major considerations discussed in this paper.

The fact that reproduction occurs more than once in the lifetime of every individual implies that the processes involved may be considered, to some extent or other, to be cyclic. The start of this cycle may be regarded for the purpose of this paper as the development and maturation of the first egg or group of eggs released into the oviduct at ovulation, generally synchronous with coitus and fertilization. It is the synchronization in the production of gametes of opposite sex, the formation of the zygote, and its subsequent nourishment and maintenance to complete independence as a replica of its parents that constitutes the framework around which reproductive physiological processes are dressed. It is implicit that the primary overt cycle is one of pregnancy, although it is well established that at puberty, and in some adults after a long period of sexual rest, reproductive activity does not necessarily result immediately in pregnancy. In many species at these times, fertility is preceded by a small number of short recurrent cycles which seem to be necessary to integrate the components of the reproductive process (Perry & Rowlands 63). The fact that all adult females living in the natural state are fertile when environmental conditions are favorable is often overlooked when discussing the literature dealing with laboratory species such as the rat, mouse, hamster, guinea pig, and rabbit; domesticated animals like the cow, sheep, and pig; and of course man. These species can be considered as exceptional examples because their reproduction is clearly modified by their unnatural environment: social in the case of man and the physical one of captivity for the others. Under these conditions many facts have been accumulated and generalizations made to fit them with little consideration of the relevance of these generalizations either to other species or to the domesticated species in their natural habitat. In a wild population a nonpregnant female is either juvenile, senile, or a failure, and mechanisms must operate to safeguard the species against any of these occurring too frequently.

Regardless of whether the cycle is one of pregnancy or not, the first necessity for mammals is the manipulation of conditions such that the two sexes are brought together at a time when each can produce mature gametes. In a continuously breeding male, production of gametes is no problem since, although each section of a seminiferous tubule may undergo cyclic activity (Leblond & Clermont 39), different sections are out of phase and the stored spermatozoa in the epididymis or vas deferens are continually augmented. A female mammal is primarily limited by the time it takes to mature the relevant numbers of eggs, although some reduction in time is achieved by development of successive groups of follicles as seen in sheep (Smeaton & Robertson 78) and mice (Pedersen 59). However, there is no shortage of oocytes (Baker 6), and follicles up to a certain stage are always present so that the final maturation changes can occur in a few hours or days. The presence of mature follicles implies secretion of estrogenic hormones which in turn affect the secondary reproductive organs and induce an hormonal state of receptivity, or estrus. The egg is released from the follicle for fertilization, except in tenrecs (Strauss 80) and the shrew, *Blarina brevicauda* (Pearson 60), in which intrafollicular fertilization has been described. After fertilization the zygote enters the uterus, implants, and develops into an embryo which is nurtured until the end of gestation. After parturition the young mammal sucks milk until it can be independent. The recurrence of estrus starts the next reproductive cycle of the mother, and reproductive ability of the young is reached at puberty. This cycle may be considered to be the basic one of pregnancy. The timing of each event is usually correlated either with other events in that or the next cycle, or with internal and external cues which act via the hypothalamus and the gonadotropic hormones. Nevertheless, the cycle can be varied or even interrupted at several of the stages, and the present task is to discover what factors normally determine how and where a cycle should be altered. The phraseology is recognized to be teleological in parts but has been retained to avoid any dullness caused by repetition of the few unambiguous grammatical constructions available.

Many of the possibilities suggested below agree in outline with those put forward by Conaway (18). There is a plethora of data on the ecological factors appertaining to reproductive patterns in mammals (see Sadleir 73) but a new appraisal is not a wasted exercise. Considerable attention is given in this review to rodents of the suborder Hystricomorpha as many species have been studied by the authors in field- and laboratory-bred individuals. The close relationship and relatively long zoogeographical isolation of these species in the New World could be expected to provide a situation where all things once may have been fairly equal and any differences now apparent may be the direct effects of different ecological conditions. Mossman's (53) suggestion that these animals would be interesting or bizarre has been amply substantiated.

PREGNANCY

Since it is postulated that pregnancy is the primary aim of a female mammal, some of the factors that influence the time from mating to parturition are considered. In most mammals the mating to parturition interval is a species characteristic. In polytocous animals the gestation length may be proportional to the size of the litter in utero (Rowlands 70), and, as in all biological samples, there are individual variations. Parturition must occur at a time when conditions are most likely to be optimal for survival of the young, but it is often forgotten that, because the gestation length is fixed for each species, mating must be effected beforehand by a time equivalent to the length of gestation. The mating period itself may be determined by environmental conditions which may not be the same as those necessary for the rearing of the young. It is obviously impossible to tell whether the gestation length was the fixed parameter and the environmental cues existing at the time were incorporated into the mating season stimuli, or whether the gestation length was the variable upon which selection acted. Both processes may have been involved, and at this point in time we can only look at individual frames from a continuous event and try to find the sequence and story of the film.

It is generally accepted that gestation length in mammals is correlated with (a) body weight of the mother and (b) the degree of development of the young at birth. Thus large mammals have longer pregnancies than small mammals (hamster, 16 days; elephant, 22 months) and animals which produce poorly developed young at birth have a shorter pregnancy (e.g. rat, 21–23 days) than do those which produce well-developed young (guinea pig, 68 days). There are correlations between birth weight and gestation length (e.g. Huggett & Widdas 35) which have resulted in formulas for general application. The Huggett & Widdas formula is $W = a(t - t_o)$, in which W = bodyweight at birth, a = fetal growth rate, t = gestation length, and t_o is the intercept on the x axis. Values of t_o for different gestation lengths are given and are based on results from several species for which a range of samples was available. Examination of the plots of Huggett & Widdas shows that there are three groups of animals which do not conform closely to the formula: cetaceans, hystricomorph rodents, and the primates.

Cetaceans have a rapid fetal growth rate and a shorter gestation length (only 16 months in the sperm whale) than one would expect for the size of the mother and the low maternal/fetal bodyweight ratio at birth. Only one well-developed young is carried, but this is true of other mammals with a relatively longer pregnancy for their size and equally well-developed young (e.g. artiodactyls such as the giraffe and camel which give birth after a 15-month pregnancy). The marine environment may be a relevant factor in permitting heavier young to be carried, but this cannot be the whole answer since other marine mammals such as seals and manatees, and some terrestrial mammals like the guinea pig, produce large, well-developed young.

However, in spite of the precocity of young guinea pigs, who can survive independently from the mother within a few days of birth, the guinea pig and other hystricomorphs do not conform to the Huggett & Widdas formula. The guinea pig is generally considered to have a long pregnancy (68 days) for its size (1000 g) but most other hystricomorphs have a gestation length of 90 days or more. Thus the chinchilla, which weighs only 500 g, gives birth after 111 days, a gestation equivalent to that of the pig. Moreover, although the young chinchilla is fully furred and has open eyes, it cannot survive without suckling until it is at least 3 weeks old. The argument that hystricomorphs have long pregnancies because they have precocious young is also refuted by the offspring of the 150 g tuco-tuco *(Ctenomys talarum)* and the 200 g degu *(Octodon degus)*. At birth, after more than 100 days and 90 days of gestation respectively, the young of both species have closed eyes and only the guard hairs have penetrated through the skin. The resemblance to adults is noticeable after about 2 weeks, but the neonates are clearly not as well developed at birth as are those of the coypu *(Myocastor coypus,* 132 days), plains viscacha *(Lagostomus maximus,* 154 days), North American porcupine *(Erethizon dorsatum,* 213 days), and guinea pig (68 days), in spite of the extended gestation. Neither can respective litter sizes be used as extenuating circumstances. Tuco-tucos and degu do have litters of 1–6 young, but so do coypu and guinea pigs.

Of the hystricomorphs mentioned the guinea pig has the shortest pregnancy; it has been postulated (Weir 88) that guinea pigs have been selectively bred for a shorter gestation during their long domestication of at least 400 years. However, it was subsequently found (Rood & Weir 69) that other caviids and echimyids had shorter pregnancies. The wild guinea pig *(Cavia aperea),* believed to be the ancestor of *C. porcellus* (George & Weir 28), gives birth after 61 days; *Microcavia australis* and *Galea musteloides,* after 54 and 52 days respectively. Thus it appears that caviids (at least of the subfamily Caviinae) have a short gestation compared with other hystricomorphs although well-developed young are produced. The same is true for the echimyid genus *Proechimys,* whose members litter after 62 days (Maliniak & Eisenberg 43, Weir 99). Since echimyids are considered to be generalized hystricomorphs, it would appear that the ancestral tendency for the group was to have a short pregnancy.

If this is so, the question arises as to why the other hystricomorphs evolved a long pregnancy. It seems likely that the long pregnancy was the basic hystricomorph trait and may have been necessary because environmental conditions precluded mating and littering in the same season. Competition was probably minimal in the Eocene and Oligocene and the species could afford to run the risk of the females being pregnant for long periods during which time mother and embryo(s) would be vulnerable. The recent (geologically) arrival of the cricetine rodents and the introduced European hare in South America, as well as persecution by man, will probably be the death knell of many hystricomorphs because their low reproductive rates preclude rapid recruitment. It is perhaps significant that both families with short gestation length are those which have been, and are, preyed upon. Caviids are the major item of diet for most of the predators in southern South America. A faster recruitment rate may have been a necessity; caviids and echimyids certainly seem to be reproductively the most efficient of the hystricomorphs (Rood & Weir 69, Weir 99, 100).

If, therefore, it is postulated that the long pregnancy is an hystricomorph characteristic which has been altered only because of certain environmental pressures (predation), one must consider the factors that may have influenced the development of a long pregnancy in the basal hystricomorph stock. The earliest unequivocal fossil hystricomorphs are found in the Deseadan formation in Patagonia (Wood & Patterson 108). It is possible that much of the hystricomorph radiation occurred from this area in or after the Oligocene. The climatic conditions of the time may have been harsh for long periods, and one can realize the advantages of having a pregnancy which delayed the birth of the young until conditions were better suited for their survival. The evolution of young fully furred at birth suggests that the climate may have been inclement even during better seasons. Why this precocity was not complete as in some artiodactyls and why there was a slow fetal growth rate rather than a delay of implantation (see below) is a mystery and indicates the homogeneity of the Hystricomorpha. In view of the controversy over the possible relationships of the New and Old World hystricomorphs (Wood 107, Landry 38), the finding of a long pregnancy of 110 days in the brush-tailed porcupine, *Atherurus africanus* (Rahm 67), is interesting. It is a great pity that more is not known about the African hystricomorphs, but the presence of a vaginal closure membrane and a long estrous cycle (33 days) in the African porcupine, *Hystrix cristata* (Weir 91), may be significant.

The third group of mammals that does not fully conform to the Huggett & Widdas (35) formula is the Primates. This is particularly noticeable in the prosimians; for example, the mouse lemur, *Microcebus murinus,* weighing only about 100 g, has a 60-day pregnancy, and the slow loris, *Nycticebus coucang,* has one of 174 days. The New World squirrel monkey, *Saimiri sciureus,* has a 6-month pregnancy and the much larger Old World chimpanzee *Pan troglodytes* gives birth after 7½ months. Thus, the Primates also seem to have a group tendency towards long pregnancies, but the difference between the Primates and Hystricomorpha is that the former give birth to immature young which are dependent until such time as they have reached maturity. Since only one or two young are born, maternal care and upbringing must be efficient to perpetuate the species even on a one for one basis. A high degree of social organization, increased brain capacity, and longevity have presumably contributed to the success of this group.

That a long pregnancy is a "designed" consequence in the Primates and the Hystricomorpha is further indicated by the occurrence in both groups of a hemoendothelial, labyrinthine placenta (Mossman 52, Amoroso 2). It would be expected that such a system would increase the efficiency of the placenta in terms of transfer of materials and thus hasten the rate of development of the fetus. This is obviously not the case and reflects once again that embryological parameters may be a useful indication for taxonomic considerations (Mossman 52, Mossman & Conaway 54, Fischer & Mossman 27).

Pregnancy in mammals is established by the hormonal activity of the corpus luteum as its secretion (progesterone) enables the endometrium to accept the early blastocyst for implantation. As a general rule, the lifespan of the corpus luteum is prolonged by gestation and in many species of carnivores, artiodactyls, and myomorph rodents it remains morphologically distinct and retains its physiological activity until parturition. In others, such as equids, some primates, probably man, and hystricomorph rodents, the corpus luteum does not outlast gestation and in an extreme case (the mare) it has disappeared by the end of the second month of the

11-month gestation. The need for progesterone beyond this time in this species is suggested by the unusual occurrence of follicular development, ovulation, and the formation of secondary corpora lutea in succession over another 2–3 month period, after which all trace of luteal activity has disappeared. If progesterone is required during the second half of gestation it is probably supplied by the placenta of the mare, as has been demonstrated in some other species. The stratagem of ovulation followed by corpus luteum formation during pregnancy appears to be unique to equids. Ovulation without luteinization occurs in the plains viscacha (Weir 93) in pregnancy, but progesterone requirements are met in this species in a manner similar to that used in other hystricomorphs, namely the development of accessory corpora lutea formed from unovulated follicles at the start of or during pregnancy (see page 153).

Delayed Implantation

Not all long pregnancies are due to very slow fetal growth rates. The phenomenon of delayed implantation in Eutherian mammals has been considered and investigated by many workers (see Enders 21). The time of fertilization of the egg and the moment of implantation of the blastocyst may be separated by many months during which period the blastocysts lie dormant in the uterine lumen. This phenomenon is common among the Mustelidae (Carnivora), but such a delay in implantation is found in so many species of different orders that it has clearly arisen independently several times. At first sight a list of some of the animals exhibiting delayed implantation (seals, badger, armadillo, stoat, roe deer, bear) suggests that there is no common factor. Once implantation has occurred, the rest of the pregnancy is equivalent to that in species of the same order which have no delay. Thus it appears that this is another device for separating the breeding period and the season of birth. The question is, why should this have been necessary, especially when closely related species which may even occupy a similar habitat have not evolved a similar reproductive stratagem.

One such example is that of the stoat *(Mustela erminea)* and the weasel *(Mustela nivalis),* species which are solitary and have extensive home ranges, often in the same areas. The female stoat mates in June or July but does not give birth until the following April or May, while the female weasel mates at about the same time and gives birth to two litters before the following spring. The gestation period is about 42 days and mating recurs at the end of the first lactation (6–8 weeks). The young weasels are mature by the following breeding season. The young male stoat also matures in the year after his birth, but the young female stoat reaches sexual maturity and mates in the nest before she is weaned at about 6–8 weeks of age (Gulamhusein 30). It is not known whether or not the active male is the father, but development of this early precocity ensures impregnation before the litter is dispersed. It also indicates why a delay of implantation may be necessary to ensure that somatic growth is sufficient to bear the stress of pregnancy. The mechanism may then have become established for all females of the species.

A similar explanation does not seem valid for other species since none is known to experience the early sexual maturity of the young female stoat. However, for most of such species there are various behavioral or climatic reasons which can be used to explain the implantation delay. As with all mammals it is essential that the young are born at a time optimal for survival, so the question reverts to what are the factors that determine the mating time? In the colonial seals, mating occurs in the same period as littering and growth of the young, that is, when the sexes are together on land. Thus the best time for birth is also the best (and only) time for mating.

A similar coincidence of seasons may have been evolved in the other boreal species which exhibit delayed implantation. Certainly it allows flexibility in the period of delay according to the conditions, but one must then consider what conditions cause the delay to end and implantation to take place. These have not yet been determined precisely for any species with delayed implantation, but the timing of events suggests it to be a photoperiodic response. Mead (47) found that blinded western spotted skunks *(Spilogale putorius)* did not implant, while others kept in an increased photoperiod (14L:10D) implanted a month earlier than usual. This suggests that the light effect is not mediated via the pineal as photoperiodic control of estrus is in the ferret (Herbert 33). The striking differences in reproductive patterns between the western forms of *Spilogale putorius,* which exhibit delayed implantation, and the eastern forms, which do not, have led Mead (45, 46) to suggest that the various subspecies should be regrouped into two distinct species.

Lighting changes alone were not effective in inducing implantation in the European badger, *Meles meles* (Canivenc 14), but a decrease in photoperiod and temperature resulted in attachment about 6 months before the usual time of December (Canivenc, Bonnin-Laffargue & Lajus-Boue 17). Although exogenous progesterone does not affect the delay, the induction of a new set of corpora lutea by treatment with exogenous gonadotropins results in implantation, indicating that the inadequate hormonal activity of the corpora lutea is caused by insufficient secretion of gonadotropin (Canivenc, Bonnin-Laffargue & Lajus 15). However, similarly induced corpora lutea in the marten, *Martes martes,* became atretic like those of the delay (Canivenc, Bonnin-Laffargue & Lajus-Boue 16), suggesting a lack of luteotropic factors. Lincoln & Guinness (41) produced some evidence that the time of implantation in the roe deer was not significantly affected by alteration of the photoperiod, and a fixed interval of delay was suggested.

A nonfunctional corpus luteum during delay is not associated with all the species in which this phenomenon is known (Enders 21). In the armadillo, *Dasypus novemcinctus* (Enders 22), and the roe deer, *Capreolus capreolus* (Short & Hay 75), the corpus luteum appears well developed during the delay and does not change in weight or histological appearance after implantation, although some increase in progesterone secretion was noted in the armadillo (Labhsetwar & Enders 37). It has been suggested (Short & Hay 75) that estrogens secreted near the time of implantation synergize with progesterone to effect implantation, as in the mouse (Finn 26). However, bilateral ovariectomy during the delay causes implantation in the armadillo (Buchanan, Enders & Talmage 11).

Although many species have evolved a similar system of delaying implantation to achieve flexibility with environmental cues, the precise methods of control are not uniform. It is therefore likely that the causative factors which initiate implantation also vary from species to species.

The occurrence of a similar stratagem, embryonic diapause, in some members of the macropodid, or kangaroo-like, marsupials is also indicative of the advantages of having an instant replacement mechanism. In all marsupials except the swamp wallaby, *Protemnodon bicolor,* gestation is shorter than the length of one estrous cycle and there appears to be an endocrine equivalence of the nonpregnant and pregnant states. Suckling activity of the young in the pouch causes ovarian inactivity in nonmacropodid marsupials and the corpus luteum from the newest ovulation remains nonfunctional in macropodids. The blastocyst from the mating at that estrus remains in the uterus until the young leaves the pouch naturally or by disaster. The inhibition of the corpus luteum is probably effected by oxytocin and can be overcome by exogenous progesterone. Thus there is no inhibition of the ovulatory mechanism in macropodids, a fact that may be important for species in which the sexes are widely dispersed and which often experience adverse conditions. It is possible, however, that factors other than the suckling inhibition are involved. In those species such as the tammar wallaby *(Protemnodon eugenii)* and the quokka *(Setonix brachyurus)* which enter the nonbreeding season with a pouch young, loss of the pouch young during anestrus does not cause development of the diapausing blastocyst. Such development does not start until the following breeding season, a delay of up to 11 months (Berger 7), presumably because of lack of pituitary stimulation. Administration of exogenous gonadotropins of eutherian origin has not been successful in stimulating the corpus luteum and ending the diapause (Sharman 74). The many differences between metatherian and eutherian mammals have led Sharman (74) to suggest that viviparity arose separately in these stocks after their derivation from a common oviparous ancestor.

Delayed Fertilization

It is clear from its ubiquity and the foregoing discussion that the development of delayed implantation has involved changes only in the timing of normal reproductive processes, thus conferring some flexibility upon events in utero and a precise correlation with environmental conditions.

Many vespertilionid bats who live in temperate climates and depend on a supply of insects for their food have also evolved a timing schedule to connect their reproductive processes to their habit of hibernating when a regular supply of food is not available. They have also instituted some unique physiological mechanisms. Mating occurs in late summer but ovulation does not take place until the following spring, when the one egg that is released is fertilized by spermatozoa stored within the uterus of the female throughout the previous 6-month period. The tubo-uterine junction remains firmly closed throughout hibernation and spermatozoa do not penetrate the oviduct (Fallopian tube) during torpor. Furthermore, the testes become completely regressed and aspermic, but in the late summer the cauda epididymis becomes greatly enlarged and serves as a reservoir of spermatozoa which can be used in spring to inseminate any females that escaped mating in the previous autumn. Thus we have the situation in which mammalian spermatozoa, which normally do not survive in either the male or the female tract for more than a few days, remain fertile over several months (Racey 64). Electron micrographs of the stored uterine sperm show that there may be attachment of the sperm heads to the uterine epithelium, so a transfer of nutrients can be postulated (Racey & Potts 66). Nourishment for spermatozoa would also seem to be provided by the epithelium of the cauda epididymis.

It might be thought likely that ovulation does not occur at the time of mating because there is no follicle at a suitable stage of maturation, but in fact there is a Graafian follicle available (presumably necessary to effect behavioral estrus) which grows very slowly during hibernation and remains viable throughout the winter. The proximate mechanism which maintains the follicle appears to be the storage of glycogen by the granulosa cells, particularly those of the cumulus oophorus (Wimsatt 105). The hormonal conditions whereby this state of affairs is controlled are as yet unknown, but Racey (65) has shown that if pipistrelle bats are deprived of food in a cold environment after implantation has taken place, the actual gestation period is extended by a period similar to that of the induced torpor. When food is available, gestation is increased at low temperatures (5°C) and shortened at high temperatures (30–35°C). These findings indicate mechanisms that can effect variation of the fetal growth rate, thus adding a fine control to the coarser control of delayed ovulation. The two factors involved, food supply and temperature, are naturally related and the separate effects of the two are difficult to distinguish. The reproductive changes appear to be a consequence of the hibernating habit (Wimsatt 106), but which came first, sperm storage, the overwintering follicle, or changes in fetal growth rate, cannot be determined. No such deviations have occurred in other hibernators such as the dormouse or hedgehog. It would seem likely that this pattern of reproduction has arisen only once in this particular group of temperate bats. There appears to be no species of bat which one can suggest is evolving towards a stratagem of delayed fertilization, although the tropical African bat, *Eidolon helvum,* is reported to exhibit delayed implantation (Mutere 56). No evolutionary or ecological reason can be given to explain this circumstance, should it be confirmed.

BREEDING SEASON

A necessary preliminary to fertilization in mammals is the union of the two sexes at a time when both are capable of producing gametes to initiate conception. The female gamete, or ovum, is made available for fertilization at known intervals depending on the periodicity of ovulation, but obviously its occurrence is not synchronous in all individuals of the same species. In the ideal situation, therefore, all males should be capable of mating with fertile results at all times; this occurs in some male primates (man) and carnivores (cat, dog) in which spermatozoa are continuously produced because of asynchronous activity of different regions of the seminiferous tubules (see page 141). However, a large number of other species, such as insectivores, rodents, artiodactyls, and some carnivores (e.g. mustelids), maintain full testicular activity only seasonally. The extent to which gonadal activity is reduced during the so-called quiescent period in seasonal breeders varies from a comparatively slight reduction in spermatogenesis with no obvious difference in gross size or condition of the testis, as in the ram, to a complete cessation with subsequent loss of testicular weight, as in wild myomorph rodents, sciuromorphs, and mustelids. In seasonally breeding female mammals, ovarian regression is more uniformly complete and it is rare for an ovary of such a species to contain even small Graafian follicles during the anestrous period. An exception occurs in some hystricomorphs in which mature Graafian follicles are present during anestrus [chinchilla (Weir 88, 90, 91), plains viscacha (Weir 93), agouti (Weir 94), acouchi (Weir 95)].

Environmental temperature is known to affect spermatogenesis, and infertility associated with reduced output in summer months, as in the ram, is thereby explained. But in most species the effect is mediated by photoperiod (Sadleir 73) and in some by food supply (Smythe 79). For example, ferrets can be made to breed in the winter by being exposed to extra light (Allanson, Rowlands & Parkes 1), and there are numerous reports of animals undergoing a 6-month shift of breeding season after transportation from one hemisphere to the other. The female of the species may be affected by the same environmental factor as the male, but the male must be more sensitive to that factor since spermatogenesis and steroidogenesis must be initiated early enough for spermatozoa to accumulate in the epididymis and for the accessory glands to respond to increasing amounts of circulating androgens. Thus in deer testosterone production and spermatogenesis begin some 6 weeks before the breeding season, or rut (Lincoln, Youngson & Short 42). The female appears to need less effort metabolically to prepare for breeding, and yet it is likely that the male has followed the female in becoming a seasonal breeder. There are no species in which females can breed throughout the year and do not do so because the males are seasonally aspermatogenic, but there are many in which the males are always fertile and females are seasonal (e.g. chinchilla, plains viscacha). It is easy to see that a female would not wish to be pregnant during inclement conditions when death would mean loss of the young as well as of her own reproductive potential. Clearly, the factors causing cessation of breeding are more critical than those initiating the breeding season. Why the male should wish to follow suit and become seasonal, thus imposing limits on breeding, is harder to understand, although there are many species in which this situation appears to be arising (e.g. sheep).

Segregation of the sexes at certain periods of the year is a common feature of some artiodactyls, but it usually occurs during the period of sexual rest and the reunion of the sexes is heralded by elaborate behavioral patterns. These may be simple patterns involving vocalization by the male (red deer), or more complex, like those shown by seals and Uganda kob (Buechner, Morrison & Leuthold 12) in which territorial behavior and harem collection are important. The marking of territory by means of scent glands (Mykytowycz 57) can be considered a reproductive stratagem for ensuring possession by the most virile males since the skin glands are generally influenced by androgens (Ebling 23). A marked territory indicates ownership to other males and can serve to attract females.

Apart from the obvious uses of specialized glands and urine for territory marking, some species use these media as an intraspecific hormonal system. It has long been known that insects use a pheromonal system to affect reproduction (e.g. the queen bee substance, Butler, Callow & Johnston 13) and similar systems have been suggested in mammals except that mammalian pheromones are generally considered to be active by olfaction rather than ingestion (Bruce 10). Work in this field was initiated by the report of Lee & Boot (40) that grouped female mice experienced nonsynchronous estrus. Whitten (102) showed that when such mice were introduced to a male, mating did not occur with equal frequency on the next four nights as would be expected for a 4-day estrous cycle, but that there was a peak of matings on the third night. This result indicated that the presence of a male was inducing a synchrony of estrus in about 75% of the females. A similar effect has been shown in deer mice *(Peromyscus)* and is caused by a species-specific constituent in the urine of the male (Whitten 103). The value of this effect in natural conditions when the females would rarely be away from males and nonpregnant is difficult to understand. With a 4-day estrous cycle a 1- or 2-day delay cannot make much difference; however, these observations may be laboratory artifacts. The results of Vandenbergh and his colleagues on the time that mice achieve puberty in different social environments are interesting because, if one accepts the fact that once pregnancy has occurred the female should remain in that state, the onset of puberty becomes important in terms of recruitment to the breeding population. If female mice are housed with, or exposed to the urine of, adult males, they reach puberty sooner than do similar mice housed with adult females or coevals (Vandenbergh, Drickamer & Colby 84, Vandenbergh 83). This would appear to be a very sensible arrangement to encourage the earliest possible breeding. The effect is presumably mediated via the hypothalamus, which does not respond until the endogenous hormonal environment is indicative of somatic and sexual competence.

The importance of an early puberty has been stressed with regard to another pheromonal phenomenon, that which occurs in the cuis *(Galea musteloides)* (Rood & Weir 69, Weir 96, 100). Unlike mice and deer mice, cuis do not exhibit regular cycles if they are isolated, either in groups or singly, from males. Being related to the guinea pig, cuis have a vaginal closure membrane and its perforation at estrus is easily detected. The incidence of spontaneous opening is very low (less than 13%) but all females show vaginal opening within 3 days of the introduction of the male into the same cage. The need for direct contact indicates a tactile requirement and the effect is believed to be instigated by the behavior pattern of chin-rump following and then augmented by ingestion of the secretions of a submandibular (chin) gland, which is hypertrophied in the male (Weir 96, 100). These animals will breed all year in the wild, except when conditions are inclement, so the significance of this mechanism during a normal breeding season is obscure. However, if breeding stops because of bad weather an early resumption is needed, especially as the inherent periodicity of estrus is 21 days (Rood & Weir 69). In laboratory conditions, and presumably in the wild also, puberty and pregnancy can be initiated as early as 13 days of age, and conceptions at 17 days of age are not uncommon. The attentions of an adult male are clearly important for encouraging females to breed. However, one questions why it is necessary to have a behavior pattern and a pheromone system when a related and sympatric species, *Microcavia australis,* seems to manage only with a chin-rump follow pattern. It seems possible that the extra refinement of a chin gland in *Galea* could be due to the similar distribution of *Microcavia.* The secretions of a chin gland may be the characteristic necessary for a female *Galea* to recognize that a male *Galea* is doing the chasing rather than a *Microcavia* male. There is evidence that although a female *Galea* will respond to a male with no chin gland she takes longer to do so and is more reluctant to mate (Weir 100). No stimuli other than the direct contact with a male are effective in evoking estrus in the cuis, unlike the findings for *Microtus ochrogaster* in which cage changes and cleaning can act as triggers, although the male is more reliable (Richmond & Conaway 68).

Another pheromonal effect, also mediated by the urine of the male, has been described for mice and some deer mice. If a recently impregnated female is exposed to a strange male before implantation, the implantation is blocked and the female returns to estrus (Bruce 9, 10). It is again difficult to visualize any use for this mechanism in the wild, unless it acts as a means for encouraging outbreeding in populations of family groups. Laboratory rats and some strains of laboratory mice do not exhibit the Whitten and Bruce effects (Whitten 104), so their significance to wild populations must remain questionable.

Only one group of mammals outside the Rodentia has been studied for pheromonal influences, and this is the Primates. Michael & Keverne (49) showed that male rhesus monkeys were stimulated to sexual activity by the vaginal secretions of the intact female at estrus or of the ovariectomized female treated with estrogens. Anosmia abolished the response which was present even when the male was blinded. A similar stratagem was found in baboons. Although normally other senses, particularly vision, are important, it is believed that many mammalian species can use olfactory cues to detect reproductive condition (Signoret 76).

THE NONPREGNANT CYCLE

Should pregnancy fail for any reason or a litter be lost soon after birth, it is a matter of urgency for small short-lived, highly predated species to become reimpregnated as quickly as possible. Predatory species and large herbivores are not subject to such selective pressures and can afford to wait for the next normal estrus of the same or following breeding season. The emergency is met by the institution of a cycle of recurring periods of estrus regulated by ovarian activity. This activity is most readily demonstrable in adult females of the commonly available domestic species. In the wild such cyclical activity is discernible in histological sections of the ovary of animals, such as myomorph rodents, the elephant, and some artiodactyls, at the onset of a period of breeding. They are usually regarded as sterile cycles and are recognizable by the simultaneous appearance of two, three, or more sets of corpora lutea of different ages. Each cycle consists of a follicular phase comprising the growth and maturation of one or more Graafian follicles, and a luteal phase representing the development of the spent or ovulated follicle(s), the corpus luteum. The sex hormones secreted during the two phases of the ovarian cycle, estrogen (follicular phase) and progesterone (luteal phase), provide the necessary hormonal stimulus for the accompanying estrous cycles.

It is generally recognized (see Brambell 8, Perry & Rowlands 63, Perry 62) that there are three types of cycle in nonpregnant mammals. The first relates to those animals which do not ovulate spontaneously (e.g. rabbit, cat, many mustelids) and in which the follicular phase is terminated by degeneration without the formation of any luteal tissue. If mating takes place ovulation follows and a corpus luteum develops. In the event of a sterile mating, a condition of pseudopregnancy is produced and the life of the corpus luteum is either shorter (rabbit) or equal (dog, ferret) to the duration of that of the corpus luteum of pregnancy. The other two types of cycle are found in animals in which ovulation occurs spontaneously; the difference between them relates to the degree of secretory activity of the corpus luteum. The corpus luteum of myomorph rodents (rat, mouse, hamster) and tree shrews (Conaway & Sorenson 19) is small and nonfunctional, but after mating it enlarges and secretes progesterone. If the mating is sterile the corpus luteum regresses after 9–12 days, that is, pseudopregnancy occurs, whereas following conception the corpus luteum remains functional throughout pregnancy. In the third type of cycle a fully functional corpus luteum is formed after the spontaneous ovulation regardless of whether the mating is sterile or fertile. This type of cycle is found in the guinea pig and occurs commonly among carnivores, ungulates, and primates. In these species the corpus luteum secretes progesterone which inhibits the return of estrus and ovulation. The life span of the corpus luteum is characteristic for all members of the same species.

In the majority of mammalian species the cycles persist while females are isolated from males. Following mating, the luteal phase becomes greatly extended by the formation of the corpus luteum of pregnancy. The polyestrous condition, as seen in most rodents, artiodactyls, and primates, contrasts markedly with the monestrous habit of the dog and a few other species in which the single cycle, whether or not it leads to pregnancy, is followed by a period of anestrus before another cycle is initiated in the following breeding season.

It has been suggested that the guinea pig type of cycle is equivalent to that of pseudopregnancy in the mouse and that this pattern represents the basic cycle of nonpregnancy (Everett 25, Conaway 18). However, this seems unlikely to us because the corpus luteum is a relatively recent mammalian acquisition and only the follicular part of the cycle can be equated with the cycle in other vertebrates (Hoar 34). Even in viviparous fish and reptiles there is very little luteal development in postovulatory follicles; in these lower vertebrates most hormones seem to be secreted by preovulatory follicles which have undergone atresia (Amoroso & Finn 3).

One difference between mammals and other vertebrates is the number of young that are gestated. Mammals have clearly opted for smaller numbers that are more likely to survive rather than for production of large numbers of offspring, many of which will be lost before maturity. In most mammals the reduction in litter size, compatible with the spatial requirements of the placenta and embryo in the uterus, is paralleled by a reduction in ovulation number; redundancy is minimized if no fault, such as resorption or abortion of one or more embryos, occurs from one end of the process of gestation to the other. Thus it is possible that the change of function of the corpus luteum in the development of true viviparity has been necessary as an insurance or safeguarding mechanism. In this process the first step is to automatically convert the spent follicle from the nonfunctional corpus luteum, which in the hamster has been described as "merely the histological consequence of ovulation," (Greenwald 29) into a potential endocrine organ. The next step should be linked to coitus so that a prepared hormonal environment is available to ensure implantation (corpus luteum of pseudopregnancy). The life of such a converted follicle needs to be relatively short so that another ovulation can take place if fertilization or implantation fails to occur. In many mammalian species follicles do not develop if an active corpus luteum is present; therefore, a built-in lifespan of the corpus luteum of pseudopregnancy is required which can be overridden if pregnancy supervenes. The automatic conversion of the follicle into a fully functional corpus luteum in species such as the guinea pig or sheep can be considered as a device to eliminate some of the stages at which failure could occur, such as relay of the signals indicating mating. Except in marsupials, the lifespan of the normal corpus luteum is shorter than that of pregnancy, so the cyclic corpus luteum must be converted to one of pregnancy once the products of conception are in the uterus. How the ovary is informed that the uterus is pregnant is as yet unknown.

In some species, like the guinea pig and the sheep, the nonpregnant uterus acts as a luteolytic mechanism to prevent survival of the corpus luteum (Melampy & Anderson 48). The real problem appears to be that of why the cycle is so long in some nonpregnant animals. For example, the cyclic corpus luteum exists for 16 days in the guinea pig, much longer than necessary for implantation to occur (6 days). In ungulates true attachment of the blastocyst does not take place before conversion of the corpus luteum is due, but the conceptus with an expanded chorion is known to be influential in this matter (Moor & Rowson 50, 51). A long luteal phase may be necessary in carnivores such as the cat and dog, in which implantation does not occur for 2–3 weeks (Amoroso 2). A similar explanation is not available for some of the other relatives of the guinea pig such as the chinchilla and agouti, in which implantation occurs by the 8th day of gestation but the cycle is 41 and 35 days respectively (Weir 92, 94). A long cycle length seems to be typical of all of the South American hystricomorphs examined, the shortest so far known being that of the domestic guinea pig. The African porcupine, *Hystrix cristata,* also has a long estrous cycle of 33 days (Weir 89).

It is possible that in natural conditions the estrous cycle is never involved in reproductive activities and so there have never been any selective pressures on its length. Another possibility is that hystricomorphs are moving towards a situation in which the corpus luteum from the ovulation initiating the pregnancy becomes superseded by other organs that secrete the progesterone necessary for maintenance of the pregnancy. The life of the cyclic corpus luteum is then extended until such time as these other sources can take over. That there may be some basis for this suggestion is indicated by the fact that many hystricomorph species develop accessory corpora lutea, which are progestationally active (Tam 81), from unovulated follicles during pregnancy [chinchilla (Weir 87), agouti (Weir 94), acouchi (Rowlands, Tam & Kleiman 72, Weir 95), mountain viscacha (Pearson 61, Weir 97), North American porcupine (Mossman & Judas 55)]. The guinea pig, moreover, is one of the few species in which the placenta has taken over the production of progesterone from the ovaries (Amoroso & Finn 3, Illingworth & Deanesly 36).

Whether or not a similar argument can be made for the elephant, *Loxodonta africana,* which appears to have no progesterone in the corpus luteum of pregnancy or in the plasma (Hanks & Short 32), is debatable.

Postpartum Estrus

Many species are able to pack several pregnancies into that part of the year suitable for survival of young. The rapidity with which reimpregnation can be effected may be important in terms of population numbers. The original pattern would be that which precluded pregnancy by inhibiting estrus while the female was lactating. Such a mechanism is found in man and clearly prevents a concurrent pregnancy and lactation which would stress the female too greatly. This is obviously an inefficient method for many species in which the young are relatively well developed and lactation is short; in any case, the greatest demands on the mother are felt in late pregnancy when the young in utero are large, and lactation needs are by then diminishing. Moreover, what is to happen if the young are lost during lactation? As soon as a litter of pigs is naturally or artificially weaned the sow again comes into estrus and will accept the boar. The hormonal pathways by which a lactation anestrus is maintained are by oxytocic action on the corpora lutea, which presumably prevents the growth of large follicles. But large follicles are present in the ovaries of many animals when the corpus luteum of pregnancy regresses towards term, and it is not surprising to find that an ovulation occurs shortly after parturition. In gregarious species males are always at hand to take advantage of this postpartum estrus which results in a concurrent lactation and pregnancy, the very situation the female has been trying to avoid. The myomorph rodents, which produce large litters of poorly developed young that need an appreciable amount of maternal care, overcome this problem by a stratagem of delaying implantation. The blastocyst waits in the uterine lumen until another crop of follicles has been allowed to grow and secrete the estrogen which may initiate the implantation reaction, as in the mouse (Finn 26).

The inhibition of the follicles and therefore the duration of the delay is related to the number of young suckling. A similar instant replacement mechanism is found in the macropodid marsupials (Sharman 74), in which growth of the embryo is suspended for as long as a joey is suckling in the pouch. No special changes have occurred in the cycle in these marsupials because pregnancy is shorter than the length of the cycle and estrus occurs after parturition as it would have done had pregnancy not occurred. Apart from avoiding having too many pouch young at a time, the stratagem of embryonic diapause enables a kangaroo, say, to take advantage of chance meetings and mate without having to heed the effects of adverse climatic conditions on metabolic demands during pregnancy and lactation. The swamp wallaby, *Protemnodon bicolor,* is the only marsupial known in which gestation is longer than the estrous cycle, but estrus occurs a few days prepartum and the embryo resulting from mating at this time diapauses in the usual way. The European hare, *Lepus europaeus,* is the only known eutherian mammal to experience a prepartum estrus (Martinet, Legouis & Moret 44), but some small rodents often mate between delivery of the various offspring in a litter. Although one cannot comprehend the mechanics of insemination before parturition, such a stratagem is more sensible than one of an estrus occurring during or after parturition when a female would be distracted, by the mating endeavors of the male, from exercising maternal care at a critical time for survival of the young.

It is clear that several stages of the process of telescoping the reimpregnation time from after lactation to before parturition are represented by examples throughout the Mammalia. To a small prey species the advantages of rapidly becoming pregnant again are obvious. The large herbivores and carnivores do not seem to have adopted this stratagem, although many of them (e.g. antelopes) will fit two pregnancies into a single breeding season if environmental and nutritional requirements are met. Although an antelope usually gives birth to a single offspring, twins may be produced; one wonders why twins are not more frequent as this would result in greater recruitment to the population if they survived. The survival potential is likely to be the critical factor, and it is probably safer on a long term basis for a species to produce a singleton twice in a breeding season if conditions are suitable than to risk losing a twin litter in poor conditions.

Spontaneous and Induced Ovulation

In agreement with many other authors (e.g. Conaway 18) we feel that animals cannot be separated into those that ovulate spontaneously and those that ovulate reflexly, as there is clearly a continuum. The degree to which either pattern is characteristic of a species possibly reflects selective pressures and, although there are several species in which both patterns can be found, it is not clear which represents the primitive pattern.

Everett (25) and Conaway (18) believe that spontaneous ovulation mechanisms have arisen independently several times during the evolution of mammals and that that ovulation dependent upon stimuli such as mating is the ancestral type. This does not seem probable to us, as ovulation in all other vertebrates, with the possible exception of some birds (Erickson & Lehrman 24), appears to be spontaneous. A system for induced ovulation would emancipate the female from cyclic behavior and would provide freedom to mate almost as soon as a male appeared. The advantage of this would be understandable for species in which the sexes are widely separated (e.g. cat), but not for those which are gregarious (e.g. rabbit, some voles). Induced ovulation is not correlated with diet, habit, or habitat. The physiological pathways by which the changes could occur are numerous; it is likely that different steps have been involved in different species. In spontaneously ovulating species there is a set hormonal pattern and luteinizing hormone (LH) is secreted at a particular time to entrain the ovulatory processes. If mating occurs before the LH surge it is obvious that the neural stimulation of coitus could affect the hypothalamic centers and cause an early LH release as described in the rat (Aron et al 4) and chinchilla (Weir 101), both of which normally ovulate spontaneously.

An induced ovulation mechanism implies a prolonged period of estrus. This is usually effected by growth of several follicular waves rather than by an extended life of one particular set, except possibly in the case of the ferret. Similar waves of follicular growth occur in the ovary of spontaneous ovulators such as the cow

(Smeaton & Robertson 78), mouse (Pedersen 59), and guinea pig (Rowlands 71), so the transition would be easily achieved. There is as yet no known reason why some follicles continue to mature while others degenerate or become atretic, but ovulation is clearly the exceptional fate for a follicle. The products of atretic follicles may be interstitial tissue or accessory corpora lutea, both of which are secretory. This is again analogous with the situation in nonmammalian vertebrates in which hormones are derived from unovulated follicles.

Polyovulation

Natural polyovulation or superovulation is a rare process in mammals, although standard in nonmammalian vertebrates. It should not be confused with polyovuly, the occurrence of several oocytes within a single follicular envelope, which is found in several mammalian species, particularly in juvenile animals. Large numbers of eggs are produced at any one time by fishes and amphibians because of the difficulties inherent in external fertilization and the numbers of young lost before reaching maturity. The development of internal fertilization and viviparity in some fishes and reptiles has resulted in the production of relatively small litters. The same trend is found in mammals. In the tenrec, *Hemicentetes ecaudatus,* and some marsupials as many as 20–40 embryos have been described at any one time (see Tripp 82); however, the armadillo, *Dasypus novemcinctus,* regularly exhibits polyembryony whereby several offspring are produced from a single fertilized egg (Hamlett 31). In the pronghorn, *Antilocapra americana,* seven or eight ovulations occur, but twin births are the rule because embryos are lost during the preimplantation stage of the expanded blastocyst (O'Gara 58).

In some elephant shrews (family Macroscelididae) the ovulation number is very much greater than the litter number. *Elephantulus myurus* gives birth to only two young but up to 60 eggs are ovulated from each ovary (van der Horst & Gillman 85). Some elephant shrews, however, produce normal numbers of eggs. There is a gradation rather than a complete separation between species which do exhibit polyovulation and those that do not (Tripp 82). The mature follicles in polyovulating animals are smaller than those of other elephant shrews; this is explicable on the physical grounds of getting them all in the ovary. There is no obvious correlation with the ecology; one can only assume that polyovulation represents some relict mechanism in ovulation procedures. There is no selection at the stage of fertilization, and all the eggs are fertilized. The specific litter size is achieved because only one part of each horn of the uterus is competent to accept a single blastocyst. The first zygote to reach this area excludes all others. This clearly is the reason why the fertilized egg reaches the uterus at an earlier stage of cleavage (4-cell) in elephant shrews than in any other species. Polyovulation can be excused in these shrews as a reptilian legacy, and the fact that all the eggs are fertilizable indicates that the correct mechanisms are acting at that stage.

Until recently the elephant shrews held the record for the largest number of eggs ovulated by a mammal, but in 1971 the strange case of the plains viscacha *(Lagostomus maximus),* an hystricomorph rodent related to the chinchilla, was reported (Weir 93). Ovulation in the plains viscacha appears to be spontaneous and at each ovulatory episode 200–800 eggs are produced. This has been confirmed in several animals, and indeed should have been deduced from the structure of the ovaries except that such a prodigality was almost beyond belief for a mammal. The surface of the ovary is deeply invaginated and the mature follicle is only 300 μm in diameter, small for the size of the animal (average bodyweight 3000 g). The combination of these factors clearly permits the development and ovulation of large numbers of eggs. However, the reason for this specialization is beyond speculation at the present time.

Unlike the elephant shrew eggs, only about 10% of viscacha eggs appear to be fertilized, but whether this is a defect of the male or the female gametes is unknown. No more than eight blastocysts implant, but only the one which implants nearest the cervix in each horn survives to term; the rest are resorbed between 26 and 35 days of pregnancy. The surviving embryo is that blastocyst which implanted first, perhaps an indication of the evolutionary pathway that the elephant shrews took. However, in many other aspects of its reproductive physiology the plains viscacha is a typical hystricomorph rodent (e.g. complete interstitial implantation and amniogenesis by cavitation) and can be considered highly evolved. Moreover, instead of getting zygotes to the uterus and implanted at the earliest possible stage, the plains viscacha appears to be the only hystricomorph so far studied in which there is a delay in implantation.

Thus, although the large number of eggs ovulated by tenrecs and some elephant shrews can be thought of as reminiscent of their reptilian origins, such a postulate does not appear to account for the excessive loss of eggs in the viscacha. There is no indication of such a trend in any of the other known hystricomorph species. The largest known litters (of 13) are those of the domestic guinea pig, which has almost certainly been selectively bred for a large litter size (Rood & Weir 69), and the coypu. It is not yet known whether the viscacha starts with a greater number of oocytes than do other mammals or whether the rate of atresia of follicles is less. The former seems more likely as the viscacha resembles other hystricomorphs (see page 153) in forming accessory corpora lutea from unovulated follicles. Thus at each ovulation over 1000 eggs are lost by ovulation and conversion into accessory corpora lutea, regardless of any others that may degenerate. No obvious advantages seem to occur by such a performance. The number of spermatozoa produced by the male seems normal and, although the low fertilization rate perhaps indicates that not all the eggs are mature, there is still an excess of fertilized eggs compared with the two young that are normally born (Weir 98). It is difficult to detect any unusual factor in the natural habitat of the viscacha that could account for these peculiarities. The viscacha is one of the few mammalian species indigenous to the pampas of Argentina. Apart from man the adults have no natural predators, although the pampas fox *(Dusicyon)* and the boa constrictor *(Constrictor constrictor)* may lie in wait for young animals. Viscacha appear to have lived in similar conditions in large burrow systems for centuries; such a constant environment would constitute little justification for a selection mechanism. The production of large numbers of eggs cannot be a direct legacy of reptilian ancestry because the change from the macrolecithal egg type of the monotremes and nonmammalian vertebrates to the microlecithal egg of eutherian mammals has occurred and because other aspects of the reproductive physiology typical of hystricomorphs have evolved. At present there is no answer as to why the viscacha ovulates so many eggs, and the question of how it does so could be an embarrassment to the currently postulated theories of ovulation.

CONCLUSIONS

The foregoing discussion has indicated that there are no obvious patterns of mammalian reproduction that can be correlated with either taxonomy or habitat.

The members of any one species can be expected to conform to the same pattern within the limits set by individual variation and local climatic variants. But this is an implicit tenet of our understanding of the species. The interest really focuses on the problem of which reproductive strategy or which individual stratagems can be considered representative of any higher taxa. Only for monotypic genera is it possible to be certain that all species of a genus have been investigated. If two species of a genus are found to be different in any respect then that particular factor is not a generic characteristic. For example, not all voles of the genus *Microtus* are induced ovulators, and one therefore cannot consider this pattern to be representative of all microtines. However, in most other features these voles do behave reproductively in very similar ways. This is not so for the mustelids, in which the variation of stratagems is much greater. Thus the stoat *(Mustela erminea)* has molded its reproduction around the phenomenon of delayed implantation, the ferret *(Mustela putorius)* and the weasel *(Mustela nivalis)* have ignored this device, and the mink *(Mustela vison)* has adopted delayed implantation as well as a system of superfetation. Although delayed implantation is more common in the Mustelidae than any other family, it is obviously not a definitive characteristic.

It is to be expected that as one considers a higher taxon the number of species with reproductive features in common should increase, but of course the proportion of the constituent taxa that have been investigated becomes less. The family Chinchillidae is one of the few for which members of all three known genera (two of them monotypic) have been studied; each displays completely different characteristics, the most bizarre being that of polyovulation in the plains viscacha (see p. 156) and the use of only one ovary in the mountain viscacha (Pearson 61). In spite of these peculiarities they show features that appear to be typical of the suborder Hystricomorpha. Other mammals may exhibit a long pregnancy for their size, a long estrous cycle, the tendency to form accessory corpora lutea, and even a type of vaginal closure membrane, but their combination and representation is minimal. The occurrence of all these characteristics in nearly all the hystricomorphs so far studied seems remarkable. The finding of a closure membrane, long estrous cycle, and a complex ovary in the African porcupine is perhaps one of the best indicators that the group is homogenous and should not be split into Hystricomorpha (Old World species) and Caviomorpha (New World species).

Therefore, although there may be very broad generalizations that cover the reproductive stratagems of related groups of mammals, there are few, if any, taxonomic reasons to explain the distribution of the unusual reproductive features. All reproductive stratagems known today must be considered successful or they would not be represented. But times and circumstances are changing and it is perhaps possible to indicate those patterns which seem to us flexible enough to cope with future fluctuations. It is often thought that marsupials represent a stage towards the evolutionary excellence of eutherian mammals, but they in fact represent an equal level of achievement on a different line. If anything, marsupials are more successful reproductively than are eutherians (Sharman 74). The ubiquity and variation (though very slight) of the myomorph rodents, particularly the murids and cricetids, is at first difficult to explain, but they are enormously efficient in numbers and frequency of young produced. The small scale on which most of these rodents live and their short generation time is probably the reason for their conspicuous speciation. Competition from other species has most likely been great and evolutionary success in being able to radiate quickly has probably been thrust upon them. An indication of the reproductive potential of myomorphs is found in the frequently reported population explosions of some species. The reasons for the disruption of the balance between production and survival are not clearly known; most workers have concentrated on the declines, often rapid, of crowded populations. A build-up of numbers is associated with features such as plentiful food, but how animals know that this will occur is not clear.

How animals obtain this apparent advance knowledge of how to order their reproductive lives is one of the most intriguing puzzles. Those species that link their reproductive timing to environmental features (Sadleir 73) can equally well be caught out by unusual climatic conditions, such as cold weather in the late spring and early summer, as can those who have a fixed reproductive pattern that must be completed once it has started.

We have tried to show how some of the stratagems observed can be related to the urgency of reproduction, but, of course, we cannot know what were the determining events during the evolution of mammals. We can only hope that a survey of the species available today reveals some of the stages through which species may have passed. It appears to us that from a basic reptilian type of pattern, mammalian reproduction was faced with several possible ways to develop, and that each species has evolved some or all of the way along some paths but not others. The combination of the progress along each facet represents the reproductive pattern achieved by any one species at the present time. Although some of our speculations may be valid, we shall never be certain of the origin or necessity of the various strategies which have been discussed.

Literature Cited

1. Allanson, M., Rowlands, I. W., Parkes, A. S. 1934. Induction of fertility and pregnancy in the anoestrous ferret. *Proc. Roy. Soc. B* 115:410–21
2. Amoroso, E. C. 1956. Placentation. *Marshall's Physiology of Reproduction,* ed. A. S. Parkes, 2:Chap. 15, 127–311. London: Longmans Green. 3rd ed. 880 pp.
3. Amoroso, E. C., Finn, C. A. 1962. Ovarian activity during gestation, ovum transport and implantation. *The Ovary,* ed. S. Zuckerman, 1:Chap. 9, 451–537. New York: Academic. 619 pp.
4. Aron, C., Asch, G., Roos, J., Luxembourger, M. M. 1964. Influence de l'intensité du stimulus copulatoire sur declenchment, par le coit, de phenomè-

nes de luteinisation ou ovulationes, chez la Ratte. *C. R. Soc. Biol.* 158:126–9

5. Asdell, S. A. 1964. *Patterns of Mammalian Reproduction.* Ithaca, New York: Cornell Univ. Press. 2nd ed. 670 pp.

6. Baker, T. G. 1972. Gametogenesis. *Acta Endocrinol. Copenhagen Suppl.* 166:18–41

7. Berger, P. J. 1966. Eleven-month "embryonic diapause" in a marsupial. *Nature London* 211:435–6

8. Brambell, F. W. R. 1956. Ovarian changes. *Marshall's Physiology of Reproduction,* ed. A. S. Parkes, 1:Chap. 5, 397–542. London: Longmans Green. 3rd ed. 688 pp.

9. Bruce, H. M. 1959. An exteroceptive block to pregnancy in the mouse. *Nature London* 184:105

10. Bruce, H. M. 1970. Pheromones. *Brit. Med. Bull.* 26:10–13

11. Buchanan, G. D., Enders, A. C., Talmage, R. V. 1956. Implantation in armadillos ovariectomized during the period of delayed implantation. *J. Endocrinol.* 14:121–8

12. Buechner, H. K., Morrison, J. A., Leuthold, W. 1966. Reproduction in the Uganda kob with special reference to behavior. *Symp. Zool. Soc. London* 15: 69–88

13. Butler, C. G., Callow, R. K., Johnston, N. C. 1961. The isolation and synthesis of queen substance, 9-oxodec-*trans*-2-enoic acid, a honey bee pheromone. *Proc. Roy. Soc. B* 155:417–32

14. Canivenc, R. 1966. A study of progestation in the European badger (*Meles meles* L.). *Symp. Zool. Soc. London* 15:15–26

15. Canivenc, R., Bonnin-Laffargue, M., Lajus, M. 1967. Preuve expérimentale de la réactivation lutéale chez le Blaireau européen (*Meles meles* L.). *C. R. H. Acad. Sci. Ser. D* 264:1486–9

16. Canivenc, R., Bonnin-Laffargue, M., Lajus-Boue, M. 1969. Induction de nouvelles générations lutéales pendant la progestation chez la martre européenne (*Martes martes* L.). *C. R. H. Acad. Sci. Ser. D* 269:1437–40

17. Canivenc, R., Bonnin-Laffargue, M., Lajus-Boue, M. 1971. Réalisation expérimentale précoce de l'ovoimplantation chez le blaireau européen (*Meles meles* L.) pendant la periode de latence blastocystaire. *C. R. H. Acad. Sci. Ser. D* 273:1855–7

18. Conaway, C. H. 1971. Ecological adaptation and mammalian reproduction. *Biol. Reprod.* 4:239–47

19. Conaway, C. H., Sorenson, M. W. 1966. Reproduction in tree shrews. *Symp. Zool. Soc. London* 15:471–92

20. Dawson, M. R. 1967. Fossil history of the families of recent mammals. *Recent Mammals of the World,* ed. Sydney Anderson, J. Knox-Jones, Jr., 12–53. New York: Ronald Press. 453 pp.

21. Enders, A. C., Ed. 1963. *Delayed Implantation.* Chicago: Chicago Univ. Press. 318 pp.

22. Enders, A. C. 1966. The reproductive cycle of the nine-banded armadillo (*Dasypus novemcinctus*). *Symp. Zool. Soc. London* 15:295–310

23. Ebling, F. J. 1963. Hormonal control of sebaceous glands in experimental animals. *Advan. Biol. Skin* 4:200–19

24. Erickson, C. J., Lehrman, D. S. 1964. Effect of castration of male ring doves upon ovarian activity of females. *J. Comp. Physiol. Psychol.* 58:164–6

25. Everett, J. W. 1971. The mammalian female reproductive cycle and its controlling mechanisms. *Sex and Internal Secretions,* ed. W. C. Young, 1:Chap. 8, 497–555. Baltimore: Williams and Wilkins. 3rd ed. 704 pp.

26. Finn, C. A. 1971. The biology of decidual cells. *Advan. Reprod. Physiol.* 5:1–26

27. Fischer, T. V., Mossman, H. W. 1969. The fetal membranes of *Pedetes capensis* and their taxonomic significance. *Am. J. Anat.* 124:89–116

28. George, W., Weir, B. J. 1972. Chromosome studies in some members of the family Caviidae (Mammalia: Rodentia). *J. Zool.* 168:81–9

29. Greenwald, G. S. 1968. Failure of hypophysectomy to affect regression of cyclic hamster corpus luteum. *J. Reprod. Fert.* 16:495–7

30. Gulamhusein, A. P. 1973. *Reproductive studies in the Mustelidae.* PhD thesis. Univ. London. 250 pp.

31. Hamlett, G. W. D. 1932. The reproductive cycle in the armadillo. *Z. Wiss Zool. Abt. A* 141:143–54

32. Hanks, J., Short, R. V. 1972. The formation and function of the corpus luteum in the African elephant, *Loxodonta africana. J. Reprod. Fert.* 29:79–89

33. Herbert, J. 1972. Initial observations on pinealectomized ferrets kept for long periods in either daylight or artificial illumination. *J. Endocrinol.* 55: 591–7

34. Hoar, W. S. 1965. Comparative physiology: hormones and reproduction in fishes. *Ann. Rev. Physiol.* 27:51–70

35. Huggett, A. St. G., Widdas, W. F. 1951. The relationship between mammalian foetal weight and conception age. *J. Physiol. London* 114:306–17

36. Illingworth, D. V., Deanesly, R. 1972. Maintenance of pregnancy by synthetic progestagens in guinea-pigs ovariectomized before implantation; progesterone-binding protein and placental secretion. *J. Endocrinol.* 54:435–44

37. Labhsetwar, A. P., Enders, A. C. 1968. Progesterone in the corpus luteum and placenta of the armadillo, *Dasypus novemcinctus. J. Reprod. Fert.* 16:381–7

38. Landry, S. O. 1957. The interrelationships of New and Old World hystricomorph rodents. *Univ. Calif. Publ. Zool.* 56:1–118

39. Leblond, C. P., Clermont, Y. 1952. Definition of the stages of the cycle of the seminiferous epithelium in the rat. *Ann. NY Acad. Sci.* 55:548–73

40. Lee, S. van der, Boot, L. M. 1955. Spontaneous pseudopregnancy in mice. *Acta Physiol. Pharmacol. Neer.* 4:442–3

41. Lincoln, G. A., Guinness, F. 1972. Effect of altered photoperiod on delayed implantation and moulting in roe deer. *J. Reprod. Fert.* 31:455–7

42. Lincoln, G. A., Youngson, R. W., Short, R. V. 1970. The social and sexual behaviour of the red deer stag. *J. Reprod. Fert. Suppl.* 11:71–103

43. Maliniak, E., Eisenberg, J. F. 1971. Breeding spiny rats, *Proechimys semispinosus,* in captivity. *Int. Zoo Yearb.* 11:93–8

44. Martinet, L., Legouis, J.-J., Moret, B. 1970. Quelques observations sur la reproduction du lièvre européen (*Lepus europaeus* Pallas) en captivité. *Ann. Biol. Anim. Biochim. Biophys.* 10:195–202

45. Mead, R. A. 1968. Reproduction in eastern forms of the spotted skunk (genus *Spilogale*). *J. Zool.* 156:119–36

46. Mead, R. A. 1968. Reproduction in western forms of the spotted skunk (genus *Spilogale*). *J. Mammal.* 49:373–90

47. Mead, R. A. 1971. Effects of light and blinding upon delayed implantation in the spotted skunk. *Biol. Reprod.* 5:214–220

48. Melampy, R. W., Anderson, L. L. 1968. Role of the uterus in corpus luteum function. *J. Anim. Sci.* 27, *Suppl.* 1:77–96

49. Michael, R. P., Keverne, E. B. 1968. Pheromones in the communication of sexual status in primates. *Nature London* 218:746–9

50. Moor, R. M., Rowson, L. E. A. 1966. The corpus luteum of the sheep: functional relationship between the embryo and the corpus luteum. *J. Endocrinol.* 34:233–9

51. Moor, R. M., Rowson, L. E. A. 1966. The corpus luteum of the sheep: effect of the removal of embryos on luteal function. *J. Endocrinol.* 34:497–502

52. Mossman, H. W. 1937. Comparative morphogenesis of the fetal membranes and accessory uterine structures. *Contr. Embryol. Carnegie Inst.* 26:126–246

53. Mossman, H. W. 1966. The rodent ovary. *Symp. Zool. Soc. London* 15:455–70

54. Mossman, H. W., Conaway, C. H. 1954. A new type of placentation demonstrating the phylogenetic and taxonomic significance of the fetal membranes. *Anat. Rec.* 118:431–2

55. Mossman, H. W., Judas, I. 1949. Accessory corpora lutea, luteal cell origin and the ovarian cycle in the Canadian porcupine. *Am. J. Anat.* 85:1–39

56. Mutere, F. A. 1967. The breeding biology of equatorial vertebrates: reproduction in the fruit bat, *Eidolon helvum,* at latitude 0°20'N. *J. Zool.* 153:153–61

57. Mykytowycz, R. 1970. The role of skin glands in mammalian communication. *Advan. Chemoreception* 1:327–60

58. O'Gara, B. W. 1969. Unique aspects of reproduction in the female pronghorn (*Antilocapra americana* Ord.). *Am. J. Anat.* 125:217–32

59. Pedersen, T. 1970. Follicle kinetics in the ovary of the cyclic mouse. *Acta Endocrinol. Copenhagen* 64:304–23

60. Pearson, O. P. 1944. Reproduction in the shrew (*Blarina brevicauda,* Say). *Am. J. Anat.* 75:39–93

61. Pearson, O. P. 1949. Reproduction of a South American rodent, the mountain viscacha. *Am. J. Anat.* 84:143–74

62. Perry, J. S. 1971. *The Ovarian Cycle of Mammals.* Edinburgh: Oliver & Boyd. 219 pp.

63. Perry, J. S., Rowlands, I. W. 1962. The ovarian cycle in vertebrates. *The Ovary,* ed. S. Zuckerman, 1:Chap. 5, 275–309. New York & London: Academic. 619 pp.

64. Racey, P. A. 1972. Viability of bat spermatozoa after prolonged storage in the epididymis. *J. Reprod. Fert.* 28:309–11

65. Racey, P. A. 1973. Factors affecting the length of gestation in heterothermic bats. *J. Reprod. Fert. Suppl.* 19:175–89

162   WEIR & ROWLANDS

66. Racey, P. A., Potts, D. M. 1970. Relationship between stored spermatozoa and the uterine epithelium in the pipistrelle bat *(Pipistrellus pipistrellus). J. Reprod. Fert.* 22:57–63
67. Rahm, U. 1962. L'élevage et la reproduction en captivité de l'*Atherurus africanus. Mammalia* 26:1–9
68. Richmond, M., Conaway, C. H. 1969. Induced ovulation and oestrus in *Microtus ochrogaster. J. Reprod. Fert. Suppl.* 6:357–76
69. Rood, J. P., Weir, B. J. 1970. Reproduction in female wild guinea-pigs. *J. Reprod. Fert.* 23:393–409
70. Rowlands, I. W. 1949. Post-partum breeding in the guinea-pig. *J. Hyg.* 47:281–7
71. Rowlands, I. W. 1956. The corpus luteum of the guinea-pig. *Ciba Found. Colloq. Ageing* 2:69–83
72. Rowlands, I. W., Tam, W. H., Kleiman, D. G. 1970. Histological and biochemical studies on the ovary and of progesterone levels in the systemic blood of the green acouchi *(Myoprocta pratti). J. Reprod. Fert.* 22:533–45
73. Sadleir, R. M. F. S. 1969. *The Ecology of Reproduction in Wild and Domestic Animals.* London: Methuen. 321 pp.
74. Sharman, G. B. 1970. Reproductive physiology of marsupials. *Science* 167:1221–8
75. Short, R. V., Hay, M. F. 1966. Delayed implantation in the roe deer, *Capreolus capreolus. Symp. Zool. Soc. London* 15:173–94
76. Signoret, J. P. 1970. Reproductive behaviour of pigs. *J. Reprod. Fert. Suppl.* 11:105–17
77. Simpson, G. G. 1945. The principles of classification and a classification of mammals. *Bull. Am. Mus. Natur. Hist.* 85:1–350
78. Smeaton, T. C., Robertson, H. A. 1971. Studies on the growth and atresia of the Graafian follicles in the ovary of the sheep. *J. Reprod. Fert.* 25:243–52
79. Smythe, N. 1970. Relationships between fruiting seasons and seed dispersal methods in a neotropical forest. *Am. Natur.* 104:25–35
80. Strauss, F. 1939. Die bildung der corpus luteum bei centetiden. *Biomorphosis* 1:489–544
81. Tam, W. H. 1970. The function of the accessory corpora lutea in hystricomorph rodents. *J. Endocrinol.* 48:liv–lv
82. Tripp, H. R. H. 1971. Reproduction in elephant shrews (Macroscelididae) with special reference to ovulation and implantation. *J. Reprod. Fert.* 26:149–59
83. Vandenbergh, J. G. 1973. Acceleration and inhibition of puberty in female mice. *J. Reprod. Fert. Suppl.* 19:409–17
84. Vandenbergh, J. G., Drickamer, L. C., Colby, D. R. 1972. Social and dietary factors in the sexual maturation of female mice. *J. Reprod. Fert.* 28:397–405
85. van der Horst, C. J., Gillman, J. 1940. Ovulation and corpus luteum formation in *Elephantulus. S. Afr. J. Med. Sci.* 5:73–91
86. Walker, E. P. 1964. *Mammals of the World.* Baltimore: Johns Hopkins. Vols. I & II, 1500 pp.; Vol. III, 769 pp.
87. Weir, B. J. 1966. Aspects of reproduction in chinchilla. *J. Reprod. Fert.* 12:410–11
88. Weir, B. J. 1967. *Aspects of reproduction in some hystricomorph rodents.* PhD thesis. Univ. Cambridge, England. 314 pp.
89. Weir, B. J. 1967. The care and management of laboratory hystricomorph rodents. *Lab. Anim.* 1:95–104
90. Weir, B. J. 1969. The induction of ovulation in the chinchilla. *J. Endocrinol.* 43:55–60
91. Weir, B. J. 1970. The management and breeding of some more hystricomorph rodents. *Lab. Anim.* 4:83–97
92. Weir, B. J. 1970. *Chinchilla. Reproduction and breeding techniques for laboratory animals,* ed. E. S. E. Hafez, Chap. 11, 209–223. Philadelphia: Lea & Febiger. 375 pp.
93. Weir, B. J. 1971. The reproductive organs of the female plains viscacha, *Lagostomus maximus. J. Reprod. Fert.* 25:365–73
94. Weir, B. J. 1971. Some observations on reproduction in the female agouti, *Dasyprocta aguti. J. Reprod. Fert.* 24:203–11
95. Weir, B. J. 1971. Some observations on reproduction in the female green acouchi, *Myoprocta pratti. J. Reprod. Fert.* 24:193–201
96. Weir, B. J. 1971. The evocation of oestrus in the cuis, *Galea musteloides. J. Reprod. Fert.* 26:405–8
97. Weir, B. J. 1971. Some notes on reproduction in the Patagonian mountain viscacha, *Lagidium boxi* (Mammalia: Rodentia). *J. Zool.* 164:463–7
98. Weir, B. J. 1971. The reproductive physiology of the plains viscacha, *Lagostromus maximus. J. Reprod. Fert.* 25:355–63

99. Weir, B. J. 1973. Another hystricomorph rodent: keeping casiragua *(Proechimys guairae)* in captivity. *Lab. Anim.* 7:125–34

100. Weir, B. J. 1973. The rôle of the male in the evocation of oestrus in the cuis, *Galea musteloides. J. Reprod. Fert. Suppl.* 19:419–30

101. Weir, B. J. 1973. The induction of ovulation and oestrus in the chinchilla. *J. Reprod. Fert.* 33:61–8

102. Whitten, W. K. 1956. Modification of the oestrous cycle of the mouse by external stimuli associated with the male. *J. Endocrinol.* 13:399–404

103. Whitten, W. K. 1966. Pheromones and mammalian reproduction. *Advan. Reprod. Physiol.* 1:155–77

104. Whitten, W. K. 1973. Genetic variation of olfactory function in reproduction. *J. Reprod. Fert. Suppl.* 19:403–8

105. Wimsatt, W. A. 1949. Glycogen, polysaccharide complexes and alkaline phosphatase in the ovary of the bat during hibernation and pregnancy. *Anat. Rec.* 103:564–5

106. Wimsatt, W. A. 1969. Some interrelations of reproduction and hibernation in mammals. *Symp. Soc. Exp. Biol.* 23: 511–49

107. Wood, A. E. 1955. A revised classification of the rodents. *J. Mammal.* 36: 165–87

108. Wood, A. E., Patterson, B. 1959. The rodents of the Deseadan Oligocene of Patagonia and the beginnings of the South American rodent radiation. *Bull. Mus. Comp. Zool.* 120:281–428

109. Woolley, P. 1966. Reproduction in *Antechinus* spp. and other dasyurid marsupials. *Symp. Zool. Soc. London* 15:281–94

THE EPISTASIS CYCLE:
A THEORY OF
MARGINAL POPULATIONS

❖ 4057

Michael Soulé

Department of Biology, University of California at San Diego, La Jolla, California

> *Many laws regulate variation, some few of which can be dimly seen . . .*
> Charles Darwin, *The Origin of Species* (1859)

INTRODUCTION

The subject of this speculative essay-review is the genetic structure of marginal populations. Admittedly, it is difficult to define a marginal population. A somewhat intuitive definition would be a population exposed to an extreme of one or more relevant environmental variables. In practice, though, such a nonoperational definition is of little use, and it would be better if some parameter of population performance served as a measure of marginality. One such definition would hinge on population dynamics: a marginal population is one characterized by relatively great fluctuations in numbers and a relatively high probability of extinction. (Here, as elsewhere, I am using population in the sense of a local group or deme, more or less isolated from other local groups.) Another such definition might be based on "population statics": a marginal population is one in which the individuals are relatively sparsely distributed and show effects of physiological stress, e.g. starvation, dehydration, or stunting. I prefer the dynamic definition.

Others (2, 26, 71) have discussed the distinctions between marginal and central populations, and I will not belabor them as such distinctions inevitably are arbitrary. Two points, however, should be made clear. First, not all marginal populations are peripherally located. Topographic relief can impose marginal conditions in the geographic center of a range by producing deserts, rain shadows, and various altitudinal effects. Second, not all peripheral populations are ecologically marginal.

In any case, such definitions are rarely employed with rigor. In virtually all of the studies herein discussed, marginality is assumed, not shown, by the author or me. In many cases the terms marginal and peripheral are more or less equated,

165

especially where no obvious geographic or physiographic barrier prevents range expansion.

Why are we interested in marginal conditions and populations? Stress forces biological systems to react in ways that often expose processes and mechanisms not otherwise observable. Just as Waddington (103) was able to analyze developmental canalization by employing heat shock to produce crossveinless *Drosophila,* so the population biologist hopes to learn about the roles of gene flow, inbreeding, and selection in the process of speciation by studying marginal (perturbed) populations. Marginal populations might also allow us to test the idea that gene flow is a cohesive force holding together the species and causing it to evolve as a unit (73, pp. 297–301), an idea falling into some disrepute (43, 44). Finally, such populations may yield unexpected fruits, such as information on the role of heterozygosity at both the allele and chromosomal levels.

For the purposes of this review I have not considered those studies dealing with nonmarginal, isolated populations, such as those on islands. They are mentioned only when they shed light on problems arising from analyses of marginal populations. While islands offer unique challenges to colonizers, they often differ from marginal habitats in having a qualitatively rather than quantitatively different climate and biota. The frequent maintenance of high population densities on islands is further evidence against marginality in the sense described above.

For similar reasons the human literature is rarely mentioned in this review. If humans have marginal populations in the sense defined, then there is little or no data on their genetics. In contrast there is a wealth of data on human isolates (29, 104), but these are mainly of interest in examining the interplay of inbreeding, genetic drift, and selection, not the effects of rigorous environmental conditions on variation.

Finally, one definitional matter must be clarified. When referring to morphological characters (excluding cytological ones), the term polymorphism refers to characters with clearly differentiated phenotypes, such as the color morphs of mimetic butterflies. The term is not applied to continuously varying morphological characters. Briefly stated, all polymorphism is variation, but not all variation is polymorphism.

To avoid an overly lengthy bibliography reviews are often cited in lieu of listing separate papers.

Factors Accounting for Loss of Variation in Marginal Populations

For several reasons, one of the first questions naturalists ask about a marginal population is whether it is less variable than the central area for genetic polymorphisms or morphological characters. We should be clear, therefore, on what factors might account for a diminution of variation. A nonexhaustive list of answers follows:

1. *Inbreeding*: As individuals or stands occur farther apart the pool of potential mates declines and the probability of mating with a close relative is increased.

2. *Gene flows*: Demes may be farther apart toward the margin because favorable habitat is sparsely distributed. Also, the net reproductive rate will be low. Hence,

there will be fewer propagules produced and a reduced probability of interdeme migration per propagule. Consequently, each local population is more genetically isolated.

3. *Genetic drift*: The above arguments imply a small effective population size, N_e, in marginal localities. Because of the harmonic mean effect (107), severe, periodic restriction of population size (bottlenecks) due to bad weather, famine, etc, will decrease the population far below the actual size found when optimum conditions prevail. Drift, consequently, will cause a decrease of variability, at least at the genetic level.

4. *Population effective size*: If a significant portion of morphological and genetic variation is neutral, then marginal populations should be relatively depauperate of variation. The reasoning behind this statement is based on the formulations of Kimura & Crow (60) and Kimura (59) showing that the effective number of neutral alleles in a population (n_e) is equal to the expression $4N_e\mu + 1$ when N_e is the effective size of the population and μ is the mutation rate for neutral alleles. If any one of points 1, 2, and 3 above are valid, it follows that there will be a smaller effective population in marginal localities.

5. *The niche width-variation theory*: The Ludwig (67; 72, p. 245) or niche width-variation hypothesis predicts a correlation between ecological amplitude and genetic and sometimes morphological variation. This idea is discussed in some detail below; at this stage it is mentioned only because it also predicts less variation in marginal habitats because the range of resources and habitats is typically less in a marginal environment. According to the proponents of this theory, fewer genes, gene arrangements, and less morphological variability will be maintained by selection in such a regime.

6. *Directional selection*: A marginal environment is, by definition, different from the central, optimal, environment. Whereas stabilizing selection will be the dominant mode in the latter environment, directional selection will prevail in the former. The result might be an attrition of genetic variation while the gene pool is being reorganized. The opposite could also occur. The breakdown of canalized developmental pathways could cause a release of cryptic variability under these circumstances, a possibility discussed by Guthrie (52), Soulé & Stewart (92), Huether (54), and Levin (64). So far, there is very little evidence for the occurrence of this phenomenon, though some morphometric studies in isolated and inbred human populations suggest it (79, 5, 6).

To summarize, all but the last genetic argument predict the same result in marginal populations, even when the arguments are mutually exclusive. This situation is not conducive to hypothesis testing.

ARE MARGINAL POPULATIONS LESS VARIABLE?

INVERSIONS   Table 1 lists all chromosomally polymorphic *Drosophila* species for which I was able to find sufficient data. The list shows a central-marginal or central-peripheral decline of paracentric inversion polymorphism in nearly all species. I found only a single case in which a chromosomally polymorphic species of *Droso-*

Table 1 The occurrence of marginal paucity and uniqueness of inversions in species of *Drosophila*

Species	Marginal (M) or Isolated (I)	Marginal Paucity Present (+) Absent (−)	Unique Arrangements Present (+) Absent (−)	Reference
D. rubida	I	+	+	69
D. flavopilosa	M	+	−	20
D. pavani	M	−	−	19
D. immigrans	M	+	−	18
D. subobscura	M, I	+	−	50
D. pseudoobscura	M, I	+	−	42, 26
D. persimilis	M	+	?	37, 22
D. funebris	M	+	−	39
D. nigromelanica	M	+	−	96
D. euronotus	M	+	−	95
D. robusta	M	+	−	23, 25
D. americana	M	+	−	Ref. in 28
D. pallidapenis	M	+	−	26
D. acutilabella	M	+	−	28
D. willistoni	M, I	+	+	35
D. nebulosa	M	+	−	27

phila does not show a reduction in polymorphism somewhere along the periphery of its range. In some cases the depauperization occurs along an altitudinal transect as well (26, 20). Cosmopolitan species, most of which are closely associated with human garbage or crops, are typically poor in chromosomal polymorphism throughout their ranges (27, 39). For this reason, and because marginality is not easily defined, these species (with the exception of *D. immigrans*) are not tabulated.

An interesting point to emerge from this table is the rarity of novel or unique gene arrangements in marginal populations. New arrangements appear to occur only in peripheral isolates of *D. rubida* and *D. willistoni*. The same is true for *D. birchii* (7), but the isolates may be reproductively isolated. It must be assumed that better sampling will effect some changes in these patterns.

ALLELES   Analysis of allelic variation in space yields a very different pattern from that seen for *Drosophila* inversions. In Table 2 the results are tabulated from several studies in which marginal paucity of genetic variation could be detected if it occurred. For reasons given above, island studies are not included, nor are studies of organisms for which samples from an ecological margin were not apparent [such as those of *Mus* (86) and *Spalax* (78)]. Only four out of the eleven organisms tabulated show a margin effect, and all four are species of North American vertebrates. Further, in all four it is the northern (or high altitude in the case of *Uta*) populations which show the reduction.

Table 2   The occurrence of paucity of allelic diversity in marginal populations

Species	Reduced Allelic Diversity Presence (+) Absence (−)	Reference
Drosophila robusta[a]	−	76
Drosophila pseudoobscura	−	81, 42
Drosophila obscura	−	62
Drosophila subobscura[b]	−	63
Drosophila willistoni[c]	−	3, 4
Nassarius obsoletus	−	48
Acris gryllus	+	36
Uta stansburiana	+	75
Peromyscus polionotus	+	87
Dipodomys merriami	+	55

[a]Based on genetic load.
[b]Questionable.
[c]Very slight reduction.

MORPHOLOGY   Very few morphological studies provide data bearing on relative variability in marginal populations. Mayr (72, p. 386) briefly reviews the older literature; references therein and other studies, such as the work on *Papilio dardanus* (30), often document the expected marginal loss of morphs in polymorphic species. However, what evidence there is on variation in the strict sense is not always easily interpretable. Power (80) has shown that there is no drop in morphometric variation in the Brewer blackbird in a region of rapid range expansion. Of course, a rapid range bulge is prima facie evidence against the assumption of ecological marginality. Agnew's data (1) on the perennial plant *Lysimachia volkensii* do not show a completely consistent reduction of morphological variability towards the periphery of the study site. Virtually all the other studies I have found are analyses of well-isolated populations rather than marginal or peripheral ones. No more will be said here about morphological variation.

The Pattern of Inversion Variation

THE LUDWIG HYPOTHESIS   The marginal-peripheral paucity of gene arrangements in *Drosophila* is not blessed by a paucity of explanations. Dobzhansky, da Cunha, Townsend, and their co-workers first elucidated the phenomenon in *D. willistoni* (41, 34, 38, 35, 101). These workers employed the Ludwig hypothesis as the explanation: "In general, the more polymorphic a species is, the more environments it can use or control" (41). By employing some of their data (34, 35), they were able to show a positive correlation between environmental complexity and inversion heterozygosity. Appeal was also made to the founder effect to explain the reduction in kinds of gene arrangements in peripherally isolated populations in the West Indies and Florida (35, 3). A third explanation, the gene flow hypothesis, has also been suggested to account for the reduction of inversion polymorphism in peripheral populations that are not ecologically marginal (20, 91). A fourth hypothesis, the homoselection-heteroselection theory of Carson, has also been proposed (22).

There are three criticisms one may make of the application of the classical Ludwig effect to gene arrangements in *Drosophila* species. The following three points compose the critique:

1. Too little is known about the ecology of *Drosophila* species to support this argument. For example, *D. willistoni* in the Amazonian Basin exploits many different fruits as breeding sites. If a particular inversion is a genetic device for coping with a particular niche, then some kind of correlation might be expected between the species of fruit and the genotypes of the flies ovipositing or developing in the fruit. At least certain inversions should do better on certain fruit and the geographic distribution of inversions should be related to the distribution of certain tree species. No such data have been reported. Carson (26) discussed the lack of evidence for the association of particular inversions with particular niches. This is not to say that particular gene arrangements are ecologically irrelevant. The classic field studies of Dobzhansky and others and population cage experiments are convincing of this relationship; these are reviewed by Dobzhansky (40). It has also been shown that gene arrangements from central populations fail to thrive when injected into a marginal population (32). The important distinction, however, is whether, on the one hand, these cytological polymorphisms help adapt their carriers to the average conditions a fly can expect in its lifetime or whether, on the other hand, the gene arrangements are very specific in their effects, such that the bearer of inversion *A* will more efficiently exploit niche alpha, the bearer of inversion *B*, niche beta, and so on. If the former is generally the function, then the degree of polymorphism (the number of inversions) should be independent of environmental diversity (e.g. food sources).

The proponents of the niche width-variation theory might, with some justification, rejoin that I have set up a straw man by taking their theory to a reductio ad absurdum, namely, the one gene arrangement—one niche hypothesis. One might respond that the hypothesis in this form is at least testable, and that unless the proponents can come up with a clearer statement of the genetic and physiological nature of the processes involved, the criticism is not unwarranted.

2. There are sibling species of *Drosophila willistoni (D. equinoxialis, D. tropicalis)* that occupy the ecologically diverse region from Central America to southern Brazil but apparently have lower levels of chromosomal polymorphism than *D. willistoni* (41). On the other hand, there are temperate species that have at least the number of inversions found in the tropical species; *D. montana*, for example, has thirty (97). Are the niches of, say, *D. tropicalis* and *D. equinoxialis* less wide than that of *D. montana?*

White (106) makes the same point, noting that the grasshopper, *Trimerotropis thalassica,* a species restricted to the chaparral plant *Adenostoma* in California, is polymorphic for 10 out of its 12 chromosomes. Another species in the same genus has no cytological polymorphism but occupies many habitats from Vancouver to

Central Mexico. He further states that "many such instances could be cited, and it does not seem that there is any very close correlation between the extent to which mechanisms of cytological polymorphism are developed in a species and the variety of habitats occupied."

An objection that should always be raised to this class of criticism is that it is not legitimate to employ interspecific comparisons because different phenomena could be responsible for the effects. For instance, the relevant "niches dimensions" of *D. montana* may not be the same kinds of variables as the "niches dimensions" of *D. equinoxialis*. Nevertheless, it would be nice if we had a theory of sufficient generality to account for patterns at least among closely related species.

3. Many population cage studies are not easily interpreted in terms of the Ludwig hypothesis. The laboratory environment is much more uniform than any natural environment for almost any variable imaginable. It is also manifestly different. Yet populations of *Drosophila* in such environments often tenaciously conserve the gene arrangement polymorphism of their founders (reviewed in 40), even though some loss of inversions can occur in long-term experiments (24, 65). There are several explanations for the preservation of inversions in laboratory populations. One of them is "mutual facilitation." For instance, Beardmore (9) found that the fitness measured by larval viability of a population is greater when the larvae are a mixture of karyotypes, and that this effect was not attributable to heterosis. On the other hand many workers have shown that heterosis of inversions exists and can indeed result in stable equilibria in laboratory populations (74, 25, 93, 94, 102, 33, 98). None of these results, however, supports or detracts from the hypothesis that chromosomal polymorphism is more likely to evolve in ecological generalists than in ecological specialists.

Though not dealing with inversions, some experiments by Beardmore and his co-workers (10–12, 66) are relevant in this context. They have shown that populations of *Drosophila melanogaster* retain more genetic variance and are more productive in fluctuating temperature regimes than in a constant one. Long, for example, kept populations for two years in four regimes: fluctuating temperature from 20–30°C with periods of 1, 32, and 96 days, and constant temperature at 25°C. He ranked their fitness based on productivity, competitive ability, and adaptability; it decreased in the above order. He also estimated genetic variance and found, in general, a positive correlation between fitness and variance. Though the results lacked significance, the population cycling every 24 hrs consistently outperformed the others.

These authors all interpret their results in terms of the Ludwig hypothesis, namely that temporal heterogeneity favors a greater assortment of genes. I would favor the alternative explanation that the 24 hr periodicity is the most "familiar" regime to the flies and from a selective viewpoint is the least directional. The higher fitness is simply the result of less genetic attrition and change compared to the populations in the more unnatural, and, therefore, rigorous constant and long-term fluctuation regimes. A corollary of this hypothesis is that control populations in population cage experiments should be maintained with fluctuating temperature and possibly light. The results of Thomson (99) are consistent with this.

CARSON'S HYPOTHESIS   Carson has proposed another model to account for marginal homozygosity of inversions. He argued that in the central, putatively optimal, part of the range, structurally heterozygous individuals will be fitter than individuals which are relatively more homozygous for gene arrangements because such optimum conditions place a premium on heterosis ("general vigor" and "heterotic buffering"), and heterosis is maximized by inversion heterozygosity. Carson sees an inversion as a device to capture simple luxuriance while reducing the genetic load. At the ecological margin, natural selection will tend to be more directional, thus favoring certain gene arrangements (those more fit as homozygotes) over others. A consequence of this kind of selection regime is a loss of some or most gene arrangements and a concommitant increase in the frequency of individuals homozygous for others. Because recombination is restricted in *Drosophila* inversion heterozygotes, these marginal, homoselected populations will have higher rates of recombination than will central heteroselected populations. This recombination will produce novel linear combinations of alleles and these will be instrumental in adapting the marginal population to the rigorous marginal habitat. Hence, according to Carson, while marginal homozygosity permits selection to operate more effectively, it is not adaptive per se.

The last point is often misunderstood. Carson views the relative structural homozygosity of marginal populations as an effect of directional selection. "Specific adaptations" are selected, and the by-product of their selection is structural homozygosity and a more open genetic system.

The principal difference between Carson's theory and the Dobzhansky-da Cunha version of the Ludwig hypothesis is the role of heterozygosity for gene arrangements. Carson sees them as means of maximizing general vigor via heterosis in central populations, rather than conferring specific adaptive properties. Dobzhansky and da Cunha apparently see them as supergenes with ecotypic effects, and inversion polymorphism as a mechanism by which "populations exploit a greater variety of ecological niches" (35). They do not say how this is accomplished, however. The two hypotheses are not mutually exclusive; neither are they mutually supporting.

Another distinction between the two theories is that Carson's is a theory of the early stages of speciation, while the Ludwig hypothesis is a theory of adaptation. Indeed, Carson views his theory as providing a mechanism (homoselection) for the "genetic revolution" that occurs, according to Mayr (70), as a consequence of the founder effect.

Stone et al (97) have criticized Carson's hypothesis by suggesting that the actual amount of recombination in highly polymorphic populations may be greater than that in relatively monomorphic populations since (*a*) inversions increase recombination elsewhere, and (*b*) there will be, in absolute numbers, more inversion homozygotes in central populations and consequently more recombination in central populations in absolute terms.

For reasons in part discussed by Mayr (71) and elaborated below, this criticism may be irrelevant, but it does serve to focus our attention on the essential prediction of Carson's hypothesis, namely, differential responsiveness to selection. Will flies from a marginal, inversion-poor population respond to selection faster and to a greater extent than those from an inversion-rich central population, everything else being equal? Carson (24) selected two strains of *D. robusta* for mobility toward light: a Nebraska strain monomorphic for gene arrangements (peripheral strain), and a strain from Missouri highly polymorphic for gene arrangements (central strain). In general, the Nebraska strain responded faster and to a greater extent. This phenomenon warrants further study.

There is a large amount of evidence favoring a period of structural homozygosity during speciation in *Drosophila* as predicted by Carson. This evidence is the common observation (27, 2, 15) that closely related species share virtually no inversions, even though they are often highly polymorphic. This evidence might be thought of as supporting Carson's homoselection hypothesis, since new species may often originate as ecologically marginal, probably isolated populations (71). But this observation is not inconsistent with the Ludwig hypothesis and cannot be used to support either hypothesis over the other. Besides, Wasserman (105) found several species in the *repleta* group to be polymorphic for the same inversions. Stone and his co-workers (97) found the same phenomenon in the *virilis* group. This probably means that speciation in this genus is not always accompanied by transient cytological monomorphism.

The Pattern of Allelic Variation

Although the data available are limited, it is probable that for some taxa, at least, marginal genetic depauperization is more common in inversions than in alleles. Even for the species tabulated, there is need for more sampling in peripheral areas, but as a working hypothesis I will assume that the phenomenon is real. At any rate, for some species of *Drosophila* it is clear that alleles, unlike inversions, show little or no tendency towards reduced polymorphism near ecological margins. This was first found in *D. pseudoobscura* by Prakash, Lewontin & Hubby (81).

This contrast in the geographic patterns of allelic versus chromosomal variation is nowhere clearer than in the variation in marginal and peripheral populations of *Drosophila willistoni*. Of the 50 or so inversions known (35), only 11 are known in Florida and only 6 and 8 on the islands of Cuba and Puerto Rico, respectively (100, 35). Some Brazilian populations have two or three times this number. The parallel data at the allelic level show only a slight decrease in geographical variation in levels of polymorphism (3). For the proportion of loci that are polymorphic, the mainland populations average 82.4, the island populations, 79.5. The mean per individual heterozygosity values are 18.4% and 16.9% respectively. The value for Puerto Rico is 19.6%. Another important result is the observation that the allele frequencies do not vary greatly over this entire range of populations, whereas there are very significant differences in inversion frequencies, even among the island populations.

Other species of *Drosophila* are not as well studied, but at least in *D. pseudoobscura* (81, 57) and possibly in *D. obscura* and *D. subobscura* the same pattern discordance is suggested. In the following section a theory is offered that resolves this paradox.

THE EPISTASIS CYCLE

Assumptions

The theory is best seen historically as an extension of Carson's heteroselection-homoselection theory. Unlike Carson's, however, this one (*a*) is not compatible with the Ludwig hypothesis, (*b*) relies heavily on geological time, (*c*) accounts for the inversion-allele paradox, and (*d*) stresses epistasis rather than heterosis.

REGULATORY DNA   In higher organisms an important source of genetic variation is regulatory DNA. That is, much evolutionary change in morphological, physiological, and behavioral characters is the result of recombination in and selection on the partially (middle) redundant DNA that composes from 15% (58) to 40% (61) or more of the genetic material of eucaryotes.

Much of the DNA of eucaryotes is now thought to have a regulatory and integrative function (17, 58, 16). In vitro DNA hybridization studies suggest that one category, the middle repetitive DNA, occurs as a large number of families, perhaps as many families as there are structural genes; therefore, a particular family might be responsible for controlling the function of a particular structural gene (16). The probable existence of so much regulatory DNA in higher organisms suggests the possibility that nucleotide differences within these genes are a major source of heritable genetic variation. In fact, one could argue that regulatory gene diversity and recombination of such genes are a major source, if not the major source, of genetic variance for complex, polygenic characters. More explicitly, the important phenotypic differences in a population may result from variation in the control of structural genes, e.g. exactly when and to what degree they are turned on and off. In the terminology of Britten & Davidson (17), even a single nucleotide change at a receptor gene locus may determine whether or not a producer (structural) gene is transcribed.

The impact of such genetic differences might be far greater than are allelic differences at a structural gene locus. The former differences can determine presence or absence of a gene product. The latter (assuming the allele produces a functional product) can affect only differences in the physicochemical properties of the resulting peptide.

A corollary of the above is that epistatic (nonallelic) interactions are a regulatory gene phenomenon. The so-called coadapted gene complex is, in this context, a well-integrated regulatory gene "tribe" whose members are families of middle repetitive DNA.

Some support for assuming that regulatory DNA is a major source of genetic variation comes from the few cases of electrophoretic identity between species. The most divergent of the side-blotched lizard populations in the Gulf of California is *Uta palmeri* on Isla San Pedro Mártir (90, 8). This giant, melanic, and behaviorally unique form does not have a single unique allele, nor is there fixation for any relatively rare alleles at any of the eighteen loci surveyed (unpublished data). More striking is the recent evidence that several widespread, sometimes sympatric species of marine porpoises are electrophoretically identical for at least the ten loci so far surveyed (Gary D. Sharp, personal communication). It seems that speciation may have occurred in these organisms with very little enzymatic change. These may be cases in which a change in regulation, rather than a change in proteins, is the explanation of the evolutionary differences.

The same argument can be made for the evolution of pesticide resistance in insects. In reviewing the subject Georghiou (47) shows that enhanced detoxication attributable to increased occurrence in the population of a specific enzyme is but one way. Other paths to resistance sound like alteration in gene regulation. These include reduced penetrability of the cuticle due to higher protein and lipid content and greater sclerotization, elimination of aestivation, increased irritability to and avoidance of the chemicals, and changes in habitat preference.

STRUCTURAL DNA   What then is the evolutionary role of DNA coding for protein, the structural DNA? I suggest that most evolutionary changes in enzymes and other proteins account for long-term, fine adjustments, but are relatively unimportant in the events normally associated with rapid selection responses of polygenic characters.

Consider evolutionary changes in body size following colonization of a new habitat. Such changes can be accomplished genetically in at least two ways: 1. amino acid substitutions having very specific enhancing effects or 2. slight changes in the rate of synthesis of an enzyme or hormone already present. The former may be thermodynamically more efficient, and will be selected for when and if it occurs. In terms of probabilities, however, the second, "quick and dirty" method is more likely, since all that is required is a combination of regulatory genes that produce the appropriate changes in transcription, and this might arise by recombination relatively frequently.

FITNESS   In this context, the genetic contributions to fitness can be thought of in terms of three categories: 1. the production of messenger RNA, 2. epistasis, and 3. heterosis. Ignoring environmental influences, the first two categories control developmental and physiological processes. That is, they determine the character state or phenotype for a particular characteristic, as well as the integration of characters throughout the organism. Heterosis, on the other hand, contributes to fitness in a totally different and less important way. Haldane (53) pointed out that heterozygosity improves biochemical efficiency or vigor. Many mechanisms have been proposed to explain the greater fitness of heterozygous individuals (85, 84, 68, 13, 88). Irrespective of the mechanism, however, the effect of heterosis, unlike the first two genetic categories, is, on the average, general rather than specific. Figure 1 illustrates these relationships in an oversimplified way, showing that most individuals would survive without the heterotic "frosting on the cake."

The parameters that *Drosophila* geneticists call fitness characters, e.g. viability, fecundity, rate of development, and biomass, are very general, organism-wide phenomena. These phenomena are the sorts of things usually improved by heterozygosity. Naturalists, however, often think of fitness in terms of a normal distribution of character states, the putative optimum being near the mean. Any

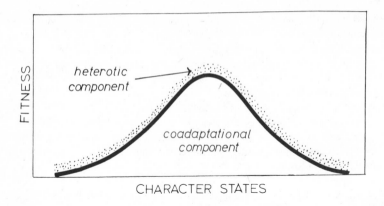

Figure 1 The "cake frosting" model of the heterotic contribution to fitness in natural populations. In general, extreme character states (phenotypes) lower one's fitness, but even relatively unfit individuals are benefited by heterozygosity since it increases "vigor."

enhancement of these means by heterozygosity (such as the mean pH, leg length, or scale size) would decrease rather than increase fitness. Nonetheless, one might argue that heterozygosity in such characters might still enhance the fitness of the population by increasing canalization and thus reducing the variance of morphological and physiological characters. More individuals would be near the mean as a result. I doubt this effect exists in natural populations. The literature suggests that this occurs only in highly, and usually artificially, inbred strains.

CHROMOSOMES   The above assumptions, if more or less correct, would force us to conclude that alleles for structural genes would respond to selection in a manner very different from that of a section of chromosome having both structural genes and the regulatory elements. This point is the central one of the argument. Directional selection favoring individuals with an extreme character state would favor certain chromosome segments because they included regions of regulatory and structural DNA interacting in such a way as to produce the extreme phenotype. Such selection should reduce the variety of gene combinations (e.g. inversions) in the gene pool. In general, however, such directional selection would not reduce the amount of structural gene polymorphism because this presumably is the basis of much heterosis and is adaptive regardless of the character state. Depending on the strength of the selection, structural alleles might be lost by virtue of tight linkage between them and their regulatory elements. In any case, the rate of loss of allelic polymorphisms should be slower under directional selection than the rate of loss of coadapted gene complexes.

Applications and Predictions

MARGINAL POPULATIONS   Ecologically marginal populations will become depleted of epistatic combinations, including inversions. The reason for attrition is that directional selection will favor new or rare character states in marginal localities, and these are determined by a minority of the available inversions or other epistatic systems of alleles.

On the other hand, structural gene alleles at polymorphic loci will tend to be conserved in marginal populations. This assumes that the principal contribution to fitness of many alleles at polymorphic loci is via heterosis because, according to this theory, heterosis will be beneficial in all environments. Marginal populations are, nevertheless, prone to severe reduction in numbers and can experience intermittent drift. Just a trickle of gene flow can, however, restore lost alleles; but, if the organism has poor dispersal powers, then marginal, isolated demes are expected to be allelically depauperate.

Such a theory would explain why marginal populations of *Drosophila* apparently retain most of their allelic diversity whereas marginal populations of the less vagile vertebrate species are depauperate (Table 2). Another explanation for the maintenance of species-wide allelic polymorphisms in marginal populations of *Drosophila* is their regeneration by mutation following loss by genetic drift. This possibility is discussed below.

The Ludwig hypothesis also predicts a loss of inversions in marginal populations. How do these two theories differ? According to the epistasis theory, new coadapted gene combinations will gradually evolve and increase in number in marginal populations. The number of such epistatic combinations that persist is independent of niche width because their function is to insure that development and physiology produce and maintain the optimum phenotypic expression. Such optima exist regardless of niche width. Recombination continually produces new allelic arrays. The probability that a new and advantageous array will be recognized and conserved in a marginal population is relatively high. For one thing, the chances of occurring as a homozygote are enhanced in a small, inbred population. For another, the genetic background will be relatively uniform. This means that a particular DNA sequence will have a relatively constant effect on fitness.

Drosophila willistoni is at the present time the most completely analyzed species in which chromosomal and allelic polymorphisms are well known in many peripheral populations. Ayala and his co-workers (3, 4), as discussed above, found that peripheral populations retained most allelic polymorphisms but were depauperate for chromosomal inversions. They speculate that there was a founder effect in all the peripheral populations at the allelic and chromosomal levels, but that the attrition of alleles due to bottlenecking was reversed by mutation-induced regeneration of alleles in situ. Furthermore, they contend that all the detectable new alleles that have arisen de novo in all of the island populations have exactly the same electrophoretic properties as those that were lost.

I believe that an alternative explanation is both simpler and more consistent with findings in other studies. Briefly, this explanation holds that:

1. The insular and Floridian populations are relatively young, at least as compared to the South American population.
2. The founder stocks were large enough or immigration was (and is) frequent enough to rapidly establish all of the presently occurring alleles and many more inversions than now persist.

Table 3 *Drosophila willistoni* heterozygote inversion frequencies in some mainland and island populations. Some values are averages [a]

Chromosome and Inversion	Locality						
	Costa Rica	Buenos Aires	Florida	Jamaica	Puerto Rico	St. Lucia	Caripe (Venezuela)
XL None	57	100	100	99	99	100	56
A	5	–	–	1	1	–	8
B	–	–	–	–	–	–	5
C	24	–	–	–	–	–	–
D	–	–	–	–	–	–	–
C & D	9	–	–	P[b]	–	P	3
D & H	–	–	–	–	–	–	3
Others	–	–	–	–	–	–	33
XR None	57	100	91	97	92	91	43
A	9	–	1	–	–	–	18
B	–	–	–	1	–	5	3
C	14	–	–	P	–	P	27
D	29	–	–	2	–	5	20
C & D	–	–	–	–	–	–	3
IIL None	8	39	46	59	70	81	19
A or B	43	–	30	32	30	7	25
A & B	8	9	–	–	–	–	17
C	3	–	–	–	–	–	–
D or E	24	35	36	10	6	–	33
D & E	5	–	–	–	–	–	15
F	57	25	–	11	12	13	49
H	–	14	–	–	–	–	1

Table 3 (continued)

Chromosome and Inversion	Locality						
	Costa Rica	Buenos Aires	Florida	Jamaica	Puerto Rico	St. Lucia	Caripe (Venezuela)
IIR None	35	87	54	56	60	90	48
A	38	–	12	15	–	1	14
B	13	–	–	1	9	4	1
C	–	–	1	–	–	–	1
E	32	12	38	37	32	5	41
F	–	–	–	–	2	–	–
III None	13	18	38	55	71	35	26
A-1	–	–	2	–	–	–	3
A or B	5	86	43	–	1	29	24
A & B	–	–	–	–	–	–	2
C	11	–	–	3	–	–	–
D	8	–	–	–	–	–	18
F	70	2	2	–	–	–	31
H	19	2	–	–	–	2	24
I	–	–	–	–	–	–	–
J	3	46	35	43	28	28	19
K	3	–	–	–	–	–	–
L	43	–	–	–	–	28	15
L & M	27	–	–	–	–	–	2
V	–	–	–	–	–	–	7

[a]Data references: 100, 34, 38, 35.
[b]P indicates presence of an inversion with indeterminable frequency.

3. The present paucity of inversion types is the result of selection favoring certain inversions and eliminating others.

4. Therefore, the high levels of protein polymorphism were not regenerated by mutation, but rather have been conserved because of the inherent advantage of individuals that are heterozygous at these loci.

The evidence favoring this hypothesis is indirect, but worthy of consideration.

First, we must examine the geographic distribution of inversions. If the present pattern of distribution were due to accidents of colonization, we might expect considerable heterogeneity among the islands even if selection were operating after colonization. In Table 3 are data showing the inversions in two likely source regions (Costa Rica and Venezuela) and in four widely separated islands, all of which have large areas of rain forest. Florida, with the most northerly population of the species, and Buenos Aires at the southern margin are also included. If anything, the data suggest similarity rather than dissimilarity of inversions among the islands. Especially interesting are the inversions on chromosome III. Inversion III-J is frequent on all the islands and Florida, and is less frequent in the source populations. Inversion III-J and III-A (or III-B) are dominant in Florida and both are very common inversions at the southern margin as well. These data do not offer much support for the founder hypothesis. It seems equally probable that the absent inversions were introduced but not successful.

The obvious similarities in the Floridian and Argentinean populations bear on the earlier discussion of the Ludwig hypothesis. Rather than inversions having very specific ecotypic functions, this amphitropical correlation suggests that inversions capture allelic complexes that enhance fitness in relation to broad climatic conditions. The reason there are fewer "subtropical" than "tropical" inversions may be related to the recentness of occupation of the marginal environment as discussed below.

Next, we consider the allelic polymorphisms. Ayala, Powell & Dobzhansky (3) surveyed 24 loci in 4 mainland and 6 island populations, but they tabulated the results for only 16. Altogether there are 47 alleles at these 16 loci, none of which are unique on the islands. It is difficult to see how (a) many of the alleles now present in these island populations could have been lost due to drift, then (b) regenerated by mutation, and (c) reestablished in their original equilibrium frequencies, without (d) a single new allele appearing that is not present on the mainland.

It is not as though the establishment of new alleles is a rare event in island populations following a colonization bottleneck. In our laboratory we have been studying electrophoretic variation in several groups of island lizards. Table 4 presents some unpublished data showing allele frequencies in some closely related species in the *roquet* group of *Anolis* in the West Indies.

A similar pattern is seen in all the *roquet* group species, but to save space I have tabulated data for only three. These three, furthermore, are a very closely related group (49), and until recently *blanquillanus* was considered a subspecies of *bonairensis*. There is compelling evidence (49, and unpublished) that *A. luciae* is the descendant of the ancestral population for the whole group and for *blanquillanus* and *bonairensis* specifically. Four alleles absent in all other members of the *roquet* group occur in the two derived species.

Table 4 Unique alleles in island populations of closely related lizards

Species	Locality	N	Number of Loci Surveyed	P[a]	H[b]	A[c]
Anolis luciae	St. Lucia Island	30	22	41	9.5	–
Anolis blanquillanus	La Blanquilla Island	25	22	14	4.9	3
Anolis bonairensis	Bonaire Island	30	22	18	2.9	2[d]

[a]percentage of loci polymorphic including rare alleles.
[b]percentage heterozygosity based on actual count of heterozygotes.
[c]number of unique alleles (see text).
[d]one unique allele shared with *A. blanquillanus*.

Others have published similar findings. An island population of *Sigmodon hispidus* has five unique alleles (56). This population could not have originated prior to Pleistocene time.

It could be argued that these unique insular alleles are "relict alleles" that have been lost or are very rare in the source populations. This argument would receive support if (*a*) there were ever alleles very rare on the mainland and common on an island, or if (*b*) a particular allele occurred on several well-separated islands, but not on the mainland, or if (*c*) the morphology of island animals gave evidence of being relictual as was once widely believed (e.g. 21). Many recent workers, emphasizing the parallel adaptive trends in island mammals (31, 46, 14), birds (51), and reptiles (89, 82, 83), have rejected the relict hypothesis of island evolution (cf. 8).

From these admittedly circumstantial evidences, I would conclude that the apparent absence of any unique alleles in the West Indian populations of *D. willistoni* supports the hypothesis that these are, in fact, relatively young populations that were never depauperate for allelic polymorphisms.

CENTRAL POPULATIONS New gene arrays with potentially useful epistasis are continually arising in central populations, but selection will rarely be able to operate on them or increase their frequency before they are broken up by recombination or lost by drift. One reason for their failure to thrive in a large, outbreeding population is that they will rarely occur as homozygotes, as Mayr (70) pointed out for alleles long ago. A second reason is that fitness of such an array will depend to a large extent on the genetic background. Genomes are so heterogeneous in central populations that fitness of a particular array could vary greatly. On the other hand, inversions and other chromosomal changes that are protected from recombination, and especially those which produce heterotic heterozygotes from the start, are exceptions. These should not show a different rate of appearance in central and marginal populations, everything else being equal.

TIME At any given time, most marginal populations or demes are young. Probably less than one in a million persist long enough to achieve genetic isolation and to emerge as a successful, expanding species. Most either become extinct or lose their nascent identity by merging with others. This recency is likely the reason for the observation that marginal populations rarely have novel chromosomal or allelic characteristics. The central body of the species will be much older and will have produced a greater diversity of genetic tools.

Epling (45) recognized this nearly two decades ago, but his emphasis on the importance of time in explaining the level of inversion polymorphism has not been appreciated. The following quotes are from his comments:

> The greatest degree of structural heterozygosity in *D. robusta* coincides with the central parts of the hardwood forest of the eastern United States—homogeneous forest substantially unchanged since Miocene or long before.

> Greatest heterozygosity of *D. pseudoobscura* has two centers—one coinciding with the Sierran forest, the other with the montane forests of central Mexico and Guatemala. These are later and different derivatives of the Arcto-Tertiary forest.

> The area of greatest structural homozygosity is geographically central, in the southern Basin and Range province . . . and coincides with vegetation and environments that have developed during late Cenozoic time from the Madro-Tertiary geoflora. The greater heterozygosity is associated in both instances with older environments, and the greater homozygosity with the younger that are ecologically peripheral.

From this we can predict that time, rather than ecological conditions or niche breadth, is the essential determinant of the diversity of epistatic combinations, including inversions. The one qualification is that the populations in question must have similar effective numbers, since bottlenecks eliminate gene complexes, and their rate of origin depends on effective population size.

THE CYCLE   New epistatic sequences will continue to arise and be conserved in any population so long as the population persists. Eventually mutation and recombination produce so much extant variation that new, potentially useful combinations can rarely be unmasked unless they are highly heterotic from the start. It is possible, but by no means certain, that at this point a population is unlikely to produce evolutionary novelty and adapt successfully to changing conditions. The number of epistatic combinations will, however, be maximized, and fitness, as operationally measured by biomass or density, will remain high in a geologically and evolutionarily static environment.

A concomitant of high population density is relatively high rates of propagule production or gene flow. Mayr (73) has invoked gene flow as a cohesive force in a species for the reason that "good mixer" genes would be selected for and these would retard the divergence of marginal populations. A simpler explanation of the cohesive effect of gene flow (if such an effect exists) is that it maintains genetic diversity at such a high level that new, potentially beneficial, epistatic combinations will have a low probability of occurring as homozygotes before they are broken up or drift out.

The cycle of epistatic buildup can begin anew in marginal or isolated populations in which drift and selection have reduced the diversity of such genetic arrays (26, 71). A novel array here can be expressed in most of the individuals that have it because of the relative uniformity of genetic background. A new gene arrangement has a relatively high probability of incorporation because of this recognition.

RECAPITULATION

The available dàta on genetic variation in marginal and peripheral populations suggest the existence of two patterns. First, marginal populations of *Drosophila* show a relative paucity of structural (inversion) polymorphism. Second, marginal populations of vertebrates show a relative paucity of allelic polymorphisms, but *Drosophila* do not. Assuming the reality of these patterns, we must explain why *Drosophila* lose inversions but retain alleles at their margins (the "*Drosophila* paradox"), and why vertebrates tend toward allelic monomorphism at their margins in contrast to *Drosophila.*

To answer the first question, it is argued that chromosomes and chromosome segments respond to directional selection by increasing in frequency if they help to produce the appropriate character states in the stressful environment and decreasing in frequency if they do not. Any novel environment, therefore, will cause a transient reduction in the diversity of coadapted gene combinations such as inversions. On the other hand, it is argued that a function of alleles (presumably, the less common ones) at polymorphic loci is to increase "vigor" via heterosis. Since "vigor" is advantageous in the margins as well as at the center, there will be a tendency for the preservation of allelic polymorphism.

To answer the second question, it is suggested that less vagile organisms tend to be allelically depauperate towards their geographic margins because of genetic drift and reduced immigration. This implicitly assumes there is no fundamental difference in genetic systems between *Drosophila* and terrestrial vertebrates.

Finally, it is argued that marginal paucity of inversions is unrelated to niche breadth or the variety of resources exploited. Given enough time, all populations should tend towards maximum organization of their gene pools as this is manifested by epistatic gene combinations such as inversion polymorphism and linkage disequilibrium. At any given time most marginal populations are simply too young to have evolved many novel gene combinations or balanced heterotic systems.

ACKNOWLEDGMENTS

Suh Yung Yang, Francisco Ayala, and Charles McKinney are thanked for their constructive suggestions. Many of the ideas in this paper arose during several delightful discussions with Christopher Wills.

This review is an outgrowth of studies supported by grants from the National Science Foundation.

Literature Cited

1. Agnew, A. D. Q. 1968. Variation and selection in an isolated series of populations of *Lysimachia volkensii* Engl. *Evolution* 22:228–36
2. Ayala, F. J., Mourao, C. A., Perez-Salas, S., Richmond, R., Dobzhansky, Th. 1970. Enzyme variability in the *Drosophila willistoni* group. I. Genetic differential among sibling species. *Proc. Nat. Acad. Sci. USA* 67:225–32
3. Ayala, F. J., Powell, J. R., Dobzhansky, Th. 1971. Enzyme variability in the *Drosophila willistoni* group. II. Polymorphisms in island and continental populations of *Drosophila willistoni. Proc. Nat. Acad. Sci. USA* 68:2480–83

4. Ayala, F. J., Powell, J. R., Tracey, M. L., Mourao, C. A., Perez-Salas, S. 1972. Enzyme variability in the *Drosophila willistoni* group. IV. Genic variation in natural populations of *Drosophila willistoni. Genetics* 70:113–39

5. Bailit, H. L. 1966. Tooth size variability, inbreeding and evolution. *Symposium on the Biology of Human Variation. Ann. N. Y. Acad. Sci.* 134: 616–23

6. Bailit, H. L., Dewitt, S. J., Leigh, R. A. 1968. The size and morphology of the Nasioi dentition. *Am. J. Phys. Anthropol.* 28:271–88

7. Baimai, V. 1970. Chromosomal polymorphism in *Drosophila birchii. J. Hered.* 61:23–34

8. Ballinger, R. E., Tinkle, D. W. 1972. Systematics and evolution of the genus *Uta* (Sauria: Iguanidae). *Misc. Pub. Mus. Zool. Univ. Mich., No. 145*

9. Beardmore, J. A. 1963. Mutual facilitation and the fitness of polymorphic populations. *Am. Natur.* 97:69

10. Beardmore, J. A. 1966. Genetic information in populations. *Advan. Sci.* 23: 128–32

11. Beardmore, J. 1970. Ecological factors and the variability of gene pools. *Essays in Evolution and Genetics in Honor of Th. Dobzhansky,* ed. M. K. Hecht, W. L. Steere, 299–314. New York: Appleton, Century, Crofts. 594 pp.

12. Beardmore, J. A., Levine, L. 1963. Fitness and environmental variation. I. A study of some polymorphic populations of *Drosophila pseudoobscura. Evolution* 17:121–29

13. Berger, E. M. 1971. A temporal survey of allelic variation in natural and laboratory populations of *Drosophila melanogaster. Genetics* 67:121–36

14. Berry, R. J. 1969. History in the evolution of *Apodemus sylvaticus* at one edge of its range. *J. Zool. London* 159: 311–28

15. Bock, I. R. 1971. Intra- and interspecific chromosomal inversions in *Drosophila bipectinata* species complex. *Chromosoma* 34:206–29

16. Bonner, J., Wu, J. 1973. A proposal for the structure of the *Drosophila* genome. *Proc. Nat. Acad. Sci. USA* 70: 535–37

17. Britten, R. J., Davidson, E. H. 1964. Gene regulation for higher cells: A theory. *Science* 165:349–57

18. Brncic, D. 1955. Chromosomal variability in Chilean populations of *Drosophila immigrans. J. Hered.* 46:59–63

19. Brncic, D. 1957. Chromosomal polymorphism in natural populations of *Drosophila pavani. Chromosoma* 8: 699–708

20. Brncic, D. 1970. Studies on the evolutionary biology of Chilean species of *Drosophila. Evolutionary Biology Suppl.,* ed. M. K. Hecht, W. L. Steere, 401–436. New York: Appleton, Century, Crofts. 594 pp.

21. Brown, W. L. 1957. Centrifugal speciation. *Quart. Rev. Biol.* 32:247–77

22. Carson, H. L. 1955. The genetic characteristics of marginal populations of *Drosophila. Cold Spring Harbor Symp. Quant. Biol.* 20:276–87

23. Carson, H. L. 1956. Marginal homozygosity for gene arrangement in *Drosophila robusta. Science* 123: 630–31

24. Carson, H. L. 1958. Response to selection under different conditions of recombination in *Drosophila. Cold Spring Harbor Symp. Quant. Biol.* 23: 291–305

25. Carson, H. L. 1958. The population genetics of *Drosophila robusta. Advan. Genet.* 9:1–40

26. Carson, H. L. 1959. Genetic conditions that promote or retard the formation of species. *Cold Spring Harbor Symp. Quant. Biol.* 24:87–103

27. Carson, H. L. 1965. Chromosomal morphism in geographically widespread species of *Drosophila. The Genetics of Colonizing Species,* ed. H. G. Baker, G. L. Stebbins, 503–37. New York: Academic. 588 pp.

28. Carson, H. L., Heed, W. B. 1964. Structural homozygosity in marginal populations of Nearctic and Neotropical species of *Drosophila* in Florida. *Proc. Nat. Acad. Sci. USA* 52:427–30

29. Cavalli-Sforza, L. L., Bodmer, W. F. 1971. *The Genetics of Human Populations.* San Francisco: Freeman. 965 pp.

30. Clarke, C. A., Sheppard, P. M. 1963. Interactions between major genes and polygenes in the determination of the mimetic patterns of *Papilio dardanus. Evolution* 17:404–13

31. Cook, L. M. 1961. The edge effect in population genetics. *Am. Natur.* 95: 295–307

32. Cordeiro, A. R., Salzano, F. M., Marques, V. B. 1960. An interracial hybridization experiment in natural populations of *Drosophila willistoni. Heredity* 15:35–44

33. Crumpacker, D. W. 1968. Uniform heterokaryotypic superiority for viability in a Colorado population of

Drosophila pseudoobscura. Evolution 22:256–61

34. Da Cunha, A. B., Dobzhansky, Th. 1954. A further study of chromosomal polymorphism in *Drosophila willistoni* in its relation to environment. *Evolution* 8:119–34

35. Da Cunha, A. B., Dobzhansky, Th., Pavlovsky, O., Spassky, B. 1959. Genetics of natural populations. XXVIII. Supplementary data on the chromosomal polymorphism in *Drosophila willistoni* in its relation to the environment. *Evolution* 13:389–404

36. Dessauer, H. C., Nevo, E. 1969. Geographic variation of blood and liver proteins in cricket frogs. *Biochem. Genet.* 3:171–88

37. Dobzhansky, Th. 1948. Genetics of natural populations. XVI. Altitudinal and seasonal changes produced by natural selection in certain populations of *Drosophila pseudoobscura* and *Drosophila persimilis. Genetics* 33:158–76

38. Dobzhansky, Th. 1957. Genetics of natural populations. XXVI. Chromosomal variability in island and continental populations of *Drosophila willistoni* from Central America and the West Indies. *Evolution* 11:280–93

39. Dobzhansky, Th. 1965. "Wild" and "domestic" species of *Drosophila*. See Ref. 27, 533–46

40. Dobzhansky, Th. 1970. *Genetics of the Evolutionary Process.* New York: Columbia. 505 pp.

41. Dobzhansky, Th., Burla, H., Da Cunha, A. B. 1950. A comparative study of chromosomal polymorphism in sibling species of the *willistoni* group of *Drosophila. Am. Natur.* 84:229–45

42. Dobzhansky, Th., Hunter, A. S., Pavlovsky, O., Spassky, B., Wallace, B. 1963. Genetics of natural populations. XXXI. Genetics of an isolated marginal population of *Drosophila pseudoobscura. Genetics* 48:91–103

43. Ehrlich, P. R., Raven, P. H. 1969. Differentiation of populations. *Science* 165:1228–32

44. Endler, J. A. 1973. Gene flow and population differentiation. *Science* 179:243–50

45. Epling, C. 1958. Statement made during discussion of a paper delivered by H. L. Carson at *Cold Spring Harbor Symp. Quant. Biol.* 23:305–6

46. Foster, J. B. 1964. Evolution of mammals on islands. *Nature* 202:234–35

47. Georghiou, G. P. 1972. The evolution of resistance to pesticides. *Ann. Rev. Ecol. Syst.* 3:133–68

48. Gooch, J. L., Smith, B. S., Knupp, D. 1972. Regional survey of gene frequencies in mud snail *Nassarius obsoletus. Biol. Bull.* 142:36–48

49. Gorman, G. C., Atkins, L. 1969. The zoogeography of Lesser Antillean *Anolis* lizards. An analysis based upon chromosomes and lactic dehydrogenases. *Bull. Mus. Comp. Zool., Harvard Univ.* 138:53–80

50. Gotz, W. 1967. Untersuchungen uber den chromosomalen Strukturpolymorphismus in Kleinasia—tischen und persischen Populationen von *Drosophila subobscura* Coll. *Mol. Gen. Genet.* 100:1–38

51. Grant, P. R. 1965. The adaptive significance of some size trends in island birds. *Evolution* 19:355–67

52. Guthrie, R. D. 1965. Variability in characters undergoing rapid evolution, an analysis of *Microtus* molars. *Evolution* 19:214–33

53. Haldane, J. B. S. 1954. The statics of evolution. *Evolution as a Process,* ed. J. Huxley, A. C. Hardy, E. B. Ford, 109–21. London: Allen & Unwin. 367 pp.

54. Huether, C. A. 1968. Exposure of natural genetic variability underlying the pentamerous corolla constancy in *Linanthus androsaceus* ssp. *androsaceus. Genetics* 60:123–46

55. Johnson, W. E., Selander, R. K. 1971. Protein variation and systematics in kangaroo rats (genus *Dipodomys*). *Syst. Zool.* 20:377–405

56. Johnson, W. E., Selander, R. K., Smith, M. H., Kim, Y. J. 1972. Biochemical genetics of sibling species of the cotton rat (*Sigmodon*). *Stud. Genet.* VIII:297–305, *Univ. Tex. Publ. 7213*

57. Jones, J. S. 1973. Protein polymorphism in a high altitude population of *Drosophila pseudoobscura*. Unpublished

58. Judd, B. H., Shen, M. W., Kaufman, T. C. 1972. The anatomy and function of a segment of the *X* chromosome of *Drosophila melanogaster. Genetics* 71: 139–56

59. Kimura, M. 1968. Genetic variability maintained in a finite population due to mutational production of neutral and nearly neutral isoalleles. *Genet. Res.* 11:246–69

60. Kimura, M., Crow, J. F. 1964. The number of alleles that can be maintained in a finite population. *Genetics* 49:725–38

61. Kohne, D. E. 1970. Evolution of higher organism DNA. *Quart. Rev. Biophys.* 3:327–75

62. Lakovaara, S., Saura, A. 1971. Genetic variation in natural populations of *Drosophila obscura*. *Genetics* 69:37–84

63. Lakovaara, S., Saura, A. 1971. Genic variation in natural populations of *Drosophila subobscura*. *Hereditas* 69: 77–82

64. Levin, D. A. 1970. Developmental instability and evolution in peripheral isolates. *Am. Natur.* 104:343–53

65. Lewontin, R. C. 1958. Studies on heterozygosity and homeostasis. II. Loss of heterosis in a constant environment. *Evolution* 12:494–503

66. Long, T. 1970. Genetic effects of fluctuating temperature in populations of *Drosophila melanogaster*. *Genetics* 66: 401–16

67. Ludwig, W. 1950. Zur Theorie der Konkurrenz. Die Annidation (Einnischung) als fünfter Evolutions faktor. *Neve Ergeb. Probleme Zool., Klatt-Festschrift* 1950:516–37

68. Manwell, C., Baker, C. M. A. 1970. *Molecular Biology and the Origin of Species*. Seattle: Univ. Wash. Press. 394 pp.

69. Mather, W. B. 1964. Speciation in *Drosophila rubida*. *Evolution* 18:10–11

70. Mayr, E. 1954. *Change in Genetic Environment and Evolution*, ed. J. Huxley, A. C. Hardy, E. B. Ford, 157–80. London: Allen & Unwin. 367 pp.

71. Mayr, E. 1959. Where are we? *Cold Spring Harbor Symp. Quant. Biol.* 24: 1–14

72. Mayr, E. 1963. *Animal Species and Evolution*. Cambridge: Belknap Press, Harvard Univ. Press. 747 pp.

73. Mayr, E. 1970. *Population, Species, and Evolution*. Cambridge: Belknap Press, Harvard Univ. Press. 453 pp.

74. Mayr, E., Stresemann, E. 1950. Polymorphism in the chat genus *Oenanthe* (Aves). *Evolution* 4:291–300

75. McKinney, C. O., Selander, R. K., Johnson, W. E., Yang, S. Y. 1972. Genetic variation in the side-blotched lizard. *Stud. Genet.* VII:307–18, *Univ. Tex. Publ. 7213*

76. Nair, P. S. 1969. Genetic load in *Drosophila robusta*. *Genetics* 63:221–28

77. Nevo, E., Gorman, G., Soulé, M., Yang, S., Clover, R., Jovanović, V. 1972. Competitive exclusion between insular *Lacerta* species (Sauria, Lacertidae): Notes on experimental introductions. *Oecologia* 10:183–90

78. Nevo, E., Shaw, C. R. 1972. Genetic variation in a subterranean mammal, *Spalax ehrenbergi. Biochem. Genet.* 7: 235–41

79. Niswander, J. D., Dhung, S. C. 1965. The effects of inbreeding on tooth size in Japanese children. *Am. J. Hum. Genet.* 17:390–98

80. Power, D. M. 1971. Range expansion of Brewer's blackbird—phenetics of a new population. *Can. J. Zool.* 49:175

81. Prakash, S., Lewontin, R. C., Hubby, J. L. 1969. A molecular approach to the study of genic heterozygosity in natural populations. IV. Patterns of genic variation in central, marginal and isolated populations of *Drosophila pseudoobscura. Genetics* 61:841–58

82. Schoener, T. 1969. Size patterns in West Indian *Anolis* lizards. I. Size and species diversity. *Syst. Zool.* 18:386–401

83. Schoener, T. 1970. Size patterns in West Indian *Anolis* lizards. II. Correlations with the sizes of particular sympatric species—displacement and convergence. *Am. Natur.* 104:155–74

84. Schwartz, D. 1971. Genetic control of alcohol dehydrogenase—a competition model for regulation of gene action. *Genetics* 67:411–25

85. Schwartz, D., Laughner, W. J. 1969. A molecular basis for heterosis. *Science* 166:626–27

86. Selander, R. K., Johnson, W. E. 1973. Genetic variation among vertebrate species. *Ann. Rev. Ecol. Syst.* 4:75–91

87. Selander, R. K., Smith, M. H., Yang, S. Y., Johnson, W. E., Gentry, J. B. 1971. Biochemical polymorphism and systematics in the genus *Peromyscus*. I. Variation in the old field mouse (*Peromyscus polionotus*). *Stud. Genet.,* VI:49–50, *Univ. Tex. Publ. 7103*

88. Sing, C. F., Brewer, G. J. 1971. Evidence for non-random multiplicity of gene products in 22 plant genera. *Biochem. Genet.* 4:297–320

89. Soulé, M. 1966. Trends in the insular radiation of a lizard. *Am. Natur.* 100: 47–64

90. Soulé, M. 1967. Phenetics of natural populations. I. Phenetic relationships of insular populations of the side-blotched lizard. *Evolution* 21:584–91

91. Soulé, M. 1971. The variation problem: The gene flow-variation hypothesis. *Taxon* 20:37–50

92. Soulé, M., Stewart, B. R. 1970. The "niche-variation" hypothesis: A test and alternatives. *Am. Natur.* 104:85–97

93. Spassky, B., Spassky, N., Pavlovsky, O., Krimbas, M. G., Dobzhansky, Th. 1960. Genetics of natural populations. XXVIII. The magnitude of the genetic load in populations of *Drosophila pseudoobscura. Genetics* 45:723–40

94. Sperlich, D. 1966. Equilibria for inversions induced by X-rays in isogenic strains of *Drosophila pseudoobscura*. *Genetics* 53:835–42

95. Stalker, H. D. 1964. Chromosomal polymorphism in *Drosophila euronotus*. *Genetics* 49:669–87

96. Stalker, H. D. 1964. The salivary gland chromosomes of *Drosophila negromelanica*. *Genetics* 49:883–93

97. Stone, W. S., Guest, W. C., Wilson, F. D. 1960. The evolutionary implications of the cytological polymorphism and phylogeny of the *virilis* group of *Drosophila*. *Proc. Nat. Acad. Sci. USA* 46:350–61

98. Stone, W. S., Wheeler, M. R., Johnson, F. M., Kojima, K. 1968. Genetic variation in natural island populations of members of the *Drosophila nasuta* and *Drosophila ananassae* subgroups. *Proc. Nat. Acad. Sci. USA* 59:102–9

99. Thomson, J. A. 1964. Genetic differentiation within laboratory populations of *Drosophila pseudoobscura*. *Genetica* 35:270–86

100. Townsend, J. I. 1952. Genetics of marginal populations of *Drosophila willistoni*. *Evolution* 6:428–42

101. Townsend, J. I. 1958. Chromosomal polymorphism in Caribbean Island populations of *Drosophila willistoni*. *Proc. Nat. Acad. Sci. USA* 44:38–42

102. Vann, E. 1966. The fate of X-ray induced chromosomal rearrangements introduced into laboratory populations of *Drosophila melanogaster*. *Am. Natur.* 100:425–49

103. Waddington, C. H. 1953. Genetic assimilation of an acquired character. *Evolution* 7:118–26

104. Ward, R. H. 1972. The genetic structure of a tribal population, the Yanomama Indians. V. Comparison of a series of genetic networks. *Ann. Hum. Genet.* 36:21–43

105. Wasserman, M. 1960. Cytological and phylogenetic relationships in the *repleta* group of the genus *Drosophila*. *Proc. Nat. Acad. Sci. USA* 46:842–59

106. White, M. J. D. 1957. Some general problems of chromosomal evolution and speciation in animals. *Surv. Biol. Prog.* 3:109–47

107. Wright, S. 1939. Statistical genetics in relation to evolution. *Actual. Sci. Ind.* 802:1–64

CHARACTER CONVERGENCE ❖ 4058

Martin L. Cody
Department of Biology, University of California, Los Angeles, California

CONVERGENCE VERSUS DISPLACEMENT

Part of our ecological heritage are the abundant examples of related species that are serially arrayed along some resource gradient such as geographic space (15), altitude (16), habitat type (10), foraging height (31), foraging site (39), or food type (28). Recent reviews of these examples are available (15, 33). The species involved are usually, though not necessarily, taxonomically related; occassionally allogeneric or even allofamilial species are found in displacement patterns (33, 61). These observations imply rather strongly that resource use is determined to a large extent by interspecific competition. About two decades ago impetus was given to this notion by the discovery that ecologically similar species may be more dissimilar in some character value in sympatry than in allopatry, a phenomenon called character displacement (5, 29).

Character displacement was initially associated with just morphological character values, but we now recognize that ecological and behavioral traits may be subject to the same variation. This is especially likely where large-scale niche readjustments occur following large-scale changes in the competitive environment. Some of the most convincing evidence for the role of competitors in determining relative character states comes from observed differences in resource use between species-rich (competitor-rich) mainlands and species-poor (competitor-poor) islands. A common observation is that island colonists show expanded use of resources when released from competitor pressure. Most commonly species use a broader range of habitats on depauperate islands, less commonly they expand their distributions of foraging heights, and least commonly they may display on islands feeding behaviors atypical of the mainland populations. Finally, the most compelling evidence that (*a*) interspecific competition determines the relative positions of species on a resource gradient and that (*b*) species are often not physiologically tied nor in any other way fixed at some absolute position on the resource gradient comes from what I call "crossovers" in character displacement. In these situations species pairs in one relative position on the gradient in one place occur elsewhere with these positions reversed. Figure 1 gives examples, using as resource gradients time of breeding, habitat type, body size, and bill size [equivalent to prey size, (28)].

189

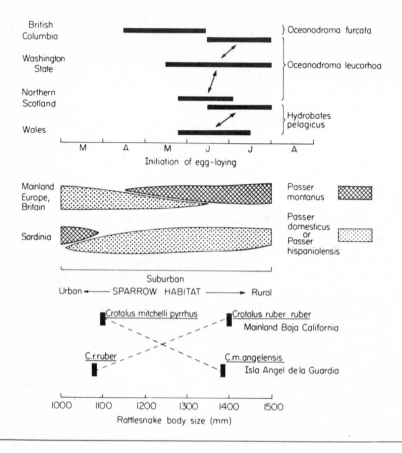

Figure 1 Crossovers in character displacement occur in (*a*) the breeding seasons of small oceanic petrels, (*b*) the habitats of sparrows *Passer*, and (*c*) body size in rattlesnakes.

The concept of and evidence for character displacement seem now to be widely accepted [notwithstanding recent doubts of the purity of commonly cited examples, (23)]. From both intuition and our knowledge of the mechanics of competition we expect that selection will limit the extent to which different species overlap in resource use, and so displacement patterns, while satisfying, are hardly surprising. But there is an alternative to character displacement, which is more recent in conception and perhaps less readily appreciated, and that is character convergence. To many ecologists the idea that selection may actually favor increased similarity between species in character value until they eventually coincide is contraintuitive and contradictory to what has been since the fifties a blanket application of the so-called "Volterra-Gause" principle. Coexistence, it is said, is assured only between species that occupy different niches. Yet there are both theoretical grounds on which to expect the evolution of convergence under certain circumstances, and empirical, field evidence to suggest that such evolution has quite often occurred. In some cases the convergent species apparently coexist in a stable association, in others the convergence is unstable with the replacement of one species by the other. This chapter reviews briefly the theoretical reasoning and at length the empirical documentation of the character convergence phenomenon; it has been adapted from a broader work that treats convergence as a predictable alternative to displacement when resources are low in abundance or predictability (15, Ch. VI).

SOME THEORETICAL CONSIDERATIONS

We must next distinguish between various characters of a phenotype that might be involved in character convergence. As by far the greater proportion of existing information on this topic comes from birds, I will use a hypothetical bird example. Convergence may take place at the level of a basic, morphological character of great ecological significance, such as bill length. As bill lengths become more similar under the influence of natural selection (see below), the converging species take increasingly similar prey size distributions and become ecologically more like a single species. Such evolution will most likely result in the competitive exclusion of one of the species by the other (although it is possible that the persistence of the two could be extended by some configurations of resource distributions over time and space). The theoretical argument for this sort of evolution follows in this section, and examples are given in a later section.

Alternatively two species with similar but not identical bill lengths may contact each other in a region of marginal sympatry or in intermediate habitats. Now overlap in the prey size distributions of the two will be great, and they must forage over large areas where they co-occur. This arrangement will be inefficient, and natural selection may favor the evolution of convergently similar signals so that the two respond to each other as they would normally respond only to conspecifics. This would result in a system of spatial organization such as territoriality between two species rather than within just one, or in a social system involving two or more species rather than one. Interspecific convergence in such signals (appearance, voice, smell) can occur independently of convergence in morphology, and indeed may occur simultaneously with divergence in morphological characteristics [as in the woodpeckers *Dinopium* and *Chrysocolaptes* on Palawan, (11)]. A theoretical basis for the evolution of convergent signals between species is given in the following section and is documented later with many examples. Convergence in morphology (e.g. bill length) can likewise occur without convergence in signals; this is especially likely when one of the two species has an overall advantage (higher fitness) where they co-occur, and when the ultimate outcome is the replacement of one species by the other rather than their coexistence. Convergence in signals, on the other hand, is indicative of a mutual benefit to both species of an integrated interspecies resource division system, and while one species may outcompete the other in some habitats, the two will be balanced competitively in intermediate habitats and coexist there. We can predict that the extent of the coexistence over the habitat gradient will be inversely proportional to the resource predictability: extensive where predictability is low, narrow where it is high. Thus the rather precise altitudinal replacements of species recorded by recent investigators (16, 61) seem to be characteristic of tropical regions of presumed high resource predictability and absent from temperate latitudes of greater environmental uncertainty.

Important inroads have already been made into the theory of the evolution of convergence (41). The framework of these investigations is a resource gradient R (abscissa) and species' "utilization curves" (40) which are distributions of the extent to which or efficiency with which each species uses each resource on the gradient. Utilization curves are more or less normal in shape. Suppose a species with resource utilization intermediate between established species 1 and 2 were to invade. How should the utilization curve of the invading species evolve to maximize fitness and to minimize the effects of competition from its neighbors on the resource gradient? The question is answered using strategic analysis (35, 36). The phenotype of the invader can be represented in the plane α_1 (niche overlap with species 1) and α_2 (niche overlap with species 2) by a point, depending on its position on the resource gradient. The invader has maximum fitness when it minimizes competition with its neighbors, where $K - \alpha_1 X_1 - \alpha_2 X_2 = C$ measures this competition load. This is the equation of a straight line in the $\alpha_1 \alpha_2$ plane (treating X_1 and X_2, the population sizes of the neighbors, as constants, as also is the carrying capacity K of the invader). Thus the optimal phenotype for the invading species is given by the intersection of the fitness set with the adaptive function $K - \alpha_1 X_1 - \alpha_2 X_2 = C$ such that C is maximized. When species 1 and 2 are well separated on the resource gradient, competition (or $K - C$) is minimized when the invader is positioned between 1 and 2; evolution then procedes until maximal divergence is reached. But when species 1 and 2 are already closely packed on the resource gradient, C is maximized as the invader converges toward the neighbor i which yields the lower product $\alpha_i X_i$. In this latter case the fitness set is convex, whereas in the former instance it is concave. The critical quantity that distinguishes between concave and convex fitness sets is D/H, the ratio of the distance D between modes of the resident's utilization curves and their standard deviation H.

An elaborate extension of this argument has recently been presented by Roughgarden (55). The invasion of a guild of residents (with a set of serially arrayed utilization curves) is more likely to be successful if (a) the invader has low niche overlap with the residents, (b) the invader has an advantage in carrying capacity ($K_{invader}/K_{residents} > 1$), and ($c$) the utilization curves are leptokurtic rather than platykurtic, for a given variance of utilization over resources. Roughgarden identifies as an "invasion barrier" to successful colonization a range of niche overlaps α between invader and residents because of which invasion cannot succeed. Invasion is possible, however, for α values below and above this barrier; "invasion from above" becomes increasingly possible (over a larger range of α below and up to 1.0) as the invader's edge in carrying capacity increases. Thus in any particular community we could expect to see a bimodal distribution of niche overlap values with a large mode centered around $\alpha = 0.5$ and a smaller mode at much higher α representing invasions from above. Interestingly enough, just such α distributions have been found (15), and are illustrated in Figure 2. The long term persistence of high niche overlap species pairs in such communities is still in question.

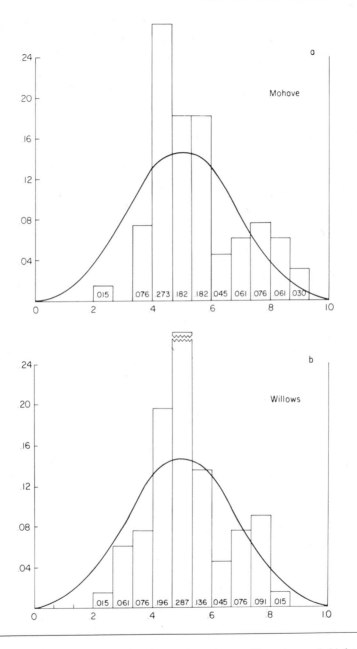

Figure 2 Bimodal distributions of niche overlap *a* in two 12-species scrub bird communities in seasonal environments.

Persistence is expected, however, in *Anolis* lizards where body size parallels overall food abundance and where both initial resident species and invading species may simultaneously become smaller following a successful colonization (56).

INTERSPECIFIC TERRITORIALITY

General Consideration

A logical extension of territoriality systems is that pairs or larger sets of species that are ecologically extremely similar may defend interspecific as well as intraspecific territories. Very often such species pairs are geographic replacements with only marginal overlap; others are species with quite distinct habitat preferences, which meet in intermediate habitat.

Orians & Willson (51) were first to assemble cases of interspecific territoriality between bird species and to generalize about the conditions under which the phenomenon might be expected. These conditions reduce to the following three situations (11) in which the usual options of interspecific divergence in resource use might not be available: (*a*) the habitat is simple (implying a narrow span of resource types), (*b*) the species involved are food specialists (in Wrightian terms food specialists occupy the summits of steep-sided adaptive peaks), (*c*) many other species are present and limit the extent to which increased species packing can take place. A stable coexistence of two species in a zone of marginal overlap implies that the resource or niche space can be stretched, or alternatively that niches can be contracted and still permit the persistence of all species involved. This increased species packing may not be possible in cases where, for historical, chance, or other reasons, maximal species packing already occurs. Tropical locations are possible candidates a priori for such type *c* situations, as are the centers of radiation or speciation for particular taxonomic groups (wrens in the neotropics, shrikes in Africa).

A Quantitative Argument

It is useful to ask the question: at what level of ecological similarity will selection favor nonoverlapping territories between different species? Clearly this will occur when the benefits of interspecific territoriality, including a more concentrated food supply spread over a smaller area, slightly more than offset the advantages of a wider foraging area. This can be seen in the following argument, which depends for its validity on the premise that individuals spend time necessarily but unproductively in traveling within the territory.

Territorial birds spend much of their time feeding young; such time has two components: actual foraging time and time spent traveling between the nest and foraging site. This latter is called traveling time and is a factor of economic importance in bird time budgets. Three kinds of evidence support this contention: (*a*) territories are often rounded as opposed to elongate in uniform habitats (17, 27) and may even approach hexagonals (22), (*b*) nests are often positioned close to the territory center (6, 17, 27, 52), (*c*) the young of species that travel long distances for food grow more slowly than those of similar species with light traveling requirements and are disadvantaged because of this (14).

We assume that natural selection favors production of the largest possible number of young and that food is limiting. The number of young that can be reared by a pair of birds will thus be proportional to 1. the fraction of their time spent foraging and 2. the abundance of food. The number of young must also be affected by 3. the size of the area over which the birds can forage; while there must be a law of diminishing returns for larger territory sizes, I assume that the population is sufficiently dense that the two are directly proportional. Representing the number of young by n, we have

$$n = kt_f FA = k(p - t_t) FA \qquad\qquad 1.$$

where territory size A, food density F, and the proportion of time the parents spend feeding young p, equal traveling time t_t plus foraging time t_f. The k is a constant of proportionality. In the first case we let two competing species maintain completely overlapping territories. The expression

$$n_1 = k'r^2 F(p - k''r/\sqrt{2}) \qquad\qquad 2.$$

gives the number of young one species can raise simultaneously in the presence of the other species. Territory size A is written πr^2 and π is absorbed into the constant k'. Traveling time is approximated in proportion to the radius of a circle enclosing half the area of the territory, i.e. proportional to the distance travelled in an average foray if the territory is uniformly searched $[A/2 = \pi(r/\sqrt{2})^2]$. The k'' is a second constant of proportionality. A similar expression describes the number of young raised by the second species. Now suppose these two species are interspecifically territorial (no territory overlap between them); that is, the same number of pairs of each species is present but now the size of their territories is reduced by half. For the above species A becomes $A/2$ and the radius of a circle enclosing half the new territory size is $r/2$ $[A/4 = \pi(r/2)^2]$. Food resources now cover a smaller area but are increased in density by a factor $(1 + \alpha)$, where α is an index of niche overlap or diet similarity. Thus traveling time is reduced, though this economic advantage may be offset by a reduction in the total food available to $AF(1 + \alpha)/2$. The number of young which can be raised under these new circumstances is

$$n_2 = k'r^2 (1 + a) F(p - k''r/2)/2 \qquad\qquad 3.$$

$$= n_1 \left[\frac{(1 + a) (p - k''r/2)}{2(p - k''r/\sqrt{2})} \right] \qquad\qquad 4.$$

If n_2 is greater than n_1 there is an advantage to interspecific territoriality. The condition is satisfied when the term in brackets in Equation 4 is greater than unity. Note that when $\alpha = 1$ this term is greater than unity for any value of r, but that increasingly smaller values of α require increasingly larger values of r for this to be so. More precisely, write n_2/n_1 as N and p/k'' as K to obtain

$$N = (1 + a)(K - r/2)/2(K - r/\sqrt{2})$$ 5.

The $r - \alpha$ plane is divided into two zones by this line $N = 1$, which has the equation

$$r \doteq K(1 - a)/(1 - a/2)$$ 6.

The intercepts of this "isocline" on the ordinate α and abscissa r in Figure 3a are $(0, 1)$ and approximately $(K, 0)$. Pairs of species with α and r values between the isocline and the origin derive no advantage from interspecific territoriality, whereas species with values beyond this line do. Figure 3b shows a plot of α versus r for examples of known interspecifically territorial bird species pairs. A logarithmic scaling of r is used to produce a correspondence to Figure 3a. Thus K is not a constant, but a function of r and increases with r. Now $K = p/k''$; it is unlikely that p increases with larger territories, more reasonably p would decrease due to larger defense times for larger areas. More likely k'' is decreasing with increasing r, perhaps reflecting a nonuniform use of larger territories (more feeding activity toward the center).

PARTIAL INTERSPECIFIC TERRITORIALITY

The Intermediate Situation

Interspecific territoriality may not be an all-or-none phenomenon and intermediate cases may occur in which ecologically rather less similar species exhibit only partial territorial segregation. Such interactions might be quite subtle in that little or nothing in the way of overt aggression or convergently-similar territory defense signals occurs. Two species might simply evolve a mutual avoidance reaction to each other's originally intraspecific spacing signals; such interactions are detectable and measurable.

One way of showing the existence of subtle interspecific spacing is to contrast two ways of measuring interspecific habitat α_H. Two species which co-occur in the same ten-acre patch will each occur at certain grid points not covered by the other, numbering p_{11} and p_{22}, and will each hold some points in common $-p_{12}$. A measure of their spatial overlap is $\alpha_{H1} = p_{12}/(p_{11} + p_{12})$ or $p_{12}/(p_{22} + p_{12})$, depending on whether $p_{11} < p_{22}$ or $p_{22} < p_{11}$ respectively, by convention. But alternatively we can measure at each grid point vegetational characteristics that are important in habitat selection (e.g. vegetation height, vegetation density). The two species can be represented as clusters of points in n space if n such habitat variables are included. Their overlap in this space can be measured exactly as before to give α_{H2}. Notice that $\alpha_{H2} \geq \alpha_{H1}$, but that if there is no interaction between the two species $\alpha_{H1} = \alpha_{H2}$. Thus the deviation from equality $(\alpha_{H2} - \alpha_{H1})$ measures the strength of the interaction between the two species and varies from 0 (no interaction) to 1 (the interaction customary within a territorial species, with no habitat difference but no spatial overlap).

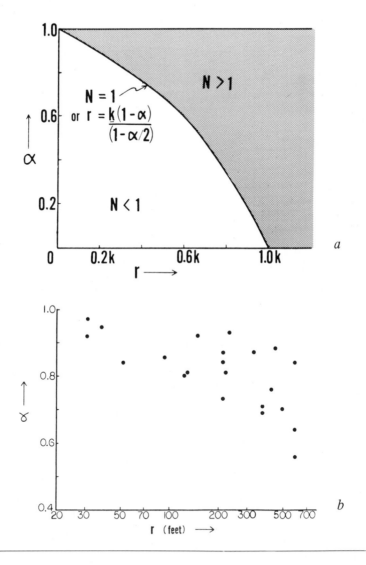

Figure 3 (a) Species pairs are predicted candidates for interspecific territoriality when their measured values of territory radius *r* and niche overlap *a* put them in the cross-hatched part of the *r–a* plane. (b) gives all known interspecifically territorial bird species pairs.

Four finch species of a grass-sagebrush site in Wyoming illustrate these interactions. In Table 1 the first of the paired figures at each matrix entry is actual spatial overlap between the species pair, the second is the overlap in the vegetational characteristics of their territories. In three species listed ($\alpha_{H2} - \alpha_{H1}$) is close to zero, but between these species and the savannah sparrow (*Passerculus sandwichensis*) significant interactions occur, with values for the difference between the two overlap measures of 0.663, 0.523, and 0.749. Thus it appears that the savannah sparrow tolerates little spatial overlap with the other three finches with which it co-occurs, but excludes them (or vice versa) from its territories. In spite of the magnitude of the interaction, no overt aggression between the savannah sparrow and the other species was observed. Apparently rather subtle avoidance reactions suffice to segregate species.

Table 1 Interactions between the finches of a Wyoming grass-sagebrush habitat. The first figure represents spatial overlap in territories; the second, overlap in the vegetational characteristics of the same territories. Large differences between the pairs of overlap values represent interactions between the species pairs. B = Brewer's sparrow *Spizella breweri*, V = vesper sparrow *Pooecetes gramineus*, S = savannah sparrow *Passerculus sandwichensis*, W = white-crowned sparrow *Zonotrichia leucophrys*.

	B	V	S	W
B	100/100	95.3/94.9	0/66.3	91.2/100
V		100/100	1.5/76.4	100/100
S			100/100	0/52.3
W				100/100

The Blue Ridge Finches

The Blue Ridge extends southeast from the main east-west ridge of the San Gabriel Mountains in southern California at around 8400 ft elevation and overlooks Los Angeles and its eastern satellite cities. The climax vegetation is chiefly jeffrey pine (*Pinus jeffreyi*) and white fir (*Abies concolor*), but the southern side of the ridge burned in 1953 and is now under a successional chaparral with scattered pines. This chaparral is composed of the same plant genera (*Ceanothus, Arctostaphylos, Ribes*) as the putative climax chaparral of lower elevations (from sea level up to about 7000 ft) with which it is contiguous. In spite of this continuity just three of the 20–30 bird species of the lower elevation chaparral breed on Blue Ridge, and instead its avifauna is drawn from locally adjacent sagebrush, coniferous forest, and streamside willows.

The role of large, ground-feeding, litter-scratching, deep-billed omnivores, the "towhee niche," is taken up by two species, the rufous-sided or spotted towhee *Pipilo erythrophthalmus* and the brown towhee *P. fuscus,* from sea level to at least 5000 ft. Above this elevation *fuscus* drops out, perhaps excluded in a diminishing resource space by competition with *erythrophthalmus.* But at 8000 ft *erythrophthalmus* occurs only in willow thickets; on the Blue Ridge the towhee niche is co-occupied by the fox sparrow, *Passerella iliaca,* and the green-tailed towhee, *Chlorura chlorura* (both 6½ in. in body size), the former being recruited from the willows and the latter from the sagebrush. Both species were common in the 11.65 acres of successional chaparral mapped on the ridge; the towhee occurred in 83 of 203 50 X 50 ft squares censused, the fox sparrow in 108.

A second species group or guild comprises three smaller finch species. In the lowlands black-chinned sparrow *Spizella atrogularis,* rufous-crowned sparrow *Aimophila ruficeps,* and sage sparrow *Amphispiza belli* form a trio similar in ecology and body size (5¼, 5¼, 5 in. respectively). They are characterized by similar feeding height distributions divided between the ground and the lower bush foliage. On Blue Ridge two of these three are replaced by chipping sparrow *S. passerina* and Oregon junco *Junco oreganus,* and black-chinned sparrow is retained (4¾, 5¼, 5¼ in.). While the trio coexists apparently quite comfortably in lowland chaparral, it does not at 8400 ft.

The five common finches on Blue Ridge, an ecological duo and a trio, quite obviously have rather different habitat preferences. The territories of all five were mapped in May–July 1971, and the occupancy by each species of each 50 X 50 ft patch of the study area was determined. Each patch was also characterized by (*a*) its cover by chaparral of height 0–1, 1–3, 3–5, and > 5 ft in height and (*b*) its number and total height of larger pines and firs that escaped the fire or grew up since 1953. The four variables most obviously involved in habitat segregation, (*a*) proportion of patch covered with brush less than 1 ft high, (*b*) proportion of patch covered with brush greater than 3 ft high, (*c*) number of trees in the patch, and (*d*) total height of all trees in the patch, were entered as variables in a discriminant function analysis. From the resultant species distributions along this linear combination of the habitat variables the habitat overlap between species pairs was calculated. This overlap, α_{H2}, is the right-hand figure of the entries in the matrix of Table 2.

Table 2 Interactions of finches on Blue Ridge, San Gabriel Mountains, southern California. Left-hand figure is spatial overlap (a_{H1}), right-hand figure overlap in vegetation (a_{H2}) in the territories. Figure below is Mahalanobis' generalized distance statistic, D^2, from discriminant function analyses incorporating four habitat variables discussed in text. F = fox sparrow, G = green-tailed towhee, B = black-chinned sparrow, C = chipping sparrow, J = Oregon junco.

	G	B	C	J
F	0.391/0.755 0.366	0.633/0.700 0.339	0.375/0.563 0.650	0.500/0.663 0.707
G	–	0.367/0.561 0.539	0.458/0.658 0.263	0.482/0.734 0.235
B		–	0.000/0.516 0.976	0.041/0.489 0.880
C			–	0.286/0.822 0.090

Thus there are certainly differential habitat preferences which separate the territories of the finches in this habitat. But when the territories of the five are actually plotted their spatial overlap (α_{H_1}, the left-hand figures in Table 2) is in several cases far less than the differences in habitat preferences would account for. Furthermore, ($\alpha_{H_2} - \alpha_{H_1}$) is great between just those species already labelled ecologically similar, the two large litter-scratching species and the three smaller ground-and-low-brush finches. The size of the interaction averages 0.466 ($n = 4$) between species within the two guilds and only 0.177 ($n = 6$) between pairs across the two groups.

There is also direct evidence for a behavioral interaction between the fox sparrow and the green-tailed towhee, and the mechanism responsible for this is apparently an evolution of song similarity. Normally one has no difficulty in distinguishing the songs of these two species, for the towhee's is shorter with a terminal trill and the sparrow's a long, fluty melodic jumble of notes. Although some songs of each of the two species on Blue Ridge are the typical sort, some are not, and even with practice the identity of the vocalist may be confused. Thus there seems to have evolved an intermediate song, which causes overt interspecies interactions. When an intermediate song recorded from one of the species is played back in the territory of the other, the territory owner approaches the recorder with great agitation, singing with great intensity. The two species have also been observed to fight at territory margins.

The songs of the three smaller finches also show a basic similarity. Each has a song which consists of a single note and each trills this note for about 2½ seconds. The note is rather different in structure between the chipping sparrow and junco, but is repeated at the same pitch. In black-chinned sparrows the note is either slurred down (like the chipping sparrow's) or slurred up (like the Oregon junco's), but is repeated at an increasing rate. More direct evidence is not yet available on their interspecific interactions.

CHARACTER CONVERGENCE

The Hypothesis

Consideration of partial interspecific territoriality on the Blue Ridge and discussion of the probable mechanism of its operation, song convergence, lead us now to signal character convergence (11, 46). The hypothesis and term were produced independently in these two papers, but the emphasis in the earlier is on the evolution of convergent signals to facilitate gregariousness, and in the later to facilitate spacing by aggression. Here I will restrict the term character convergence to the latter situation and reserve the term "social mimicry" for the former phenomenon.

The character convergence hypothesis states that interspecific territoriality is sometimes associated with, and presumably developed in parallel with, a convergence between species in those characteristics used to defend the territory; that is, appearance and voice. Males would thus fail to distinguish in territorial defense between other males of their own and males of the convergently similar species, and the two would divide space as a single species. Interbreeding is prevented by the retention of species-specific recognition cues used by the females to select mates of their own species [mate selection is usually the female's choice in birds (50)].

Attempted hybridization and its attendant ecological disadvantages are thereby avoided.

Character convergence may evolve wherever conditions favor interspecific territoriality, as discussed above. As the advantages to spatial separation in ecologically-similar species are mutual, evolution may proceed simultaneously in both species toward new convergent phenotypes. Alternatively it may be a one-sided change, especially likely when the advantage in morphological character convergence is one-sided (see Chilean bird species, below). That such an arrangement has evolved many times and in many different animal groups is shown below. It is only surprising that interspecific territoriality often occurs with no such convergence in signals between participants. But notably these cases almost always involve congeners (51, but see 9, 18), and thus it is difficult to distinguish signal convergence from genetic affiliation. For this and other reasons (11) we are likely to identify only the most obvious examples of signal convergence. We can note that signal character convergence is a convergence in the stimulus part of the interaction. It is quite possible that evolution can also produce a convergence in the receptor part of the interaction, such that a behavioral reaction now results from not just the reception of one type of (intraspecific) stimulus but from two different stimuli, both intraspecific and interspecific. The flow chart of these possibilities given in Figure 4 serves as a summary.

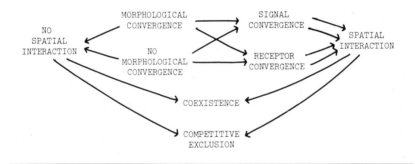

Figure 4 Flow chart of alternative selective forces and outcomes involving character convergence.

The Evidence

Both appearance and vocalization may provide the signals by which birds defend territories. First we treat convergence in appearance.

Two species groups provide only circumstantial evidence. These are the southeast Asian woodpeckers (Picidae) and the African bushshrikes *Malaconotus* and *Chlorophoneus* (11). The woodpeckers include several pairs of genera, such that one genus per convergent species pair comes from each of two taxonomically distinct sections of the family. Sympatric species in these distinct genera show remarkable similarities in appearance. Most striking are the species of *Dinopium* (*javense* and *benghalense*) and *Chrysocolaptes* (*lucidus*), difficult to separate in the field (Plate 1 in 11) despite taxonomic divergence (20, but see 3 for an argument for taxonomic similarity). The convergence in appearance extends to coincidence of coloration (red, yellow, black, and white) and patterning, which varies clinally from north to south in India and Ceylon. The only subspecies which are distinct (not convergent) are those in allopatry: *D. b. dilutum* and *D. b. benghalense* (part) in northwest India and West Pakistan, and six subspecies of *C. lucidus* on the Philippines. The characters which remain different in the convergent species are malar stripe and bill color, both associated by limited evidence (48, 60) with mate selection in this family. Thus *Dinopium* has a dark bill and *Chrysocolaptes* a pale one, except in some allopatric subspecies.

The shrikes provide a similar picture with quite exact similarities in appearance linking pairs of species, one from each of the two genera (color plate in 26; the authors propose merging the two genera into *Malaconotus*). The convergently similar pairs are those that occur together in a habitat type, and the species within genera are separated by habitat preference. Thus savannah, rain forest, and montane forest each support a pair of taxonomically distinct (at a level near the genus) but visually similar species. In the savannah indirect evidence associates one of the pair with more open and the other with less open habitat, and although they are widely, almost completely, sympatric most collecting localities produce only one of the pair. Local competitive exclusion may therefore take place. Voice and eye color might provide the necessary species-specific recognition cues to mate selectors.

Convincing evidence of the association of like appearance with interspecific territoriality was recently provided for some Mexican finches (15). The widespread North and Central American *Pipilo erythrophthalmus* and the south Mexican endemic *P. ocai* co-occur in brushy, forest-edge habitats and hybridize everywhere (57) except on Cerro San Felipe, Oaxaca, where a third species, *Atlapetes brunneinucha*, is also present. *A. brunneinucha* and *P. ocai* are extremely similar in appearance, the more striking because of their bright green, chestnut, black, and white plumage. The two are interspecifically territorial on Cerro San Felipe and behave ecologically as a single species there. Only the former occurs within the forest and only the latter in the drier, more open habitat lower on the mountain. Differences in voice may aid mate selection. It was concluded that selection would act against *ocai-erythrophthalmus* hybridization by selecting against the *ocai* that lost their *brunneinucha*-like appearance and hence their profitable interaction with *A. brunneinucha*.

A. brunneinucha ranges widely through Central America to South America showing considerable geographic variability (7). It sometimes occurs with other ground-scratching finches that it closely resembles in appearance. In Veracruz, for example, it is found with three saltators, *Saltator atriceps*, *S. maximus*, and *S. coerulescens*, and the sparrow *Arremon aurantiirostris* (64). At least two of the saltators also vary geographically in some of the characteristics, e.g. throat color, in which they resemble *brunneinucha*. These finches, plus some others, may be involved in a character convergence complex which extends throughout Central America into South America.

Unusual similarity in appearance in unrelated sympatric species appears to be not uncommon. In some cases the species vary geographically in parallel with each other, accentuating species interdependence and inviting an explanation in terms of character convergence. Wallace (63) came upon such a case on the island of Buru in the Malay Archipelago, where an oriole (*Oriolus*) and the friarbird *Philemon moluccensis* (of different families) look identical. On the nearby island of Ceram the two appear quite different from the Buru birds (and are taxonomically distinguished), but again resemble each other quite precisely. This relationship between the two taxa extends to the New Guinea representatives, which J. Diamond (personal communication) has observed in some detail. The two species are found in exactly the same habitats, in the same altitudinal range, and feeding at the same heights—the crowns of trees. They also take the same types of food. He reports that they are virtually indistinguishable in the field, for as well as being similar in appearance they have similar behaviors and apparently a similar social structure. The two species are not territorial in the sense that most temperate bird species are [as Karr (31) has described for Panamanian forest birds], but tolerate each other, both intra- and interspecifically, only beyond a certain distance when they feed together. The convergence seems, therefore, to have evolved in response to very similar ecologies promoting interspecific aggression, rather than to have been a result of selection for (Batesian) mimicry as first postulated by Wallace.

There are several other cases of possible character convergence in appearance that are worth mentioning. The North African wheatear *Oenanthe hispanica* changes subspecies where it meets *O. desertii* to very closely resemble the latter. Wheatears have been reported to be interspecifically territorial on wintering grounds (59). Terborgh (personal communication) has commented upon an amazing resemblance between a tanager *Hemispingus sp. nov.* and a warbler *Basileuterus luteoviridis* in the Cordillera Vilcabamba between 3000 m and 3500 m in Peru. On the Micronesian island of Palau a white-eye *Rukia palauensis* (Zosteropidae) and a warbler *Psomathia annae* (Sylviidae) look "amazingly similar" (2). Character convergence could also account for the confusing similarities in neotropical flycatchers of the genera *Myiozetetes, Pitangus, Conopias,* and *Megarhynchus,* the vireos (genus *Vireo*) of North and Central America, and the European wagtails of the *Motacilla flava-cinerea-alba* groups.

Just as the New World hummingbirds Trochilidae are commonly interspecifically territorial (9, 18), their African counterparts, sunbirds Nectariniidae, exhibit the same phenomenon. *Nectarinia tacazze* and *N. famosa* have nearly identical diets and defend interspecific territories above 3450 m in Kenya (8). The two species are alike in shape and proportion and also in voice. Color differences between the males exist but, as in hummingbirds, do not preclude interspecific aggression.

Character convergence in voice is appropriately discussed next, for on Cerro San Felipe the two towhees *P. ocai* and *P. erythrophthalmus* are also interspecifically territorial and show similarities in their songs. Each has a number of song types, with a broader range in *ocai,* and most songs are easily assigned to one or the other species. However, just as with the fox sparrow and towhee on Blue Ridge, some song types of each are "intermediate" and cannot be attributed to either except by observing the vocalist (15). This may have been a recent development, as song similarity was not noticeable in the 1940s (57). Thus *brunneinucha-ocai* of like appearance and *erythrophthalmus-ocai* of like voice show nonoverlapping territories, even though their habitat requirements where mapped are identical. But the species *brunneinucha-erythrophthalmus* differ in both appearance and voice, and should therefore overlap in territories; they do (15). Is there any disadvantage to this overlap? It appears that there is, and that a reduction in available food supply is compensated by an increased territory size over that measured in pairs which by chance do not overlap. Of five complete *brunneinucha* territories mapped, three did not overlap with *erythrophthalmus* and averaged (5500 + 5400 + 4400)/3 = 5100 sq ft, one overlapped partially and measured 8400 sq ft, and the last was completely enclosed by an *erythrophthalmus* territory and was 18,400 sq ft in area.

Character convergence in voice is also known in the wrens *Thryothorus* and the meadowlarks *Sturnella*. In each case the associated interspecific territoriality has been documented (21, 34). *Thryothorus felix* and *T. sinaloa* sing identical songs and show no territorial overlap on mainland Mexico. Only *T. felix* occurs on the Tres Marias Islands, where it sings a different song and where its pattern of facial striping resembles that of mainland *sinaloa*. Mainland *felix,* however, shows a different facial pattern, indicating that this characteristic might be involved in mate selection, diverging in sympatry and converging in allopatry as the song type does exactly the opposite. The meadowlarks *Sturnella magna* of eastern North America and *S. neglecta* of western North America are rather more difficult to interpret, as their song is learned rather than innate (34). Thus selection for character convergence here is tantamount to selection for an indiscriminacy in the source of the learned song.

Two other cases worth mentioning are in European warblers Sylviidae. *Hippolais icterina* and *H. polyglotta* are geographic replacements and are interspecifically territorial where they overlap on a northwest-southeast front across eastern France (19). Their songs there are very similar. Likewise *Phylloscopus bonelli* and *P. sibilatrix* have very similar songs; these species are extensively allopatric, but occur together over large areas in central Europe (4). A similar situation is described for the titmice *Parus montanus* and *P. palustris* (62). The former shows a greater resemblance in song to the latter, with which it shares habitat and food (see also 32), than with its own subspecies which occurs at higher elevations.

Song Variation in Passerine Birds

In many passerine species a great deal of variation in song exists within individuals, within populations, and among populations of the same species. This sort of variation is particularly common among wrens Troglodytidae, New World and Old World thrushes Mimidae and Turdidae, and among finches Fringillidae and Emberizidae. All have quite generalized feeding habits. Individuals of a species encounter different sound environments in different parts of their individual, population, and species ranges (42). They encounter parallel differences in competitive environments. Possibly a partial territorial exclusion amongst different species through the mechanism of partial song convergence is widespread amongst passerines and accounts for some of the song variability. Similar reasoning might account for the sort of song mimicry observed in starlings (*Sturnus vulgaris*) and mockingbirds (*Mimus polyglottos*), which incorporate into their song repertoires parts of the songs of other species with which they share habitats.

Distributional Patterns in Chilean Birds

Despite similarities in the ranges of habitat types and altitudes between Chile and California, the distribution of bird species differs considerably between the two countries (12). Most striking is the wider use of habitat types in Chile by bird species; even drastic changes of habitat structure produce few if any new species. In order to see new species one must change latitude or, less productively, altitude as species replacements occur by geographic area. The scrub and open country habitats are well packed with species, rather more densely than are their Californian counterparts. Thus the picture is one of habitats well packed with species which persist through habitat changes, a pattern akin in this latter characteristic to islands. Central Chile, from where these census data come, is in fact an isolated area. It must have a slow, island-like colonization rate by differentiated species in trans-Andean populations, but lacks, because of its large area, the high extinction rates of islands.

In Chile there are very few groups of both ecologically similar and sympatric species, for most congeners are allopatric there. The genera *Muscisaxicola* and *Cinclodes* and the family Rhinocryptidae are exceptions; in these natural selection appears to have pared down within-habitat diversity from a higher number of taxonomic species to a lower number of ecological species by producing interspecific convergences in bill size (12, 15). The *Muscisaxicola* are tundra flycatchers and reach maximum ecological diversity in northern Chile; *Cinclodes*, which are dipper-like birds, reach maximum diversity on the streams and lakes in central Chile; and the rhinocryptids are most abundant in the *Nothofagus* and *Cryptocarya* forests in the south. Away from these areas of maximum resource space, which permit maximal species coexistence, the number of taxonomic species can exceed the number which the habitat can support as separate ecological entities, and bill convergences result. Limited evidence reveals that local competitive exclusion is resulting from the convergences, which in turn produces a bird species turnover with habitat more typical of a continent (12).

Character Convergence in Other Animal Groups

A good instance of character convergence in insects appears to be the dragonflies *Micrathyria eximia* and *Nephepeltia phryne* (D. Paulson, personal communication). These are interspecifically territorial, visually oriented, and look identical in the field.

Many coral-reef fish, particularly those of the families Chaetodontidae, Pomacentridae, Ballistidae, and Labridae, are territorial. Some of these are extremely brightly colored, but others are not; in some species the young are more brightly colored and more aggressive than the adults. Some species, particularly pomacentrids, have been recorded as interspecifically territorial, but the interpretation of this behavior has been either that it is a misdirected form of intraspecific agonistic behavior (37), or that it is simply defense of eggs and nest against potential predators (37, 53).

Pomacentrus jenkinsi defends its territory against all intruders (53), but more interesting is the behavior of *P. flavicauda* on the Great Barrier Reef. This species excluded 38 species of reef fish from its territories, but ignored another 16. The 38 that were excluded varied considerably in appearance, behavior, and taxonomy, but held in common among themselves and with *flavicauda* the fact that they were algae-eaters, while the 16 permitted within the territory were all carnivores (38). The association of the excluded species and their diet similarity with the territory holder is described above for interspecifically territorial bird species. This behavior is apparently achieved without convergence in appearance, and qualifies therefore as convergence in receptor; at least the receptor-effector system has been modified to include responses to signals that would normally trigger only conspecifics. In fact many reef fish are remarkably alike in their color patterns. Aquarium triggerfish reacted aggressively towards two other species (37), one of which was similar in appearance and the other similar in shape. An even more clear-cut example of convergence in signals has recently been reported in freshwater territorial fish on the Olympic Peninsula (25).

Signal character convergence apparently occurs in the salamanders *Plethodon jordani* and *Desmognathus ochrophaeus* in the Great Smoky Mountains (11). Different subspecies on different mountains possess in common a red "ear" patch or red forelegs, although not all individuals in the population are so endowed. The two forms are also alike in general body coloration and occur in the same habitat at the same altitudes. Where the two species do not coincide in range or altitude the incidence of these red markings drops appreciably or altogether. A possible explanation involving Batesian mimicry has received extensive testing but no confirmation (30, 49). Intraspecific aggression has been observed in both species and it seems reasonable to attribute their coincident markings to selection for interspecific spacing. Stevan Arnold (personal communication) finds these species lying with their distinctively colored forequarters extended out of their burrows, and rarely more than one individual per log. The same interaction may occur between *Plethodon cinereus* and *Desmognathus ochrophaeus* (different subspecies) further north, which possess a red mid-dorsal stripe in common, and between salamander species of the genus *Chiropterotriton* in Mexico. Indeed, it may be widespread. Of particular interest are the many amphibians in which the young are colorful and aggressive and the adults dull and placid. It appears that, for many amphibians, sites for larval and juvenile development are more widely used in common than are habitats and feeding places for the adults. In such cases the young of several species may share a simple and indivisible habitat (e.g. tree bole) and may profitably become interspecifically aggressive and spatially assorted.

Character convergence in signals has not been reported for lizards, although because of visual signalling among brightly colored males it could be expected to occur. A possible instance involves similar dewlap coloration in Puerto Rican anolis lizards, *Anolis cooki* and *A. cristatellus* (Roughgarden, personal communication).

Finally, in mammals the only possibility of signal character convergence I have encountered is the resemblance between the ground squirrels *Spermophilus lateralis* and *S. leucurus,* including their close relatives, and the chipmunks *Eutamias.* These are remarkably similar in general coloration and both have prominently striped sides; lack of facial striping in the ground squirrels distinguishes them from the chipmunks. There are broad areas of sympatry between the two species groups, and at least two chipmunk species and one of the ground squirrels can be found in the same habitat. Insufficient information exists about their behavioral interactions to discuss the situation further. Lastly, an instance of morphological character convergence has been reported in the weasels *Mustela*, in which two species are most similar in body size where they co-occur (54).

SOCIAL MIMICRY

The Hypothesis

Studies of interspecific social behavior in neotropical birds have resulted in the conclusion that natural selection may favor within mixed species groups the evolution of signal convergence to promote gregariousness (43–46); this is "social mimicry." Moynihan's (44) explanation for flock formation and cohesion is largely a mechanistic one involving the reactions of "circumference" species to individuals or groups of "nucleus" species. The ultimate factors involved in flock formation, particularly in flocks of different species, were not discussed, although it is safe to assume that selection has favored flocking in birds and that it is ecologically advantageous.

The Ecological Advantage to Flocking in Birds

Sociality is extremely common among birds. Indeed, the species that do not participate in some sort of flocking behavior at some time during the season or during their lifespan may well be in the minority. A common pattern in many temperate bird species is territoriality during the breeding season and flock formation at other times. In tropical forests flocks may be seen throughout the year; individuals leave the flocks for breeding schedules and rejoin them after reproductive efforts cease. Not only do individuals of single species form flocks, but in many different situations several species, even dozens, participate in this behavior together.

There are potential adaptive advantages of various sorts to flocking. Mixed species aggregations around clumped, temporarily abundant food supplies form for obvious reasons. Flocks may serve some function in dominance-subordinance hierarchies (37), and may also present predators with a more difficult task through "confusion effects" (24, 65). But mixed species flocks feeding on small, dispersed food items present the most challenging problem. In wintering finches in the Mohave Desert flock formation appears to be a mechanism to regulate "return time" or time intervals between successive visits to a particular point in the habitat. With renewing food supplies each location in the habitat replenishes its stock after the flock has passed by, but it is only worth revisiting after a certain time interval. Flocks can move over the habitat in such a way that the return time mean is regulated to match food renewal rates and the return time variance is minimized. Individuals which forage independently have no knowledge of the history of prior visits to any one feeding site and thus have no way of avoiding short return times except by trial and error. Thus independent individuals would be selected against (13).

The general concept of return time regulation by flock formation as a strategy for exploiting renewing food supplies may be of more general applicability. The tropical forest flocks of insectivores are of course utilizing renewing insect food, and their paths may show considerable regularity (47).

Social Mimicry in Flocking Birds

Neotropical bird species which flock together often tend to resemble each other in appearance, in spite of the fact that the flock members are of different species from different families. Thus lowland forest flocks in Panama which are composed of antbirds (Formicariidae), a tanager (Thraupidae), and a warbler (Parulidae) show considerable uniformity. Montane bush flocks in western Panama contain a preponderance of species with black and/or yellow markings, as well as finches, tanagers, and warblers. In the humid temperate forest of the northern Andes the members of mixed species flocks are predominantly either blue or blue and yellow (tanagers and honeycreepers Coeribidae), while in the same habitat further south species are commonly blue or blue-grey above with chestnut or buff underparts (tanagers, honeycreepers, and a plush-capped finch Catamblyrhynchidae) (46). Another instance of such convergence on a grand scale comes from lowland rain forests in Peru. Of 26–30 species of antbirds, eleven in four different genera are understory foragers that frequently join mixed species flocks. The males of eight of these species are uniformly dull blue-grey in color, the other three are charcoal; they are distinguished only by small black and white flecks of embellishments (Terborgh, personal communication). The similarities among species are far too striking to be mere coincidence, and an argument invoking common selection for crypticity is rendered unlikely by the fact that the females of all eleven species are similarly reddish- and greyish-browns, but are much more readily distinguished than the males. The semisocial coots (*Fulica*), which are intra- and interspecifically aggressive at short range and show very little divergence in appearance, may be a further example (12).

Convergence to Facilitate Both Aggression and Aggregation

One of the most striking cases of social mimicry is that of the Panamanian grassland finches (43), in which the characteristics of an all black plumage with a white flash on the wing are retained across three genera, *Volatinia, Sporophila,* and *Oryzoborus.* These form mixed species flocks and, at least some species in some combinations, are interspecifically territorial (N. G. Smith, personal communication). Thus the convergent appearance might serve equally well to separate breeding individuals as to promote flocking outside the breeding season. The tyrant flycatchers *Muscisaxicola* are a second example of the same phenomenon. All species are similar in appearance and are distinguished chiefly by a dab of color on the back of the head. They are not only interspecifically territorial during the breeding season, but in the months preceding the establishment of territories they form mixed species flocks in the Andean foothills (12, 58). Again, the similarity in appearance could function equally well in promoting interspecific spacing at one season as in facilitating flocking at another.

CONCLUSION

Species that converge in appearance, voice, and/or morphology are apparently responding to selection by restricted resources unable to support them as separate ecological entities. This is particularly apparent in the genus *Muscisaxicola*, in which the number of "ecological species" matches the altitudinal range available to the genus better than the number of taxonomic species. Similarly, the towhee niche in the San Gabriel Mountains supports two species at lower elevations, one at higher elevations, and at still higher elevations two species co-occur but behave ecologically more like one.

Flock formation is similarly associated with food resources that are low in abundance and variety. Mohave Desert finch flocks become larger and develop from single species to mixed species aggregations as the winter season progresses and food supplies dwindle (13). Food supplies of sparse, small items which are not defendable but are renewing over time are also best exploited by flocks. Such flocks in the neotropics promote social mimicry, which facilitates their formation and coherence.

Displacement patterns, the result of selection for divergence in ecological characteristics particularly associated with the ways in which food is gathered, evolve only on resources that are neither superabundant nor extremely low in either predictability or abundance. Towards either extreme of resource abundance selection no longer favors divergence. Superabundant resources are exploited opportunistically, but when resources are in short supply relative to the number of species (or individuals) character convergence is evolved. The result is that taxonomically different species may behave ecologically as a single species, either in mixed species flocks in the nonbreeding season or by defending territories interspecifically in the breeding season.

ACKNOWLEDGMENTS

I thank Jared Diamond for reading an early draft of this manuscript. The original field work reported in the paper has been generously supported by the National Science Foundation.

Literature Cited

1. Albrecht, H. 1969. Behavior of four species of Atlantic damselfishes from Colombia, South America (*Abudefduf saxatiles, A. taurus, Chromis multilineata, C. cyanea;* Pisces, Pomacentridae). *Z. Tierpsychol.* 26:662–76
2. Baker, R. H. 1951. The avifauna of Micronesia, its origin, distribution and evolution. *Univ. Kans. Natur. Hist. Mus. Publ.* 3:1–359
3. Bock, W. 1963. Evolution and phylogeny in morphologically similar groups. *Am. Natur.* 97:265–85
4. Brémond, J. C. 1970. Recherche experimentale sur les composantes assurants la spécificite du chant chez *Phylloscopus* *bonelli. Abstr. Congr. Int. Ornithol., 15th.* 1970:73
5. Brown, W. L., Wilson, E. O. 1956. Character displacement. *Syst. Zool.* 5:49–64
6. Cade, T. J. 1960. Ecology of the peregrine falcon and gyrfalcon populations in Alaska. *Univ. Calif. Publ. Zool.* 63:151–290
7. Chapman, F. M. 1923. Mutation among birds of the genus *Buarremon. Bull. Am. Mus. Natur. Hist.* 48:243–78
8. Cheke, R. A. 1971. Feeding ecology and significance of interspecific territoriality of African montane sunbirds (Nectariniidae). *Rev. Zool. Bot. Afr.* 84:50–64

9. Cody, M. L. 1968. Interspecific territoriality among hummingbirds. *Condor* 70: 270, 271
10. Cody, M. L. 1968. On the methods of resource division in grassland bird communities. *Am. Natur.* 102:107–47
11. Cody, M. L. 1969. Convergent characteristics in sympatric populations: A possible relation to interspecific territoriality. *Condor* 71:222–39
12. Cody, M. L. 1970. Chilean bird distribution. *Ecology* 51:455–64
13. Cody, M. L. 1971. Finch flocks in the Mohave Desert. *J. Theor. Pop. Biol.* 2: 142–58
14. Cody, M. L. 1973. Coexistence, coevolution and convergent evolution in seabird communities. *Ecology* 54:In press
15. Cody, M. L. 1973. *Bird Communities.* *Monogr. Pop. Biol.* Princeton: Princeton Univ. Press. In press
16. Diamond, J. M. 1972. *The Avifauna of Eastern New Guinea.* Cambridge, Mass.: Publ. Nuttall Ornithol. Club, No. 12
17. Drury, W. H. Jr. 1961. Studies on the breeding biology of the horned lark, water pipit, lapland longspur and snow bunting on Bylot Island, Northwest Territories, Canada. *Bird Banding* 32:1–46
18. Dunford, C., Dunford, E. 1972. Interspecific aggression of resident broadtailed and migrant rufous hummingbirds. *Condor* 74:479
19. Ferry, C., Deschaintre, A. 1966. *Hippolais icterina* et *polyglotta* dans leur zone de sympatrie. *Abstr. Congr. Int. Ornithol.* 1966:57, 58
20. Goodge, W. R. 1972. Anatomical evidence for phylogenetic relationships among woodpeckers. *Auk* 89:65–85
21. Grant, P. R. 1966. The coexistence of two wrens of the genus *Thryothorus*. *Wilson Bull.* 78:266–78
22. Grant, P. R. 1968. Polyhedral territories of animals. *Am. Natur.* 102:75–80
23. Grant, P. R. 1972. Convergent and divergent character displacement. *Biol. J. Linn. Soc.* 4:39–68
24. Grinnell, J. 1903. Call notes of the bushtit. *Condor* 5:85–87
25. Hagen, D. W., Moodie, G. E. E., Moodie, P. F. 1972. Territoriality and courtship in the Olympic mudminnow *(Novumbra hubbsi)*. *Can. J. Zool.* 50: 1111–15
26. Hall, B., Moreau, R. E., Galbraith, I. C. 1966. Polymorphism and parallelism in the African bushshrikes of the genus *Malaconotus*(including *Chlorophoneus*). *Ibis* 108:161–82
27. Harris, R. D. 1944. The chestnut-collared longspur in Manitoba. *Wilson Bull.* 56:105–15

28. Hespenheide, H. 1971. Food preference and the extent of overlap in some insectivorous birds, with special reference to the Tyrannidae. *Ibis* 113:59–72
29. Hespenheide, H. 1973. *Ecological Inferences from Morphological Data.* *Ann. Rev. Ecol. Syst.* 4:213–29
30. Huheey, J. E., Brandon, R. A. 1961. Further notes on mimicry in salamanders. *Herpetologica* 17:63, 64
31. Karr, J. R. 1971. Structure of avian communities in selected Panama and Illinois habitats. *Ecol. Monogr.* 41:207–33
32. Lack, D. L. 1966. *Population Studies of Birds.* Oxford: Oxford Univ. Press
33. Lack, D. L. 1971. *Ecological Isolation in Birds.* Oxford: Blackwell
34. Lanyon, W. E. 1957. The comparative biology of the meadowlarks *(Sturnella)* in Wisconsin. *Publ. Nuttall Ornithol. Club.* no. 1:1–67
35. Levins, R. 1962. Theory of fitness in a heterogeneous environment. I. The fitness set and adaptive function. *Am. Natur.* 96:361–78
36. Levins, R. 1968. *Evolution in Changing Environments.* *Monogr. Pop. Biol.* 2. Princeton: Princeton Univ. Press
37. Lorenz, K. 1966. *On Aggression.* New York: Bantam
38. Low, R. M. 1971. Interspecific territoriality in a pomacentrid reef fish, *Pomacentrus flaviċauda* Whitley. *Ecology* 52:648–54
39. MacArthur, R. H. 1958. Population ecology of some warblers of northeastern coniferous forests. *Ecology* 39:599–619
40. MacArthur, R. H. 1970. Species packing and competitive equilibria for many species. *J. Theor. Pop. Biol.* 1:1–11
41. MacArthur, R. H., Levins, R. 1967. The limiting similarity, convergence and divergence of coexisting species. *Am. Natur.* 101:377–85
42. Marler, P. R. 1960. Bird songs and mate selection. In *Animal Sounds and Communication*, ed. W. E. Lanyon, W. N. Tavolga. Wash. DC: AIBS
43. Moynihan, M. 1960. Some adaptations which help to promote gregariousness. *Proc. Congr. Int. Ornithol., 12th.* 1958: 523–41
44. Moynihan, M. 1962. The organization and probable evolution of some mixed species flocks of neotropical birds. *Smithson. Misc. Collect.* 143(7):1–140
45. Moynihan, M. 1963. Interspecific relations between some Andean birds. *Ibis* 105:327–39
46. Moynihan, M. 1968. Social mimicry: Character convergence versus character displacement. *Evolution* 22:315–31

47. Nicholson, E. M. 1932. *The Art of Bird Watching.* London: Witherby
48. Noble, G. K. 1936. Courtship and sexual selection of the flicker *(Colaptes auratus luteus).* Auk 53:269–82
49. Orr, L. P. 1967. Feeding experiments with a supposed mimetic complex in salamanders. *Am. Midl. Natur.* 77: 147–55
50. Orians, G. H. 1969. On the evolution of mating systems in birds and mammals. *Am. Natur.* 103:589–603
51. Orians, G. H., Willson, M. F. 1964. Interspecific territories of birds. *Ecology* 45:736–45
52. Pitelka, F. A., Tomich, P. Q., Triechel, G. W. 1955. Ecological relations of jaegers and owls as lemming predators near Barrow, Alaska. *Ecol. Monogr.* 25:85–117
53. Rasa, O. A. E. 1969. Territoriality and the establishment of dominance by means of visual cues in *Pomacentrus jenkinsi* (Pisces:Pomacentridae). *Z. Tierpsychol.* 26:825–45
54. Rosensweig, M. 1968. Anecdotal evidence for the reality of character convergence. *Am. Natur.* 102:491, 492
55. Roughgarden, J. 1973. On invading a guild of coexisting species. Unpublished
56. Schoener, T. W. 1969. Models of optimal size for solitary predators *Am. Natur.* 103:277–313
57. Sibley, C. G. 1950. Species formation in the red-eyed towhees of Mexico. *Univ. Calif. Publ. Zool.* 50:109–94
58. Smith, W. J., Vuilleumier, F. 1971. Evolutionary relationships of some South American ground tyrants. *Bull. Mus. Comp. Zool.* 141(5):179–268
59. Stresemann, E. 1950. Interspecific competition in chats. *Ibis* 92:148
60. Tanner, J. T. 1942. *The Ivory-billed Woodpecker. Res. Rep. No. 1, Nat. Audubon Soc.*
61. Terborgh, J. 1971. Distribution on environmental gradients: Theory and a preliminary interpretation of distributional patterns in the avifauna of the Cordillera Vilcabamba, Peru. *Ecology* 52:23–40
62. Thönen, W. 1962. Studien uber die Mönchsmeise. *Ornithol. Beob.* 59: 103–72
63. Wallace, A. R. 1869. *The Malay Archipelago.* London: Macmillan
64. Wetmore, A. 1943. The birds of southern Veracruz, Mexico. *Proc. US Nat. Mus.* 93:215–340
65. Wynne-Edwards, V. C. 1962. *Animal Dispersion in Relation to Social Behavior.* Edinburgh: Oliver and Boyd

ECOLOGICAL INFERENCES ❖ 4059
FROM MORPHOLOGICAL DATA

Henry A. Hespenheide[1]

Biological Sciences Group, University of Connecticut, Storrs, Connecticut

To some extent ecologists have differentiated into two groups: one whose primary concern is theory, another whose primary concern is collection of data. Fretwell's engaging analysis of research strategies in ecology (19:x–xix) suggests that specialization as theorist or data collector increases prestige, which is almost certainly favored by selective biology faculties in the current era of restricted budgets and PhD oversupply. One corollary of such specialization has been the tendency of theoreticians to test models, not by direct field observation of ecological systems, but by measurement of museum specimens for a morphological characteristic indicating the organism's ecology. Certain types of ecologic analysis over large biogeographic areas can probably be done in no other way short of several man-lifetimes of study (46, 57, 74). How valid is this methodology?

The purpose of this review is to scrutinize critically some of the studies in which ecological generalizations have been made or tested on the basis of morphological measurements and to examine the validity of the relationship assumed between morphology and ecology. As is demonstrated in the next section, this area of concern involves a wide range of phenomena, some recently and well reviewed elsewhere. I therefore concentrate attention on two related aspects of this topic: the relationship of prey size to predator size and the phenomenon of character displacement understood in terms of niche width, separation, and overlap.

ECOLOGICALLY RELEVANT MORPHOLOGICAL CHARACTERS

Let me specify at the outset that I deal only with mensural characteristics such as size and shape, rather than such characteristics as color and pattern. The latter are ecologically important enough to be treated separately, as in Cody's review (9) of

---

[1]Present address: Department of Biology, University of California, Los Angeles, California 90024.

the behavioral aspects of character convergence or in the relationship between palatability and color pattern in aposematic coloration and mimicry (11).

A variety of mensural characters have been used at one time or another to make ecological inferences. These are usually from either trophic or locomotory append-ages, or often body size considered by itself. A variety of ecological-size correlations are involved, depending on the choice of character. Choice of a body part is compli-cated by allometry (24, 31, below; review in 20, 21). In the case of trophic append-ages the feeding method of the group of organisms in question will determine what is relevant: for studies of birds, bill length is relevant (or width or depth—23, 43, 69, 99); of lizards, jaw length (78 and included references); of parasitic wasps, ovipositor length (27, 67); of filter feeding fish, distance between gill rakers (7); of certain phytophagous insects, perhaps body size itself (29, 33, 42). In most cases the size of the organ is assumed to be related to the size of the prey, or alternatively to the distance of the food source/prey item from the predator (50, 87, 27, 67). Relative size and shape of locomotory appendages usually are discussed for their relation to the method of foraging for food; for example, the relative lengths of wing and tarsus in birds (18, 19, 44–46, 64) and of ear and wing in bats (16, 17, 57, 89). Wing size and dispersal ability, as well as egg number, are correlated in some insects (67).

Body size is related to a variety of ecological characteristics. We have already mentioned that body size may be closely related to prey size for certain types of organisms; in addition to wood-boring and seed-predator beetles, parasitic insects which completely devour single hosts likely fall in this category (67), as would predatory wasps for which prey capture is chemically arranged and limits on prey size are set by the ability to subdue the prey prior to stinging or to carry it to the nest afterward — both likely functions of total size (see below). I have found (31) that body weight of birds is even a better predictor of mean prey size than is bill size, which might be expected to give best results. Reproductive rates of organisms are complexly related to body size. In many organisms, especially fish, amphibians, and reptiles, egg number within and between species is directly proportional to body size (65, 90, references in 80, review in 91). Pianka (66, also 5, 26) has recently discussed reproductive rates, generation time, and body size in terms of the concepts of r- and K-selection. As in the wasps mentioned above, plant seed size and number are closely related to dispersal method (26, reviewed in 42). Metabolic rate of organisms is a function of weight to the 2/3 or 3/4 power (47, 83, references in 80), as are certain types of energetic efficiency (72, 100) and speed of movement (5, 36). Home range or territory size in birds (1, 75), mammals (39, 56), and lizards (92) is a function of body size within a particular food type, the slope and/or position of the regression varying among food types. These last three are interrelated in ways discussed below. Social dominance may be a function of body size, both within (19) and between species (61), larger individuals being dominant.

This variety of ecological-morphological correlations could be developed along any of these several lines, but I have chosen to concentrate on the relation of prey size to predator size and its implications for character displacement, food being a, perhaps *the,* critical niche dimension.

RELATIONS OF PREY SIZE TO PREDATOR SIZE

The assumption that large predators take large prey and small predators take small prey is basic to discussions of character displacement and models of community size distributions. Indeed, one may block out large regions of a prey size/predator size graph with relatively little empirical data, but largely on the basis of energetic and similar considerations (Figure 1). There are two basic series of strategems in feeding, defined in terms of size relationships: to be a relatively larger predator than your prey (e.g. birds feeding on insects), or a relatively smaller parasite than your host (e.g. internal parasites, insects on plants). As indicated in Figure 1, predators very much larger than their prey are usually filter feeders, the boundary between those predators collecting individual items (hunters) and those filter feeding set by energetic (benefit–cost) considerations (77, 80). Some predators are able to take prey larger than themselves by using chemical poisons (spiders, wasps) or by hunting socially (pack hunters as army ants and carnivores). Parasites include a large relative size range as well: from insect parasites of insects or insect predators of seeds (42,

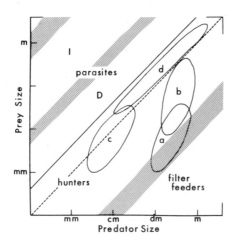

Figure 1 Generalized pattern of the relative sizes of prey and predator (both scales, log length). Dotted line indicates prey and predator are equal in size; above the line prey is larger than predator; below it, predator is larger. Solid line marks upper limit of parasite size relative to host; fuzzy areas indicate transitions/overlap between adjacent strategies. Letters indicate: D, host killed by parasite; I, host injured by single parasite individual. Ellipses indicate areas occupied by a. insectivorous birds and insect prey (31); b. hawks and owls and all prey (12, 75); c. solitary wasps and insect prey (34); d. pack hunters (army ants, carnivorous mammals).

67), in which a single parasite kills its host and is as close to the size of its host as its energetic efficiency will allow, to viruses or insects on plants, in which a single parasite individual has a negligible effect on the host and a single host individual will support large populations of the parasite.

Specific correlations of prey size to predator size have now been demonstrated for the prey of mammals (70), raptorial birds (75), fish-eating birds (31, after 2), insectivorous birds (4, 31), frugivorous birds (50), seed-eating birds (6, 28, 50, 60, 68, 98), insectivorous lizards (65, 76, 78, 81), insectivorous wasps (15, 34, 49, 55), insectivorous flies (35), and wood-boring beetles (29, 33). Two observations with respect to these correlations will be important to our subsequent discussion of niche width and overlap and their influence on character displacement:

1. Organisms of different foraging method but feeding on the same prey type show different increases of mean prey size with increased predator size (31, and included references). This is a necessary consequence of the energetic cost–benefit analysis of the prey capture process (54, 77, 80, and included references).

2. Organisms of the same foraging type feeding on different prey taxa show different increases of mean prey size with increased predator size among different prey types. This fact emerges from current studies on the prey of insectivorous insect predators (34, 35) and is shown in Table 1. Although most data are for length, the few data for weight substantiate the conclusion based on length. Analysis of data for a single bird predator (swift, *Chaetura spinicauda;* 32) also shows that different prey taxa have different mean sizes in the diet, which is interpreted as evidence that different prey taxa are differentially easy or difficult to capture in terms of energetic cost. Neither this nor the previous observation is remarkable or unexpected, but both become important in the following discussion of character displacement. The relation between the size and taxonomic identity of prey taken by generalized predators is complex, as suggested by 2 above, and may be a severe qualification of studies of organisms with mixed diets where this fact is not taken into account. In

Table 1 Regression coefficients (slope, b) for prey of predatory insects

	Pompilidae, Sphecidae[a]						Asilidae[b]		
	Length (mm)			Weight (mg)			Length (mm)		
	n	b	se	n	b	se	n	b	se
Spiders	859	0.743	0.026	91	2.222	0.150			
Coleoptera	99	0.719	0.108				36	0.295	0.088
Diptera	29	1.019	0.050				119	0.594	0.076
Hemiptera	6	0.389	0.107				15	0.398	0.128
Homoptera	22	1.244	0.156				42	0.205	0.093
Hymenoptera	6	0.404	0.148				94	0.279	0.079
Lepidoptera							17	0.440	0.176
Orthoptera	19	1.280	0.295	14	1.491	0.100	58	0.652	0.124

[a]Data from 34; reanalyzed from 15, 49, and others.
[b]Data from 35.

the discussion which follows I talk as if all predators prefer only a single prey type (and therefore size), assuming that different taxon–size relationships can be made equivalent mathematically, as is likely (32).

CHARACTER DISPLACEMENT

History and Definitions

Grant (24) has recently reviewed the historical development of the concept of character displacement and critically examined examples in the literature. Lack's essay (101) on ecological isolation and coexistence attributes to Huxley (40) the origin of the idea that differences in size imply ecological differences, but Lack includes size differences in a more general discussion of the history of ideas about the role of competition in producing community structure (a subject with which this review concludes). Lack's view is that a combination of geographic range, habitat choice, and food, under which he includes foraging station and method as well as food type and size, specify the niche of an organism vis-a-vis competitors; that is, these three are the fundamental niche dimensions of at least birds (30, 37, 52; for lizards, 65). Discussions of character displacement (as well as of niche width and overlap) often suffer from the myopia that Lack avoids; size differences are only one component of one dimension of the niche, albeit an important one. A complete definition of an organism's niche in evolutionary terms with respect to competitors almost certainly involves only a small number of dimensions [to differ slightly from Lack, perhaps only geography, time (especially for nonperennial organisms, but also for nocturnal versus diurnal forms), feeding site (habitat or host plant), feeding method (the "guild" concept of Root, 69), and food itself; see also 25, review in 93].

A review of character displacement within the context of the competitive niche dimensions of an organism is useful in evaluating recommendations of Grant (24) concerning the redefinition of the term character displacement. Brown & Wilson (8) originally defined the term to include increased morphological differences (i.e. divergence) between sympatric populations of two species; these differences were supposed to have evolved to facilitate either (a) reproductive isolation, (b) ecological isolation, or both together. Schoener (74) suggested that differences evolved under ecological pressures be termed character difference, but the term character displacement has been used primarily in the ecological sense in the literature. Character convergence (9) and character release (review in 73) were subsequently described for situations in which competition, its presence or absence, also influenced the morphology of organisms. Grant (24) recommends that both convergence and divergence be termed displacement since both include morphological changes as the result of competition, but the two are quite different phenomena. Displacement as defined by Grant compares a species' new morph size only with its previous size, independent of the direction of the change toward (convergence) or away from (divergence) the competing species. It seems clear that Brown & Wilson (9) equate displacement and divergence and are thinking about the change relative to the competitor. In most situations the result of convergence is to increase competition and facilitate exclusion of one of the competitors [9, 51, 53, 77; except in the interspecific flocks of Moynihan (63) in which plumage and behavior, not size or ecological characters, facilitate social interaction]. Since displacement (divergence) favors coexistence and convergence favors exclusion, and both result in morph changes opposite in direction, it seems useful to retain both terms.

Character Displacement and Competition

Discussions of character displacement have almost invariably invoked competition as the raison d'etre (37, 101), but there have been some situations observed in nature in which niches seemed more carefully spaced (displaced) than required (2, 37:419). There is a growing body of evidence and theory (19) that bird populations are controlled by winter densities, with seasonally low carrying capacity, K_w, and that breeding populations are well below high summer carrying capacities, K_s. For example, the bills of finches are adapted for their winter food (seeds) rather than for their summer food (insects), circumstantially suggesting winter conditions are critical. Yet breeding birds are neatly spaced with respect to their resources (30, 101) as if competition were important. How do we resolve this seeming paradox?

Let us suppose we have two breeding bird species with very similar requirements along some resource continuum (Figure 2). Let us also suppose that both are kept well below maximum possible summer density K_s by winter mortality and that not even high clutch size and multiple nestings bring either species up to K_s. Figure 2 suggests they will diverge even in the absence of strict food limitation and consequent competition, not because resources are overutilized in the region of overlap between the two species, but because they are underutilized in the area above the upper species on the continuum. The upper species should move up the continuum to exploit opportunistically the greater availability of unused resources (represented by the relative lengths of the solid vertical lines), even though there is enough to support it in situ. In the sense that the lower species reduces resource levels in the area of overlap below those in areas to the upper side of the upper species there is an element of competition involved, although not in the sense that one of the two species could not survive if the competition were not alleviated. This divergence therefore has elements of both character displacement and competitive release, as discussed above, but the existence of the competitive release can only be determined by a careful study of resource abundances and their relative use. This rough model for opportunistic character displacement would explain divergence unpredicted by considerations of food limitation, but would simultaneously invalidate displacement as proof of control of breeding population size by food limitation. If divergence occurs both under conditions of limited food and abundant food then it cannot be used as evidence for the former. Ashmole (2) has come to similar conclusions from considerations of sizes of local faunas.

Character Displacement and Niche Theory

With quantitative interpretation of the niche (53), character displacement between two different species can be seen to be a compound of interactions among at least

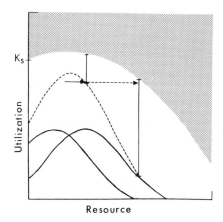

Resource

Figure 2 Pressure for character displacement in order to exploit unused resources. Open area below K_s indicates resources available. Dashed curve represents total use of the resources by two species indicated by solid curves. Height of vertical lines indicates amount of unused resources. Pressure on upper (right hand) species to move further to the right is provided primarily by the possibility of exploiting greater unused resources (dotted arrow), rather than by competition (solid arrow).

three different niche properties: 1. the direction of morphological change(s) from noncompetitive situations [most of the possibilities are covered by Grant (24)], 2. the magnitude of the change (38, 74, review in 24) [the difference between means along a niche continuum (D of MacArthur & Levins, 53), often expressed as a ratio of the larger to the smaller species], and 3. the niche width of the species (measurable by the variance or standard deviation of a normal distribution; H of MacArthur & Levins). The interaction of these three determines the amount of overlap of the two species' utilization curves (α of MacArthur & Levins), the overlap being the measure of competitive interaction and, therefore, of the pressure for displacement. The effects of differences in means and of niche width on α are considered independently in some detail.

NICHE PARAMETERS, SIZE, AND OVERLAP

Overlap and Difference in Means

MacArthur (52:40ff) has described the expected behavior of the measure of overlap α for two organisms using a continuously variable resource in which (*a*) the amount of resource used by each is identical and is normally distributed; (*b*) the niche widths, measured as the standard deviation (σ), are identical; and (*c*) the distance between the means is varied over a distance *d*. In this case, $\alpha = \exp(-d^2/4\sigma^2)$. If *d* is measured in standard deviations and the two curves are moved slowly apart from $d = 0$, α falls slowly, then more rapidly, then most rapidly at $d = \sqrt{2}\sigma$, and then more slowly again as $\alpha \longrightarrow 0$.

If the relationship between prey size and predator size had the same slope and position for all prey and all predator types then a predator of a certain mean size would always specify prey individuals of a certain mean size. We would then be free to compare size ratios of pairs of congeners or other potential competitors and discuss the observed differences among these ratios as equivalent numbers. Figure 3 shows, however, that this is impossible if the slope of the increase in prey size with predator size differs for different predator or prey types, as we have demonstrated they do (similar implications for niche width and its relation to morpohological variability are discussed below). I have previously discussed this in detail (31) for the observation that bird predators of different foraging methods show different prey size/predator size relationships, and the argument is easily extended to the second observation for predators that forage similarly but take different prey taxa (Table 1). The situation is complicated by considerations of niche width, but it can be shown (31) that the amount of overlap (α) in prey size differs for identical predator/ predator size ratios when the slope of prey size/predator size differs. If the amount of overlap determines the extent of character displacement (52, 53), then some groups of predators can achieve a particular optimal overlap for a small difference in predator size (Figure 3, curve *a;* ratio of midrange predator sizes = 1.57), whereas others require a much larger difference in size (Figure 3, curve *b;* ratio of midrange predator sizes = 1.98). This consequence of different prey size/predator size rela-

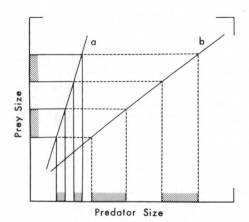

Figure 3   The effect of different relations of prey size to predator size (lines *a* and *b*) on separation of means and variability of predators. Spacing of means and variability of prey identical. See text.

tionships explains at least part of the difficulty of authors of earlier studies in generalizing about the nature of size ratios among close competitors (2, 24, 38, 74). The relationship of α to d, the difference between two species' means along a resource continuum, can rather simply incorporate the different relations of prey size to predator size by multiplying d by the coefficient b, the slope of the prey size/predator size function for a particular predator foraging method (1, above) or type of prey (2, above). The behavior of this coefficient under varying conditions of predator and prey type is currently under study (34, 35).

Niche Width and Overlap

Theoretical discussions of overlap and the limits to the similarity of competing species have usually been made on the assumption of equal niche widths of the species considered (52, 53). Pianka (65) has reasoned that if two species with different niche widths interact, the species with the narrower niche will be more adversely affected, presumably because more individuals of the species with the narrower niche will be exposed to competition than individuals of the species with the broader niche. MacArthur (52) implies that the effect is equal on the two species by giving the formula for α as $\exp(-d^2/2[\sigma_{small}^2 + \sigma_{large}^2])$. Roughgarden (73), in a paper on the evolution of niche width, reasons that larger individuals will require more energy and will therefore have larger niches than smaller individuals. He then measures overlap by comparing the area of overlap to the whole area used by each species; since the area of overlap is the same for both species but the larger species has a larger utilization curve and the smaller a smaller curve, he finds the effective overlap will be greater on the species with the smaller niche.

We can measure the overlap on species i by species j, α_{ij}, and that of i on j, α_{ji}, by the formulas of MacArthur & Levins (53)

$$a_{ij} = \Sigma p_i p_j / \Sigma p_i^2 \text{ and } a_{ji} = \Sigma p_i p_j / \Sigma p_j^2$$

where p_i is the proportionate utilization by species i of an interval of some continuously variable resource such as prey size (31). We determine the effect of different variances in the utilization curves by placing two hypothetical utilization curves a given distance apart along a continuum, fixing the variance of one, and then systematically altering the variance of the other (Figure 4). The niche width $B = 1/\Sigma p_v^2$ of the variable species v increases with increased variance and decreases with decreased variance. The values of α for the effect of the variable species on the fixed (α_{fv}) and of the fixed species on the variable (α_{vf}) change as shown in Figure 5 for three situations of different fixed variances and separations of means. By definition the curve with the smaller variance must have a greater competitive effect on the species with the larger variance than vice versa, as measured by α since if $\Sigma p_j^2 > \Sigma p_i^2$, then $\alpha_{ij} > \alpha_{ji}$, because $\Sigma p_i p_j$ is constant. For a fixed difference between means the difference between the pairs of values of α (α_{fv} and α_{vf}) in Figure 5 is greater as the difference in the variances increases, although less strongly so as the variable variance becomes smaller than the fixed variance (Figure 6). Within a region where F, defined as the ratio of the larger to smaller variance, is less than about

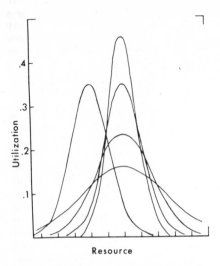

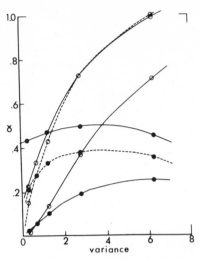

Figure 4 Hypothetical relationship of resource use by two organisms. Means of curves at distance of two units; variance of left hand curve = 1.11 units; variance of right hand curve is varied to determine effect of differences in variance on overlap (Figure 5).

Figure 5 Relation of α to differences in variances of two utilization curves at a fixed distance (Figure 4). Open circles = $\alpha_{variable,\ fixed}$, closed circles = $\alpha_{fixed,\ variable}$. Upper pair of curves (solid lines), distance = 2 units, variance = 1.11; middle pair (broken lines), distance = 2, variance = 0.83; lower pair (solid lines), distance = 3, variance = 0.83.

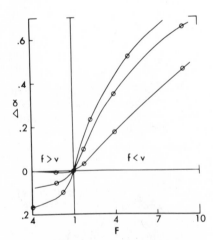

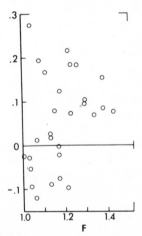

Figure 6 Difference in α as a function of difference in variance (F = higher/lower variance); $\alpha_{high,\ low}$ is always greater than $\alpha_{low,\ high}$. Data from Figure 4; curves to the right of $F = 1$ in the same relationship as in Figure 5.

Figure 7 Values of difference in α for pairs of flycatchers of high and low variance. Vertical scale is $\alpha_{high,low}$ − $\alpha_{low,\ high}$. Negative values indicate differences in the opposite direction from that predicted by Figure 6.

1.5, the two parts of the curve relating the difference in α's to F are relatively symmetrical. Figure 7 shows empirical data for comparisons of α_{ij} with α_{ji} for overlap in flycatcher prey sizes when the two birds compared differ in the variance in prey size (reanalyzed from 31). For values of F less than 1.2 differences in α are as often in the wrong direction as in the correct direction (that is, the species with the lower variance has a lesser rather than a greater competitive effect; 9 of 17 comparisons), but for F greater than 1.2, almost always positive (10 of 11 comparisons); the poor conformation to expectations when F is less than 1.2 almost certainly reflects the sampling error in estimating the true variances of the distributions.

Values of α must take into account the carrying capacities K of the species compared to have any meaning (53). As mentioned above, Roughgarden (73) also adjusted the size of his utilization curves for lizards to account for the greater energetic needs of the larger lizards, but this is doubtfully justifiable. Although larger predators do need more energy, they are also taking larger prey items. In fact, Schoener (75) has theorized that if prey weight increases as predator weight W is raised to the 0.9–1.0 power, and energy needs increase only as about $W^{\sim 0.7}$, then larger predators should actually take fewer prey per unit time than smaller ones. It can be shown that larger flycatchers have fewer prey items per stomach than do smaller flycatchers (unpublished work), although this is not true for swifts, for which the coefficient relating prey size to predator size must be less than $W^{.75}$. If unit prey size utilization curves are used to calculate α for comparisons between pairs of flycatcher species, and if the values of α are separated into $\alpha_{small, large}$ and $\alpha_{large, small}$, the two resulting clouds of points are completely overlapping showing no effect of size per se (data from 31, reanalyzed).

The relative effect of a larger competitor on a smaller competitor is in fact a compound of at least four factors: (a) the relative energy needs, (b) the relation of prey size to predator size, (c) the relative abundance of resources along the prey size continuum, and (d) the intensity of utilization of habitat space (K for a given resource level) since home range/territory size is also an increasing function of body size, lowering the per-individual intensity of use of space by large animals with respect to smaller ones (75, and above; it is unclear whether this is a simple function of prey density). To date, no study known to me has tried to relate all four of these to estimate the true values of α with respect to food size. Again it must be emphasized that food size is only one component of one dimension of the niche, as discussed above, and complete competition coefficients must include spatial, temporal, and other components even more difficultly quantifiable than prey size (reviews in 10, 51, 93).

Niche Width and Morphological Variability

It has been reasonable to many to suppose that if there is a relationship between prey size and predator size then organisms which have generalized food habits ought to be more variable morphologically than those which are specialized. This is a corollary of a more general "niche variation hypothesis" that includes nonmorphological variation as well (95, review in 94). With respect to morphological variation Van Valen (94) showed that island birds which left competitors behind on the mainland were more variable in bill measurements on the island than on the mainland (6 of 12 comparisons with the probability of $F < 0.05$; all others in the hypothesized direction). Willson (97) subsequently compared a number of temperate and tropical birds on the suggestion (48) that tropical species were more specialized, but she found no significant differences. Soulé & Stewart (86) compared African birds with mixed diets to those with more specialized diets and found no differences in variability between the sets within sexes, although those with more variable diets showed significant sexual dimorphism in one to three characters (see below).

If we return to Figure 3 we note that one of the consequences of different slopes of prey size to predator size for different foraging or prey types is that one may obtain the same range of prey sizes for a small range of predator sizes if the slope is large (curve *a*) as one may obtain from a large range of predator sizes if the slope is small (curve *b*). Conversely, if the range of predator sizes (i.e. the morphological variability) is the same for both foraging or prey types, the one with a steeper slope for prey size ought to yield a wider niche width for prey size than another with a lesser slope. Table 2 compares the slope of prey/predator size, mean niche width of prey size, and coefficients of variation for bill or wing length for three groups of insectivorous birds. The result is chaotic; despite high variability and high slope, vireos have no greater prey niche width than flycatchers, and swallows have the largest prey niche width despite the lowest slope and lowest variability.

Unfortunately, a direct relationship of prey variability to predator variability with different slopes of prey/predator size will obtain only if the error variances of the regressions are zero or equal for all foraging/prey types. Van Valen & Grant (95) and Roughgarden (73) pointed out that there are two components to niche width: within phenotype (the error variance of the regression) and between phenotypes (the regression). The total prey variance (niche width) will be the sum of these two. One may therefore attain the same niche width with respect to prey size by having low within-phenotype and high between-phenotype components, or by having high within-phenotype and low between-phenotype components. At least the within-phenotype component of niche width varies with environmental conditions. Emlen (13, 14) has predicted that predator niche width with respect to prey is an inverse function of prey abundance, a prediction verified for a Central American swift, *Chaetura spinicauda* (32) and experimentally (references in 80). Returning to Table

Table 2 Relation of niche width and morphological variability to relation of predator size with prey size

	slope, $\log_{10}$ prey size/predator size[a]	variance (mean) in prey size[a]	C.V., bill/ wing length[b]
Vireos	0.0275	0.384	0.055 – 0.070 (b)
Flycatchers	0.0088	0.385	0.021 – 0.045 (b)
Swallows	0.0060	0.505	0.027 – 0.034 (w)

[a]From Hespenheide (31).
[b]C.V. = Coefficient of Variation (σ/mean); data from Fretwell (19 – wing length, w) or Willson (97 – bill length, b).

2 we find that the three types of bird predators are listed in order of decreasing prey abundance, and therefore perhaps also in order of increasing within-phenotype niche width. However, the variability in prey size must also be affected by the energy expenditure per capture (54, 77, 80), which is highest for the flycatchers, less for the swallows, and least for the vireos. The difficulty of relating the three variables in Table 2 (prey variability, predator variability, and prey/predator size slope) appears to be due to the inability to specify the within-phenotype component to prey variability, presumably largest for the swallows and smallest for the vireos. This is presently under study (34, 35), but reanalysis of data for swifts (32) indicates all of the variance in prey size can be assigned within- rather than between-phenotypes over a small sample of birds ($n = 4$; prey items, $n = 60$).

One ambiguous case relating morphological variability to niche width may be mentioned. Study of wood-boring beetles of the genus *Agrilus* (29, 33) shows that large species bore larger branches as larvae, smaller species bore smaller twigs and branches. In some species the size of the beetles increases directly with branch diameter (33), but analysis of variance on five species in nine rearing combinations, sexes considered separately or together, shows no simple relation of morphological variability and niche width, although the problem deserves further study.

Roughgarden (73) has considered the population genetics of the evolution of morphological variability in the particular situation of colonists on islands released from competition. He concludes that asexual organisms may reach a new optimum phenotype distribution rapidly; sexually reproducing organisms, slowly if at all. Neither Roughgarden, in discussing sexually reproducing species, nor I have reviewed the growing literature on ecological sexual dimorphism in size, which is certainly relevant to the topic of intraspecific variation and niche width. Sexual dimorphism in birds (62) in an ecologically important sense seems to occur most commonly at the two ends of the food density spectrum: birds whose prey is in very low abundance [raptors (12, 88) and woodpeckers (84, 101); see also Fretwell's discussion (19)] and island species which lack competitors and presumably thereby have more food available (84, 101). As evidence, dimorphism in owls (12) increases with owl size as major prey taxa simultaneously switch from insects (more abundant prey) to vertebrates (less abundant prey). In island lizards (78) size dimorphism decreases with increased fauna size.

Roughgarden's genetic argument (73) for difficulty in arranging morphological variability in sexual species predicts such variation may be the exception in contrast to alternative nonmorphological mechanisms for attaining essentially identical ends (54, 95). In view of differences in the coefficients of variation for bill characters in Table 2 (and for other data in 19:193, and 97), the original data of Van Valen (94), and the patterns of occurrence of sexual dimorphism just discussed, the dismissal (85, 86) of differential morphological variability as a consequence of different optimal niche widths seems premature at best. More precise knowledge of prey/predator size relationships should help direct the search for such variability. This discussion of niche width has emphasized the relationship of morphological variability to food size, but other types of morphological variation associated with foraging method may also be related to other components of niche width (e.g. 18, 19; but see 3).

COMMUNITY STRUCTURE AND MORPHOLOGY

Size Frequency Distributions

The sizes of organisms, whether individuals or species, that make up a community or a fauna usually conform more or less closely to a log-normal distribution (31, 32, 39, review in 82). This would be expected (39) if competition favored (*a*) a mean size for a given type of organism, (*b*) a fixed ratio of sizes among organisms inhabiting niches larger or smaller than the mean, and (*c*) proportional size changes, if becoming either larger or smaller were equally easy to achieve. Hutchinson & MacArthur (39) attempted to derive a theoretical model of ecological size distributions based on the probabilities of selecting particular numbers and combinations of contiguous, equal-sized habitat patches among *r* different types from an environment in which the patch types were equally abundant and distributed at random. Their attempt was tested with data on mammal faunas and was not successful in detail, probably because of the assumptions of equal abundance and random distribution of patch types. It seems, however, that a successful model could be developed along these lines. Their suggestion that the extinction of (nonhunted) larger animals may be due to randomization of habitat types by human land-use patterns is insightful.

In empirical studies of the size distributions of organisms the means and variances of the distributions have been shown or hypothesized to be affected by (*a*) amount of moisture and length of growing season (insects, 32, 82), (*b*) energetics of food gathering in relation to climate and food availability (birds, 19), (*c*) the effects of predators (cladocera, 7; molluscs, 96; insects, 31), and (*d*) the availability of prey (starfish, 59; carnivores, 58; birds, 69, 79). Comparisons have been made over latitudinal (58, 79, 82) and climatic (31, 32, 82, 96) gradients, among different taxa (32, 82), and with respect to height above ground relative to dispersal mechanisms (32). The discussion of the relation of prey size to predator size has obvious implications for the size distributions of the predator guilds which exploit them, some of which are under study (34, 35).

Morphology and Faunal Composition

The ultimate correlations of morphology and ecology will include not only size distributions but also functional anatomy as an index of foraging methods and of major differences in food type for whole communities. One approach is characterization of microcommunities involved in the exploitation of single hosts or resources (67; Root, unpublished). Most extant studies are attempts to describe and ecologically compare biogeographically disparate faunas (16, 41, 46, 57). This level of analysis should eventually indicate to what extent the ecological structure of communities is dependent or independent of historical/taxonomic constraints.

Finally, two other areas of research deal with biogeographic patterns of morphological variation and its correlation with ecology, but are only mentioned here. Geographic variation in body size within species, generalized in several biogeographic rules, has been subjected to reinterpretation (58, but cf 71; see review, 22). Island patterns in sizes of organisms also have been studied in detail recently (78, and included references; review in 24).

ACKNOWLEDGMENTS

I thank G. A. Clark for bibliographic help and for critically reading a draft of the manuscript, F. A. Streams for reading the manuscript and making useful suggestions, and Martin Cody for making helpful comments on some of the ideas. J. R. Karr, F. G. Stiles, and M. F. Willson allowed me to see unpublished manuscripts. Computational assistance was given by the University of Connecticut Computer Center, in turn supported by NSF Grant GJ-9.

Literature Cited

1. Armstrong, J. T. 1965. Breeding home range in the nighthawk and other birds: its evolutionary and ecological significance. *Ecology* 46:619–29
2. Ashmole, N. P. 1968. Body size, prey size, and ecological segregation in five sympatric tropical terns (Aves: Laridae). *Syst. Zool.* 17:292–304
3. Banks, R. C. 1970. On ecotypic variation in birds. *Evolution* 24:829–31
4. Betts, M. M. 1955. The food of titmice in oak woodland. *J. Anim. Ecol.* 24: 283–323
5. Bonner, J. T. 1965. *Size and Cycle.* Princeton: Princeton Univ. Press. 219 pp.
6. Bowman, R. I. 1963. Evolutionary patterns in Darwins finches. *Occas. Pap. Calif. Acad. Sci.* 44:107–40
7. Brooks, J. L., Dobson, S. I. 1965. Predation, body size, and composition of plankton. *Science* 150:28–35
8. Brown, W. L. Jr., Wilson, E. O. 1956. Character displacement. *Syst. Zool.* 5: 49–64
9. Cody, M. L. 1973. Character convergence. *Ann. Rev. Ecol. Syst.* 4:189–211
10. Colwell, R. K., Futuyma, D. J. 1971. On the measurement of niche breadth and overlap. *Ecology* 52:567–76
11. Cott, H. B. 1940. *Adaptive Coloration in Animals.* London: Methuen. xxxii + 508 pp.
12. Earhart, C. M., Johnson, N. K. 1970. Size dimorphism and food habits of North American owls. *Condor* 72: 251–64
13. Emlen, J. M. 1966. The role of time and energy in food preference. *Am. Natur.* 100:611–17
14. Emlen, J. M. 1968. Optimal choice in animals. *Am. Natur.* 102:385–89
15. Evans, H. E. 1970. Ecological-behavioral studies of the wasps of Jackson Hole, Wyoming. *Bull. Mus. Comp. Zool.* 140:451–511
16. Fenton, M. B. 1972. The structure of aerial-feeding bat faunas as indicated by

ears and wing elements. *Can. J. Zool.* 50:287–96
17. Fleming, T. H., Hooper, E. T., Wilson, D. E. 1972. Three Central American bat communities: structure, reproductive cycles, and movement patterns. *Ecology* 53:555–69
18. Fretwell, S. D. 1969. Ecotypic variation in the non-breeding season in migratory populations: a study of tarsal length in some Fringillidae. *Evolution* 23:406–20
19. Fretwell, S. D. 1972. *Populations in a Seasonal Environment.* Princeton: Princeton Univ. Press. xxiii + 217 pp.
20. Gould, S. J. 1966. Allometry and size in ontogeny and phylogeny. *Cambridge Biol. Rev.* 41:587–640
21. Gould, S. J. 1971. Geometric similarity in allometric growth: a contribution to the problem of scaling in the evolution of size. *Am. Natur.* 105:113–32
22. Gould, S. J., Johnston, R. F. 1972. Geographic variation. *Ann. Rev. Ecol. Syst.* 3:457–98
23. Grant, P. R. 1972. Bill dimensions of the species of *Zosterops* on Norfolk Island. *Syst. Zool.* 21:289–91
24. Grant, P. R. 1972. Convergent and divergent character displacement. *Biol. J. Linn. Soc.* 4:39–68
25. Green, R. H. 1971. A multivariate statistical approach to the Hutchinsonian niche: bivalve molluscs of central Canada. *Ecology* 52:543–56
26. Harper, J. L., Lovell, P. H., Moore, K. G. 1970. The shapes and sizes of seeds. *Ann. Rev. Ecol. Syst.* 1:327–56
27. Heatwole, H., Davis, D. M. 1965. Ecology of three sympatric species of parasitic insects of the genus *Megarhyssa* (Hymenoptera: Ichneumonidae). *Ecology* 46:140–50
28. Hespenheide, H. A. 1966. The selection of seed size by finches. *Wilson Bull.* 78: 191–97
29. Hespenheide, H. A. 1969. Larval feeding site of species of *Agrilus* (Coleop-

tera) using a common host plant. *Oikos* 20:558–61
30. Hespenheide, H. A. 1969. *Niche overlap and the exploitation of flying insects as food by birds.* PhD thesis. Univ. Pennsylvania. xix + 82 pp.
31. Hespenheide, H. A. 1971. Food preference and the extent of overlap in some insectivorous birds, with special reference to the Tyrannidae. *Ibis* 113:59–72
32. Hespenheide, H. A. Three bird predators and their aerial prey. Unpublished
33. Hespenheide, H. A. Patterns in the use of single plant hosts by wood-boring beetles. Unpublished
34. Hespenheide, H. A. Statistical properties of the prey of solitary wasps (Hymenoptera: Pompilidae, Sphecidae). Unpublished
35. Hespenheide, H. A., Lavigne, R., Dennis, D. S. Statistical properties of the prey of flies of the family Asilidae. Unpublished
36. Hocking, B. 1953. The intrinsic range and speed of flight of insects. *Trans. Roy. Entomol. Soc. London* 104:223–345
37. Hutchinson, G. E. 1958. Concluding remarks. *Cold Spring Harbor Symp. Quant. Biol.* 22:415–27
38. Hutchinson, G. E. 1959. Homage to Santa Rosalia or Why are there so many kinds of animals? *Am. Natur.* 93:145–59
39. Hutchinson, G. E., MacArthur, R. H. 1959. A theoretical ecological model of size distributions among species of animals. *Am. Natur.* 93:117–25
40. Huxley, J. 1942. *Evolution: The Modern Synthesis.* New York: Harper. 645 pp.
41. James, F. C., Karr, J. R. 1973. Computers and the ecology of tropical bird communities. *Mus. News.* In press
42. Janzen, D. H. 1969. Seed eaters versus seed size, number, toxicity and dispersal. *Evolution* 23:1–27
43. Johnson, N. K. 1966. Bill size and the question of competition in allopatric and sympatric populations of dusky and gray flycatchers. *Syst. Zool.* 15:70–87
44. Keast, A. 1968. Competitive interactions and the evolution of ecological niches as illustrated by the Australian honeyeater genus *Melithreptus* (Meliphagidae). *Evolution* 22:762–84
45. Keast, A. 1970. Adaptive evolution and shifts in niche occupation in island birds. *Biotropica* 2:61–75
46. Keast, A. 1972. Ecological opportunities and dominant families, as illustrated by the Neotropical Tyrannidae

(Aves). *Evol. Biol.* 5:229–77
47. Kendeigh, S. C. 1970. Energy requirements for existence in relation to size of bird. *Condor* 72:60–65
48. Klopfer, P. H., MacArthur, R. H. 1960. Niche size and faunal diversity. *Am. Natur.* 94:293–300
49. Kurczewski, F. E., Kurczewski, E. J. 1968. Host records for some North American Pompilidae (Hymenoptera) with a discussion of factors in prey selection. *J. Kans. Entomol. Soc.* 41:1–33
50. Leck, C. F. 1971. Overlap in the diet of some Neotropical birds. *Living Bird* 10:89–106
51. Levins, R. 1968. *Evolution in Changing Environments.* Princeton: Princeton Univ. Press. 120 pp.
52. MacArthur, R. H. 1972. *Geographical Ecology: Patterns in the Distribution of Species.* New York: Harper & Row. xviii + 269 pp.
53. MacArthur, R. H., Levins, R. 1967. The limiting similarity, convergence, and divergence of coexisting species. *Am. Natur.* 101:377–85
54. MacArthur, R. H., Pianka, E. R. 1966. On optimal use of a patchy environment. *Am. Natur.* 100:603–9
55. Mason, L. G. 1965. Prey selection by a non-specific predator. *Evolution* 19:259–60
56. McNab, B. K. 1963. Bioenergetics and the determination of home range size. *Am. Natur.* 97:133–40
57. McNab, B. K. 1971. The structure of tropical bat faunas. *Ecology* 52:352–58
58. McNab, B. K. 1971. On the ecological significance of Bergmann's rule. *Ecology* 52:845–54
59. Menge, B. A. 1972. Competition for food between two intertidal starfish species and its effect on body size and feeding. *Ecology* 53:635–44
60. Morris, D. 1955. The seed preferences of certain finches under controlled conditions. *Avicult. Mag.* 61:271–87
61. Morse, D. H. 1970. Ecological aspects of some mixed-species foraging flocks of birds. *Ecol. Monogr.* 40:119–68
62. Morse, D. H. 1971. The insectivorous bird as an adaptive strategy. *Ann. Rev. Ecol. Syst.* 2:177–200
63. Moynihan, M. 1968. Social mimicry; character convergence versus character displacement. *Evolution* 22:315–31
64. Osterhaus, M. B. 1962. Adaptive modifications in the leg structure of some North American warblers. *Am. Midl. Natur.* 68:474–86
65. Pianka, E. R. 1969. Sympatry of desert

lizards *(Ctenotus)* in western Australia. *Ecology* 50:1012–30

66. Pianka, E. R. 1970. On *r*- and *K*-selection. *Am. Natur.* 104:592–97
67. Price, P. W. 1972. Parasitoids utilizing the same host: adaptive nature of differences in size and form. *Ecology* 53: 190–95
68. Pulliam, H. R., Enders, F. 1971. The feeding ecology of five sympatric finch species. *Ecology* 52:557–66
69. Root, R. B. 1967. The niche exploitation pattern of the blue-gray gnatcatcher. *Ecol. Monogr.* 37:317–50
70. Rosenzweig, M. L. 1966. Community structure in sympatric carnivora. *J. Mammal.* 47:602–12
71. Rosenzweig, M. L. 1968. The strategy of body size in mammalian carnivores. *Am. Midl. Natur.* 80:299–315
72. Rosenzweig, M. L., Sterner, P. W. 1970. Population ecology of desert rodent communities: body size and seed-husking as bases for heteromyid coexistence. *Ecology* 51:217–24
73. Roughgarden, J. 1972. Evolution of niche width. *Am. Natur.* 106:683–718
74. Schoener, T. W. 1965. The evolution of bill size differences among sympatric congeneric species of birds. *Evolution* 19:189–213
75. Schoener, T. W. 1968. Sizes of feeding territories among birds. *Ecology* 49: 123–41
76. Schoener, T. W. 1968. The *Anolis* lizards of Bimini: resource partitioning in a complex fauna. *Ecology* 49:704–26
77. Schoener, T. W. 1969. Models of optimal size for solitary predators. *Am. Natur.* 103:277–313
78. Schoener, T. W. 1970. Size patterns in West Indian *Anolis* lizards. II. correlations with the sizes of particular sympatric species—displacement and convergence. *Am. Natur.* 104:155–74
79. Schoener, T. W. 1971. Large-billed insectivorous birds: a precipitous diversity gradient. *Condor* 73:154–71
80. Schoener, T. W. 1971. Theory of feeding strategies. *Ann. Rev. Ecol. Syst.* 2: 369–404
81. Schoener, T. W., Gorman, G. C. 1968. Some niche differences among three species of Lesser Antillean anoles. *Ecology* 49:819–30
82. Schoener, T. W., Janzen, D. H. 1968. Some notes on tropical versus temperate insect size patterns. *Am. Natur.* 101:207–24
83. Schmidt-Nielsen, K. 1970. Energy metabolism, body size, and problems of scaling. *Fed. Proc. Fed. Am. Soc. Exp. Biol.* 29:1524–32
84. Selander, R. K. 1966. Sexual dimorphism and differential niche utilization in birds. *Condor* 68:113–51
85. Soulé, M. 1970. A comment on the letter by Van Valen and Grant. *Am. Natur.* 104:590–91
86. Soulé, M., Stewart, B. R. 1970. The "niche variation" hypothesis: a test and alternatives. *Am. Natur.* 104:85–97
87. Stiles, F. G. Ecology, flowering phenology, and hummingbird pollination of some Costa Rican *Heliconia.* Unpublished
88. Storer, R. W. 1966. Sexual dimorphism and food habits in three North American accipiters. *Auk* 83:423–36
89. Tamsitt, J. R. 1967. Niche and species diversity in neotropical bats. *Nature* 213:784–86
90. Tilley, S. G. 1968. Size-fecundity relationships and their evolutionary implications in five desmognathine salamanders. *Evolution* 22:806–16
91. Tinkle, D. W., Wilbur, H. M., Tilley, S. G. 1970. Evolutionary strategies in lizard reproduction. *Evolution* 24: 55–74
92. Turner, F. B., Jennrich, R. I., Weintraub, J. D. 1969. Home ranges and body size of lizards. *Ecology* 50: 1076–81
93. Vandermeer, J. H. 1972. Niche theory. *Ann. Rev. Ecol. Syst.* 3:107–32
94. Van Valen, L. 1965. Morphological variation and width of ecological niche. *Am. Natur.* 99:377–89
95. Van Valen, L., Grant, P. R. 1970. Variation and niche width reexamined. *Am. Natur.* 104:589–90
96. Vermeij, G. J. 1972. Intraspecific shore-level size gradients in intertidal molluscs. *Ecology* 53:693–700
97. Willson, M. F. 1969. Avian niche size and morphological variation. *Am. Natur.* 103:531–42
98. Willson, M. F. 1971. Seed selection in some North American finches. *Condor* 73:415–29
99. Willson, M. F. 1972. Seed size preference in finches. *Wilson Bull.* 84:449–55
100. Willson, M. F., Harmeson, J. C. 1973. Seed preference and digestive efficiency of cardinals and song sparrows. *Condor.* In press
101. Lack, D. 1971. *Ecological Isolation in Birds.* Cambridge, Mass.: Harvard Univ. Press. xi + 404 pp.

ECOLOGY OF FORAGING BY ANTS

❖ 4060

C. R. Carroll[1]
Department of Biology, University of Chicago, Chicago, Ill.

D. H. Janzen
Department of Zoology, University of Michigan, Ann Arbor, Mich.

Ants (Formicidae)[2] have received much attention by taxonomists, ethologists, and biochemists (10, 16, 20, 133, 134, 175). Little attention has been given to the ecology of their food gathering, a major selective force in the evolution of their morphology, ethology, and biochemistry. In this review we wish to collate and reinterpret some of the widely scattered observations on ant foraging ecology.

First and foremost, we must recognize that selection in this system is not operating at the level of the individual ant; as Wilson (174) has so properly emphasized, colony fitness is the only near approximation we have of worker ant foraging "success." Second, ant colonies forage primarily for particulate and widely scattered food items. Sugar-rich exudate from extra-floral nectaries and their homopteran animal counterparts is the only stationary and immediately renewable food harvested by many species of ants. Third, most ant species are scavengers on animal matter. In addition, they range from active predators to seed gatherers, including extreme generalists such as swarm-raiding army ants (Ecitoninae, Dorylinae) to specialists on termites (*Megaponera* and *Termitopone*), centipedes (*Amblypone*), and other ants (Cerapachyinae). A few are even obligatory foragers on the vegetative parts of specific plants (Attini, *Azteca, Pseudomyrmex, Pachysima*).

In Table 1 we list food characteristics of major importance in selection for colony and worker foraging patterns. None of these characteristics is likely to dominate the entire foraging ecology of an ant species because the characteristics are not mutually exclusive and because ants harvest food types that require mixed foraging strategies in space and seasonal time. For example, desert seed-harvesting ants obtain large

---

[1]Present address: Department of Ecology and Evolution, State University of New York, Stony Brook, N.Y.

[2]Taxonomy used here follows Brown (20) as much as possible.

numbers of seeds directly from the plant at some times, widely scattered seeds off the ground at other times, and scavenge for dead insects simultaneously (39, 95, 137, 155). In north temperate zone forests, wood ants (*Formica rufa* group) prey heavily on insects in late spring and early summer, but feed mainly on homopteran honeydew in late summer (66, 134, 154).

KINDS OF FOOD

The important kinds of food for ants (mobile insect prey, nests of other social insects, seeds, vegetative parts of plants, dead animal matter) place certain constraints on foraging ecology. We consider these kinds here, and in later sections examine foraging from the viewpoint of intracolony economics.

Hunting Mobile Insects

Mobile insects are protein- and calorie-rich resources that are energetically costly to harvest and of uncertain location in time and space. Even in the tropics there is a strong seasonal component to insect prey availability (51, 75, 81, 82). Many edible prey can escape easily from substrate-bound ants owing to their limited vision, small size, and reliance on direct capture by mandibles and sting alone. Almost all truly predaceous ants supplement their carnivore diet by scavenging and/or collecting secretions from Homoptera.

Very few ant species are exclusively hunters of mobile insects, and these species tend to be far from the representative ant in hunting technology. Dacetine ants (Myrmicinae) are highly predaceous (171), with most species preying on Collembola and other soft-bodied litter organisms (21); *Strumigenys, Smithistruma,* and other genera have greatly elongated mandibles with specialized cocking mechanisms and "trigger" hairs lining the inner mandible edge (21). By preying on abundant soil

Table 1   Food characteristics of importance in selection for colony and worker foraging patterns. Citations refer to exemplary studies that deal at least in part with the characteristic.

1. A "worker-load" of food occurs as widely spaced units or strongly clumped (8, 119, 137).
2. The difference between the resource value (calories + elements + strategic materials) of each worker-load of food and the cost to the colony of obtaining it (50, 52, 147).
3. The traits of food items as compared with the short and long term needs of the colony (37).
4. Storage properties of food items within the nest (39, 155).
5. Longevity of unharvested food items (137).
6. Timing of food appearance with respect to labor availability (37, 127).
7. Size of foraging range with respect to the balance of search and ownership costs against yield (53, 54).
8. Probability distribution of the means and variances of the food characteristics described in this table.
9. Time required to convert food to ants (15, 16).
10. Physical environmental heterogeneity of foraging range (17, 39).

arthropods (59, 128–130), dacetines appear to be specialized for short prey handling and location times. The pantropical *Odontomachus* (Ponerinae) has mandibles highly convergent with dacetine mandibles. However, *Odontomachus* is 7 to 12 mm in length and it is likely that the abundant soil and litter microfauna is unavailable to them. *O. haematodis,* a pantropical species, preys on large and live insects. Even this specialist predator, which we have frequently seen in Costa Rica with large insects as prey, tends coccids (Homoptera) for their sugar-rich exudates and may even construct shelters over them (48). Even the neotropical *Ectatomma* (Ponerinae), which are able but generalized predators on single large insects such as *Trigona* bees visiting sugar baits, also tend Homoptera (23). For a colony of their biomass, the ants mentioned above have a small number of large workers, have nests that are scarcely more than cavities in the ground, litter, or fallen stems, and may be viewed as effective predators only at a prey density that is far above that for diminishing returns by more generalized foragers.

Army ants (Ecitoninae, Dorylinae) are generalized predators that greatly extend the size and behavioral range of live arthropod prey by foraging in large groups (175). They may lose as much as half the colony through foraging mortality over a two-month period, but take in an immense amount of prey in exchange. To various degrees, they lack permanent nests (111, 122), and it is very unlikely that an army ant colony could sustain itself from a fixed nest site. As with the solitary foraging predators mentioned in the previous paragraph, army ants harvest so exclusively from the "cream" of the arthropod biomass that they appear not to be food limited. This is, of course, not true if efficiency is taken into account in harvest economics. Most species also raid other social insect nests (111), an exceptionally concentrated food source.

Raids on Nests of Social Insects

Social insect nests are rich sources of food for ants. They are relatively sessile, often perennial at the individual and population level, chemically conspicuous, and well protected from other types of predators. They may even be directly (*Termitopone* continually raiding the same termite nest) or indirectly (*Eciton* catching only the larvae from a vespid wasp nest) renewable resources. That portion of the ant colony's resource budget that could have been expended on searching may be diverted evolutionarily to overcoming the nest defense. It is somewhat surprising that other social insects can survive at all in ant-rich habitats. Stingless bees (*Trigona, Melipona*) and army ants seem to be the only conspicuous social insects that are free of nest predation by ants. Stingless bees are notorious for very well-developed chemical defense (175), which should be the only truly effective defense against a numerically superior attacker. Tropical wasps with open-faced nests (comb) are under constant threat of attack (47) and *Mischocyttarus drewseni* coats the nest pedicel with an ant-repellant chemical (85). That most tropical social wasps have arboreal and closed nests may well be in major part a defense against foraging ants.

Neotropical army ants (Ecitonini) frequently prey on other ant nests. Rettenmeyer (111) found that "all species of army ants studied captured more [broods of] ants than any other kind of prey." However, they catch few adult ants. Some ants show an explosive escape behavior that is triggered by army ant pheromones. Trails of the army ant *Neivamyrmex nigrescens* cause a "startle reaction by workers of *Camponotus* and other ants preyed on by army ants" (122). In the dry forests of Guanacaste, Costa Rica, the nests of a common stem-dwelling ant, *Camponotus planatus,* are often raided by army ants when the stem is large and near the ground; when a worker army ant enters the nest, each worker *C. planatus* picks up a larva or cocoon and rushes out of the nest; these workers escape. If air from an aspirator containing army ants (*Labidus* sp.) is blown over a *C. planatus* colony in the laboratory, the same reaction occurs (R. Carroll, unpublished). On the other hand, Old World tropical army ants (Dorylini) avoid trees occupied by the very aggressive formicine *Oecophylla longinoda* (140, 146). *O. longinoda* is very abundant in African lowland forest (117, 160) and thus a major part of the ant biomass in the forest is not harvested by army ants. Some much less conspicuous ponerine ants, such as *Simopelta,* and cerapachyine ants, such as *Cerapachys, Phyracaces,* and *Sphinctomyrmex,* make group raids on ant nests and sometimes catch adults as well as larvae (55, 165, 166). Wilson (165) notes that the raided colony often survives the raid with a substantial part of the brood and worker force unharmed; this might well be a place where there is true selection for a "prudent predator."

Both specialized and generalized ants prey on termites in the nest. *Leptogenys, Termitopone,* and *Megaponera* mount small raids against termite colonies all times of the year, but only harvest a small amount from each nest visited (123, 162). *Pheidole megacephala* is an important predator on West African savanna termites, especially following heavy rains when the walls of the termite nests are soft and exits have been made for the sexual forms (123). It is perplexing that ants in general (and especially army ants) do not prey more heavily on termite colonies, as a very large number of species harvest individual workers if caught by themselves. The major deterrents to the ants probably lie in the defensive ability of the termite soldier castes (in Costa Rica, one *Nasutitermes* soldier placed by itself on an ant-acacia is able to repel 10 to 40 worker ants before it exhausts its supply of defensive secretions; Janzen, unpublished) and in the improbability of the evolution of foraging behavior that encompasses both complex group raiding from termite nests and more generalized foraging when termites are absent.

The temperate zone "slave-making" ants (e.g. *Polyergus rufescens* and *Formica sanguinea*) obtain a major part of their worker force from colonies of other *Formica* (135, 136, 158), which is a behavior we might expect to evolve where the growing season is short and success in resource harvest is directly related to the number of workers foraging for the colony. There is the possibility for an interesting feedback between the density of slave-making colonies and slave colonies at a site. As the frequency of raids on slave colonies increases, the slave colonies should be weakened not only by the loss of workers but by the increasing foraging competition from their workers that are working for the other colony. Slaving activities will therefore have to be adjusted downward so as to maintain an optimal density of host colonies. In short, the slaver is explicitly leaving part of its prey population unharmed, just as we would expect of a long-established parasite. We may also expect strong selection for slave-making species which have small colonies of very aggressive and organizationally competent workers, as with *Harpagoxenus* (156); the end result of such a chain might be a parasitic queen who simply enters the host nest and takes over the extant foraging system, e.g. *Teleutomyrmex* queens in *Tetramorium* nests (175).

Strangely, intercolony aggression among ants only occasionally involves predation, and then usually when alternate foods are unavailable and territorial boundaries are not yet established (e.g. during a temperate zone spring, 37, 45).

In summary, ant predation on social insects appears to be divided into two adaptive peaks. Huge colonies (Dorylinae, Ecitoninae) attempt to capture the entire prey colony and are continually on the move. Very small colonies (in comparison to their prey colonies) take small fractions of the colony, probably forage within the same small area for a long time, and otherwise behave in a manner we would expect of a parasite.

Seed Collecting

Seeds with an oily covering may be fed on by almost any kind of ant (including some army ants, 111, 160); whether the seed itself is eaten depends on its hardness and chemistry. When the seed bears an external oil body (elaiosome), a wide variety of ants are customarily the dispersal agents and do not eat the seed (7); from a foraging standpoint, the elaiosome is simply a dead insect analogue. In this section we consider only those cases where the ant harvests seeds to eat.

Seed predation occurs in many genera of ants, most prominently in the Myrmicinae in dry habitats (*Pheidole*, 30–32, 94, 158; *Pogonomyrmex*, 29; *Veromessor*, 8, 137; *Messor*, 39). Seeds have high nutrient values (high lipid and nitrogen content, 4, 78, 91) and should be almost as valuable as are insect prey. However, they can also be stored in quantities adequate to lower the need for continual foraging (39) and may be metabolized for free water (39, 155). We know that many seeds contain large amounts of toxic compounds (74, 78) and the question then becomes how do seed-harvesting ants deal with them. Ants prey on many species of seeds (95, 99, 137, 155); Cole (27, 28) recorded 29 species of seeds in the diet of *Pogonomyrmex occidentalis* at one site. There are several strategies available. (*a*) They may select from the nontoxic end of the spectrum. Such seeds are likely to be small and very seasonal in abundance, as they are likely involved with predator satiation as a plant reproductive syndrome (74). Grass seeds are low in toxins (78) and sometimes constitute a major part of the diet of ant seed predators (27, 28, 95, 158). However, ants are unlikely to be able to survive solely on grass seeds as many harvester ants live in grass-poor sites where they subsist on the very large store of dicot annual seeds in the soil (137) or blown into the depressions between sand dunes (39). (*b*) They may eat only a nontoxic part of the seed and the embryo is a possible candidate. However, there are no records of such fine resource partitioning. In a certain sense, eating the elaiosome and ignoring the seed, as does a Californian *Pogonomyrmex* with *Dendromecon* seeds (Bullock, unpublished), is such an event. (*c*) Toxic seeds may be eaten in small quantities or only when other seeds are not available (presumably at some loss in fitness to the ant) as when *Veromessor* progressively take more *Plantago* seeds as the supply of others is depleted (137). Seed mixtures may be very important in this context; it may well be that the colony can eat toxic seeds only at some low proportion of the total seed intake (stored), and this proportion becomes the colony growth rate-limiting process. Seed digestion by *Veromessor pergandei* has high potential for seed detoxification (155). The ants apply a secretion from the gaster to the seeds, which results in softening and degradation of the seed. (*d*) Ants may have a generalized internal detoxification mechanism for seed toxins, but there is no evidence for this.

Seed harvesting strategies clearly require discriminatory behavior by the individual ant. *Pogonomyrmex badius* workers apparently form a chemical search image for certain seeds, as a single worker tends to collect one species of seed repeatedly (99). *Veromessor pergandei* switch abruptly from *Plantago* seeds to the seeds of other desert annuals when they become available after a desert rain (137).

Scavenging

Most worker ants are highly opportunistic in scavenging dead insects, bird droppings, animal waste, fruit fragments, vertebrate carcasses and scraps left by larger predators, etc. At the one end of the spectrum, we have specialists that take an occasional insect; even *Veromessor pergandei* worker-loads may be as much as 21.5% items other than seeds (137) (though here it may be that the ant has chemotactically confused an insect part with a seed, 155). At the other end, we have a large number of genera containing species that subsist almost entirely on scavenged material (e.g. *Tetramorium, Monomorium, Solenopsis, Pheidole, Myrmica, Leptothorax, Iridomyrmex, Lasius,* etc). As the food items obtained by scavengers are generated by a large number of processes, they are of highly unpredictable food value and detailed spatial location. However, their general abundance should vary with habitat to the degree that scavenging ant workers or species can be expected to specialize in the area over which they forage. We should expect intense competition for scavenged food items in the most unpredictable habitats, and a good deal of ecological release by scavenging ants when competitors are removed (as in island ant faunas).

Scavenging ants may be expected to maximize the number of searchers and minimize the cost per searcher, as well as to develop mechanisms for rapid concentration of nest-mates at a food item. Species that scavenge usually have many small workers per colony, well-developed chemical communication systems, and often display conspicuous aggression over baits. When they have polymorphic workers (e.g. *Camponotus, Pheidole, Solenopsis*), the smaller workers usually do the foraging and large workers arrive well after it has been located by the colony (16, 22, 152, 153).

Harvesting from Plants and Homoptera

Direct foraging on plants has two very different components. On the one hand, we have the very specialized members of the tribe Attini (Myrmicinae) that feed entirely on fungi, and on the other hand, there are many species of ants involved in protection of plants as a result of the plant feeding the ant. Indirectly, many ants feed on plant sap by harvesting "honeydew" from Homoptera. In all three cases, there is extensive and detailed coevolution of the ants with their specific food sources.

Attini, and especially *Atta* and *Acromyrmex,* may be viewed as subdivided cows (just as army ants may be viewed as subdivided wolves). These two genera harvest very specific subsets of the community's leaves and grow fungi on them (25, 93, 115, 116, 150). Their foraging patterns thus come under the same selective forces as any large ruminant confronted by variably toxic and species-rich forage. They are also highly territorial, and not surprisingly exhibit foraging patterns that appear to be evolutionarily designed so as to minimize the proportional reduction of desirable food plants within their colony's territory. Other attine genera collect things such as caterpillar feces or fruits as substrate for their fungi (150) and are hardly more than somewhat specialized scavengers.

Ants that aid plants can be divided into three types. Those involved in obligatory mutualisms with the plant (72, 73, 77, 79, 80, 83, 84) have contracted their foraging range almost entirely to the host surface and have the most localized food supply of any ant species. They undoubtedly exhibit the highest ant biomass per unit area under long-term foraging of any ant species (see especially 83). The nectar and food bodies produced by the plant are taken by other ants when the myrmecophyte is unoccupied by its usual occupant (73). In this sense, their food is very generalized. However, highly specialized behavior is required by the obligate occupant to retain and nurture the food supply. At least three myrmecophytic ants (*Pseudomyrmex, Azteca, Pachysima*) have become so aggressive that even a potential food item, such as a dead insect, is attacked and thrown off the plant (73, 79).

A very large number of tropical ant species obtain sugar (and perhaps amino acids; Baker & Opler, personal communication) from extra-floral nectaries. These structures have evolved independently on many plant families (e.g. Passifloraceae, Leguminosae, Bignoniaceae, Orchidaceae, Bixaceae, Euphorbiaceae, etc). The selective pressure for nectary production has been the direct predation and "scarecrow" effect against the plant's herbivores (B. Carroll, unpublished; 72), though in the case of groundnuts, the ants appear to function as pollinators and in aiding in burial of the fertilized flower (42). This food source is extremely generalized and extra-floral nectar is eaten by practically any ant that encounters it. Detailed chemical analyses of extra-floral nectar have shown so far only glucose and fructose (72), and glucose, fructose, and sucrose (86). At the level of the plant and the individual nectary, this food is highly predictable in space, time, and quality. It is not surprising to find it the source of rigorous territorial defense (B. Carroll, unpublished; 86). From the viewpoint of the ant, extra-floral nectaries and honeydew-excreting Homoptera may be very similar, except that the latter may require more complex care, and produce protein- and lipid-rich food as well as sugar (see below).

The third form of plant aid by ants is entirely different from those listed above. Southeast Asian epiphytic myrmecophytes have hollow tubers or modified leaves that serve as nest sites for species of *Iridomyrmex* ants with the atypical behavior of sequestering their nest refuse and directly scavenged material in the nesting cavities. As this material decomposes, nutrients are taken up directly through the cavity wall or by the plant's roots that have grown into the cavity. In this system, the ants are foraging over the large area expected of a large and generalist ant colony, but concentrating part of the intake in a manner so as to grow nest sites (84).

The harvesting of plant sap via Homoptera probably dates back to the Oligocene (159) and has been extensively documented (1, 100, 134, 147, 175). Almost all groups of ants do it, from the solitary hunters of active insects (e.g. *Daceton armigerum,* 171; *Odontomachus haematodis,* 48; and even the morphologically primitive Australian Myrmeciinae are avid flower nectar collectors, 26) to those that appear almost entirely dependent on carefully tended Homoptera (e.g. *Pachysima* with their scale insects in the hollow branches of *Barteria* trees, 79; *Acanthomyops,* 179, and *Acropyga,* 148, with their aphids on roots; *Crematogaster* with its mealy-bugs in swollen acacia thorns, 63). Strickland (131) records 50 species of ants tending the mealybug *Pseudococcus njalensis* on cacao in West Africa. Wellenstein (154) maintains that honeydew makes up more than 59% of the diet of *Formica rufa,* and Way (147) points out that crops full of honeydew constitute the largest class of food items brought into the colony.

There are substantial problems with the interpretation of the above generaliza-tions. The volume of food and its frequency of intake by the colony may be a very poor measure of the reduction in colony fitness that would occur if the volume or frequency were modified. Sugar probably has its primary importance in providing the energy used by workers while they are foraging for protein- and lipid-rich prey, and caring for Homoptera; an ant colony may have a very high overhead. Ants often build protective coverings for Homoptera (110, 138, etc) and patrol them constantly to ward off other ants, parasites, and predators. Nest sites and structures may be modified or chosen to accommodate Homoptera (5, 35, 147).

Multiple-use also complicates the analysis. The more dependent the ant colony appears to be on the Homoptera, the more likely it is to be harvesting them as protein and lipid sources as well. *Pachysima,* obligate to the African tree *Barteria,* appears to obtain almost all of its food from its scale insects (79). In the ant-plant mutualism of *Azteca* and *Cecropia* (76, 80), a system that undoubtedly evolved from *Azteca* tending nectar-excreting Homoptera in hollow stems just as many other arboreal ants do now (5, 6, 43, 79, 114), the *Cecropia* tree produces glucose-rich food bodies (113) and the Homoptera do not excrete nectar but are harvested directly by the ants (Janzen, unpublished).

Homoptera are a mixed blessing from the viewpoint of the plant. While the plant may gain protection as a byproduct of the ants' protection of the Homoptera (there is even a positive effect of *Formica* on European pine plantation production despite the tending of large populations of aphids, 53, 54, 154), the plant cannot directly control the amount or location of sap removed by the ants (Homoptera). Homoptera are frequently maintained at levels where the seed crop or vigor of the plant is substantially reduced. *Lasius fuliginosus* transports young aphids to parts of the tree with the best honeydew production (147). *L. niger* transports its coccids from dying clover roots to nearby living ones (69). The aphid eggs may be tended over winter (105) and then the newly hatched aphids placed on roots so that as many as 16 generations of aphids may be produced between April and October (158). *Formica* may tend aphids on aerial parts of trees in spring, and on roots in late summer and fall (119) and thereby may maximize their drain from the tree's resource pool.

BRINGING FOOD TO THE NEST

Once a food item has been located, it may be brought directly to the nest in the mandibles or the crop. If very small (fungal spores, insect fragments), it may be retained in the mouth cavity until a mixed bag has been prepared. These cases are straightforward and of little interest in foraging economics compared to the problem of encountering a food item too large for one worker to carry. Many ants are capable of communicating information about a new food source to other members of the colony. The result is that the food is harvested (and defended) far faster than would be the case were each worker to locate the food independently. The dynamics of such an interaction are particularly critical with animals like ants, owing to their generalized food needs and home ranges that often overlap with those of severe competitors. Recruitment to a large food item seems to be involved with one or more of the following three systems.

Worker ants search for food as individuals or as groups; the former is by far the more common strategy and includes those species with highly developed mass-recruiting behavior (e.g. *Solenopsis saevissima*, 172). We do not know what the search pattern of the workers of an individual looks like with respect to the pattern of food items. At best, we can offer a few bits of information on what foraging behavior looks like.

It is not clear to what degree individual ants rely on chemical trails while searching. Workers of *Pogonomyrmex maricopa* and *P. californicus* occasionally touch the gaster to the substrate when searching for food in a new area (65), a behavior reminiscent of trail laying (followed by mass-recruitment) into new areas by *Solenopsis saevissima* workers (172). Such chemical marks as described for *Pogonomyrmex* may be of importance in aiding the worker to return to the same area (as ants do even in the absence of recruitment trails, 8, 60, 134), and in aiding the worker to avoid redundant searching. These marks may be of great importance in worker fidelity to parts of the colony's foraging range, and therefore to the worker's efficiency and colony foraging flexibility.

There is no suggestion of use of pheromones for individual foraging by many species of worker ants. A diverse group of diurnal ants orient visually (e.g. *Messor barbatus*, 121; *Cataglyphis bicolor*, 121, 151; *Formica rufa* group, 70, 71, 119, 142; *Lasius niger*, 24, 24a; *Myrmica rubra*, 143) and we have observed many tropical arboreal species acting as though they are intimately familiar with a large foraging area and are clearly not restricted to chemical trails. The good discussions of the ethology of orientation (70, 71, 175) conspicuously omit an analysis of the relationship of foraging pattern to food pattern. For example, we should expect that the frequency of visual orientation should increase with permanency of food location (the epitome being represented by extra-floral nectaries and Homoptera), use of vision in predation, worker size (bigger eyes and brain), and exposure to predators (vision being used to find one's way after evasive action, as well as for avoiding predators). Since trail pheromones should be strongly selected for ephemerality (175), the exclusive use of chemical trails to guide the individual worker to her hunting area should be restricted to short distances and colonies with many small workers. The recruited foragers of *Pheidole crassinoda* may even use a pheromone trail for a short distance from the nest to fix their angular direction, but rely on visual orientation past the end of the trail (134). Examination of such combinations of visual orientation and pheromones (e.g. 64) in foraging behavior are needed to reverse the current reductionism in ant ethology, but are extremely difficult.

Group forager species are spectacular, but constitute a very minor portion of the ant species (Ecitoninae and Dorylinae, 111, 122; Ponerinae, 62, 134, 162, 165). Tropical species of swarm-raiding army ants are extreme generalists and take many more nonsocial insect prey (i.e. more mobile and unpredictable) than do column-raiding species (111, 134). Seasonal maps of the movements of swarm-raiding army ants (122) suggest strongly that the food is comparatively uniformly distributed within a tropical forest habitat with the ant colony optimizing intake by rarely, if ever, doubling back on itself when in the migratory phase, and foraging along a new radius each day during the statary phase. Either type of army ant foraging could evolve from the other, with, for example, swarm-raiding being selected for in more seasonal lowland tropical habitats where prey densities fluctuate greatly between seasons and habitats, and column-raiding being a form of competitive specialization on the more reliably present social insect nests in wetter lowland tropical habitats. While it is not clear if group-raiding ponerines forage as a group, or are merely recruited as a group, Wilson (165) strongly suggests that the former is the case. They may even include large arthropods other than termites in their diet and raid other ant nests (e.g. *Simopelta*, 55).

Increase in Foraging Activity

An ant that returns with exceptional food may cause an increased rate of exit of foragers from the nest. If workers merely leave the nest at a higher rate, the lack of directional bias related to food location may result in some parts of the foraging range receiving very disproportionate numbers of workers owing to long-term individual worker fidelity to paths and foraging areas. Workers of *Formica obscuriventris* (in the laboratory) returning from food baits generate "excitement" in the nest that is correlated with increased foraging activity (2). However, the workers build up on a bait at a geometric rate. Either there is some information being transmitted to the colony concerning bait location or the returning workers are having a multiplicative effect on exit of foragers from the nest. As the study was done under laboratory conditions, it may also be that the large number of workers at the bait rendered that small part of the foraging arena more attractive visually or chemically.

Under natural circumstances, the new food source may be far from the nest and well hidden. If it is diffuse, such as a fresh seed fall, generalized increased foraging activity appears most economical. If it is a large item, we would expect more directional information (see later sections) to be selected for. The type of ant foraging strategies found in a habitat clearly depends on the pattern of food input. In large and generalized colonies of temperate zone forest ants, we may expect a third pattern. The workers from colonies of *Formica execta, F. rufa,* and *F. polyctena* appear to partition their colony's foraging range and not transmit information on food location to the nest (40). Much of the protein source of these ants appears to be from localized and seasonal outbreaks of caterpillars (3, 37, 67, 68, 124, 141). We need to know to what degree the foraging range is partitioned to include patches likely to contain local peaks in caterpillar density. With *Formica,* however, we are up against the usual difficulty of understanding mixed foraging strategies and budgets. These ants are notorious tenders of aphids on trees scattered through the foraging range (40, 45, 66, 119) and the high sugar input may allow superficially inefficient search and harvest methods for protein as the workers may have temporary excesses of "fuel," with nitrogenous compounds being the rate-limiting material for the colony.

Tandem Running

Wilson (169) coined the term "tandem running" for recruitment whereby a worker of *Cardiocondyla* or *Camponotus* closely follows another worker back to a piece of food that is only slightly larger than what can be carried by one worker. Sudd (134) observed that this behavior is especially frequent among *Camponotus* that "forage under the sun of a tropical noon." Only a few of the workers in the small (175) nests of *Cardiocondyla* are involved in tandem running; the pair of ants return very rapidly to the bait (169). As *Cardiocondyla* and *Camponotus* normally forage as individuals and do not recruit at a large bait, it is likely that this behavior is the intermediate between only getting those pieces of food small enough for a worker to carry to the nest, and mobilizing a large part of the colony at a large bait to harvest it slowly and aggressively exclude competitors. Tandem running may also be a way of making use of naive ants and part of their apprenticeship for learning the location of foraging areas. That this is reasonable is suggested by the occasional observation of tandem running by large ponerines moving from one fragment of a subdivided colony to another.

Recruitment with Short-Lived Pheromone Trails

A very large number of myrmicine, formicine, and dolichoderine ants lay short-lived pheromone trails to food items that are too large for one or two workers to carry (134, 175). Lasting only a few hours, such trails serve for mass or group recruiting. In the former case, the workers at the food increase gradually to an asymptote; in the latter case, a pulse of workers arrives and there is likely to be little later recruitment.

Mass recruiting characterizes large ant colonies with their biomass subdivided into many small workers (e.g. *Iridomyrmex, Paratrechina, Myrmica, Monomorium, Pheidole, Solenopsis, Crematogaster* etc). Wilson (170, 175) suggests that mass recruitment with chemical trails is most likely to evolve in large colonies with small workers, though we feel that it is a bit difficult to determine which came first, if either. There are at least five biological circumstances where selection may drive colony structure in the direction of mass recruiting. (*a*) If there are many workers, there is a large pool of recruits available without seriously depleting the nest population (provided that *numbers* of ants in the nest is more important than *biomass).* (*b*) The pheromone contribution is small by one worker, and therefore requires multiple trail-laying by many workers to produce a long trail (172), though worker ability to make trails is undoubtedly coevolved with the foraging needs of the colony. (c) In competition-rich habitats, especially where colonies do not have very distinct foraging ranges owing to substrate heterogeneity, mass recruiting of many small workers may insure a protective force adjusted to the size of the food and its required rate of removal. However, this assumes, for example, that 0.1 g of 200 small ants is a superior competitive force to 0.1 g of 20 large ants, with respect to the average challenge for the food. (d) Numerous small workers at the end of an inaccurately laid recruitment trail (as happens with fire ants, 172, 175) may well be the best method for locating moving prey. (e) Mass recruitment should be the best foraging strategy in habitats where large items of food appear erratically within a few meters of the nest.

Solenopsis saevissima, one of the fire ants, is the most thoroughly studied of the mass recruiting species (172). Small to medium sized workers search for food and each is probably very familiar with a small subset of the colony's foraging range. When a large food item is located, the worker returns to the nest by visual orientation, laying an odor trail as she goes. Exited workers follow the trail back toward the food item, but it lasts only about 50 cm. Foraging from there, they eventually locate the food and their steady passage guarantees a continuous trail. Numbers increase at the food until it is too crowded for arriving workers to feed; these workers do not lay a trail back to the nest. Each recruit at the food must make a decision as to whether to feed and add to the odor trail, and nest inhabitants must evaluate the returning workers to determine whether to forage. The equilibrium density of workers at the food is higher the richer the food and the more starved the colony. A colony with 300 workers can achieve mass foraging over a distance of only 50 cm and one 87 times as large over a distance of 150 cm. This diminishing return phenomenon should be extremely important in setting the size of a colony's territory (foraging range) but, as we cautioned earlier, may in fact be coevolved with the energetics of foraging and prey retrieval.

Group recruiting characterizes large ant colonies with their biomass subdivided into a smaller number of large workers (e.g. *Pogonomyrmex, Camponotus, Formica,* 64, 65, 119, 165, 166). According to Hölldobler's (64) analysis of this strategy for *Camponotus socius,* the worker who has located a large food source returns to the nest by visual orientation, depositing a trail substance from the hind gut as she goes. Inside the nest, she does a "waggle" dance which is correlated with the exit of a group of foragers from the nest. The worker then follows her old trail back to the food, depositing formic acid from the sting gland (as well as more hind gut materials) which the recruits follow. Owing to the rapid evaporation of the formic acid trail, a maximum of about 30 workers may arrive at the food. Some workers then become involved in leading other groups to the food, while the larger castes tend to be involved in prey dissection and transport. The group of ants arriving at the food is larger the hungrier the colony or the worker, implying that the worker communicates her opinion as to the relative desirability of the food. When the food is honey water, the foraging changes from group to individual activity; it is likely that honey water in large amounts (normally produced in small amounts continually by extra-floral nectaries or Homoptera) is so beyond the evolutionary experience of these ants that they cannot express behavioral release in order to harvest large amounts rapidly.

Persistence of hind gut pheromone trails, such as the one described above, may be very important in seasonal adjustment of the colony's searching intensity to those parts of the foraging range with the highest productivity. If the naive (new) workers use them as initial guides while learning the range, the worker force should track the local availability of prey in time. For *Formica,* Rosengran (119) reports a recruitment system nearly identical to that of *Camponotus socius.* In addition, he finds that the frequency with which naive workers are recruited to a permanent trail is proportional to the number of workers that will be constant to that trail later in the season.

Variation in the expression of group foraging behavior suggests that its value may vary with the pattern of food input to the habitat. Hölldobler & Wilson (65) found that *Pogonomyrmex badius, P. californicus,* and *P. rugosus* workers displayed striking group foraging in the laboratory, and even the ability to discriminate between a trail leading to a rich food source and one to a poor food source. However, in the field, *Pogonomyrmex californicus* and *P. rugosus* do not display group foraging even when piles of highly desirable seeds were placed as close as 5 ft from the colony entrances; the piles were found by individual foragers in all cases (8). These differences between *Pogonomyrmex* in the laboratory and field remind us of the general problem that a great deal of the ethological work with ant foraging has been done with laboratory colonies. In the laboratory the food is often present in far different amounts and kinds than in the field, closer and in an atypical pattern with respect to the nest entrance, and the colony is not subject to the competitive and physical environmental stresses that molded its genotype. We should expect an ant colony to be physiologically and behaviorally among the most plastic of all insects, and therefore its laboratory ethology to be most suspect as a basis for understanding the evolution and ecological efficiency of observed foraging patterns. This situation is not helped by assignment of the information-free adjectives "primitive" and "advanced" to traits that have a very high probability of repeated appearance and disappearance during the evolutionary history of a lineage (e.g. "The next step up the ladder of sophistication in chemical recruitment techniques . . .", 175; "The most primitive form seems to be tandem running in *Cardiocondyla . . .*", 64).

Polyethism

Division of labor within an ant colony (polyethism) is expressed in great worker morphological and behavioral variation. The former is most conspicuous when the workers are polymorphic (163) and the latter is often correlated with aging of individual workers. Presumably an increase in the efficiency of the colony, as measured by eventual production of new colonies (174), is the selective pressure determining the range and frequency distribution of worker types. Foraging economics will be only one of several major components of the colony's budget: specialization of the parts of an ant colony cannot occur nearly to the degree they can between the parts of an organism, because in an ant colony one part has to perform a number of very different functions. We face a major problem in that we know next to nothing of the specifics of the relationship between the frequency distribution of morphological and age types in the colony's foraging range, the pattern of food items, and the frequency distribution of worker types in the colony as a whole. Of course, it is clear that different types of workers perform different tasks (22, 73, 87, 122, 150, 163) but the frequency distribution of types is related to the fate of the individual colony with respect to season and local conditions (127).

Wilson (174) has made an admirable attempt to analyze caste polyethism with respect to the kinds and temporal patterns of challenge faced by the entire ant colony. He drafted a set of linear equations that describe the proportionate weight of one class as a function of 1. fitness loss when the challenge is not met, 2. frequency of the challenge, and 3. probability that the caste will successfully meet the challenge. His model predicts that if the frequency of challenges remains constant over evolutionary time, the number of specialized castes will increase to match the number of constant challenges. However, it is obvious that the extent of caste proliferation is related to the distinctness of the challenges. For example, active predators, such as large ponerines, may use the same tools (mandibles, eyes, sting) for both foraging and nest defense; here we would expect evolutionary convergence of caste types and selection for other traits in the life form that rendered these two challenges more similar from the colony's viewpoint. Wilson also predicts that as a particular caste becomes more efficient, the proportionate weight of that caste should decrease in the colony. However, the numbers of workers in the colony should place strong lower limits on this process; in colonies with a small number of large workers the turnover rate of workers may well be low enough that a caste has to be kept above a certain relative density to be able to absorb the variance in its mortality and still be present. As it stands, Wilson's ergonomics model requires one major area of modification; a model of ant colony economics must incorporate the problem that challenges are not independent events. The failure to meet one challenge is likely to change the probability that a second challenge will occur and will change the effect of ignoring some third challenge. For example, the failure to make nest repairs will increase the probability of a predation attempt on the colony. The failure to patrol a foraging territory may result in permanent loss of foraging space.

Only a very few patterns appear when we examine the literature for information on the relationship of polyethism to colony foraging ecology. In species with a large size range, the foragers generally belong to the smaller size classes (172). However, *Oecophylla longinoda* foragers are larger than the brood-tending workers (149) and *Cataglyphis* foragers are larger than the workers digging nests (151). All sizes of workers participate in all tasks on ant-acacias, as division of labor is based on age (73). We may expect the small workers to be the foragers when (*a*) there is a high chance of worker death (e.g. army ants, predator-rich habitats), (*b*) food items are very small and widely scattered, or very large and unpredictable (either way many small workers is better than a few large ones), and (*c*) food items are of very low value relative to the amount of work in obtaining them (e.g. leaf-cutter ants). These situations tend towards self-fulfilling prophecy, however. For example, worker mortality should rise as worker size declines so the outcome depends on net rather than gross gain in survivorship.

A second pattern is that foragers are often older workers (22, 73, 119, 152, 153, 172). Numerous selective pressures may lead to this colony trait: 1. Older workers have already contributed labor to the colony so as to partially amortize their cost to the colony; they are more expendable than younger workers. 2. The more expendable workers may be more effective foragers as they can afford to be more aggressive in battles over rich food sources; aggressiveness does increase with age in some species (73, 102, 152, 153). This again is a two-edged sword; presumably, increased aggressiveness with age has been selected for owing to its value to the colony and therefore the worker may also increase in value with age. 3. Older foragers are clearly going to know more about foraging ranges than are juveniles, thereby increasing the efficiency (and perhaps quality) component of foraging (119).

Spatial Foraging Specialization

If a colony structure appears whereby individual workers become very familiar with a particular fraction of the foraging range, the size of the colony's foraging range may increase greatly. A worker can spend part of her search time merely traveling to an area she knows well enough to examine very effectively. The colony's foraging range then becomes a large patchwork of intensively visited areas, mingled with large areas that are only occasionally visited (perhaps by naive young workers?). *Pogonomyrmex barbatus* workers may travel as far as 24 m from the nest along a common route before beginning to forage for seeds (95), *Iridomyrmex purpureus* may move 70 m before beginning to forage for insects (57), and *Formica fusca* may tend aphids as far as 200 m from the nest (13). In such a strategy, numerous workers commonly use the same relatively permanent trail to get to their own respective foraging area. Such trails have been recorded for Australian *Irodomyrmex detectus* (42a) and *I. purpureus* (56, 57), neotropical leaf-cutter ants (150), Israeli *Polyrachis simplex* (101), north temperate zone *Pogonomyrmex* (95), *Camponotus* (120), *Formica* (37, 40, 45, 66, 119), and many others. Even army ants (122) may be viewed as foraging in this manner, though the "highway" changes from hour to hour or day to day. Perhaps most commonly, permanent trails lead to plants bearing the colony's Homoptera (e.g. 45, 66, 134) but they may also end in hunting or scavenging areas (119). Rosengren (119) found that *Formica* main routes had some branches that ended at Homoptera colonies and some in scavenging areas; workers displayed high and increasing trail fidelity during the foraging season. Those workers that survived the Finnish winter retained much of their trail fidelity. Establishment of new trails in the spring is based not only on the old memories, but on recruitment of naive workers to those trails with the highest food yield.

The desert harvester ant, *Veromessor pergandei*, displays even more complex colony-level foraging behavior by the workers. The workers leave the nest via a single large trail, and do not begin to forage off it until they reach the distal 10 m (8). Each day the large trail rotates about 20 degrees around the nest (8, 155). There is little doubt that this is an effective foraging strategy where seeds are in low abundance, highly dispersed, and remain a long time when unharvested. When seeds become abundant through natural events or experimental manipulation, *Veromessor* workers change to individual foraging and the diameter of the foraging range decreases substantially (8).

Food Storage

We should expect food storage to be prominent in habitats where food is unpredictable from the colony's viewpoint, when needed in large amounts over a short period (as when producing reproductives), and where it is absent for long periods yet the colony has high metabolic costs such as in hot deserts and tropical dry seasons.

Seed storage is conspicuous in deserts (29, 39) but has not been the subject of economic analysis. A critical question will be the rate of nutrient loss by the dormant seed through normal respiration, and the likelihood of attack on seed caches by other animals. We should also expect ants to treat stored seeds so that they do not germinate under the high humidity conditions of the nest.

Fat bodies in adults and larvae are a primary form of food storage. Even temperate zone mesic forest ants (*Formica,* 119; *Solenopsis,* 112) may enter the winter with large amounts of lipids in their body. Desert *Camponotus* may have greatly enlarged fat bodies (39, 175). Cannibalization of brood during a severe dry season by arboreal ants (R. Carroll, unpublished) is probably a widespread mechanism for surviving periods of low food abundance; brood cannibalization under stress is commonly observed in the laboratory (16, 92, 175). Ant larvae should be ideal food storage devices as they should have low overhead costs, put little resource into inedible cuticular structure, can contain large fat bodies, and have long starvation times (73). Eggs produced by workers and then fed to queens and larvae (trophic eggs) are known from many diverse groups of ants (16, 49, 88, 92, 150). Pharyngeal proteinaceous secretions are also produced by the workers and fed to the queens (98, 150). While such eggs and secretions may well be simply part of the colony's catabolic pathways, they may also be indicative of complex valves regulating the flow of food reserves within the colony.

Sugar is stored by specialized workers (repletes) with distended honey crops that are fated to spend their lives hanging from the roofs of the nest cavities (96). These workers have been independently evolved in the deserts of Australia, South Africa, and northern Mexico-southwestern U.S. (158). Less well-developed repletes are found in mesic temperate zone forest (*Prenolepis* and *Proformica*) and tropical rain forest (*Oligomyrmex*) (132, 175).

COMPETITION FOR FOOD

Scramble and territorial (interference) competition among ant colonies occurs largely for nest sites and food for young or mature colonies. Competition among ant colonies is complicated by various degrees of intercolony predation. In some cases, the degree of predation is strongly influenced by season.

Very few studies go beyond a general description of behavior that appears to be related to competition. The paucity of competition data is likely to be a historical consequence of the fact that current competition theory was developed in the context of free moving, nonsocial animals lacking behavioral plasticity. The detailed studies at hand (8, 11–14, 17–19, 33, 34, 41, 44, 58, 97, 103, 106–109, 140, 145, 164, 167, 168, 173, 180) describe interactions that, with few exceptions, are strongly biased towards temperate zone conditions and the constraints of habitats with low structural and ant species richness. The only reports dealing with complex tropical faunas in even a peripheral manner are Delage-Darchen (38), Room (117, 118), Leston (89), Wilson (164, 167, 168, 170), and Brown (20).

Competition for food and nest sites is very difficult to separate. Under some circumstances, number and quality of nest sites will influence the number and size of ant colonies of a particular species at a site. For example, the scarcity of logs in a patch of young temperate-zone forest may limit the number of *Camponotus* colonies (120), and in tropical deciduous forests, tree species with hard wood stems will support more arboreal ant colonies if soft wood stems are added to the tree crown (R. Carroll, unpublished). However, just as the competitive suitability of a seed germination site varies with the amount of reserves in the seed, the suitability of a nest site in the face of competition should vary with the amount of food the ant colony can harvest from that point. From the viewpoint of the ant, nest sites may be scarcer in habitats with lower harvestable productivity. Wilson (167) states that "Perhaps the most important single ultimate factor regulating colony size in the rain forest is limitation of nest space." While small colony size may be an adaptation to small nest cavities, there is a distinct possibility that small colonies are a consequence of food harvesting specialization in a predictable and competitor-rich environment. Further, reliance on nest limitation ignores the question of why there are not a small number of large but highly subdivided ant colonies in Wilson's rain forest site; this life form is common in tropical ants. A large arboreal twig-inhabiting ant colony may be able to survive while nesting only those twigs with short half-lives if there is a high food input from the surrounding foliage.

It may be possible to decide in relatively simple environments if the density of a particular species of ant is set by food or nest sites. In more complex environments, however, such statements are valid only in local time and space. To define the critical niche axis with respect to limiting resources, it is necessary to make a species by species analysis that takes into consideration the patch size for each limiting resource, the reproductive output within each patch, and the migration between patches. The entire subject has been grossly oversimplified in the ant ecology literature. For a review and analysis of the general problem see Vandermeer (139).

Interaction Among Founding Queens

A queen that is founding a new colony may produce her first workers from food reserves in her body or she may forage for food. In the former (claustral) type of founding, the queen typically builds a small nest chamber and does not open the chamber until the first workers are produced. Claustral colony founding is found among some Myrmeciinae, Ponerinae, and many Myrmicinae, and is typical in Formicinae and Dolichoderinae. In general, the claustral queen is rather defenseless and her body is large relative to her workers.

Nonclaustral colony founding may involve groups of queens (60, 83) but colony founding by multiple queens usually involves claustral queens (104, 144, 157, 161, 175). When there is strong competition for food it is likely that each nonclaustral queen will have greater fitness if she has her own foraging area, thereby reducing the average foraging distance and the time away from the brood [see Smith (125) for a similar argument as applied to squirrel territories]. Where there is an advantage to queen cooperation, as in building a worker force large enough to protect an acacia, then there may be a group of queens at the same site even if they have to forage (83). Claustral founding queens contain their food supply and do not compete for food at that stage. However, the energy expended to find a suitable mate and a nest site may reduce the number of workers critical for colony survival. Under these circumstances, there is selection for unrelated conspecific queens to pool their food reserves. Once the mixed colony begins growth, aggression is expected among the queens for colony ownership. We may expect group founding to occur frequently among claustral queens in food-poor environments or seasons. We also note that in the sense of traditional competition models, the sign of the competition coefficient will here depend on colony age in relation to food abundance in an odd sort of manner; with less food and more difficulty in locating a nest site, the more likely is cooperation.

Killing Founding Queens

Founding queens are usually killed by worker ants and often workers of the same species are the worst offenders (73, 83, 90, 106). This is a case of a strong colony displacing a very weak one; the queens are relatively defenseless and full of food reserves (they may serve as prey) so that the risk taken by a worker in dispatching a searching or founding queen is minimal. There should be strong selection favoring the behavior of killing infringing colonies at an early stage. The probability of tropical arboreal founding queens being collected in specific tree crowns is inversely related to the size of conspecific colonies in the crown and independent of other species' colony sizes (R. Carroll, unpublished).

Territoriality

Strong separation of foraging grounds is a major consequence of killing founding queens and competition among colonies. When the food supply is sparsely distributed and unpredictable in time and location, the average foraging distance and search time per worker is long. Defense of the perimeter of the large foraging grounds becomes costly relative to the yield lost by each ant infringing from another colony. Here, we expect considerable overlap of foraging grounds and defense of nest sites only. This pattern is strongly suggested by the large number of worker ant species that may be taken with any very small and repeated sample method at the same site in a tropical forest.

Territoriality is correlated with the following circumstances: (a) The colony foraging area is stable and contains a rich food supply (45, 119, 175). (b) The weight of a worker is small relative to the weight of the colony. (c) The physical structure of the environment is simple (178).

No published studies experimentally analyze the conditions under which we may expect aggressive territoriality. Aggressive tropical species in environments with much structural complexity invariably seem to have a strongly defended territory containing plants with honeydew-producing Homoptera [e.g. *Crematogaster* in West African cacao plantations (117, 118); tea plantations in Asia (35); and neotropical deciduous forests (R. Carroll, unpublished)]. The same pattern holds for the temperate species of the *Formica rufa* group (13, 45). Since honeydew supplies much of the food for adult workers, the activity of workers can be greatly increased over that which could be maintained on a diet of insect prey. It is undoubtedly for this reason that the huge territories of *Formica polyctena* (37) can be defended.

Worker territorial aggressiveness on foraging grounds seems to be not only species-specific but to change during the growth of the colony, with the intensity of food needs of the colony, and with the seasons. Aggressiveness is a byproduct of the balance of a trade-off between the risk of losing workers plus the energetic drain of agonistic behavior, and the yield in resources. The death of a forager is more costly to small colonies than large ones, and the probability of winning in a multiworker aggressive encounter is smaller. It is not surprising that aggressiveness as displayed on foraging grounds decreases dramatically away from foraging grounds, even for very aggressive species such as *Oecophylla* and *Crematogaster brevispinosa* (R. Carroll, unpublished). In the spring (13, 45), *Formica* workers are much more aggressive towards foragers infringing on the colony's territory than at other times. Aggression towards other *Formica* colonies decreases in the summer and involves more ritualized combat (37). In early spring, the workers are all old survivors from the previous year. Their days are limited and correspondingly they might be expected to attempt more risk-taking foraging behavior. The sexual brood is also increasing at this time and thus demands for protein are increasing. Territories established at this time will pay maximal long-term investments. Much of the foraging activity at this time involves collecting honeydew because the spring population of insects is still low, and therefore early spring intraspecific colony encounters include a larger component of predation.

As physical structure in an environment decreases, the probability of encounters between colonies for the same food increases. To the extent that the environment is food-rich, colonies may be large and foraging distances small. Under these circumstances, aggressive territorial defenses should be expected and competitive displacement should be common (33, 61, 178).

Competitive Displacement

There is a wealth of descriptive information concerning species displacement (9, 12, 33, 46, 61, 126, 176) when ant colonies of various species compete. For example, Haskins & Haskins (61) and Crowell (33) have documented the rate at which *Iridomyrmex humilis* has displaced *Pheidole megacephala* on Bermuda, and Erickson (46) has followed the displacement of several native species by *I. humilis* for 6 yr in an old field in southern California. Bhatkar et al (9) present a detailed account of the mechanism of competitive displacement of *Lasius neoniger* by the imported fire ant. The fire ant colony recruits huge numbers of workers to the conflict, and ultimately overwhelms the *Lasius* colony with greater numbers. The outcome is usually the complete destruction of the *Lasius* colony and consumption of the brood. However, several mounds of *Lasius* may act in concert to repel a larger colony of the fire ant. *Lasius alienus* and *Lasius niger* compete for juices of new bracken fronds. *L. alienus* approaches underground and *L. niger* on the surface; attacks commonly occur between the two at the frond base. Both species usually lose in aggressive encounters with *Tetramorium caespitum* (17). In nature *L. alienus* tends to be distributed in dry heath and *L. niger* in wet heath. When *Tetramorium* is present in moderately wet heath the two *Lasius* species do not overlap. When *Tetramorium* is removed, *L. niger* spreads into drier heath and overlaps with *L. alienus* (44). Thus, the presence of *Tetramorium* prevents much food competition from occurring between the two *Lasius* species.

In interspecific competition, a common result is the evolution of the species into minimally overlapping local distributions (109). For example, habitat selection by searching founding queens (177) may partly result in minimally overlapping distributions (17, 44). A considerable amount of spatial separation occurs among North American species of *Lasius* that relates to differences in foraging behavior. For example, E. Goldstein (personal communication) noted that four species of *Lasius* occurring in Connecticut are divisible into two pairs of species associated with differing amounts of canopy cover. Furthermore, each pair of species consists of one surface forager and one subterranean forager. Stabaev & Reznikova (126) describe two types of interactions among Siberian *Formica* that clearly relate to competition for food. *F. pratensis* nests above ground and *F. subpilosa* nests below ground. The colonies have very similar activity patterns unless the colony of *F. subpilosa* is within the territory of *F. pratensis*. Under these circumstances the activity period of *F. subpilosa* shifts to minimize overlap with *F. pratensis.* In the other pattern, the above-ground nesting species *F. uralensis* searches more thoroughly in the area where foraging zones overlap with the underground-nesting *F. picea* than they do elsewhere. For example, when *F. picea* is absent the foragers of *F. uralensis* find hidden baits less often than when *F. picea* is present.

The remarkably orderly changeover in tropical regions from a nocturnal to a diurnal fauna (175) is very likely related to avoiding food competition. For example, many of the nocturnal species of *Camponotus* in the neotropics have nest sites that appear indistinguishable from the nest sites of diurnal *Camponotus* (R. Carroll, unpublished). Since the length of time a dead insect persists as free prey in the tropics is on the order of minutes, diurnal and nocturnal species scavenging this kind of food are feeding largely on different resources. Nectar-producing plants likewise may have a distinct diurnal and nocturnal ant contingent that visits them; however, some aggressive species such as *Monacis* and *Pseudomyrmex* on *Acacia* (73) will patrol the nectar producing area on a 24 hr schedule.

The aggressive component in ant behavior interacts in very complex ways with the environment. For example, in the deciduous forests of Costa Rica, *Cremato-gaster brevispinosa* lives in cavities in tree trunks and dead stems. If there is a large colony, it is sometimes the only species in the tree canopy. However, the size of the colony depends on the availability of homopterans and a trunk cavity in which to tend them during the long dry season when other sources of food are scarce. Woodpecker damage to dead stems increases during the dry season and many stem-dwelling ant colonies are damaged by fire during the dry season. It is during this time when colonies of other species are suffering from nest damage that *C. brevispinosa* actively evicts their colony fragments and captures the brood whenever possible. Isolated trees often have only *C. brevispinosa*. However, if the trunk cavity begins to heal, these ants no longer have a place to maintain their homopteran food resources and their colony size begins to decline. They then become increasingly less successful competitors as a colony and other ant species begin to reinvade the canopy. This cycle may start again if a new cavity appears in the trunk (R. Carroll, unpublished). Aggressive processes require much finer scale examination than they have received in the past. For example, woodpecker and flicker damage to nests of *F. rufa* are extremely common (36, 45). If larger nests are selected by woodpeckers, we have a mechanism that leads to higher colony density among populations of these aggressive species.

ACKNOWLEDGMENTS

This study was supported by NSF GB-7819 and the Organization for Tropical Studies. B. Carroll provided high quality technical assistance, and S. Carroll provided social encouragement.

Literature Cited

1. Auclair, J. L. 1963. Aphid feeding and nutrition. *Ann. Rev. Entomol.* 8: 439–90
2. Ayre, G. L. 1969. Comparative studies on the behavior of three species of ants (Hymenoptera: Formicidae). II. Trail formation and group foraging. *Can. Entomol.* 101:118–28
3. Ayre, G. L., Hitchon, D. E. 1968. The predation of tent caterpillars, *Malacosoma americana* (Lepidoptera: Lasiocampidae) by ants. *Can. Entomol.* 100:823–26
4. Baker, H. G. 1972. Seed weight in relation to environmental conditions in California. *Ecology* 53:997–1010
5. Baker, J. A. 1934. Notes on the biology of *Macaranga* spp. *Gard. Bull.* 8:63–68
6. Bequaert, J. 1922. Ants of the American Museum Congo Expedition, a contribution to the myrmecology of Africa. IV. Ants in their diverse relations to the plant world. *Bull. Amer. Mus. Nat. Hist.* 45:333–583
7. Berg, R. Y. 1972. Dispersal ecology of *Vancouveria* (Berberidaceae). *Am. J. Bot.* 59:109–22
8. Bernstein, R. A. 1971. The ecology of ants in the Mojave desert: their interspecific relationships, resource utilization and density. PhD thesis. Univ. Calif., Los Angeles. 129 pp.
9. Bhatkar, A., Whitcomb, W. H., Buren, W. F., Callahan P., Carlyse, T. 1972. Confrontation behavior between *Lasius neoniger* and the imported fire ant. *Environ. Entomol.* 1:274–79
10. Blum, M. S., Brand, J. M. 1972. Social insect pheromones: their chemistry and function. *Am. Zool.* 12:553–76
11. Brian, M. V. 1952a. The structure of a dense natural ant population. *J. Anim. Ecol.* 21:12–24
12. Brian, M. V. 1952b. Interaction between ant colonies at an artificial nest-site. *Entomol. Mon. Mag.* 88:84–88
13. Brian, M. V. 1955. Food collection by a Scottish ant community. *J. Anim. Ecol.* 24:336–51
14. Brian, M. V. 1956. Segregation of species of the ant genus *Myrmica*. *J. Anim. Ecol.* 25:319–37
15. Brian, M. V. 1957. The growth and development of colonies of the ant *Myrmica*. *Insectes Soc.* 4:177–90
16. Brian, M. V. 1965. *Social Insect Populations*. New York: Academic. 135 pp.
17. Brian, M. V., Hibble, J., Kelly, A. F. 1966. The dispersion of ant species in a southern English heath. *J. Anim. Ecol.* 35:281–90
18. Brown, E. S. 1959a. Immature nutfall of coconuts in the Solomon Islands. I. Distribution of nutfall in relation to that of *Amblypelta* and of certain species of ants. *Bull. Entomol. Res.* 50:97–134

19. Brown, E. S. 1959b. Immature nutfall of coconuts in the Solomon Islands. II. Changes in ant populations, and their relation to vegetation. *Bull. Entomol. Res.* 50:523–58

20. Brown, W. L. Jr. 1973. A comparison of the hylean and Congo-west African rain forest ant faunas. In *Tropical Forest Ecosystems in Africa and South America: A Comparative Review,* ed. B. J. Meggers, E. S. Ayensu, W. D. Duckworth, 161–85. Wash. DC: Smithsonian

21. Brown, W. L. Jr., Wilson, E. O. 1959. The evolution of the Dacetine ants. *Quart. Rev. Biol.* 34:278–94

22. Buckingham, E. N. 1911. Division of labor among ants. *Proc. Amer. Acad. Arts Sci.* 46:425–507

23. Carroll, C. R., Carroll, B. L. 1973. The ecology of *Ectatomma tuberculatum.* Unpublished

24. Carthy, J. D. 1951a. The orientation of two allied species of British ant. I. Visual direction finding in *Acanthomyops (Lasius) niger. Behaviour* 3:304–18

24a. Carthy, J. D. 1951b. The orientation of two allied species of British ant. II. Odour trail laying and following in *Acanthomyops (Lasius) fuliginosus. Behaviour* 3:304–18

25. Cherrett, J. M. 1972. Some factors involved in the selection of vegetable substrate by *Atta cephalotes* in tropical rainforest. *J. Anim. Ecol.* 41:647–60

26. Clark, J. 1934. Notes on Australian ants, with descriptions of new species and a new genus. *Mem. Nat. Mus. Victoria* 8:5–20

27. Cole, A. C. 1932a. The ant, *Pogonomyrmex occidentalis,* Cr., associated with plant communities. *Ohio J. Sci.* 32:10–20

28. Cole, A. C. 1932b. The relation of the ant, *Pogonomyrmex occidentalis,* Cr., to its habitat. *Ohio J. Sci.* 32:133–46

29. Cole, A. C. 1968. *Pogonomyrmex* harvester ants: a study of the genus in North America. Knoxville: Univ. Tenn. Press. 222 pp.

30. Creighton, W. S. 1950. The ants of North America. *Bull. Mus. Comp. Zool.* 104:1–585

31. Creighton, W. S. 1966. The habits of *Pheidole ridicula* Wheeler with remarks on habit patterns in the genus *Pheidole. Psyche* 73:1–7

32. Creighton, W. S., Creighton, M. P. 1959. The habits of *Pheidole militicida* Wheeler (Hymenoptera: Formicidae). *Psyche* 66:1–12

33. Crowell, K. L. 1968. Rates of competitive exclusion by the Argentine ant in Bermuda. *Ecology* 49:551–55

34. Culver, D. 1972. A niche analysis of Colorado ants. *Ecology* 53:126–31

35. Das, G. M. 1959. Observations on the associations of ants with coccids of tea. *Bull. Entomol. Res.* 50:437–48

36. De Bruyn, G. J., Goosen-De Roo, L., Hubregtse-van den Berg, A. I. M., Feijen, H. R. 1972. Predation of ants by woodpeckers. *Ekol. Pol.* 20:83–91

37. De Bruyn, G. J., Mabelis, A. A. 1972. Predation and aggression as possible regulatory mechanisms in *Formica. Ekol. Pol.* 20:93–101

38. Delage-Darchen, B. 1971. Contribution a l'etude ecologique d'une savane de Cote d'Ivoire (LAMTO) les fourmis des strates herbacee et arboree. *Biol. Gab.* 7:462–96

39. Delye, G. 1968. Recherches sur l'ecologie, la physiologie et l'ethologie des fourmis du Sahara. Theses. Univ. d'Aix-Marseille

40. Dobrzanska, J. 1958. Partition of foraging grounds and modes of conveying information among ants. Warsaw: *Acta Biol. Exper.* 18:55–67

41. Dobrzanska, J. 1966. The control of the territory by *Lasius fuliginosus. Acta Biol. Exper.* 26:193–213

42. Doku, E. V., Karikari, S. K. 1971. Role of ants in pollination and pod production of Bambarra groundnut. *Econ. Bot.* 25:357–62

42a. Duncan-Weatherby, A. H. 1953. Some aspects of the biology of the mound ant *Iridomyrmex detectus* (Smith). *Aust. J. Zool.* 1:178–92

43. Duviard, D., Segeren, P. 1972. La colonisation d'un myrmecophyte, le parasolier, par *Crematogaster* spp. (Myrmicinae) en Cote d'Ivoire forestiere. Cote d'Ivoire: *ORSTOM* 24 pp.

44. Elmes, G. W. 1971. An experimental study on the distribution of heathland ants. *J. Anim. Ecol.* 40:495–500

45. Elton, C. 1932. Territory among wood ants (*Formica rufa* L.) at Picket Hill. *J. Anim. Ecol.* 1:69–76

46. Erickson, J. M. 1971. The displacement of native ant species by the introduced Argentine ant, *Iridomyrmex humilis* Mayr. *Psyche* 78:257–66

47. Evans, H. E., Eberhard, M. J. W. 1970. *The Wasps.* Ann. Arbor: Univ. Mich. Press. 265 pp.

48. Evans, H. C., Leston, D. 1971. A ponerine ant (Hymenoptera: Formicidae) associated with Homoptera on cocoa in Ghana. *Bull. Entomol. Res.* 61:357–62

49. Freeland, J. 1958. Biological and social patterns in the Australian bulldog ants of the genus *Myrmecia. Aust. J. Zool.* 6:1–18

50. Gentry, J., Stiritz, K. L. 1972. The role of Florida harvester ant, *Pogonomyrmex badius,* in old field mineral nutrient relationships. *Environ. Entomol.* 1:39–41

51. Gibbs, D. G., Leston, D. 1970. Insect phenology in a forest cocoa-farm locality in west Africa. *J. Appl. Ecol.* 7: 519–48

52. Golley, F. B., Gentry, J. B. 1964. Bioenergetics of the southern harvester ant, *Pogonomyrmex badius. Ecology* 45: 217–25

53. Gösswald, K. 1954. Uber die Wirtschaftlichkeit des Masseneinsatzes der Roten Waldameise. *Z. Angew. Zool.* 145–85

54. Gösswald, K. 1958. Neue Erfahrungen uber Einwirkung der Roten Waldameise auf den Massenwechsel von Schadinsekten sowie einige methodische Verbesserungen bei ihrem praktischen Einsatz. *Proc. Int. Congr. Entomol. 10th, Montreal* 4:567–71

55. Gotwald, W. H. Jr., Brown, W. L. Jr. 1966. The ant genus *Simopelta. Psyche* 73:261–77

56. Greaves, T. 1971. The distribution of the three forms of the meat ant *Iridomyrmex purpureus* in Australia. *J. Aust. Entomol. Soc.* 10:15–21

57. Greenslade, P. J. M. 1970. Observations on the inland variety (*V. viridiaeneus* Viehmeyer) of the meat ant *Iridomyrmex purpureus* (Frederick Smith). *J. Aust. Entomol. Soc.* 9:227–31

58. Greenslade, P. J. M. 1971. Interspecific competition and frequency changes among ants in Solomon Islands coconut plantations. *J. Appl. Ecol.* 8:323–52

59. Greenslade, P. J. M., Greenslade, P. 1968. Soil and litter fauna densities in the Solomon Islands. *Pedobiologia* 7: 362–70

60. Haskins, C. P., Haskins, E. F. 1950. Notes on the biology and social behavior of the archaic ponerine ants of the genera *Myrmecia* and *Promyrmecia. Ann. Entomol. Soc. Am.* 43:461–91

61. Haskins, C. P., Haskins, E. F. 1965. *Pheidole megacephala* vs. *Iridomyrmex humilis* in the Bahamas. *Ecology* 46: 736–40

62. Hermann, H. R. Jr. 1968. Group raiding in *Termitopone commutata* (Roger). *J. Georgia Entomol. Soc.* 3: 23, 24

63. Hocking, B. 1970. Insect associations with the swollen thorn acacias. *Trans. Roy. Entomol. Soc. London* 122: 211–55

64. Hölldobler, B. 1971. Recruitment behavior in *Camponotus socius. Z. Vergl. Physiologie* 75:123–42

65. Hölldobler, B., Wilson, E. O. 1970. Recruitment trails in the harvester ant *Pogonomyrmex badius. Psyche* 77: 385–99

66. Holt, S. J. 1955. On the foraging activity of the wood ant. *J. Anim. Ecol.* 24: 1–34

67. Horstmann, K. 1970. Untersuchungen uber den Nahrungserwerb der Waldameisen (*Formica polyctena* Foerster) im Eichenwald. I. Zu sammensetzung der Nahrung, Abhängigkeit von Witterungsfaktoren und von der Tageszeit. *Oecologia* 5:138–57

68. Horstmann, K. 1972. Untersuchungen uber den Nahrungserwerb der Waldameisen (*Formica polyctena* Foerster) im Eichenwald. II. Abhängigkeit von Jahresuerlauf und vom Nahrungsangebot. *Oecologia* 8:371–90

69. Hough, W. S. 1922. Observations on two mealybugs, *Trionymus trifolii* Forbes, and *Pseudococcus meritimus* Ehrh. *Entomol. News* 33:171–76

70. Jander, R. 1957. Die optische Richtungsorientierung der Roten Waldameise (*Formica rufa*). *Z. Vergl. Physiol.* 40:162–238

71. Jander, R. 1963. Insect orientation. *Ann. Rev. Entomol.* 8:95–114

72. Janzen, D. H. 1966. Coevolution of mutualism between ants and acacias in Central America. *Evolution* 20:249–75

73. Janzen, D. H. 1967. Interaction of the bull's-horn acacia (*Acacia cornigera* L.) with an ant inhabitant (*Pseudomyrmex ferruginea* F. Smith) in Eastern Mexico. *Univ. Kans. Sci. Bull.* 47:315–558

74. Janzen, D. H. 1968a. Seed eaters versus seed size, number, toxicity, and dispersal. *Evolution* 23:1–27

75. Janzen, D. H. 1968b. Differences in insect abundance and diversity between wetter and drier sites during a tropical dry season. *Ecology* 49:96–110

76. Janzen, D. H. 1969a. Allelopathy by myrmecophytes: the ant *Azteca* as an allelopathic agent of *Cecropia. Ecology* 50:146–53

77. Janzen, D. H. 1969b. Variation in behavior among obligate acacia-ants from the same colony. *J. Kans. Entomol. Soc.* 42:58–67

78. Janzen, D. H. 1971. Seed predation by animals. *Ann. Rev. Ecol. Syst.* 2: 465–92

79. Janzen, D. H. 1972. Protection of *Barteria* (Passifloraceae) by *Pachysima* ants (Pseudomyrmecinae) in a Nigerian rain forest. *Ecology* 53:885–92

80. Janzen, D. H. 1973a. Dissolution of mutualism between *Cecropia* and its *Azteca* ants. *Biotropica.* In press
81. Janzen, D. H. 1973b. Sweep samples of tropical foliage insects: effects of seasons, vegetation types, elevation, time of day, and insularity. *Ecology.* In press
82. Janzen, D. H. 1973c. Sweep samples of tropical foliage insects: description of study sites, with data on species abundances and size distributions. *Ecology.* In press
83. Janzen, D. H. 1973d. Evolution of polygynous obligate acacia-ants in western Mexico. *J. Anim. Ecol.* In press
84. Janzen, D. H. 1973e. Epiphytic myrmecophytes: mutualism by ants feeding plants. *Biotropica.* In press
85. Jeanne, R. L. 1972. Social biology of the neotropical wasp *Mischocyttarus drewseni. Bull. Mus. Comp. Zool.* 144:63–150
86. Jeffrey, D. C., Arditti, J., Koopowitz, H. 1970. Sugar content in floral and extrafloral exudates of orchids: pollination, myrmecology and chemotaxonomy implications. *New Phytol.* 69:187–95
87. King, R. L., Walters, F. 1950. Population of a colony of *Formica rufa melanotica* Emery. *Proc. Iowa Acad. Sci.* 57:469–73
88. Le Masne, G. 1953. Observations sur les relations entre le couvain et les adultes chez les fourmis. *Ann. Sci. Nat.* 15:1–56
89. Leston, D. 1970. Entomology of the cocoa farm. *Ann. Rev. Entomol.* 15:273–94
90. Levieux, P. S. 1971. Mise en evidence de la structure des nids et de l'implantation des zones de chasse de deux especes de *Camponotus* a l'aide de radio-isotopes. *Insectes Soc.* 18:29–48
91. Levin, D. A. 1973. The oil content of seeds: an ecological perspective. Unpublished.
92. Markin, G. P. 1970. Food distribution within laboratory colonies of the Argentine ant, *Iridomyrmex humilis* (Mayr). *Insectes Soc.* 17:127–58
93. Martin, M. M., Carman, R. M., MacConnell, J. G. 1969. Nutrients derived from the fungus cultured by the fungus-growing ant *Atta colombica tonsipes. Ann. Entomol. Soc. Am.* 62:1386, 1387
94. McColloch, J. W., Hayes, W. P. 1916. Preliminary report on the life economy of *Solenopsis molesta* Say. *J. Econ. Entomol.* 9:23–38
95. McCook, H. C. 1879. *The natural history of the agricultural ant of Texas. A monograph of the habits, architecture, and structure of Pogonomyrmex barbatus.* Philadelphia: Acad. Nat. Sci. 311 pp.
96. McCook, H. C. 1882. *The Honey Ants of the Garden of the Gods, and the Occident Ants of the American Plains.* Philadelphia: Lippincott. 188 pp.
97. Morisita, M. 1941. Interrelations between *Formica fusca japonica* and other species of ants on a tree. *Kontyu* 15:1–9
98. Naarman, H. 1963. Untersuchungen uber Bildung und Weitergabe von Drusensekreten bei *Formica* (Hymenoptera: Formicidae) mit Gilfe der Radioisotopenmethode. *Experientia* 19:412, 413
99. Nickle, D. A., Neal, T. M. 1972. Observations on the foraging behavior of the southern harvester ant, *Pogonomyrmex badius. Fla. Entomol.* 55:65, 66
100. Nixon, G. E. J. 1951. *The Association of Ants with Aphids and Coccids.* London: Common. Inst. Entomol. 36 pp.
101. Ofer, J. 1970. *Polyrhachis simplex,* the weaver ant of Israel. *Insectes Soc.* 17:49–82
102. Otto, D. 1958. Uber die Arbeitsteilung im Staate von *Formica rufa rufo-pratensis minor* Gossw. und ihre verhaltensphysiologischen Grundlagen: Ein Beitrag zur Biologie der Roten Waldameise. *Wiss. Abh. Deut. Akad. Landwirtschaftswiss. Berlin* 30:1–169
103. Pickles, W. 1940. Fluctuations in the populations, weights, and biomasses of ants at Thornhill, Yorkshire, from 1935 to 1939. *Trans. Roy. Entomol. Soc. London* 90:467–85
104. Poldi, B. 1963. Studi sulla fondazione dei nidi nei Formicidi I. *Tetramorium caespitum* (L.). *4th Congr. UIEIS Pavia* 12:132–99
105. Pontin, A. J. 1960a. Observations on the keeping of aphid eggs by ants of the genus *Lasius. Entomol. Mon. Mag.* 96:198, 199
106. Pontin, A. J. 1960b. Field experiments on colony foundation by *Lasius niger* (L.) and *L. flavus* (F.). *Insectes Soc.* 7:227–30
107. Pontin, A. J. 1961. Population stabilization and competition between the ants *Lasius flavus* (F.) and *L. niger* (L.). *J. Anim. Ecol.* 30:47–54
108. Pontin, A. J. 1963. Further considerations of competition and the ecology of the ants *Lasius flavus* (F.) and *L. niger* (L.). *J. Anim. Ecol.* 32:565–74
109. Pontin, A. J. 1969. Experimental transplantation of nest-mounds of the ant

Lasius flavus (F.) in a habitat containing also *L. niger* (L.) and *Myrmica scabrinodis* Nyl. *J. Anim. Ecol.* 38:747–54

110. Rau, P. 1934. Notes on the behavior of certain ants of St. Louis County, Mo. *Trans. Acad. Sci. St. Louis* 28:207–15

111. Rettenmeyer, C. W. 1963. Behavioral studies of army ants. *Kans. Univ. Sci. Bull.* 44:281–465

112. Ricks, B. L., Vinson, S. B. 1972. Changes in nutrient content during one year in workers of the imported fire ant. *Ann. Entomol. Soc. Am.* 65: 135–38

113. Rickson, F. R. 1971. Glycogen plastids in Müllerian body cells of *Cecropia peltata*—a higher green plant. *Science* 173:344–47

114. Ridley, R. W. 1910. The symbiosis of ants and plants. *Ann. Bot. London* 24: 457–83

115. Rockwood, L. 1973a. Distribution, density, and dispersion of two species of *Atta* (Hymenoptera: Formicidae) in Guanacaste Province, Costa Rica. *J. Anim. Ecol.* In press

116. Rockwood, L. 1973b. The effects of seasonality and host plant selection on foraging of two species of leaf-cutting ants (*Atta*) in Guanacaste Province, Costa Rica. Unpublished

117. Room, P. M. 1971. The relative distributions of ant species in Ghana's cocoa farms. *J. Anim. Ecol.* 40:735–51

118. Room, P. M. 1972. The fauna of the mistletoe *Tapinanthus bangwensis* (Engl. & K. Krause) growing on cocoa in Ghana: relationships between fauna and mistletoe. *J. Anim. Ecol.* 41:611–21

119. Rosengren, R. 1971. Route fidelity, visual memory and recruitment behaviour in foraging wood ants of the genus *Formica. Acta Zool. Fennica* 133:1–106

120. Sanders, C. J. 1970. The distribution of carpenter ant colonies in the spruce-fir forests of northwestern Ontario. *Ecology* 51:865–73

121. Santschi, F. 1911. Observations et remarques critiques sur le mecanisme de l'orientation chez les fourmis. *Rev. Suisse Zool.* 19:303–38

122. Schneirla, T. C. 1971. *Army Ants: a Study in Social Organization.* San Francisco: Freeman. 349 pp.

123. Sheppe, W. 1970. Invertebrate predation on termites of the African savanna. *Insectes Soc.* 17:205–18

124. Sitowski, L. 1924. Strzygonia choino'wka (*Panolis flammea* Schiff.) i jej pasorzyty na ziemiach polskich. *Szecs II. Roczn. Nauk Rol. 12.* 18 pp.

125. Smith, C. C. 1968. The adaptive nature of social organization in the genus of tree squirrels *Tamiasciurus. Ecol. Monogr.* 38:31–63

126. Stebaev, I. V., Reznikova, J. I. 1972. Two interaction types of ants living in steppe ecosystem in south Siberia, USSR. *Ekol. Pol.* 20:103–9

127. Steyn, J. J. 1954. The pugnacious ant (*Anoplolepis custodiens* Smith) and its relation to the control of citrus scales at Letaba. *Mem. Entomol. Soc. S. Africa* 3:1–96

128. Strickland, A. H. 1944. The arthropod fauna of some tropical soils. *Trop. Agr. Trinidad* 21:107–14

129. Strickland, A. H. 1945. A survey of the arthropod soil and litter fauna of some forest reserves and cacao estates in Trinidad, British West Indies. *J. Anim. Ecol.* 14:1–11

130. Strickland, A. H. 1947a. The soil fauna of two contrasted plots of land in Trinidad, British West Indies. *J. Anim. Ecol.* 16:1–10

131. Strickland, A. H. 1947b. Coccids attacking cacao in West Africa. *Bull. Entomol. Res.* 38:497–523

132. Stumper, R. 1961. Radiobiologische Untersuchungen uber den sozialen Nahrungshaushalt der Honigameise *Proformica nasuta* (Nyl). *Naturwissenschaften* 48:735, 736

133. Sudd, J. H. 1960. The foraging method of Pharaoh's ant, *Monomorium pharaonis* (L.). *Anim. Behav.* 8:67–75

134. Sudd, J. H. 1967. *An Introduction to the Behavior of Ants.* New York: St. Martin's. 200 pp.

135. Talbot, M. 1967. Slave raids of the ant *Polyergus lucidus* Mayr. *Psyche* 74: 299–313

136. Talbot, M., Kennedy, C. H. 1940. The slave-making ant, *Formica sanguinea subintegra* Emery, its raids, nuptial flights, and nest structure. *Ann. Entomol. Soc. Am.* 33:560–77

137. Tevis, L. Jr. 1958. Interrelations between the harvester ant *Veromessor pergandei* (Mayr) and some desert ephemerals. *Ecology* 39:695–704

138. Tynes, J. S., Hutchins, R. E. 1964. Studies of plant-nesting ants in east central Mississippi. *Am. Midl. Natur.* 72: 152–56

139. Vandermeer, J. H. 1972. Niche theory. *Ann. Rev. Ecol. Syst.* 3:107–32

140. Vanderplank, F. L. 1960. The bionomics and ecology of the red tree ant, *Oecophylla* and its relationship to the coconut bug. *J. Anim. Ecol.* 29: 15–33

141. Studi ed esperienze pratiche di protezione biologica. 1965. *Min. Agr. For.,* Roma, Collana Verde 16. 414 pp.

142. Voss, C. 1967. Uber das formensehen der roten waldameise (*Formica rufagruppe*). *Z. Vergl. Physiol.* 55:225–54

143. Vowles, D. M. 1950. Sensitivity of ants to polarized light. *Nature, London* 165:282, 283

144. Waloff, N. 1957. The effect of the number of queens of the ant *Lasius flavus* (Fab.) on their survival and the rate of development of the first brood. *Insectes Soc.* 4:391–408

145. Waloff, N., Blackith, R. E. 1962. The growth and distribution of the mounds of *Lasius flavus* (Fabricius) in Silwood Park, Berkshire. *J. Anim. Ecol.* 31:421–37

146. Way, M. J. 1954. Studies of the life history and ecology of the ant *Oecophylla longinoda* Latreille. *Bull. Entomol. Res.* 45:93–112

147. Way, M. J. 1963. Mutualism between ants and honeydew producing Homoptera. *Ann. Rev. Entomol.* 8:307–44

148. Weber, N. 1944. The neotropical coccid-tending ants of the genus *Acropyga* Roger. *Ann. Entomol. Soc. Am.* 37:89–122

149. Weber, N. 1949. The functional significance of dimorphism in the African ant, *Oecophylla. Ecology* 30:397–400

150. Weber, N. A. 1972. Gardening ants: the attines. *Am. Phil. Soc. Phil. Mem.* Vol. 92. 146 pp.

151. Wehner, R., Menzel, R. 1969. Homing in the ant *Cataglyphis bicolor. Science* 164:192–94

152. Weir, J. S. 1958a. Polyethism in workers of the ant *Myrmica. Insectes Soc.* 9:97–127

153. Weir, J. S. 1958b. Polyethism in workers of the ant *Myrmica* (Part II). *Insectes Soc.* 5:315–39

154. Wellenstein, G. 1952. Sur Ernahrungsbiologie der Roten Waldameise (*Formica rufa* L.). *Z. Pflanzenkr.* 59:430–51

155. Went, F. W., Wheeler, J., Wheeler, G. C. 1972. Feeding and digestion in some ants (*Veromessor* and *Manica*). *BioScience* 22:82–88

156. Wesson, L. G. 1939. Contribution to the natural history of *Harpagoxenus americanus* (Hymenoptera: Formicidae). *Trans. Am. Entomol. Soc.* 65:97–122

157. Wheeler, W. M. 1906. On the founding of colonies by queen ants, with special reference to the parasitic and slave-making species. *Bull. Am. Mus. Nat. Hist.* 22:33–105

158. Wheeler, W. M. 1910. *Ants, Their Structure, Development and Behavior.* New York: Columbia Univ. Press. 663 pp.

159. Wheeler, W. M. 1914. The ants of the Baltic amber. *Schr. Phys. Okon. Ges. Konigsberg* 55:1–142

160. Wheeler, W. M. 1922. Ants of the American Museum Congo Expedition, a contribution to the myrmecology of Africa. VII. Keys to the genera and subgenera of ants. VIII. A synonymic list of the ants of the Ethiopian region. IX. A synonymic list of the ants of the Malagasy Region. *Bull. Amer. Mus. Nat. Hist.* 45:631–710, 711–1004, 1005–55

161. Wheeler, W. M. 1933. *Colony-Founding Among Ants, With an Account of Some Primitive Australian Species.* Cambridge: Harvard Univ. Press. 179 pp.

162. Wheeler, W. M. 1936. Ecological relations of ponerine and other ants to termites. *Proc. Am. Acad. Arts Sci.* 71:159–243

163. Wilson, E. O. 1953. The origin and evolution of polymorphism in ants. *Quart. Rev. Biol.* 28:136–56

164. Wilson, E. O. 1958a. Patchy distributions of ant species in New Guinea rain forests. *Psyche* 65:26–37

165. Wilson, E. O. 1958b. The beginnings of nomadic and group-predatory behavior in the ponerine ants. *Evolution* 12:24–31

166. Wilson, E. O. 1958c. Observations on the behavior of the cerapachyine ant. *Insectes Soc.* 5:129–40

167. Wilson, E. O. 1959a. Some ecological characteristics of ants in New Guinea rain forests. *Ecology* 40:437–47

168. Wilson, E. O. 1959b. Adaptive shift and dispersal in a tropical ant fauna. *Evolution* 13:122–44

169. Wilson, E. O. 1959c. Communication by tandem running in the ant genus *Cardiocondyla. Psyche* 66:29–34

170. Wilson, E. O. 1961. The nature of the taxon cycle in the Melanesian ant fauna. *Am. Natur.* 95:169–93

171. Wilson, E. O. 1962a. Behavior of *Daceton armigerum* (Latreille), with a classification of self-grooming movements in ants. *Bull. Mus. Comp. Zool.* 127:401–22

172. Wilson, E. O. 1962b. Chemical communication among workers of the fire ant *Solenopsis saevissima* (Fr. Smith). I. The organization of mass-foragers. *Anim. Behav.* 10:134–47

173. Wilson, E. O. 1965. Trail sharing in ants. *Psyche* 72:2–7

174. Wilson, E. O. 1968. The ergonomics of caste in the social insects. *Am. Natur.* 102:41–66

175. Wilson, E. O. 1971. *The Insect Societies.* Cambridge: Belknap. 548 pp.

176. Wilson, E. O., Brown, W. L. Jr. 1958. Recent changes in the introduced population of the fire ant *Solenopsis saevissima* (Fr. Smith). *Evolution* 12: 211–218

177. Wilson, E. O., Hunt, G. L. Jr. 1966. Habitat selection by the queens of two field-dwelling species of ants. *Ecology* 47:485–87

178. Wilson, N. L., Dillier, J. H., Markin, G. P. 1971. Foraging territories of imported fire ants. *Ann. Entomol. Soc. Am.* 64:660–65

179. Wing, M. W. 1968. Taxonomic revision of the nearctic genus *Acanthomyops.* *Cornell Univ. Mem.* 405 pp.

180. Yasuno, M. 1965. Territory of ants in the Kayano grassland at Mt. Hakkoda. *Sci. Rep. Tohoku Univ. Ser. 4* 31: 195–206

FOLK SYSTEMATICS IN RELATION TO BIOLOGICAL CLASSIFICATION AND NOMENCLATURE

❖ 4061

Brent Berlin

Department of Anthropology, University of California, Berkeley, California

Folk systematics as a field of study is concerned with the elucidation of those general principles which underlie prescientific man's classification, naming, and identification of living things. The subject is part of the more inclusive area of folk science, the aim of which is to describe the nature of primitive knowledge of the natural world. In this paper, I attempt to summarize recent findings in the study of folk classification and point out what I believe to be some close relationships between folk systematics and western science. I hope that these data will be of interest to readers concerned with the historical and philosophical aspects of biological classification as a system for organizing our experience of natural history.

MAJOR AREAS OF STUDY IN FOLK SYSTEMATICS

It appears useful to recognize three major areas of study in folk systematics, each tied closely to the others. These areas may be referred to as classification, nomenclature, and identification. In the study of classification, one is concerned with discovering those principles by which classes of organisms are naturally organized in the preliterate mind. Nomenclatural studies are devoted to the description of linguistic principles of naming the conceptually recognized classes of plants and animals in some particular language. The area of identification deals with the study of those physical characters utilized when assigning a particular organism to a particular recognized class.

Although all three topics should ideally be considered in a complete description of a particular society's folk systematics, no such study has yet been completed and would require many man-years of collaborative effort on the parts of ethnographers, biologists, and psychologists. Research into the nature of folk biological classifica-

259

tion and nomenclature has been carried sufficiently far in recent years to allow one to describe several general principles which apparently underlie most, if not all, systems of folk biological classification. Field work on the problem of folk biological identification proceedures is, with one important exception (22), almost nonexistent.

THE BASIS OF FOLK BIOLOGICAL CLASSIFICATION

One of the best documented findings of folk systematics is that prescientific man's classification of his biological universe is highly systematic and quite developed. The principles which form the basis of folk biological classification seem to be ones which arise out of the recognition of groupings of organisms formed on the basis of gross morphological similarities and differences. Only rarely is classification based primarily on functional considerations of the organisms involved, such as, for example, their cultural utility. Less than half of the named folk generic classes of plants in the folk botany of the Tzeltal, a group of Mayan horticulturalists with whom my collaborators and I have been working for several years, can be shown to have any cultural significance whatsoever (6, 10). My current studies among the Aguaruna Jívaro of the rain forests of north-central Peru suggest the same findings (5). The primitive natural systematist is apparently as much concerned with bringing classificatory order to his biological universe as is his western counterpart.

Perhaps more surprising to the western biologist are recent field data which continue to suggest that the objective biological discontinuities recognized by primitive man are, for the most part and with explainable exceptions, identical at some level with those recognized by western science (3, 9, 12, 13). I believe that these findings, to be partially documented below, can be interpreted as support for the view held by the few remaining conservative taxonomists concerning the "reality of species" and are contrary to the relativistic position I once espoused myself (7).

FOLK TAXONOMY

The fundamental organizing principle of folk biological classification—the result partially, perhaps, of the large numbers of classes of organisms involved—is taxonomic, whereby recognized groupings (hereafter called *taxa*) of greater and lesser inclusiveness are arranged hierarchically (9, 24). It should be noted that the taxa which occur as members of the same folk ethnobiological category are always mutually exclusive. Furthermore, it now appears that in natural folk taxonomies most taxa are members of just five ethnobiological categories that are logically comparable to the ranks of western systematics. These are the *unique beginner, life form, generic, specific,* and *varietal.* A sixth category, tentatively called *intermediate* and containing taxa which fall hierarchically between the life form and generic categories, may be established with further research and as additional data on folk systematics become available.

The *unique beginner* is a distinctive category in that it has but one member, that being the taxon which includes all other taxa. The terms *living things* or *plants and animals* are often used to refer to this taxon in American English folk biology.

Members of the category *life form* represent the broadest, most encompassing classification of organisms into groups that are apparently easily recognized on the basis of numerous gross morphological characters. Taxa of this category are invariably few in number, usually somewhere between five and ten, and among them include the majority of all taxa of lesser rank. Such terms as *tree, vine, herb, fish,* and *bird* refer to examples of commonly recognized life form taxa in most folk taxonomies.

In contrast to life form taxa, which refer to the largest groupings of organisms distinguished by multiple characters, members of the ethnobiological category *generic* refer to the smallest discontinuities in nature which are easily recognized on the basis of large numbers of gross morphological characteristics (12, 13). Folk generic taxa are the most numerous in any folk taxonomy that has been more or less fully described, yet their numbers appear to be within the range of 500 to 800 in any actual system. Examples of folk generic taxa in American English folk botany would be those classes referred to by the names *hickory, maple, tulip tree,* and *cottonwood,* all of which are included in the life form *tree.*

Taxonomically, the majority of all generic taxa in any natural folk taxonomy are included in one of the recognized life form taxa. There are, nonetheless, generic classes which are aberrant in some fashion or another, which prohibits their inclusion in one of the major life form classes. In Tzeltal, the cactus *pehtak (Opuntia* sp.) is one such example. Possessing characteristics unlike any other grouping of plants in the area inhabited by the Tzeltal, it is considered a conceptual isolate. Aberrancy of a generic may, at times, be due to the fact that it possesses characteristics of two life form taxa simultaneously. In Aguaruna Jívaro, for example, members of the generic taxon *úwi (Clusia* sp.) are considered neither to be kinds of *númi* 'tree' nor kinds of *dáek* 'liana,' by virtue of the simultaneous tree-like and liana-like stem habit found in members of this class, a commonly seen strangler.

Finally, the majority of all generic taxa in folk taxonomies are monotypic and include no taxa of lesser rank. Polytypic generic taxa almost invariably refer to those classes of organisms which are important culturally.

Taxa which occur as members of the *specific* and *varietal* ethnobiological categories differ from both life form and generic taxa in several respects, the most important of which appears to be that such taxa are conceptually distinguished on the basis of very few morphological characters. As will be seen in the section on nomenclature, a single, multivalued character, such as color or size, is often sufficient to differentiate two or more folk specifics of the same folk genus.

Generally, specific taxa in folk taxonomies occur in sets of two or three members. It is quite rare for a set of specific taxa to exceed ten; those that do are invariably organisms of supreme cultural significance. Varietal taxa, as might be expected, are rare in all folk taxonomies.

Examples of specific taxa in American English folk botany would be those categories labeled by such names as *white oak* and *sugar maple.* Varietal taxa may be seen in the names *baby lima bean* and *butter lima bean.*

At the opening of this section on folk taxonomy, I mentioned the possibility of recognizing a sixth ethnobiological category, termed *intermediate,* which is comprised of taxa that fall between the life forms and generics. As such, intermediate taxa taxonomically include two or more generic taxa. It now appears that such intermediate forms are relatively rare in folk taxonomies and, as pointed out in an earlier paper (8), when such taxa are found, they most commonly are not labeled by an habitual expression. The rarity of intermediate taxa in folk systematics, but more importantly, the fact that they are not named, casts doubt as to whether our current knowledge empirically justifies establishing an ethnobiological category of this rank.

THE BASIS OF FOLK BIOLOGICAL NOMENCLATURE

Recent research into the nature of folk biological nomenclature reveals that the naming of plants and animals in folk systematics is essentially identical in all languages and can be described by a small number of nomenclatural principles. While a detailed linguistic discussion of these principles has appeared elsewhere (9, 10), a brief summary is presented here.

There is a fairly close correspondence between the linguistic form of a name for some folk biological taxon and its ethnobiological rank. Linguistically, two basic types of names for plants and animals can be recognized in folk systematics. For lack of more original terminology, these forms can be referred to as primary and secondary names. Primary names occur as labels, almost without exception, for generic and life form taxa and, for the unique beginner, when this latter taxon is named (but see below). Secondary names are generally restricted to taxa of lesser rank, namely, the specific and varietal forms.

Nomenclatural Properties of Generic Names

Generic taxa form the basic core in any folk taxonomy. The labels for taxa of this category are also fundamental and are among the first words in folk ethnobiological lexicon learned by children in preliterate societies (26). The botanist H. H. Bartlett, discussing the development of modern botanical nomenclature, noted that ". . . the concept of genus must be as old as folk science itself" (2, p. 341), and provided an essentially nomenclatural definition of the concept. For Bartlett, a folk genus is any class of organisms ". . . which is more or less consciously thought of as the smallest grouping requiring a distinctive name" (2, p. 356).

Etymologically, it is often impossible to provide linguistic analysis of generic names, a fact that should not be surprising since such names are generally quite ancient. When analysis is possible, it is often the case that the name is descriptive of some quality of the class of organisms to which it refers. In Tewa, an American Indian language of the southwestern US, the white fir, *Abies concolor,* is known as *tenyo,* literally, 'large tubes,' presumably due to the hollow stems used in pipes (21).

Onomatopoeia is also important in the formation of many generic names, especially of animals such as birds and frogs whose distinctive calls are often quite characteristically represented (23).

A final linguistic feature of generic names which appears to be widespread in many languages is the use of the generic plus some modifier to refer to some taxon that is conceptually related to the class indicated by the generic name alone. Often the modifier is an animal name as, for example, in Tzeltal where one finds many such pairs. Typical is the pair *ishim* 'corn' and *ishim ahaw,* literally, 'snake's corn' (*Anthurium* spp.), the latter formed on the basis of the presumed similarity of the mature spadix in many members of *Anthurium* to an ear of corn. In English, one finds such pairs as *oak, poison oak; apple, horse apple* (also known as *Bodark* in some dialects); *cabbage, skunk cabbage; cypress, false cypress; orange, mock orange;* and many others.

It should be pointed out that none of these superficially binomial expressions are seen as conceptually subordinate to their monomial counterparts. Thus, *skunk cabbage* is not a kind of *cabbage* nor is *poison oak* a kind of *oak.* Each simply shares some characters which are seen to be similar to the monomially designated form. This point is discussed by the California botanist Edward Lee Greene in his important *Landmarks of Botanical History* in describing the early nomenclatural writings of Theophrastus. Greene notes that many of Theophrastus' generic names are linguistically complex expressions, several of which appear to be derived from monomial generics, e.g. *Calamos* 'reed grass' (*Arundo* spp.) and *Calamos Euosomos* 'sweet flag' (*Acornus calamus*). There is no doubt in Greene's mind, however, that Theophrastus meant the two taxa as distinct genera:

> It is not imaginable that a botanist of Theophrastus' ripe experience and great attainments should think those large grass-plants and the sweet-flag to be of the same genus. Beyond doubt, however, the name Calamos Euosomus did originate in the notion that arundo and acornus are next of kin; for, however unlike they are as to size, foliage, and other particulars, there is a remarkably close similarity in their rootstocks, these being of almost the same size, form and color in the two. The gatherers of roots and herbs, as we know, looked first of all to the 'roots' of things, and these were their first criteria of plant relationships. To these it should be perfectly natural to place the sweet-flag alongside arundo, the true [Calamos] by its closely imitative "root," and then on account of the aromatic properties of the root to call the plant [Calamos Euosomos] (20, p. 123).

Nomenclatural Properties of Life Form Names

As with generic taxa, members of the ethnobiological category, *life form,* are invariably marked by primary linguistic expressions. It is often the case that these names are linguistically unanalyzable, suggesting some antiquity. On the other hand, in many languages spoken by preliterate peoples, it is not uncommon to find that an identical linguistic expression for some generic taxon also occurs as the label for the life form class as well. Such a term, with two distinct but semantically related meanings, is linguistically *polysemous.* An example of life form–generic name polysemy can be seen in Klamath, an Indian language once spoken in Oregon, where the term *k'osh* (*Pinus* sp.) is used to refer to pines as well as to the general life form taxon *tree* (18).

Of the several possible explanations for such a nomenclatural feature, the one most appealing to me suggests that, over time, the name of the most salient or culturally important generic class has become elevated to life form status. This view receives support from the work of Almstedt (1) who has done research among the Digueño, a small group of Indians of Southern California. She reports that the term *isnyaaw* 'live oak' (*Quercus agrifolia*) is also used for the concept of *tree* in general.

This species is of critical importance to these people; it has the widest distribution of any major tree and is the most generally available source of edible acorns. Early historical linguistic research indicates that the Digueño lacked a term for tree until relatively recently. For Almstedt, "... it seems logical that the name *isnyaaw* should be used for tree when the need arose" (1, p. 13).

In many Indian languages of the American Southwest, the term for cottonwood, the only deciduous tree which is widely distributed outside the major forests, is also used for *tree*, as well (19, 27). A recent linguistic survey by Demory (15) shows that in several languages of the Hokan family one finds life form and generic polysemy as a common occurrence. In each case, the generic name which refers to trees of major cultural importance in that particular geographic area is used for the the more general concept as well. In these cases, the range is wide, including such diverse forms as juniper, sugar pine, live oak, and broad leaf maple.

There is at least some evidence to suggest that an identical nomenclatural development took place in Indo-European, the ancestral language from which most of the major languages of Europe are thought to have evolved. Buck, in an extensive study of synonyms in the major languages of this stock, notes that a commonly widespread group of words for *tree* can be traced etymologically to an Indo-European word "... which probably denoted a particular kind of tree, namely the oak" (11, p. 48). The most conclusive evidence in this regard can be found in Paul Friedrich's detailed and authoritative treatment of the proto Indo-European taxonomy of trees. His conclusions are stated here in detail:

> ... It seems probable that the primitive, arboreally oriented PIE [Proto Indo-European] distinguished several species of oak by distinct morphs, and that *ayg-, *perkʷ-, and *dorw- served in this way. As the oak and mixed-oak forests were reduced and contracted, and as the speakers of the PIE dialects migrated into their new homelands—two simultaneous processes during the second and third millennia—the denotations of the *dorw- reflexes shifted to "wood, tree, hardness" and yet other referents. ... It is quite possible that even in PIE times the main name for the oak—a sort of *Urbaum*—was occasionally or dialectically applied to 'tree' in general. Within pre-Homeric Greek δρῦς and δρυός could denote either 'oak' or 'tree' with disambiguation through social or literary context. By Classical Greek times the meaning had narrowed to the original PIE 'tree.' In more recent centuries the identical process has been documented in Germanic, where *eik* shifted from 'oak' to 'tree' in Icelandic—oaks being virtually absent in that country (17, p. 146).

Nomenclatural Properties of Specific and Varietal Names

Linguistically, the structure of specific names in folk systematics is regularly binomial (with one singular, but explainable exception). Formally, the generic name is modified by an adjective which usually designates some obvious morphological character of the plant class such as color, texture, size, location, or the like. Examples such as *sakil ishim* 'white corn' and *tsahal ishim* 'red corn,' in Tzeltal, typify the binomiality of specific names.

It is perhaps an unintentional bit of western systematic ethnocentrism to attribute the "invention" of our current binomial system of nomenclature to Linnaeus (or to

Bauhin) if in so doing one is suggesting a radical break with folk tradition. It is more close to the facts to observe that Linnaeus and his predecessors formally codified a system of nomenclature present in the folk systematics of earliest prescientific man and still recognized in the natural folk biological systems of classification found in the languages of preliterate peoples today (25).

Monomial specific names are also found in folk taxonomies, but when such is the case, the monomial specific is usually polysemous with its superordinate generic. Invariably, such monomially designated specifics are considered to be the best known or most widely distributed members of a particular folk genus. Wyman & Harris, for the Navajo of the American Southwest, have said it is as if ". . . in our binomial system the generic name were used alone for the best known species of a genus, while binomial terms were used for all other members of the genus" (29, p. 120). Following early botanical tradition, we will refer to folk species exhibiting these nomenclatural characteristics as *type species*.

In Tzeltal, the custard apple *k'ewesh* (*Annona* spp.) includes at least three specific taxa. One, the type specific, is simply labeled *k'ewesh* (*A. cherimola*) due to its wider distribution. In Aguaruna Jívaro, this kind of specific name formation appears to be the rule with polytypic generic taxa which denote wild plants. A single example can be seen in the generic *kamanchá* (*Bactris* spp.), the most important specific member of which is also *kamanchá* due to its frequency. Among the Guaraní of Argentina, the generic taxon *Mboreví* refers to both kinds of tapirs in the area. *Mboreví* is used polysemously to designate the type species, *Tapirus terrestris* while *Mboreví hovih*, a binomial, refers to the lesser known and less prominent *M. terrestris* var. *obscura* (14).

It is particularly interesting to note that Theophrastus, considered by some botanists to be the father of western systematic botany, preserved the basic structure of folk plant names in his early nomenclatural studies. Or, as Greene has stated, Theophrastus ". . . left plant nomenclature as he found it" (20, p. 125), providing by his ethnobotanical insight historical validation of many of the structural principles suggested here. This is particularly evident in his treatment of type specific terminology. Greene notes:

> The Theophrastan nomenclature of plants is as simply natural as can be imagined. Not only are monotypic genera called by a single name; where the species are known to be several, the type-species of the genus—that is, that which is most historic—is without a specific name, at least very commonly, and only the others have each its specific adjective superadded to the generic appellation (20, p. 120).

The following examples bear out this claim.

Theophrastus	*Recent Equivalents*
Peuce	*Pinus picea*
Peuce Idaia	*P. maritima*
Peuce conophoros	*P. pinea*
Peuce paralios	*P. halepensis*
Mespilos	*Mespilos cotoneaster*
Mespilos anthedon	*Crataegus tominalis*

Varietal Names

The nomenclatural characteristics of varietal names are only trivially different from those of specific names and will be discussed here only briefly. It has been mentioned that varietal taxa are distinctly rare in natural folk taxonomies. Such names refer exclusively to those taxa of major cultural importance such as plants and, rarely, animals that have been under intense domestication and that are represented by morphologically distinct forms.

Linguistically, varietal names are formed by the addition of an attributive to the specific name. For example, in Tzeltal, beans are divided into several specific classes, one or two of which are further partioned into varietals. Thus, the specific name for the common bean, *shlumil chenek'* (*Phaseolus vulgaris*), is further divided into the two color varieties, *tsahal shlumil chenek'* 'red common bean' and *ihk'al shlumil chenek'* 'black common bean.'

Shortening of the full varietal name is, as might be expected, common in actual speech. One can often hear of *tsahal shlumil* and *ihk'al shlumil* in actual conversation, where the generic appellation has been dropped.

Nomenclatural Properties of the Unique Beginner

Typically, the most inclusive taxon in a folk taxonomy, the unique beginner, is not labeled. This is not to say that the domain of 'plant' or 'animal' is not recognized conceptually, of course, and various descriptive devices can be utilized to refer to these broad classes. In Tzeltal, the domain of plants is referred to as those things "that grow from the earth but do not move," contrasting with the domain of animals, a class of beings which "move by their own power." In many American Indian languages, the contrasting kingdoms are indicated grammatically by affixes which occur with names indicating 'animalness' or 'plantness.'

If the unique beginner is named, it is often the case that the term employed is polysemous, or at least partially so, with some life form class. In Aguaruna Jívaro, the term for 'tree' is *númi* and the domain for plants as a whole is designated by the expression *númi aídau,* literally, 'all (classes) of trees.'

Even in many modern languages, the term 'plant' may be seen to have two meanings. In Spanish, *planta* can be used to refer to the major division as a whole but its central meaning is 'herbaceous plant.' Something of this usage can still be found in English, especially in the speech of botanically naive individuals, where the primary meaning of *plant* is 'small, herbaceous, leafy thing,' excluding trees and shrubs.

Sometimes, the name for the most inclusive taxon may be a compound of two or more life form names. There is some evidence that in ancient Sumerian, the notion of 'plant' was indicated by a compound expression including the terms for 'tree,' 'grass,' and 'vegetable.' And it is well known that in Latin the terms 'tree' and 'herb' were commonly joined (*arbor et herba*) to designate the more general concept. The linguist Ullmann has noted that the term *plant,* in folk botany, at least, is quite recent indeed.

According to a recent inquiry, the modern meaning of 'plant' is first found in Albertus Magnus in the 13th century, whereas the French *plante* did not acquire this wider sense until 300 years later (28, p. 181).

Finally, it should be noted that in modern folk English systematics, no single common expression can be found for both biological kingdoms united. The expression *living things* is, at best, a bit stilted and may be prevalent only in the speech of those with some biological sophistication. The more common, but nonetheless fairly educated label, *plants and animals,* is a linguistic compound.

CORRESPONDENCE OF FOLK AND SCIENTIFIC CLASSIFICATION

I have attempted to point out several formal characteristics of folk systematics which appear to be widespread in actually occurring folk biological systems throughout the world. I believe many of these features of classification and nomenclature can be found in modern western systematics, which is, at least partially, a development of folk systematics.

But aside from these formal structural correspondences, can one also observe substantive correspondences between folk and scientific systems of classification? If such substantive correspondences exist, they might reveal aspects of the natural world which are in some sense 'natural' and which are apparently perceived as the same by persistent observers of nature everywhere.

Some field biologists have noted a rather close correspondence between scientifically recognized species and the linguistic designations given these groups by preliterate peoples (16). Conservative systematists have interpreted these findings, sparse as they are, as support for their views concerning the "reality of species."

Many anthropologists, whose traditional bias is to see the total relativity of man's variant classifications of reality, have generally been hesitant to accept such findings, which suggest some kind of universal ordering of the natural world. Some population biologists, for quite different reasons of course, have also tended to treat lightly or ignore evidence in favor of the objective nature of species—because they regard species as artificial units. My colleagues and I, in an earlier paper (7), have presented arguments in favor of the "relativist" view. Since the publication of that report more data have been made available, and it now appears that this position must be seriously reconsidered. There is at present a growing body of evidence that suggests that the fundamental taxa recognized in folk systematics correspond fairly closely with scientifically known species.

Units of Comparison

One of the difficulties in any comparison concerns the units of analysis to be considered. In the case of western systematics, the selection of the basic unit is straight forward—it must be the species. In folk systematics, it now appears useful to focus on the folk genus as the primary unit. The folk genus, it will be recalled, is the smallest linguistically recognized class of organisms that is formed, as the folk zoologist Bulmer has succinctly stated, ". . . by multiple distinctions of appearance, habitat, and behaviour" (13, p. 335). These two units, then, the scientific species and the folk genus, will be those selected as the basic taxa to be examined in any comparison of the folk and scientific systems of classification.[1]

Assuming that a detailed folk systematic study of some biological domain has been completed, that all of the folk genera have been discovered, and that their corresponding scientific species have been determined, one can recognize at least three logical types of correspondence between the two systems. These three types of correspondence will be referred to as one-to-one correspondence, over-differentiation, and under-differentiation.

The first type of mapping, one-to-one correspondence, can be observed when a single folk generic taxon refers to one and only one scientific species. The common willow *tok'oy* in Tzeltal folk botany would be in one-to-one correspondence in that it maps perfectly onto the single botanical species *Salix bonplandiana.*

Over-differentiation can be observed when two or more folk generic taxa refer to a single scientific species. As will be seen below, this type of mapping has a quite low occurrence in Tzeltal and I predict it will be rare in other folk taxonomies as well. An example would be the three Tzeltal generics, *bohch, tsu,* and *ch'ahko',* all of which denote the various shape varieties of the common bottle gourd *Lagenaria siceraria.*

Under-differentiation can be divided into two easily recognized types. Type 1 under-differentiation occurs when a single folk generic taxon refers to two or more scientific species of the same genus. The Tzeltal generic *ch'ilwet* would exemplify this type of mapping as it refers to at least five species of the genus *Lantana.*

Type 2 under-differentiation is recognized when a single folk generic refers to two or more species of two or more scientific genera. This case can be exemplified by the Tzeltal generic *tah* which refers to several species of *Pinus* as well as to at least one species of *Abies.*

Before proceeding further, it should be pointed out that the inventory of biological species utilized in any comparison are those—and only those—species which occur in the geographic area of the society being studied. For example, one may observe that a particular folk generic such as *oak* refers to one or more of the species of *Quercus* in the area inhabited by the society under study. In the absolute sense, of course, all folk systems are obviously under-differentiated when the totality of all western systematic knowledge is considered. Such an observation is trivial, however, if one is concerned with evaluating the classificatory treatment of those species for which a particular society has first-hand knowledge.

Furthermore, it is obvious that one must restrict one's comparison to those species of organisms which, because of their size, behavior, and significance, are readily

[1]A failure to recognize this important fact led to the conclusions published in Berlin, Breedlove & Raven in 1966 (7). Here, the units of comparison selected from the folk system of classification were all *terminal* taxa, regardless of ethnobiological rank, leading to the inclusion of folk generics, specifics, and varietals.

observable to the primitive natural historian. It should not be surprising if many algae and fungi are omitted from the classificatory structures of preliterate peoples, nor, for that matter, species of organisms which can be distinguished only on the basis of characters apparent with the aid of a 10X hand lens.

The materials that my colleagues and I have collected on Tzeltal folk botany, and those of Eugene Hunn on Tzeltal folk zoology (23), are the only data available, to my knowledge, where the conventions of one-to-one mapping, under-, and over-differentiation have been used in measuring the correspondence of scientific and folk taxonomies. Other research now in progress, however, will shortly be available from another society and the early findings appear to support those from Tzeltal (5). Since the Tzeltal results have been reported in greater detail elsewhere (10), I will only summarize them here.

After long-term field work, we are confident in recognizing 471 widely known generic taxa in Tzeltal folk botany. The distribution of these 471 generic forms in terms of the conventions of one-to-one correspondence, under-differentiation, and over-differentiation can be seen in Table 1.

Table 1 Correspondence of Tzeltal generic taxa with botanical species in the area (which are named in Tzeltal)

One-to-one correspondence	291
Under-differentiation, type 1	98
Under-differentiation, type 2	65
Over-differentiation	17
	$N = 471$

Table 1 reveals that a major portion of Tzeltal generics map in a one-to-one fashion onto botanical species. In our inventory of 471 generic taxa, 291, or approximately 61%, show this type of correspondence.

Only 17 generic taxa, or 3% of the inventory, are over-differentiated. In most cases, the plants involved here are important cultivated forms which show rather marked morphological differences that partially explain the occurrence of two or more generic folk names for members of the same botanical species.

While some 36% of Tzeltal generic taxa are under-differentiated, given our earlier stated conventions, it is of interest to observe that more than 2/3 these taxa are polytypic, i.e. include folk specifics. In all such cases, the folk species refer to single botanical species as well.

ACKNOWLEDGMENTS

Many of the ideas discussed here have appeared in one form or another in papers written in collaboration with my botanical colleagues, Peter H. Raven and Dennis E. Breedlove, from whom I have learned most of what little systematic botany I know. It would be unfair, however, to hold them responsible for my interpretations of the relationships of folk and scientific systematics, some of which I am sure they will find questionable if not outlandish. I am also in major debt to Paul Kay, who has provided the most revealing treatment of the formal structure of folk taxonomies yet published, and to William Geoghegan for opportunities to discuss the nature of primitive classification and its implications for broader cognitive studies. In addition to Kay's work on taxonomy, Harold C. Conklin's early research on folk botany and Ralph Bulmer's recent studies in folk zoology have been most influential in the development of my views on folk systematics to the present. Finally, I am grateful to Eugene S. Hunn who, by example, has led me to appreciate more fully what it is to be a "natural historian."

The sources of funds for this research in folk biology include the National Science Foundation, the National Institute of Mental Health, and the Language-Behavior Research Laboratory, University of California, Berkeley.

Literature Cited

1. Almstedt, R. L. 1968. Diegueño *tree:* an ecological approach to a linguistic problem. *Int. J. Am. Linguist.* 34:9–15
2. Bartlett, H. H. 1940. History of the generic concept of botany. *Bull. Torrey Bot. Club* 67:349–62
3. Berlin, B. 1972. Speculations on the growth of ethnobotanical nomenclature. *Lang. Soc.* 1:51–86
4. Berlin, B. 1973. Folk biology. In *Handbook of North American Indians,* ed. C. Sturtevant. Washington, DC: Smithsonian Inst. In press
5. Berlin, B. 1973. First ethnobotanical expedition of the University of California to the Upper Marañón River, Amazonas, Peru. *Rep. No. 1* to Nat. Inst. Mental Health, Grant 22012-01. Berkeley: Lang.-Behav. Res. Lab.
6. Berlin, B., Breedlove, D. E., Laughlin, R. M., Raven, P. H. Lexical retention and cultural significance in Tzeltal-Tzotsil ethnobotany. In *Meaning in the Mayan Languages,* ed. M. S. Edmonson. The Hague: Mouton. In press
7. Berlin, B., Breedlove, D. E., Raven, P. H. 1966. Folk taxonomies and biological classification. *Science* 154:273–75
8. Berlin, B., Breedlove, D. E., Raven, P. H. 1968. Covert categories and folk taxonomies. *Am. Anthropol.* 70:290–99
9. Berlin, B., Breedlove, D. E., Raven, P. H. 1973. General principles of classification and nomenclature in folk biology. *Am. Anthropol.* 75:214–42
10. Berlin, B., Breedlove, D. E., Raven, P. H. 1973. *Principles of Tzeltal Plant Classification: An Introduction to the Botanical Ethnography of a Mayan Speaking People of Highland Chiapas.* New York: Seminar

11. Buck, C. D. 1949. *A Dictionary of Selected Synonyms in the Principle Indo-European Languages.* Chicago: Univ. Chicago Press
12. Bulmer, R. 1970. Which came first, the chicken or the egghead? In *Echanges et Communication, Mélanges offerts á Claude Lévi-Strausse á l'occasion de son 60éme Anniversaire,* ed. J. Pouillon, P. Maranda, 1069–91. The Hague: Mouton
13. Bulmer, R., Tyler, M. 1968. Karam classification of frogs. *J. Polynesian Soc.* 77: 333–85
14. Dennler, J. G. 1939. Los nombres indígenas en Guaraní de los mamíferos de la Argentina y países limítrofes y su importancia para la systemática. *Physis* 16:225–44
15. Demory, B. 1972. The word 'tree' in the Hokan language family. *Informant.* Dept. Anthropol., Calif. State Univ., Long Beach
16. Diamond, J. M. 1966. Classification system of a primitive people. *Science* 151: 1102–4
17. Friedrich, P. 1970. *Proto-Indo-European Trees.* Chicago: Univ. Chicago Press
18. Gatschet, A. S. 1890. *The Klamath Indians of Southwestern Oregon. Contributions to North American Ethnology, Vol. II, Pt. II.* Washington, DC: Dept. Interior, US Geogr. Geol. Survey Rocky Mt. Reg.
19. Gatschet, A. S. 1899. "Real," "true," or "genuine" in Indian languages. *Am. Anthropol.* 1:155–61
20. Greene, E. L. 1909. *Landmarks of Botanical History.* Washington, DC: Smithsonian Misc. Collect.
21. Harrington, J. P., Robbins, W. W., Freire-Marreco, B. 1916. Ethnobotany of

the Tewa Indians. *Bur. Am. Ethnol. Bull. 55*
22. Hunn, E. S. 1970. *Cognitive processes in folk ornithology: the identification of gulls.* Berkeley: Lang.-Behav. Res. Lab.
23. Hunn, E. S. 1973. *Tzeltal Folk Zoology: The Classification of Discontinuities in Nature.* PhD thesis. Univ. Calif. Berkeley, Berkeley, Calif.
24. Kay, P. 1971. On taxonomy and semantic contrast. *Language* 47:866–87
25. Raven, P. H., Berlin, B., Breedlove, D. E. 1971. The origins of taxonomy. *Science* 174:1210–13
26. Stross, B. 1969. *Language Acquisition by Tenejapa Tzeltal Children.* PhD thesis. Univ. Calif. Berkeley, Berkeley, Calif.
27. Trager, G. L. 1939. "Cottonwood" = "tree": a southwestern linguistic trait. *Int. J. Am. Linguist.* 9:177–18
28. Ullman, S. 1966. Semantic universals. In *Universals of Language,* ed. J. H. Greenberg. Cambridge: MIT
29. Wyman, L. C., Harris, S. K. 1941. Navajo Indian medical ethnobotany. *Univ. N. Mex. Bull.* 366, *Anthropol. Ser.* 3.5

APPLICATION OF MOLECULAR GENETICS AND NUMERICAL TAXONOMY TO THE CLASSIFICATION OF BACTERIA

❖ 4062

T. E. Staley and R. R. Colwell

Department of Microbiology, University of Maryland, College Park, Maryland

INTRODUCTION

The classification of bacteria has traditionally been a tedious and often frustrating task undertaken by every bacteriologist at some point in his career. Classical, or alpha, taxonomies of the bacteria are based solely upon weighted phenotypic expressions; the tests used in determining phenotypes are not applied to all strains of all genera. For example, the indole or Voges-Proskauer tests are carried out for enterobacteria but not for myxobacteria, since the tests are considered important for the former and not the latter. Two new approaches to bacterial classification have been developed over the past two decades and when applied according to the tenets of polyphasic taxonomy (27) provide a solid base for developing a natural classification (24) for the bacteria. These approaches are numerical taxonomy and measurement of deoxyribonucleic acid hybridization (DNA/DNA hybridization, homology, or reassociation).

Numerical taxonomy (NT; also termed Adansonian analysis, taxonometrics, taxometrics, and computer taxonomy), as applied to the taxonomy of bacteria, involves the collection of morphological, physiological, biochemical, serological, and other data for bacterial strains. Each bit of information, i.e. each unit of data reducible to as close to a single-gene-single-enzyme representation as possible, is treated on an equal weight basis and taxa are constructed from similarity values (146) calculated for each strain pair. A set of bacterial strains is commonly examined for 100–200 features per strain. Coding methods vary from investigator to investigator, but generally speaking, the positive = 1, negative = 0, and not tested or not applicable = 3 coding is used. The $i \times m$ matrix (strain $\times$ feature) is subjected to analysis, using one of several possible similarity coefficient calculations, and the output is sorted according to highest linkage, average linkage, or similar calculations

273

so that groupings, or clusters, of strains emerge (146). The clusters can be further analyzed, if desired, to obtain the hypothetical median or mean organisms (91). Output from the computations is generally presented in the form of "trees" (phenetic dendrograms or phenograms) or similarity value "triangles" (142).

Concomitant with the development of NT of bacteria was the emergence of highly sophisticated molecular genetic procedures for assessing relatedness on the primary level of DNA base sequence. This provides a more sensitive estimate of relatedness than phenetic analysis. Polynucleotide base composition (mole % guanine + cytosine, i.e. % G + C) and genetic compatibility studies have appeared in the literature since 1961; at about that time molecular hybridization techniques were being refined and applied to taxonomic problems. Molecular hybridization makes use of the renaturation property of denatured DNA. DNA from closely related strains will almost completely reanneal (seek complementary base sequences), whereas less closely related strains will reanneal much less completely. When reannealing is nearly complete the DNAs of the two strains are believed to exhibit a high degree of homology and the strains are considered closely related genetically (102). Lack of reannealing is interpreted as an indication of low homology and distant genetic relatedness.

In this paper we sketch briefly the course of development of numerical and molecular genetic taxonomy. The present state of reliability and utility of these approaches for bacterial taxonomy is assessed and relations between the two methodologies are examined. This review is designed purposely for neither the molecular geneticist nor the numerical taxonomist, since each would find his respective area of competence too briefly covered. But we hope that those geneticists interested in taxonomy and those numerical taxonomists interested in the methods of molecular genetics will contemplate the value of combined molecular genetic-numerical taxonomic analysis for the systematics of both procaryotes and eucaryotes.

MOLECULAR GENETIC CLASSIFICATION OF BACTERIA

The classification of bacteria, on a genetic basis, has proceeded with some success along three lines: 1. nucleotide base composition; 2. genetic compatibility (transformation, transduction, conjugation, and sex-deduction); and 3. molecular hybrid formation. Review articles by Marmur et al (94), Jones & Sneath (78), and Mandel (92), and a series of papers by DeLey (34, 35, 38, 44) discuss the correlation of nucleotide base composition and genetic compatibility data with phenetic data. Since these articles appeared, considerable material on molecular hybrid formation between polynucleotides of bacterial DNA (also referred to as percent reassociation, hybridization, or homology) as applied to bacterial taxonomy has been published.

Historical Background

In the early 1960s, the phenomenon of DNA denaturation and renaturation was intensively studied (48, 50, 93, 95, 98). One of the first investigations using specific reassociation, characteristic of single-stranded DNA, for detecting base sequence homology between two bacterial strains was that of Schildkraut et al (134). By reannealing equal amounts of heavy (N^{15} deuterated) and of light (unlabeled) DNA of several *Escherichia coli* strains, followed with equilibrium centrifugation, it was possible to detect hybrid (intermediate density) molecule formation, thereby providing evidence for extensive, complementary polynucleotide sequences. This kind of analysis was extended to examine formation of hybrid molecules between species of genetically unrelated *Bacillus.* No hybrid formation was detected. However, when the same experiment was done using genetically related strains, i.e. the mutually transforming strains *Bacillus subtilis* and *Bacillus natto,* substantial hybrid formation did occur. From the hypothesis that genetic compatibility indicated sequence homology, significant hybrid formation among genera of the family Enterobacteriaceae was expected. *Escherichia coli, Shigella dysenteriae,* and *Salmonella typhimurium* were examined (134). Partial reassociation occurred between *E. coli* and *Sh. dysenteriae* reannealing mixtures, while no such homology was detected between *E. coli* and *S. typhimurium.* Negative results for the latter were attributed to a dispersion of homologous regions over the genome or to the existence of very low numbers of such regions so that they would not be detectable by analytical ultracentrifugation.

Subsequent investigations by Rownd et al (131 and Falkow et al (53), who used equilibrium centrifugation to detect DNA hybrids of intermediate density, showed that gene sequences in hybrids from an *E. coli* HFr X *Salmonella* F⁻ cross were homologous with the donor strain DNA to the extent estimated by recombinant studies, even though no DNA homology was detectable between parental strains.

To increase the sensitivity and convenience of hybrid detection, a direct method involving reannealing of single-stranded RNA of high molecular weight to denatured DNA immobilized in agar was developed by Bolton & McCarthy (9) and McCarthy & Bolton (102). Varying ranges of relative homology (percent DNA bound, relative to homologous reaction) were shown between *E. coli* B (reference strain, source of radioactive DNA fragments) and *E. coli* K–12 (101%), *Aerobacter aerogenes* 211 (51%), *Klebsiella pneumoniae* (25%), *Serratia marcescens* 4180 (7%), *Pseudomonas aeruginosa* (1%), and other organisms. *Aerobacter aerogenes* 211 was also used as a reference strain. A reciprocal relative homology of 49% for *A. aerogenes* 211 with *E. coli* B was observed.

Further refinements for separation of double and single-stranded DNA were described by Bautz & Hall (5), Gillespie & Spiegelman (60), Denhardt (46), and Warnaar & Cohen (153). All of these followed a general procedure whereby immobilization of denatured DNA onto cellulose or nitrocellulose membrane filters was followed by annealing of single-stranded DNA or ribonucleic acid (RNA) with complementary sequences on the immobilized DNA. Most of these studies were directly concerned with isolation of specific gene sequences from bacteriophage, to examine synthesis of the phage DNA in host cells, or to determine the amount of complementarity between phage DNA and RNA. The membrane filter technique has been very useful, however, in determining sequence homologies among bacterial strains.

In 1963 Nygaard & Hall (114) used nitrocellulose membrane filtration to separate RNA (synthesized using a DNA template)/DNA complexes formed in solution. This application has been employed in few bacterial taxonomic studies.

The most recently developed method for separation of duplex DNA from the single stranded form is that of Miyazawa & Thomas (109) and Bernardi (6). In this procedure, hydroxyapatite (a form of calcium phosphate) is used to specifically adsorb duplex DNA, with single-stranded DNA remaining in the column eluate. The method was applied by Brenner & Cowie (13), and Brenner et al (14–19, 21) to studies of the Enterobacteriaceae, and by Citarella & Colwell (26) to investigations of *Vibrio* spp.

Theory, Limitations, and Applicability of Molecular Genetic Methods

Whenever information obtained from one method for measurement of homology is to be compared with results from another method, the limitations of the techniques have to be assessed. Determination of base sequence homologies, or more precisely the separation of single-stranded DNA molecules (denatured) from double-stranded DNA molecules (renatured), between or among bacterial strains has been done using all available methods—heavy isotope equilibrium centrifugation, DNA template, DNA-agar, DNA-membrane, or hydroxyapatite (HA). Each of these methods has limitations, as well as advantages and disadvantages (8, 16, 45, 46, 76, 80, 102, 103, 114).

HEAVY ISOTOPE EQUILIBRIUM CENTRIFUGATION   This method was one of the first used to detect hybrid formation of bacterial DNA (40, 41, 53, 55, 97, 134). Simply stated, a reference strain is grown in a solution of heavy (deuterated) water and the radioactive isotope, N^{15}. Heavy, double-stranded, labeled DNA is subsequently isolated. The labeled, heavy DNA is mixed with an equal amount of light, double-stranded, unlabeled DNA prepared from a second strain. The mixture is denatured and reannealed. By equilibrium centrifugation in a heavy salt solution such as CsC1, three bands of different densities are revealed if hybridization occurs: light duplexes appearing as a peak of radioactivity of low density; hybrid duplexes of intermediate density; and heavy, high density duplexes.

The amount of hybrid formed is estimated roughly by calculating the difference in total area under the hybrid curve when the area from the renatured, parental DNA curve is subtracted. The result is expressed as percent of maximal hybridization. Friedman & DeLey (55) were thus able to obtain hybridization values ranging from 0–100%, with standard deviations of 1–5%. However, even when a significant amount of hybridization was suspected for certain nonhomologous reactions, values were obtained in the range 22–52%. In these cases, suspected high homology (80–100%) was confirmed (41) when the DNA-agar technique of McCarthy & Bolton (102) was applied, requiring us to question the validity of equilibrium centrifugation for assessing gene sequence homology. However, some information on homology (presumably perfect or nearly perfect complementarity) of high molecular weight molecules is obtained by the latter. Nevertheless, complete differences cannot be assumed from the low values obtained by the equilibrium centrifugation method (41).

Equilibrium centrifugation has not found widespread application in bacterial taxonomy for the following reasons: Quantitative resolution of peaks of similar densities is very difficult. Some organisms are unable to grow in a heavy (N^{15}-deuterium oxide) medium. High molecular weight, highly polymerized DNA molecules presumably must be completely identical to be resolved as hybrids. The ultracentrifugation equipment is necessary and expensive and only a few samples can be processed at one time. Thus the literature on the use of equilibrium centrifugation in bacterial systematics is scanty.

DNA TEMPLATE   In 1963 Nygaard & Hall (114) reported on a technique for detecting RNA-DNA hybrids by their retention in nitrocellulose membrane filters. The RNA-DNA hybrids formed in solution (as well as single-stranded DNA) were retained on the filters, whereas the unreacted RNA passed through the filters. Since single-stranded DNA was nonspecifically adsorbed to the filter, this technique cannot be used in detecting DNA-DNA duplexes, unless a labeled RNA, using the reference DNA as template, is synthesized in vitro. In the reports available on RNA-DNA studies (3, 108, 125, 126, 147, 154), reasonably good results were obtained by incubating labeled RNA with an equal amount of DNA and passing the mixture through nitrocellulose membrane filters. Percent homology was calculated as that amount of activity retained on the filter in the homologous or heterologous system X 100, divided by the total amount added. Values are usually normalized by dividing this figure by that obtained with the homologous system. The amount of activity specifically bound in the homologous system can be as low as 15–20% of the total activity in the reassociation mixture. Although few estimates of the precision of this procedure have been made, the results apparently are reproducible within 5% of the mean values (107).

In the RNA-DNA approach for estimation of bacterial relatedness, there are two assumptions of untested validity: 1. that the polymerase used (in most cases from *E. coli*) neither copies certain sequences preferentially nor forms a double copy of selected segments, and 2. that the polymerase does not possess species specificity (3).

The RNA-DNA method has the significant advantage over the equilibrium centrifugation technique of increased sensitivity for short sequence homologies, and it requires only microgram quantities of nucleic acid. Nevertheless, more convenient, reliable, and reproducible techniques are now available for estimating molecular genetic relatedness.

DNA-AGAR   This technique, first reported by Bolton & McCarthy (9), has undergone several modifications (69, 71, 72, 102). In general, radioactively labeled, single-stranded, sheared (200,000–300,000 mol wt) DNA is incubated with unlabeled, single-stranded, high molecular weight DNA previously immobilized in extruded (through a fine mesh screen) 4% agar. The ratio of labeled DNA in the reaction suspension to the unlabeled DNA is kept sufficiently low (1/100–1/500) to minimize reassociation of labeled fragments. Nonspecific adsorption of the labeled DNA onto the agar is usually less than 0.5% of the total radioactivity.

Labeled DNA fragments with sequences complementary to the immobilized unlabeled DNA reanneal and can be separated from the nonhomologous fragments by washing with buffer. The extent of hybridization is usually determined by elution of labeled from unlabeled fragments using low ionic strength buffers at elevated temperatures. Although there is variation in the absolute amount of bound labeled DNA, which is also dependent on conditions of incubation, the literature reports values ranging from 40–60% of the total radioactivity. The ratio of bound radioactivity to the total added is interpreted as the percent of DNA homology, the homologous reaction serving as the reference, i.e. as 100%. The values are reproducible to within 2–5% of the mean value.

The DNA-agar method offers the obvious advantages of simplicity, reproducibility, and high resolution. It has proven particularly useful for taxonomic applications. Unlike equilibrium centrifugation, homologous sequences of relatively short length (2,000–3,000 nucleotides) can be detected and small amounts of labeled DNA are required. Furthermore, thermal chromatography becomes possible, whereby the extent of mismatching of heteroduplexes is estimated by changes in thermal stabilities, Δ $Tm(e)$ values (30).

There is, however, a precaution to be observed with this technique: Homology values obtained by the DNA-agar method can be affected significantly by conditions of incubation, especially by temperature. Although most of the published results follow established procedures, it is commonly overlooked that the optimal temperature of incubation rises in proportion to the mole % G + C composition of the DNA. A general rule is to incubate at a temperature 25–30°C below the thermal denaturation midpoint (Tm) of the native DNA (93, 96). Even so, the percent homology of the DNA of two strains with similar mole % G + C composition may be overestimated (76), probably because of incomplete complementarity between reannealed duplexes due to impaired (free ended) or mispaired (looped out) base sequences.

A modification of the direct binding DNA-agar method was proposed by Hoyer et al (69) and is widely practiced by molecular geneticists. Essentially, the procedure described above is followed, except that the amount of reduction of bound label is measured by including an excess of unlabeled homologous or heterologous DNA with the labeled DNA. Thus, an inhibition of the reaction occurs, and we speak of "competition" for complementary sites on the immobilized DNA. The reduction is directly proportional to the complementarity of the labeled and unlabeled DNA preparations.

The competition procedure has several advantages over the direct method since any study of genetic relationships involving many strains will require (a) only a single, labeled, DNA preparation and (b) only a single DNA-agar preparation (batches of which can differ in annealing properties). The competition method may also be more sensitive in detecting related sequences (70, 102). However, assessment of mispairing cannot be done since few immobilized duplexes occur. Also, some nonspecific inhibition of binding may be encountered when increasing amounts (> 25 μg/ml) of unlabeled competitor DNA is used. This effect seems, however, small in relation to the actual specific DNA binding and occurs only with heterologous mixtures (81). Contaminating protein or polysaccharides in the DNA preparations may actually cause such inhibition.

DNA-MEMBRANE   In 1965 Gillespie & Spiegelman (60) published a description of a procedure for detecting RNA/DNA hybrids by annealing sheared, single-stranded RNA with high molecular weight DNA immobilized on nitrocellulose membrane filters. Denhardt (46) subsequently modified this procedure, by which

DNA/DNA hybrids on nitrocellulose membrane filters could be detected. With the inclusion of albumin in the preincubation solution used to wash the filters, non-specific adsorption of single-stranded DNA to the filter could be prevented. This modification provided a distinct advantage over the procedure of Nygaard & Hall (114) for detecting DNA complementarities between strains. The latter required the synthesis of a labeled, intermediate RNA from the reference DNA, followed by reannealing with DNA. Warnaar & Cohen (153) were successful in eliminating the nonspecific adsorption of single-stranded DNA to nitrocellulose filters by washing the filters with buffers of low ionic strength and high pH. With the elution conditions used, a 10-fold decrease in nonspecific adsorption was obtained over that of the procedure of Denhardt. More recently, Legault-Démare et al (90) observed that the nonspecific adsorption problem of single-stranded DNA can be reduced by inclusion of dimethyl sulfoxide in the reassociation reaction solution.

The theory behind hybrid formation and separation using DNA immobilized on membrane filters is, for practical purposes, identical to that of DNA-agar. High molecular weight, single-stranded DNA is deposited by filtration onto a membrane and dried. The membrane, with the adherent DNA, is placed in a vial containing the sheared, single-stranded, labeled DNA. The percent homology is calculated as the ratio of counts (radioactivity) of the immobilized DNA on the membrane filter in the heterologous system (labeled, reference DNA in solution; unlabeled, heterologous DNA on the filter) to that of the homologous system X 100. These values are normalized to relative percent binding, using percent homology of the homologous system. A very low concentration of labeled DNA in solution ($1\mu g$) and an excess of unlabeled DNA on each filter ($15-20$ μg) is used to minimize self-reassociation of labeled strands and to optimize specific adsorption of labeled and unlabeled strands. As with the DNA-agar technique, the amount of label bound in this direct binding procedure is very much dependent upon conditions of incubation. Under the usual conditions ($60-70°C$, $36-48$ hours), about $40-50\%$ of the total label will be bound, with $0.1-0.5\%$ bound nonspecifically to the control, or "blank," membrane filter. The precision has been calculated to be $1-4\%$ of the mean relative values. The "competition" technique discussed in the preceeding section is also applicable in this instance and has been extensively employed.

The advantages of the DNA-membrane are essentially the same as those for DNA-agar. Additionally, there is the advantage of increased simplicity in preparation of immobilized DNA and the ease with which the method can be carried out. This method is by far the most commonly used for molecular taxonomic studies of bacteria.

The one major problem with the DNA-membrane method is spontaneous release of immobilized DNA from filters at the incubation temperatures required for DNA of high ($> 65\%$) mole % G + C (10, 45, 90, 105, 129). This "leaching out" has been partially overcome by inclusion of formamide or urea in the reassociation reaction solutions (10, 59, 86, 105, 116, 135). Also, nonspecific adsorption of single-stranded, label DNA to the membrane can give spuriously high homologies, unless the precautions of Denhardt (46), Warnaar & Cohen (153), or Legault-Démare et al (90) are taken into consideration. For further details of this technique, Bollum (8) and DeLey & Tijtgat (45) should be consulted.

HYDROXYAPATITE   Bernardi (6) and Miyazawa & Thomas (109) reported separation of duplex DNA hybrids by specific adsorption of the hybrids onto hydroxyapatite (HA), a calcium phosphate gel. The method has recently been applied to taxonomic problems. Briefly, labeled, sheared, single-stranded DNA is reannealed in solution with unlabeled fragments similarly prepared. The solution mixture is allowed to incubate until completion of the reaction (homologous system); the solution is then passed through a column of HA. The DNA duplexes specifically adsorb onto the surface of the HA (presumably due to its strong positive surface charge arising from Ca^{2+} ions and the strong negative charge of the polynucleotide sequences arising from the phosphate residues). Single-stranded fragments pass through the column. Percent homology is calculated as in the DNA-agar and membrane methods. With HA, the reaction conditions are controlled so that homologous reactants are nearly completely reassociated (usually 85–95%), thus ruling out the possibility that the reassociation is not representative of the entire DNA molecule (16). The ratio of labeled to unlabeled DNA is normally 1/1,000–1/2,000 to minimize reassociation of labeled fragments (usually <5% of the total activity present). Zero-time binding, the amount of radioactivity bound to HA immediately after denaturization, is generally 1–2% of the total (23). The precision of the HA method is in the order of 1–2% of the mean value (149).

The advantages of the HA method are several. The quantity of DNA can be determined by Cerenkov counting, eliminating any need to filter and making the sample reusable. No immobilization of DNA is needed, thus preventing leaching out of the DNA (as can occur in the DNA agar and membrane procedures), and the amount of label that is bound is very large (see above). The quantity of unlabeled reactant involved can be determined precisely by spectrophotometric measurements. The reassociation reaction follows second order kinetics, being inversely proportional to the genome size (16). HA also readily lends itself to thermal chromatography.

The disadvantage of this method is that only a limited number of columns can be run simultaneously, even though Brenner et al (16) report on a batch procedure using up to ten samples.

MISCELLANEOUS TECHNIQUES (RENATURATION RATE AND THEORETICAL CONSIDERATIONS)   A theoretical method for predicting the maximal homology between two bacterial strains has been described by DeLey (37). The procedure entails a statistical analysis of the overlap of the mean nucleotide base composition of a pair of strains with corrections for differences in molecular weight and compositional nucleotide distribution of the chromosomal DNA. To date only one other paper using this procedure has appeared (51).

Another technique, proposed by DeLey et al (39), makes use of a spectrophotometric determination of initial renaturation rates of single-stranded DNA from which a percent homology calculation can be made. Seidler & Mandel (137) employed a similar method to calculate percent DNA homology between two bacterial strains. Taking advantage of the fact that renaturation rates of single-stranded DNA from organisms with similar polynucleotide sequences are additive, they were able to derive a formula relating any lack in rate additivity to a specific decrease in DNA homologies between two bacterial strains.

The main advantage of these two methods is that no radioactively labeled DNA need be prepared. They have been little used, however, and there may be a reduction in the accuracy of estimation of homology.

Comparison of Molecular Genetic Methods

Very poor correlation of *Xanthomonas* and *Pseudomonas* DNA homology data was obtained when the equilibrium centrifugation and DNA-agar methods were used (41). Studies have not been done in which equilibrium centrifugation data were compared with results from any of the other methods. Kingsbury et al (83), however, found very good correlation for results obtained using DNA-membrane (direct) with HA methods in their studies of *Neisseria*. Also, Brenner et al (15) were able to show significant correlation of results comparing the DNA-agar and HA methods in a study of relationships among members of the Enterobacteriaceae. DeLey et al (39) used the renaturation rate and DNA-membrane methods and found them to correlate well. Results of other workers comparing direct and competition experiments with the membrane method have also shown reasonably good correlations (56). Comparative data (15, 39, 83) are given in Table 1.

A conclusion immediately evident from a review of the literature is that there is no standard method for assessing molecular genetic relatedness. This is unavoidable in those instances where microorganisms possessing markedly different overall DNA base compositions (moles % G + C) are examined. However, it is also obvious that some workers have continued using their own methods for reassociation measurement even when other methods provide improved resolution and greater reproducibility. Clearly, if a study is to clarify taxonomic relationships, standardization of methodology must be achieved and a careful selection made of reference strains. Ideally, every study should include one or more standard reference strains so that results of different workers can better be interpreted.

Intrinsic Errors in Percent Homology Determinations

Regardless of the accuracy of a given method for assessing the extent of DNA hybrid formation, intrinsic properties of the DNA and the reaction products can be a source of serious misinterpretation of the results of reassociation reactions carried out with total cell, or "bulk" DNA.

ESTIMATION OF MISPAIRING OF BASE SEQUENCES   Reannealing of single-stranded DNA always involves the possibility that duplexes may form that are not entirely complementary. Mispairing of base sequences is more likely when strains showing marked divergence are examined. Three methods were developed expressly for the assessment of mispaired base sequences. The same methods reveal nonpaired base sequences.

The original procedure of Schildkraut et al (134) employed a phosphodiesterase from *E. coli* to eliminate single-stranded DNA sequences present in imperfect

Table 1 Comparison of homology data obtained following different methods

	DNA agar	DNA membrane	Hydroxyapatite	Renaturation Rate
Neisseria meningitidis*/*N. meningitidis*[a]		(100)+	(100)+	
Neisseria meningitidis*/*N. gonorrhoeae*[a]		78	80	
Neisseria meningitidis*/*Neisseria* strain Z[a]		88	89	
Neisseria meningitidis*/*N. sicca*[a]		45	45	
Neisseria meningitidis*/*N. flava*[a]		35	30	
Escherichia coli*/*Salmonella typhimurium* LT2[b]	38		40	
Escherichia coli*/*Aerobacter aerogenes*[b]	41		37	
Escherichia coli*/*Shigella flexneri* 24570[b]	82		87	
Proteus mirabilis*/*Escherichia coli* K12[b]	5		7	
Agrobacterium tumefaciens*/*A. tumefaciens*[c] B6		96		98
Agrobacterium tumefaciens*/*A. tumefaciens*[c] 398		45		44
Agrobacterium tumefaciens*/*A. tumefaciens*[c] 396		55		58

[a]Kingsbury et al (83).
[b]Brenner et al (15).
[c]DeLey et al (39).
* Source of labeled DNA.
+ Normalized to 100%.

duplexes. DeLey et al (43) used phosphodiesterase to assess mispairing of DNA among *Pseudomonas* species and discovered that about 5% of labeled duplex DNA was susceptible to a phosphodiesterase prepared from dogfish. Since homologous, as well as heterologous, reactions showed this effect, relative (normalized) homology values remain valid estimates of DNA homology. Phosphodiesterase has not been widely used for eliminating mispaired base sequences.

Standardization of homology determinations has been attempted through selection of more "stringent" (high) incubation temperatures for reassociation reactions. The optimal temperature for reassociation is generally accepted to be 20–30°C below the Tm for the native DNA. Higher temperatures restrict formation of all but identical, or nearly identical, DNA sequences. For DNA with an overall base composition of 50% G + C the optimal temperature is 60°C. Highly homologous strains studied using incubation temperatures of 75°C for the reaction mixtures showed expected relative binding (15). Reaction products of high thermal stability, i.e. within 1–2°C of that observed for the native DNA, were obtained. DNA preparations obtained from more distantly related strains tested under high temperatures showed a reduction of the percent relative binding. The effect was greater with decreasing relatedness. Reaction products of high temperature reassociations generally showed a smaller change in thermal stability [Δ $Tm(e)$, the temperature at which 50% of the hybrid molecules elute from the supporting matrix] than those formed at optimal temperatures. Brenner and his colleagues (14–19, 21) used this method extensively in their studies of the Enterobacteriaceae, and coined the term thermal binding index (TBI) to indicate the ratio of relative percent homology obtained under high incubation temperatures to that obtained at optimal temperatures (15, 99).

Although the above procedure has the effect of reducing the formation of duplexes of noncomplementary base sequences, it is nevertheless still difficult to quantify the amount of mispairing in duplexes formed under optimal conditions. Most investigators carry out thermal chromatography of duplexes in agar, on nitrocellulose membranes, or on HA. The linear relationship derived from examination of synthetically prepared polynucleotides (4) or deaminated DNA residues (87) indicates that a 0.7°C decrease in the $Tm(e)$ value of a hybrid duplex corresponds to approximately 1% mispaired base sequences and permits estimation of the amount of mispairing in reassociation reaction products. The linearity of this relationship is maintained to ~25 mole % of nonpaired base residues (4).

Decreased thermal stability of heteroduplex DNA can also result from related sequences containing large amounts of A-T (adenine-thymine) base pairs. This, however, is unlikely to account for all the decreased thermal stability, since bacterial DNA characteristically shows a sharp, unimodal distribution of mole % G + C around a mean base composition (44).

Homology figures from different laboratories can best be compared when at least one set of data is obtained under high temperature conditions, in addition to data being obtained under optimal temperatures. Thermal elution chromatography of reannealed duplexes is also recommended.

GENOME SIZE   Bacteria have been found to possess an amazing range of genome sizes: from 2×10^8 daltons ($\sim$300 cistrons) for *Mycoplasma gallisepticum* to 3×10^9 daltons ($\sim$4900 cistrons) for *E. coli* (36). Sometimes equivalent % homology values with reciprocal reactions (i.e. labeled DNA from strain A reannealed with unlabeled DNA from strain B, and vice versa) were not obtained (17, 71); perhaps genome size differences prevented equivalency of reciprocal reactions.

Theoretically, only genomes of approximately equal size yield equivalent reciprocal homology values. Therefore, the amount of nonreciprocity is an indication of the difference in genome size. This was shown for several strains of *E. coli,* the reciprocal reactions with a reference strain being different by 9 and 19% (17). The results were confirmed by renaturation determinations (these being 9 and 16%, respectively). Other workers (71) have reported nonreciprocal molecular genetic relationships similar to that of *E. coli.* Since genome sizes for strains of the same species may differ by 10–20%, it is essential that reference strains be carefully selected and that interpretations be suitably guarded.

EXTRACHROMOSOMAL ELEMENTS   Bulk DNA prepared from bacterial strains may include not only chromosomal DNA, but also plasmid particle (episome) DNA as well. With the discovery by Skerman et al (141) that $\sim$38% of all *E. coli* strains harbored at least one plasmid (resistance transfer factor, or RTF), it is reasonable to suspect that strains of other genera also contain extrachromosomal elements. The effect of plasmid or episomal DNA on estimates of homology between strains would depend on the amount of plasmid DNA present per cell. Guerry & Falkow (64) estimated the plasmid DNA in *E. coli* strains to be 2–4% of the total DNA and to have molecular weights of 4.0–7.0×10^7 daltons. If plasmid particles are not eliminated from reference strains (as by curing with acridine orange), low and invalid homology values would be obtained. Polynucleotide sequence relationships of bacterial plasmids (64) could be used in bacterial taxonomy if routine methods for isolating and examining them can be developed.

BASE COMPOSITION AND RENATURATION RATES   Polynucleotide base sequences containing higher mole % G + C renature at a slightly faster rate than sequences of lower mole % G + C. When the extent of homology between strains (and the genome size) is determined by either the initial renaturation rate method of DeLey et al (39) or the $Cot_{0.5}$ method of Seidler & Mandel (137), the results must be corrected for differences in renaturation rates due to different mole % G + C of the reactant DNAs. However, in more conventional methods (i.e. the direct and competitive, DNA-agar, and DNA-membrane procedures) these slight differences in renaturation rates are of little consequence, since long incubation times are employed. Reassociation reactions carried out in solution are incubated at a sufficiently high Cot[1] (usually $\sim$100 for bacterial DNA) to allow sufficient time for even the lowest mole % G + C reactants to reanneal maximally; thus the effect of different

[1] Cot is defined as the product of the nucleic acid concentration (Co) and time (t) of incubation, expressed in moles • sec • l^{-1} (23).

mole % G + C composition on homology determinations would, in this case, also be negligible. Moreover, as DeLey (37) has pointed out, the likelihood of close relationship between strains possessing DNA of vastly different mean mole % G + C composition is remote. Hence, any discrepancies in measurements of homology arising from different DNA base compositions of closely related strains would be minimal.

CHROMOSOME REPLICATION STATE   It is conceivable that homology estimates may be affected by the state of replication of the chromosome at the time of DNA isolation. Sequences of DNA immediately after replication would be present at a concentration twice that of unreplicated (unique) portions of the circular bacterial chromosome. Assuming the chromosome has one replicating fork, log-phase chromosomes in a nonsynchronous population are replicated for one third, on the average. If so, Gillis et al (61) estimated the resulting renaturation rate would be 7% too high. Where renaturation rates are used to measure homology, the rate increase would be insignificant for the homology estimates obtained (61). In addition, Seidler et al (137) showed that the state of chromosome replication of *E. coli* is dependent on the stage of growth. DNA extracted from cells grown in glucose salts medium, and which contained only one replication fork, yielded renaturation rates that were 6% higher than for cells in the stationary phase of growth. A rough calculation of the effect of rate increase on homology estimates determined by the $Cot_{0.5}$ technique showed it was well within the precision of the method (5–10%).

Replication state effect on DNA homology estimates would be further reduced if DNA were prepared from both reference and test strains at the same stage of growth, preferably the stationary phase.

REPEATED GENE SEQUENCES   The presence of extensive lengths of repeated gene sequences in bacteria has not been shown. Britten & Kohne (22, 23), applying reassociation kinetics, were unable to demonstrate fractions of bacterial DNA from *E. coli, Clostridium perfringens,* and *Proteus mirabilis* reannealing at faster rates than bulk DNA, nor did they find reassociation products with low thermal stabilities in homologous reactions. Such findings would indicate the absence of repeated gene sequences (see also 104).

Kohne (85), however, has subsequently shown that there are some DNA sequences in *E. coli* B which are similar to ribosomal RNA of *Proteus mirabilis*–1. Since the amount of DNA involved comprises about 0.3% of the bulk DNA (148), i.e. enough for 4–5 ribosomal genes, it would not seriously affect gross DNA homology estimations.

Gene Conservation and Evolution

The bacterial chromosome contains information for the synthesis of three populations of ribonucleic acid molecules: mRNA (messenger), sRNA (soluble), and rRNA (ribosomal). Although these macromolecules reflect the base composition of the DNA from which they were transcribed, the base compositions of the sRNA and rRNA are clearly distinct from that of the bulk DNA. This is not particularly surprising, except for the fact that, although the nucleotide base composition of bacterial genera ranges from 30–70 mole % G + C, the sRNA and rRNA compositional range is much narrower (50–60 mole % G + C). One obvious explanation is that those gene sequences coding for sRNA and rRNA have been conserved in the course of evolution of the bacteria (100, 101, 117, 118). This, indeed, has been shown by several workers to be so for rRNA in *E. coli* and *Proteus mirabilis* (85), *Bacillus* species (47, 49), and enteric bacteria and myxobacteria (111), and for sRNA in *E. coli* B and other enterics (62). However conservation was not considered to be absolute, judging from the thermal stability data.

With the methods now available for separating RNA, the bacterial taxonomist should be able to extend significantly the studies of ribosomal gene conservation to encompass widely diverse genera. Lines of evolution tracing back much further in time than those detectable by DNA/DNA homology techniques may thereby be elucidated.

Genera Examined to Date

The volume of literature on molecular genetic relationships among the bacteria is extensive. In the main, these reports were prepared either to provide identification of a few isolates or to classify sets of strains. The successes of the former have been considerable; the very fact of these successes has led to doubt concerning alpha taxonomy of bacteria based solely on relatively few phenetic characteristics. The accomplishments of molecular genetic study in classifying sets of strains have received less recognition, for the following reasons: First, truly representative strains for a given genus or species often do not exist or have not been designated. The strains chosen as representative of a given taxon may show great variation as the selection procedure is arbitrary. The use of single characteristics, such as nitrogen-fixation, pathogenicity, utilization of gaseous hydrogen, etc, for identification of species or genera has resulted in establishment of bacterial genera, the species within which show gene sequence homologies varying from 0–100% (2, 3, 81, 108, 140). Second, the selection of the reference strain for a homology study is critical. It is this strain with which all other isolates are compared. If it is not representative of the taxon, taxonomic conclusions drawn from the results may not be valid for the taxon. Third, it should be clear that a homology determination is not an absolute measurement, as is the case for nucleotide base composition, but is a relative estimate, subject to considerable variation depending upon the experimental parameters employed. Therefore, construction of a phylogeny of the bacteria based solely on molecular genetic evidence is not recommended at the present time. It is, however, instructive to examine the results of those homology studies employing large numbers of strains of common bacterial taxa (Table 2). The conclusions drawn from such studies will provide some predictions concerning the taxonomy of other groups of bacteria when sufficiently large numbers of strains have been analyzed.

The most thorough studies to date are those of Brenner (12), Brenner & Falkow (14), and Brenner et al (15, 17), in which members of the Enterobacteriaceae, in particular *Escherichia, Shigella,* and *Salmonella,* were examined. A DNA homology study was done in which 47 strains of *E. coli* from diverse sources,

Table 2 Polynucleotide sequence relationships among selected genera of bacteria[a]

	Lactobacillus	Streptococcus	Aerococcus	Propionibacterium	Neisseria	Sarcina	Streptomyces	Escherichia	Erwinia	Pasturella	Haemophilus	Pseudomonas	Agrobacterium	Vibrio	Myxococcus
Lactobacillus	100														
Streptococcus		100													
Bifidobacterium		2-6 (136)		9 (75)											
Aerococcus		100													
Diplococcus		1 (154)													
Gaffkya			40-95 (82)												
Propionibacterium				100											
Neisseria					100			0 (15)							
Sarcina						100									
Staphylococcus						0 (68)									
Bacillus						1 (68)									
Corynebacterium				0-7 (75)									3 (67)		
Actinomyces				2-13 (75)											
Streptomyces							100								
Nocardia							44 (151)								
Escherichia	0 (108)				2-3 (82)		2 (151)	100	11 (56)	22 (127)		30 (43)	14 (67)	5 (1)	
Shigella								80-89 (17)							
Salmonella								38 (15)	11 (56)						
Aerobacter								37 (15)							
Erwinia									100						
Proteus								6 (15)							
Pasturella								4 (15)		100					
Haemophilus											100				
Pseudomonas								4 (15)				100		6 (66)	0 (76)
Xanthomonas												31 (43)			
Aeromonas														7 (66)	
Azotobacter												40 (42)			
Rhizobium												56 (43)	52 (67)		
Agrobacterium									0 (56)				100		
Vibrio														100	
Myxococcus															100
Cytophaga															
Mycoplasma											4-6 (130)				0 (76)

[a]Numbers in parenthesis refer to references.

Shigella, and Alkalescens-Dispar strains were compared with *E. coli* K–12. The average relative relatedness of these strains was found to be 90–94%, and approximated a normal distribution. Half of the *Shigella* strains fell within the *E. coli* range. All of the Alkalescens-Dispar strains were well within the *E. coli* range.

Extensive studies of homologies between strains of different recognized genera have been made, for example *Escherichia coli/Shigella flexneri* (82–86%) (15). However, much lower estimates were found between strains of *E. coli/Salmonella typhimurium* (26–39%), *Shigella flexneri/Salmonella typhimurium* (5%), and *Proteus mirabilis/Enterobacter aerogenes* (5%) (15). DNA homologies among fifteen strains of *Erwinia* ranged between 35–100% at 40°C. *Erwinia herbicola* was found to be 29% related to *Salmonella typhimurium,* 28% related to *Klebsiella aerogenes,* and completely unrelated to *Agrobacterium tumefaciens* (56). Palleroni et al (119) showed that a number of flourescent and nonfluorescent nomenspecies of *Pseudomonas* could be placed within a single DNA homology complex, distinct from other *Pseudomonas* species. Park & DeLey (121) also found DNA "ancestral remnants" of *Pseudomonas* and *Xanthomonas* strains.

It seems logical to conclude that bacterial evolution has progressed to a point where it is not always possible to detect common (core) gene sequences, even in phenetically similar strains. It would be expected that groups of bacteria showing similar core gene sequences are most likely to be more related to each other than to groups devoid of such core sets, i.e. the higher the percentage of core DNA held in common, the more closely related the strains, on an evolutionary basis. Indeed, this conclusion seems to be supported by numerical taxonomic (NT) data, which provide an estimate of phenetic relationship. However, the size of the core gene sequence from which such NT data have been drawn has yet to be defined. In this connection, a major hurdle at this point appears to be the selection of representative strains. Reasonable taxonomic success has been achieved when representative, or reference, strains for homology studies were chosen after NT analysis; the statistically median organism (91) was hypothesized and the strain closest to it selected for homology studies. A list of genera from which reference strains have been selected is given in Table 3.

PHENETIC CLASSIFICATION OF BACTERIA

Numerical taxonomy (NT) is defined by Sokal & Sneath (146) as "the numerical evaluation of the affinity or similarity between taxonomic units and the ordering of these units into taxa on the basis of their affinities." The methods applied to taxonomic data involve conversion of the information into numerical quantities. The approaches consist of a variety of multivariate techniques. The fundamental assumptions and philosophical attitudes are amply discussed and defended by Sokal & Sneath (146).

Historical Background

The publications of Sneath (142–145) on the genus *Chromobacterium* usnered in the studies of bacterial classification using NT. Since 1960 the number of papers

Table 3 Bacterial genera from which reference strains were selected for DNA homology studies

Genera	References	Genera	References
Propionibacterium	73, 75	Klebsiella	18, 21
Lactobacillus	108, 140	Pasturella	127
Aerococcus	136	Brucella	71
Neisseria	82, 83	Haemophilus	7, 130
Streptococcus	31, 154	Achromobacter	74
Moraxella	74	Alcaligenes	74, 123
Halobacterium	113	Acetobacter	40
Sarcina	68	Xanthomonas	41, 43, 45
Clostridium	33, 73, 88,	Aeromonas	66
	89, 157	Azotobacter	42
Bacillus	150, 155	Bdellovibrio	138
Bifidobacterium	32, 132,	Rhizobium	51, 58, 67
	133	Agrobacterium	58, 67
Actinobacillus	128, 129	Pseudomonas	2, 25, 42, 43,
Actinoplanes	54		45, 66, 119,
Streptomyces	110, 115, 116,		120, 123, 128,
	151, 152		129
Actinomyces	11, 52, 158	Vibrio	1, 3, 66, 81,
Arachnia	75		149
Mycobacterium	63	Rhodopseudomonas	156
Enterobacter	21	Cytophaga	76
Escherichia	15, 17, 18,	Myxococcus	76, 77
	20, 66	Hyphomicrobium	112
Shigella	15, 17	Leptospira	65
Salmonella	15, 20	Mycoplasma	106, 107, 122,
Erwinia	18, 56		125, 126
Proteus	15, 130	Chlamydia	84
		Bedsonia	57

appearing on the theory and application of NT in microbiology has more than doubled. About 300 papers concern the taxonomy of bacteria, and others deal with viruses, yeasts, and fungi; a review for bacteria is provided by Colwell (28).

Comparison of Methods and Results

Considering only the bacteria, the development of new methods for NT has revolved around coding of characters, i.e. single-state versus multi-state characters and suitable methods for coding these, and the coefficients of association to be applied. The S_J coefficient, omitting negative matches, is the most widely used in bacterial studies. However, the simple matching coefficient has also been used, with negative matches included (139). The latter has not materially improved discrimination of clusters of related strains, but generally has had the opposite effect, i.e. merging clusters at higher levels of similarity so that cluster discrimination is very difficult to obtain.

In nearly all of the 35 bacterial genera so far studied, some clarification or improvement of the classification was achieved. Generally, conclusions concerning bacterial classification drawn from NT studies in different laboratories are in agreement. Precise correlations have not been possible because standard reference strains have not been included in every study, nor have the same tests been done on each strain. Hence, only generalized comparisons of the analyses can be made. Nevertheless, the indications are clearly that classifications based on NT are reproducible.

CORRELATION OF MOLECULAR GENETIC AND PHENETIC CLASSIFICATION

We now examine correlations between classifications obtained by these two techniques. Data from several sources (17, 27, 36, 119) are presented in Figure 1. Good linear correlation (0.84) was observed. However, the prediction of a perfect relationship, represented by the line with slope of 1.0 in Figure 1, was not obtained (slope = 1.63). For values in the range of 75–100% homology and similarity, congruence was extremely good, and the strains involved are identical, or very similar, to the reference strain used in the study. A cluster of related species occurred in the 50–75% homology and 60–75% similarity ranges. These levels are interpreted as indicating molecular genetic (above the line) and phenetic (below the line) levels of relatedness. Below 50% homology and 60% similarity are the organisms of diverse gene sequences. The gene expressions of these organisms, however, suggest closer relationships than the sequences.

Calculated slopes (m) for data from *Vibrio* strains ranging from 5–100% homology and 40–100% similarity (28, 79) varied from 1.66–1.78 and were in good agreement with the line determined by using published data ($m = 1.63$; Figure 1). The equation derived from Figure 1, expressing the relationship of percent homology determined under optimal reassociation conditions to percent similarity is: % homology = 1.63 × (% similarity) – 60.

Explanations can be offered to explain the departure from unity in Figure 1. DeLey (36) stated that "this is partially due to the limited number of unique features which are used in numerical taxonomy, but also to different nucleotide sequences of homologous genes," the latter coding for similar active sites on enzymes (isozymes) although unable to hybridize under the usual reannealing conditions. Jones & Sneath (78) also postulated that the discordance may occur because "the same phenotypic expression (e.g. protein sequence) was represented by a very different nucleotide triplet." The reverse relationship, in which homology values are higher than similarity values, could arise from homologous genes not being expressed phenetically. Previously, Heberlein et al (67) concluded from their studies of species of Rhizobiaceae, as did DeLey et al (43) from studies of *Pseudomonas* and *Xanthomonas* spp., that such did not occur since none of their homology values were higher than the corresponding similarity values. But, studies (17, 27, 36) summarized in Figure 1 conflict with this conclusion. Indeed, Brenner et al (17) found such a case of excessive discordance between the two techniques; they used well-characterized strains of *Salmonella typhimurium* and *Shigella flexneri,* finding these to be

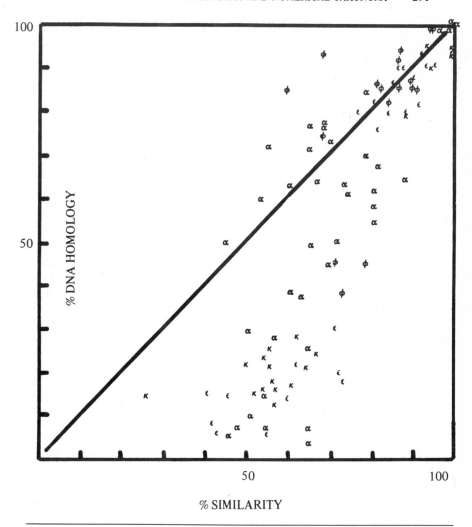

Figure 1 Correlation of % homology (DNA sequence) and % similarity (phenetic). Data are taken from Palleroni et al (119), κ; DeLey (36), α; Colwell (27), ε; and Brenner et al (17), φ; slope of linear regression line = 1.63; linear correlation coefficient = 0.84.

related to *E. coli* on a genetic basis by 45 and 85%, respectively, at 60°C (12 and 80%, respectively, 75°C). The phenetic relationships were 71 and 59%, respectively, a difference of 26% in both cases at 60°C.

Nevertheless, there are no reports indicating two strains to be genetically identical but phenetically completely unrelated, or vice versa. Thus the correlation of classifications obtained from molecular genetic and phenetic approaches is creditable.

A typical $E.$ $coli$ genome contains enough genetic material (3×10^9 daltons) to code for approximately 3×10^3 protein molecules. Assuming these proteins are enzymes and recognizing that in NT studies approximately 200 features are coded and computed, even if each of the coded features represented 10 enzymes, the genomic representation in NT analyses is incomplete. Furthermore, the similarity value may be accurate only to 10% (146). Thus the precision of numerical taxonomy needs to be improved.

Table 4 summarizes some of the recent data showing temperature effects on correlations of homology and similarity measurements. If we assume that the cause for the nonunity relationship between percent homology and percent similarity data is due to technical difficulties and not a true reflection of natural relationships, then alternative procedures for determining these values might give better congruence (unity).

High temperature incubation conditions for determining percent homology values result in an increase in the slope of the linear regression line for percent homology vs percent similarity data. (See Table 4, references 17, 118.) This suggests that either the optimal incubation temperatures were not selective enough to allow only sequences of high complementarity to anneal or that there is poor discrimination at low similarity values. It seems probable that this effect is a combination of both possibilities, in view of the intrinsic inaccuracies of each technique for distantly related strains. Should the slopes of the linear regression lines decrease, as for example, in the case of the data of Palleroni et al (119), one could postulate that the higher incubation temperatures provide a better measure of natural relationships, assuming that similarity data are truly indicative of natural relationships. The fact that increased temperature increased the slope of the linear regression lines away from unity, using the data of Palleroni et al (119) as an example, and toward unity, as in the case of the data from Brenner et al (17), is a reflection of the diversity of strains employed in each study. In the study of $Pseudomonas$ species (119), percent homology values ranged from 0–100%, being fairly evenly dispersed over this range. However, the study of the enterics (17) involved primarily $E.$ $coli$ strains showing high homologies and only a few strains of no less than 38% homology. In the latter type of study of a fairly homogeneous group of bacteria, high incubation temperatures might well be more useful for assessing natural relationships than low temperatures, but high temperatures would be of limited value in studies of widely different strains.

A further difficulty in relating molecular genetic and phenetic studies results from the different coefficients used in phenetic studies, as mentioned earlier.

LIMITS OF TAXOSPECIES AND GENOSPECIES

Ravin (124) defined a taxospecies as a group of strains sharing a high proportion of similar properties and a genospecies as a group of strains that can exchange genes. Extending the genospecies to include those strains possessing high genomic relatedness strengthens the definition. It has been proposed that for bacteria, similarity values of 70–75% delineate the taxogenus (139), with strains sharing similarity

Table 4 Effect of temperature of reassociation and numerical taxonomy methods on correlation of % homology and % similarity determinations

Reference Strain	Reassociation Temperature (C)	Numerical Taxonomy Procedure	Linear Regression Line Slope	Correlation Coefficient	Literature Reference
Pseudomonas aeruginosa (Strain 131)	72	S_J*	1.24	0.79	(119)
	80	S_J	1.30	0.80	
	72	S_{SM}++	2.14	0.85	
	80	S_{SM}++	2.39	0.85	
Escherichia coli (Strain 0128a)	60	S_J	0.84	0.46	(17)
	75	S_J	1.15	0.41	
Vibrio cholerae (NIH 35A3)	60	S_J	1.78	0.91	(27)
Vibrio parahaemolyticus (SAK 4)	60	S_J	1.73	0.81	(27)
Vibrio parahaemolyticus (MD 8657-71)	60	S_J	1.66	0.98	(79)

*S_J: similarity coefficient for positive matches only.
++S_{SM}: matching coefficient for positive and negative matches.

values of $\geq 75\%$ comprising a taxospecies. Because both the bacteriological testing procedures and the computational methods vary among laboratories, the similarity value levels for taxogenus and taxospecies may show great variation among laboratories.

Only a few combined molecular genetic and phenetic studies have been made (17, 27, 43, 63, 67, 79, 119, 120). Nevertheless, we can suggest, judging from both the published work and our own, that the homology values delineating a genus lie in the range 50–75% (43, 67) and the species level is 75–80% or greater (17, 67).

It may well be, as suggested by Cowan (29), that the bacteria form a continuous spectrum of integrating forms, thereby making any precise statement on the limits of genospecies and taxospecies wholly arbitrary. The limits given above are suggested as working guidelines.

PROSPECTS FOR BACTERIAL CLASSIFICATION

The contributions of NT to bacterial systematics have already been reviewed (28). It is clear that NT has improved the classification of many taxa. Molecular genetics permits studies of relationships at the nucleotide level, and hopefully will eventually provide an evolutionary history, i.e. phylogeny of the bacteria. For the present, the combined approaches are likely to provide major improvements in bacterial classification. We hope, however, that bacterial taxonomic studies eventually will emphasize the following procedural points: 1. a broad geographic coverage in sources of strains, 2. selection of character variables likely to represent products of one-gene-one-enzyme activity, 3. calculation of hypothetical median organisms (HMO) for any groups of strains used in an NT study, and 4. selection of reference strains similar to HMO's for hybridization studies. Bacterial taxonomy might thus become not only reproducible but also useful for information storage and retrieval.

Literature Cited

1. Anderson, R. S., Ordal, E. J. 1972. Deoxyribonucleic acid relationships among marine vibrios. *J. Bacteriol.* 109:696–706
2. Ballard, R. W., Palleroni, N. J., Doudoroff, M., Stanier, R. Y., Mandel, M. 1970. Taxonomy of the aerobic pseudomonads: *Pseudomonas cepacia, P. marginata, P. alliicola,* and *P. caryophylli. J. Gen. Microbiol.* 60:199–214
3. Basden, E. H. II, Tourtellote, M. E., Plastridge, W. N., Tucker, J. S. 1968. Genetic relationship among bacteria classified as vibrios. *J. Bacteriol.* 95:439–43
4. Bautz, E. K. F., Bautz, F. A. 1964. The influence of noncomplementary bases on the stability of ordered polynucleotides. *Proc. Nat. Acad. Sci. USA* 52:1476–81
5. Bautz, E. K. F., Hall, B. D. 1962. The isolation on T4-specific RNA on a DNA-cellulose column. *Proc. Nat. Acad. Sci. USA* 48:400–8
6. Bernardi, G. 1965. Chromotography of nucleic acids on hydroxyapatite. *Nature* 206:779–83
7. Boling, M. E. 1972. Homology between the deoxyribonucleic acids of *Haemophilis influenzae* and *Haemophilis parainfluenzae. J. Bacteriol.* 112:745–50
8. Bollum, F. J. 1968. Filter paper disk techniques for assaying radioactive macromolecules. In *Methods in Enzymology,* ed. L. Grossman, K. Moldave, 12:Pt. B (Nucleic acids), 169–73. New York: Academic
9. Bolton, E. T., McCarthy, B. J. 1962. A general method for the isolation of RNA complementary to DNA. *Proc. Nat. Acad. Sci. USA* 48:1390–97

10. Bonner, J., Kung, G., Bekhor, I. 1967. A method for hybridization of nucleic acid molecules at low temperature. *Biochemistry* 6:3650–53
11. Bradley, S. G. 1965. Interspecific genetic homology in actinomycetes. *Int. Bull. Bacteriol. Nomencl. Taxon.* 15: 239–41
12. Brenner, D. J. 1970. Deoxyribonucleic acid divergence in *Enterobacteriaceae. Develop. Ind. Microbiol.* 11:139–53
13. Brenner, D. J., Cowie, D. B. 1968. Thermal stability of *Escherichia coli Salmonella typhimurium* deoxyribonucleic acid duplexes. *J. Bacteriol.* 95: 2258–62
14. Brenner, D. J., Falkow, S. 1971. Genetics of the enterobacteriaceae. C. Molecular relationships among members of the enterobacteriaceae. *Advan. Genet.* 16:81–118
15. Brenner, D. J., Fanning, G. R., Johnson, K. E., Citarella, R. V., Falkow, S. 1969. Polynucleotide sequence relationships among members of *Enterobacteriaceae. J. Bacteriol.* 98:637–50
16. Brenner, D. J., Fanning, G. R., Rake, A., Johnson, K. E. 1969. A batch procedure for thermal elution of DNA from hydroxyapatite. *Anal. Biochem.* 28: 447–59
17. Brenner, D. J., Fanning, G. R., Skerman, F. J., Falkow, S. 1972. Polynucleotide sequence divergence among strains of *Escherichia coli* and closely related organisms. *J. Bacteriol.* 109: 953–65
18. Brenner, D. J., Fanning, G. R., Steigerwalt, A. G. 1972. Deoxyribonucleic acid relatedness among species of *Erwinia* and between *Erwinia* species and other enterobacteria. *J. Bacteriol.* 110: 12–17
19. Brenner, D. J., Fanning, G. R., Steigerwalt, A. G., Orskov, I., Orskov, F. 1972. Polynucleotide sequence relatedness among three groups of pathogenic *Escherichia coli* strains. *Infec. Immunity* 6:308–15
20. Brenner, D. J., Martin, M. A., Hoyer, B. H. 1967. Deoxyribonucleic acid homologies among some bacteria. *J. Bacteriol.* 94:486, 487
21. Brenner, D. J., Steigerwalt, A. G., Fanning, G. R. 1972. Differentiation of *Enterobactor aerogenes* from *Klebsiellae* by deoxyribonucleic acid reassociation. *Int. J. Syst. Bacteriol.* 22:193–200
22. Britten, R. J., Kohne, D. E. 1966. Nucleotide sequence repetition in DNA. *Carnegie Inst. Washington Yearb.* 65: 78–106
23. Britten, R. J., Kohne, D. E. 1968. Repeared sequences in DNA. *Science* 161:529–40
24. Cain, A. J. 1962. The evolution of taxonomic principles. *Symp. Soc. Gen. Microbiol.* 12:1–13
25. Chakrabarty, A. M., Gunsalus, I. C. 1970. Transduction and genetic homology between *Pseudomonas* species *putida* and *aeruginosa. J. Bacteriol.* 103:830–32
26. Citarella, R. V., Colwell, R. R. 1970. Polyphasic taxonomy of the genus *Vibrio:* Polynucleotide sequence relationships among selected *Vibrio* species. *J. Bacteriol.* 104:434–42
27. Colwell, R. R. 1970. Polyphasic taxonomy of the genus *Vibrio:* numerical taxonomy of *Vibrio cholerae, Vibrio parahaemolyticus,* and related *Vibrio* species. *J. Bacteriol.* 104:410–33
28. Colwell, R. R. 1973. Genetic and phenetic classification of bacteria. *Advan. Appl. Microbiol.* In press
29. Cowan, S. J. 1962. The microbial species–a macromyth? *Symp. Soc. Gen. Microbiol.* 12:433–35
30. Cowie, D. B., Szafranski, P. 1967. Thermal chromatography of DNA-DNA reactions. *Biophys. J.* 7:567–84
31. Coykendall, A. L. 1971. Genetic heterogeneity in *Streptococcus mutans. J. Bacteriol.* 106:192–96
32. Crociani, F., Scardovi, V., Trovatelli, L. D. 1970. Mannitol fermenting bifids from rumen and their DNA homology relationships. *Ann. Microbiol.* 20:99–106
33. Cummins, C. S., Johnson, J. L. 1971. Taxonomy of the clostridia: wall composition and DNA homologies in *Clostridium butyricum* and other butyric acid producing bacteria. *J. Gen. Microbiol.* 67:33–46
34. DeLey, J. 1962. Comparative biochemistry and enzymology in bacterial classification. *Symp. Soc. Gen. Microbiol.* 12:164–95
35. DeLey, J. 1968. DNA base composition and taxonomy of some *Acinotobacter* strains. *Antonie van Leeuwenhoek J. Microbiol. Serol.* 34:109–14
36. DeLey, J. 1968. Molecular biology and bacteriol phylogeny. In *Evolutionary Biology,* ed. T. Dobzhausky, M. K. Hecht, W. C. Steare, 2:103–56. Amsterdam: North-Holland
37. DeLey, J. 1969. Compositional nucleotide distribution and the theoretical prediction of homology in bacterial DNA. *J. Theor. Biol.* 22:89–116

38. DeLey, J. 1971. A molecular approach to microbial taxonomy. *J. Gen. Microbiol.* 69: Reports of Proceedings, p. ii

39. DeLey, J., Cattoir, H., Reynaerts, A. 1970. The quantitative measurement of DNA hybridization from renaturation rates. *Eur. J. Biochem.* 12:133–42

40. DeLey, J., Friedman, S. 1964. Deoxyribonucleic acid hybrids of acetic acid bacteria. *J. Bacteriol.* 88:937–45

41. DeLey, J., Friedman, S. 1965. Similarity of *Xanthomonas* and *Pseudomonas* deoxyribonucleic acid. *J. Bacteriol.* 89:1306–9

42. DeLey, J., Park, I. W. 1966. Molecular biological taxonomy of some free-living nitrogen fixing bacteria. *Antonie van Leeuwenhoek J. Microbiol. Serol.* 42: 6–16

43. DeLey, J., Park, I. W. Tijtgat, R., van Ermengem, J. 1966. DNA homology and taxonomy of *Pseudomonas* and *Xanthomonas*. *J. Gen. Microbiol.* 42: 43–56

44. DeLey, J., Schell, J. 1963. Deoxyribonucleic acid base composition of acetic acid bacteria. *J. Gen. Microbiol.* 33:243–53

45. DeLey, J., Tijtgat, R. 1970. Evaluation of membrane filter methods for DNA/DNA hybridization. *Antonie van Leeuwenhoek J. Microbiol. Serol.* 36:461–74

46. Denhardt, D. T. 1966. A membrane-filter technique for the detection of complementary DNA. *Biochem. Biophys. Res. Commun.* 23:641–46

47. Doi, R. H., Igaraski, R. T. 1965. Conservation of ribosomal and messenger ribonucliec acid cistrons in *Bacillus* species. *J. Bacteriol.* 90:384–90

48. Doty, P., Marmur, J., Eigner, J., Schildkraut, C. 1960. Strand separation and specific recombination in deoxyribonucleic acids: physical and chemical studies. *Proc. Nat. Acad. Sci. USA* 46:461–76

49. Dubnau, D., Smith, I., Morell, P., Marmur, J. 1965. Gene conservation in *Bacillus* species. I. Conserved genetic and nucleic acid base sequence homologies. *Proc. Nat. Acad. Sci. USA* 54: 491–98

50. Eigner, J. 1960. The native denatured and renatured states of deoxyribonucleic acid. PhD thesis. Harvard Univ., Cambridge

51. Elkan, G. H., Usanis, R. A. 1971. Theoretical deoxyribonucleic acid homology between strains in *Rhizobium japonicum*. *Int. J. Syst. Bacteriol.* 21:295–98

52. Enquist, L. W., Bradley, S. G. 1970. Nucleotide divergence in deoxyribonucleic acids of actinomycetes. *Advan. Front. Plant Sci.* 25:53–73

53. Falkow, S., Rownd, R., Baron, L. S. 1962. Genetic homology between *Escherichia coli* K–12 and *Salmonella*. *J. Bacteriol.* 84:1303–12

54. Farina, G., Bradley, S. G. 1970. Reassociation of deoxyribonucleic acids from *Actinoplanes* and other actinomycetes. *J. Bacteriol.* 102:30–35

55. Friedman, S., DeLey, J. 1965. "Genetic Species" concept in *Xanthomonas*. *J. Bacteriol.* 89:95–100

56. Gardner, J. M., Kado, C. I. 1972. Comparative base sequence homologies of the deoxyribonucleic acids of *Erwina* species and other *Enterobacteriaceae*. *Int. J. Syst. Bacteriol.* 22:201–9

57. Gerloff, R. K., Ritter, D. B., Watson, R. V. 1966. DNA homology between the meningopneumonitis agent and related microorganisms. *J. Infec. Dis.* 116:197–202

58. Gibbons, A. M., Gregory, K. F. 1972. Relatedness among *Rhizobium* and *Agrobacterium* species determined by three methods of nucleic acid hybridization. *J. Bacteriol.* 111:129–41

59. Gillespie, S., Gillespie, D. 1971. Ribonucleic acid-deoxyribonucleic acid, hybridization in aqueous solutions and in solutions containing formanide. *Biochem. J.* 125:481–87

60. Gillespie, D., Spiegelman, S. 1965. A quantitative assay for DNA/RNA hybrids with DNA immobilized on a membrane. *J. Mol. Biol.* 12:829–42

61. Gillis, M., DeLey, J., De Cleene, M. 1970. The determination of molecular weight of bacterial genome DNA from renaturation rates. *Eur. J. Biochem.* 12:143–53

62. Goodman, H. M., Rich, A. 1962. Formation of a DNA soluble RNA hybrid and its relation to the origin, evolution and degeneracy of soluble RNA. *Proc. Nat. Acad. Sci. USA* 48: 2101–9

63. Gross, W. M., Wayne, L. G. 1970. Nucleic acid homology in the genus *Mycobacterium*. *J. Bacteriol.* 104:630–34

64. Guerry, P., Falkow, S. 1971. Polynucleotide sequence relationships among some bacteriol plasmids. *J. Bacteriol.* 107:372–74

65. Haapala, D. K., Rogul, M., Evans, L. B., Alexander, A. D. 1969. Deoxyribonucleic acid base compositions and homology studies of *Leptospira*. *J. Bacteriol.* 98:421–28

66. Hanaoko, M., Kato, V., Amano, T. 1969. Complementary examinations of DNA's among *Vibrio* species. *Biken J.* 12:181–85

67. Heberlein, G. T., DeLey, J., Tijtgat, R. 1967. Deoxyribonucleic acid homology and taxonomy of *Agrobacterium, Rhizobium* and *Chromobacterium. J. Bacteriol.* 94:116–24

68. Herndon, S. E., Bott, K. F. 1969. Genetic relatedness between *Sarcina ureae* and members of the genus *Bacillus. J. Bacteriol.* 97:6–12

69. Hoyer, B. H., Bolton, E. T., McCarthy, B. J. 1964. A molecular approach to the systematics of higher organisms. *Science* 144:959–68

70. Hoyer, B. H., King, J. R. 1969. Deoxyribonucleic acid sequence losses in a stable streptococcal form. *J. Bacteriol.* 97:1516, 1517

71. Hoyer, B. H., McCullough, N. B. 1968. Polynucleotide homologies of *Brucella* deoxyribonucleic acids. *J. Bacteriol.* 95:444–48

72. Hoyer, B. H., McCullough, N. B. 1968. Homologies of ribonucleic acids from *Brucella ovis,* canine abortion organisms and other *Brucella* species. *J. Bacteriol.* 96:1783–90

73. Johnson, J. L. 1970. Relationship of deoxyribonucleic acid homologies to cell wall structure. *Int. J. Syst. Bacteriol.* 20:421–24

74. Johnson, J. L., Anderson, R. S., Ordal, E. J. 1970. Nucleic acid homologies among oxidase-negative *Moraxella* species. *J. Bacteriol.* 101:568–73

75. Johnson, J. L., Cummins, C. S. 1972. Cell wall composition and deoxyribonucleic acid similarities among the anaerobic coryneforms, classical propionibacteria, and strains of *Arachina propionica. J. Bacteriol.* 109:11047–66

76. Johnson, J. L., Ordal, E. J. 1968. Deoxyribonucleic acid homology in bacterial taxonomy: effect of incubation temperature on reaction specificity. *J. Bacteriol.* 95:893–900

77. Johnson, J. L., Ordal, E. J. 1969. Deoxyribonucleic acid homology among fruiting myxobacteria. *J. Bacteriol.* 98:319, 320

78. Jones, D., Sneath, P. H. A. 1971. Genetic transfer and bacterial taxonomy. *Bacteriol. Rev.* 34:40–81

79. Kaneko, T. 1973. Ecology of *Vibrio parahaemolyticus* and related organisms in Chesapeake Bay. PhD thesis. Georgetown Univ., Washington DC

80. Kennel, D. 1971. Principles and practices of nucleic acid hybridization. In *Progress in Nucleic Acid Research and Molecular Biology,* ed. J. N. Davidson, W. E. Cohn, Vol. 11. New York: Academic

81. Kiehn, E. D., Pacha, R. E. 1969. Characterization and relatedness of marine vibrios pathogenic to fish: deoxyribonucleic acid homology and base composition. *J. Bacteriol.* 100:1248–55

82. Kingsbury, D. T. 1967. Deoxyribonucleic and homologies among species of the genus *Neisseria. J. Bacteriol.* 94:870–74

83. Kingsbury, D. T., Fanning, G. R., Johnson, K. E., Brenner, D. J. 1969. Thermal stability of interspecies *Neisseria* DNA duplexes. *J. Gen. Microbiol.* 55:201–8

84. Kingsbury, D. T., Weiss, E. 1968. Lack of deoxyribonucleic acid homology between species of the genus *Chlamydia. J. Bacteriol.* 96:1421–23

85. Kohne, D. E. 1968. Isolation and characterization of bacterial ribosomal RNA cistrons. *Biophys. J.* 8:1104–19

86. Kourilsky, P., Leidner, J., Tremblay, G. Y. 1971. DNA-DNA hybridization on filters at low temperature in the presence of formamide or urea. *Biochemie* 53:1111–14

87. Laird, C. D., McConaughy, B. L., McCarthy, B. J. 1969. On the rate of fixation of nucleotide substitutions on evolution. *Nature* 22:149–54

88. Lee, W. H., Rieman, H. 1970. Correlation of toxic and nontoxic strains of *Clostridium botulinum* by DNA composition and homology. *J. Gen. Microbiol.* 60:117–23

89. Lee, W. H., Rieman, H. 1970. The genetic relatedness of proteolytic *Clostridrim botulinum* strains. *J. Gen. Microbiol.* 64:85–90

90. Legault-Démare, J., Desseaux, B., Heyman, T., Séror, S., Ress, G. P. 1967. Studies on hybrid molecules of nucleic acids. I. DNA-DNA hybrids on nitrocellulose filters. *Biochem. Biophys. Res. Commun.* 28:550–57

91. Liston, J., Wiebe, W., Colwell, R. R. 1963. Quantative approach to the study of bacterial species. *J. Bacteriol.* 85:1061–70

92. Mandel, M. 1969. New approaches to bacterial taxonomy: perspective and prospects. *Ann. Rev. Microbiol.* 23:239–74

93. Marmur, J., Doty, P. 1961. Thermal renaturation of deoxyribonucleic acids. *J. Mol. Biol.* 3:585–94

94. Marmur, J., Falkow, S., Mandel, M. 1963. New approaches to bacterial tax-

onomy. *Ann. Rev. Microbiol.* 17: 329–72

95. Marmur, J., Lane, D. 1960. Strand separation and specific recombination in DNA's. I. Biological studies. *Proc. Nat. Acad. Sci. USA* 46:453–61

96. Marmur, J., Rownd, R., Schildkraut, C. L. 1963. Denaturation and renaturation of DNA. *Progr. Nucl. Acid Res., Mol. Biol.* 1:231–300

97. Marmur, J., Schildkraut, C. L., Doty, P. 1961. The reversible denaturation of DNA and its use in studies of nucleic acid homologies and the biological relatedness of microorganisms. *J. Chim. Phys. Physicochim. Biol.* 58:945–55

98. Marmur, J., Schildkraut, C. L., Doty, P. 1961. Biological and physical chemical aspects of the reversible denaturation of deoxyribonucleic acids. In *The Molecular Basis of Neoplasia, Ann. Symp. Fund. Cancer Res., 15th,* ed. K. Houston. Austin: Texas Univ. Press

99. Martin, M. A., Hoyer, B. H. 1966. Thermal stabilities and specificites of reannealed animal deoxyribonucleic acids. *Biochemistry* 5:2706–13

100. McCarthy, B. J. 1965. The evolution of polynucleotides. *Progr. Nucl. Acid Res., Mol. Biol.* 4:129–60

101. McCarthy, B. J. 1967. Arrangement of base sequences in deoxyribonucleic acid. *Bacteriol. Rev.* 31:215–29

102. McCarthy, B. J., Bolton, E. T. 1963. An approach to the measurement of genetic relatedness among organisms. *Proc. Nat. Acad. Sci. USA* 50:156–64

103. McCarthy, B. J., Church, R. B. 1970. The specificity of molecular hybridization reactions. *Ann. Rev. Biochem.* 39: 131–50

104. McCarthy, B. J., McConaughy, B. L. 1968. Related base sequences in the DNA of simple and complex organisms. I. DNA/DNA duplex formation and the incidence of partially related base sequences in DNA. *Biochem. Genet.* 2:37–53

105. McConaughy, B. L., Laird, C. D., McCarthy, B. J. 1969. Nucleic acid reassociation in formamide. *Biochemistry* 8:3289–95

106. McGee, Z. A., Rogul, M., Wittler, R. G. 1967. Molecular genetic studies of relationships among *Mycoplasma,* L-forms and bacteria. *Ann. NY Acad. Sci.* 143:21–30

107. McGee, Z. A., Rogul, M., Falkow, S., Wittler, R. G. 1965. The relationship of *Mycoplasma pneumoniae* (Eaton agent) to *Streptococcus* MG: applica-

tion of genetic tests to determine relatedness of L-forms and PPLO to bacteria. *Proc. Nat. Acad. Sci. USA* 54:457–61

108. Miller, A. III, Sandine, W. E., Elliker, P. R. 1972. Deoxyribonucleic acid homology in the genus *Lactobacillus.* *Can. J. Microbiol.* 17:625–34

109. Miyazawa, Y., Thomas, C. A. Jr. 1965. Composition of short segments of DNA molecules. *J. Mol. Biol.* 11:223–37

110. Monson, A. M., Bradley, S. G., Enquist, L. W., Cruces, G. 1969. Genetic homologies among *Streptomyces violaceoruber* strains. *J. Bacteriol.* 99: 702–6

111. Moore, R. L., McCarthy, B. J. 1967. Comparative study of ribosomal ribonucleic acid cistrons in enterobacteria and myxobacteria. *J. Bacteriol.* 94:1066–74

112. Moore, R. L., Hirsch, P. 1972. Deoxyribonucleic acid base sequence homologies of some budding and prosthecate bacteria. *J. Bacteriol.* 110: 256–61

113. Moore, R. L., McCarthy, B. J. 1969. Base sequence homology and renaturation studies of the deoxyribonucleic acid of extremely halophilic bacteria. *J. Bacteriol.* 99:255–62

114. Nygaard, A. P., Hall, B. D. 1963. A method for the detection of RNA-DNA complexes. *Biochem. Biophys. Res. Commun.* 12:98–104

115. Okanishi, M., Akagawa, H., Umezawa, H. 1972. An evaluation of taxonomic criteria in streptomyces on the basis of deoxyribonucleic acid homology. *J. Gen. Microbiol.* 72:49–58

116. Okanishi, M., Gregory, K. F. 1970. Methods for determination of deoxyribonucleic acid homologies in *Streptomyces.* *J. Bacteriol.* 104:1086–94

117. Pace, B., Campbell, L. L. 1971. Homology of ribosomal ribonucleic acid of *Desulfovibrio* species with *Desulfovibrio vulgaris.* *J. Bacteriol.* 106:717–19

118. Pace, B., Campbell, L. L. 1971. Homology of ribosomal ribonucleic acid of diverse bacterial species with *Escherichia coli* and *Bacillus stearothermophilus.* *J. Bacteriol.* 107:543–47

119. Palleroni, N. J., Ballard, R. W., Ralston, E., Dondoroff, M. 1972. Deoxyribonucleic acid homologies among some *Pseudomonas* species. *J. Bacteriol.* 110:1–11

120. Palleroni, N. J., Dondoroff, M. 1971. Phenotypic characterization and deox-

yribonucleic acid homologies of *Pseudomonas solanacearum. J. Bacteriol.* 107:690–96

121. Park, I. W., DeLey, J. 1967. Ancestral remnants in deoxyribonucleic acid from *Pseudomonas* and *Xanthomonas. Antonie van Leeuwenhoek J. Microbiol. Serol.* 33:1–16

122. Peterson, M., Pollock, M. E. 1969. Deoxyribonucleic acid homology and relative genome size in *Mycoplasma. J. Bacteriol.* 99:639–44

123. Ralston, E., Palleroni, N. J., Dondoroff, M. 1972. Deoxyribonucleic acid homologies of some so-called *"Hydrogenomonas"* species. *J. Bacteriol.* 109:465, 466

124. Ravin, A. W. 1963. Experimental approach to the study of bacterial phylogeny. *Am. Natur.* 97:307–18

125. Reich, P. R., Somerson, N. L., Hybner, C. J., Chanock, R. M., Weissman, S. M. 1966. Genetic differentiation by nucleic acid homology. I. Relationships among *Mycoplasma* species of man. *J. Bacteriol.* 92:302–10

126. Reich, P. R., Somerson, N. L., Rose, J. A., Weissman, S. M. 1966. Genetic relatedness among mycoplasmas as determined by nucleic acid homology. *J. Bacteriol.* 91:153–60

127. Ritter, D. B., Gerloff, R. K. 1966. Deoxyribonucleic acid hybridization among some species of the genus *Pasteurella. J. Bacteriol.* 92:1838–39

128. Rogul, M., Brendle, J., Haapala, D. K., Alexander, A. D. 1968. DNA homologies among *Pseudomonas pseudomallei, Actinobacillus mallei* and phenotypically similar organisms. *Bacteriol. Proc.* 68:19

129. Rogul, M., Brendle, J. J., Haapala, D. K., Alexander, A. D. 1970. Nucleic acid similarities among *Pseudomonas pseudomallei, Pseudomonas multivorans,* and *Actinobacillus mallei. J. Bacteriol.* 101:827–35

130. Rogul, M., McGee, Z. A., Wittler, R. G., Falkow, S. 1965. Nucleic acid homologies of selected bacteria, L-forms and mycoplasma species. *J. Bacteriol.* 90:1200–3

131. Rownd, R., Falkow, S., Baron, L. S. 1962. Genetic transfer and molecular hybrid formation between *E. coli* K–12 and *Salmonella typhosa. Abstr. Biophys. Soc. Ann. Meet., 6th, Washington, D.C.*

132. Scardovi, V., Trovatelli, L. D., Zani, G., Crociani, F., Matteuzzi, D. 1971. Deoxyribonucleic acid homology relationships among species of the genus *Bifidobacterium. Int. J. Syst. Bacteriol.* 21:276–94

133. Scardovi, V., Zani, G., Trovatelli, L. D. 1970. Deoxyribonucleic acid homology among the species of the genus *Bifidobacterium* isolated from animals. *Arch. Mikrobiol.* 72:318–25

134. Schildkraut, C. L., Marmur, J., Doty, P. 1961. The formation of hybrid DNA molecules and their use in studies of DNA homologies. *J. Mol. Biol.* 3: 595–617

135. Schmeckpeper, B. J., Smith, K. D. 1972. Use of formamide in nucleic acid reassociation. *Biochemistry* 11:1319–26

136. Schultes, L. M., Evans, J. B. 1971. Deoxyribonucleic acid homology of *Aerococcus viridans. Int. J. Syst. Bacteriol.* 21:207–9

137. Seidler, R. J., Mandel, M. 1971. Quantitative aspects of DNA renaturation: base composition, state of chromosome replication and polynucleotide homologies. *J. Bacteriol.* 106:608–14

138. Seidler, R. J., Mandel, M., Baptist, J. N. 1972. Molecular heterogeneity of the bdellovibrios: Evidence of two new species. *J. Bacteriol.* 109:209–17

139. Silvestri, L., Turri, M., Hill, L. R., Gilardi, E. 1962. A quantitative approach to the systematics of actinomycetes based on overall similarity. *Symp. Soc. Gen. Microbiol.* 12:333–60

140. Simonds, J., Hansen, P. A., Lakshmanan, S. 1971. Deoxyribonucleic acid hybridization among strains of Lactobacilli. *J. Bacteriol.* 107:382–84

141. Skerman, F. J., Formal, S. B., Falkow, S. 1972. Plasmid-associated enterotoxin production of *Escherichia coli* isolated from humans. *J. Infec. Immunity* 5:622–24

142. Sneath, P. H. A. 1956. Cultural and biochemical characteristics of the genus *Chromobacterium. J. Gen. Microbiol.* 15:70–98

143. Sneath, P. H. A. 1957. Some thoughts on bacterial classification. *J. Gen. Microbiol.* 17:184–200

144. Sneath, P. H. A. 1957. The application of computers to taxonomy. *J. Gen. Microbiol.* 17:201–26

145. Sneath, P. H. A. 1960. A study of the bacterial genus *Chromobacterium. Iowa State J. Sci.* 34:243–500

146. Sokal, R. R., Sneath, P. H. A. 1963. Principles of numerical taxonomy. San Francisco: Freeman. 359 pp.

147. Somerson, N. L., Reich, P. R., Walls, B. G., Chanock, R. M., Weissman, S. M.

1966. Genetic differentiation by nucleic acid homology. II. Genotypic variations within two *Mycoplasma* species. *J. Bacteriol.* 92:311–17

148. Spiegelman, S., Yankovsky, S. A. 1965. In *Evolving Genes and Proteins*, ed. V. Bryson, H. J. Vogel. New York: Academic

149. Staley, T. E., Colwell, R. R. 1973. Polynucleotide sequence relationships among Japanese and American strains of *Vibrio parahaemolyticus. J. Bacteriol.* 114:916–27

150. Takahashi, H., Saito, H., Ikeda, Y. 1966. Genetic relatedness of sporeforming bacilli studied by the DNA agar method. *J. Gen. Appl. Microbiol.* 12:113–18

151. Tewfik, E. M., Bradley, S. G. 1967. Characterization of deoxyribonucleic acids from streptomyces and nocardiae. *J. Bacteriol.* 94:1994–2000

152. Tewfik, E. M., Bradley, S. G., Kuroda, S., Wu, R. Y. 1968. Studies on deoxyribonucleic acids from streptomycetes and nocardiae. *Develop. Ind. Microbiol.* 9:242–49

153. Warnaar, S. O., Cohen, J. A. 1966. A quantitative assay for DNA/DNA hy-brids using membrane filters. *Biochem. Biophys. Res. Commun.* 24:554–58

154. Weissman, S. M., Reich, P. R., Somerson, N. L., Cole, R. M. 1966. Genetic differentiation by nucleic acid homology. IV. Relationships among Lancefield groups and serotypes of streptococci. *J. Bacteriol.* 92:1372–77

155. Welker, N. E., Campbell, L. L. 1967. Unrelatedness of *Bacillus amyloliguefaciens* and *Bacillus subtilis. J. Bacteriol.* 94:1124–30

156. Witkin, S. S., Gibson, K. D. 1972. Ribonucleic acid from aerobically and anaerobically grown *Rhodopseudomonas spheroides:* comparison by hybridization to chromosomal and satellite deoxyribonucleic acid. *J. Bacteriol.* 110:684–90

157. Wu, J. I., Riemann, H., Lee, W. H. 1972. Thermal stability of the deoxyribonucleic acid hybrids between the proteolytic strains of *Clostridium botulinum* and *Clostridium sporogenes. Can. J. Microbiol.* 18:97–99

158. Yamaguchi, T. 1967. Similarity in DNA of various morphologically distinct actinomycetes. *J. Gen. Appl. Microbiol.* 13:63–71.

STRUCTURE OF FOLIAGE ✦4063
CANOPIES AND PHOTOSYNTHESIS

Masami Monsi

Department of Botany, University of Tokyo, Tokyo, Japan

Zenbei Uchijima

Division of Meteorology, National Institute of Agricultural Sciences, Tokyo, Japan

Takehisa Oikawa

Department of Botany, University of Tokyo, Tokyo, Japan

INTRODUCTION

All living things on the earth, including plants, rely principally upon the photosynthate produced by plants for their daily food, and accordingly are strongly affected by the variation of plant photosynthesis over the globe. The distribution of solar energy with latitude determines to a great extent the geographic variation of photosynthetic activity of plants. Its latitudinal change in turn sets broad geographic limits to the different forms of terrestrial life, affecting the energy flow and the cycle of materials in ecosystems.

On the other hand, the morphological and physiological characteristics of plants are thought to result from their evolutional adaptation to environmental conditions during the geological past. The morphological features as characterized by the geometrical structure (or architecture) of plant canopies have a great influence upon the processes of action and reaction between plants and their environment through the modification and interception of fluxes of radiation, heat, carbon dioxide, etc. Consequently it is obvious that the canopy structure is determinant of the photosynthetic productivity of plant canopies. The canopy structure as well as the physiological properties of leaves with respect to photosynthesis and respiration, therefore, can play an important role in the competition between plants.

Since the relation between the photosynthetic activity and the structure of plant canopies was elucidated by Monsi & Saeki (100) in 1953, a great number of both theoretical and experimental studies have been done on this problem. Studies in this field have been greatly stimulated by the activities of the International Biological

301

Program (IBP). An understanding of the physical and biological processes involved in canopy photosynthesis helps us to obtain high crop yields toward solving food problems and to clarify the distribution of primary production over the globe related to the development of biome.

In this review, our emphasis is on the geometrical structure of the plant canopies as systems for the capture and conversion of solar radiation energy. We also attempt to stress the need for comprehensive research on phytometrics and to show the applicability and feasibility of modeling and computer simulation to aid research on canopy photosynthesis. Numerous excellent reviews and symposium volumes concerning the physical and physiological aspects of canopy photosynthesis have already been published (7, 26, 91, 93, 94, 101, 110, 136, 177) and are recommended to the reader.

PLANT CANOPIES AS SYSTEMS FOR CAPTURE OF RADIATION ENERGY

Although the importance of canopy structure has long been recognized among plant and crop ecologists, the quantitative measurement of the canopy structure of plant stands has started relatively recently. In 1953 Monsi & Saeki (100) pointed out in their epoch-making paper that the inclination of leaves forming plant canopies plays a decisive role in the interception of light by canopies and in canopy photosynthesis. Thereafter, the influence of canopy structure upon dry-matter production by plants has been investigated by many workers (12, 14, 15, 19, 23, 35, 53, 56, 62, 76, 80, 96, 102, 118, 119, 135, 138, 139, 147, 148, 151, 162, 168, 172).

Productive Structure

The productive structure of a plant community according to Monsi & Saeki (100) denotes a complex composed of the distribution of plant elements in the canopy and the microenvironmental factors influencing plant photosynthesis. Of these factors, radiation intensity, temperature, wind, and CO_2 concentration are of most importance. Comprehensive measurements of the productive structure were made with grassland canopies (Monsi & Saeki, 100), weeds (100), crop canopies (72, 73, 131, 132), forests (74, 75, 121, 122, 142), and submerged water plants (60). In spite of this wide variety of plant species, the productive structures observed seem to be classified into the grass- and forb-types. The grass-type is characterized by maximum leaf area in the middle of the canopy, and light attenuation is more gentle than that in the forb-type with the maximum leaf area density in the upper 7th and 8th tenths of the canopy, starting from the ground. The productive structure of submerged water plants was nearly a forb-type.

Comparison of the productive structures in each plant revealed that the ratio of photosynthesizing plant elements to nonphotosynthesizing plant elements varies from 0.22 for forests to 1.16 for submerged water plants. Tadaki et al (142) observed with *Abies veitchii* that the ratio of photosynthesizing elements to nonphotosynthesizing elements decreases gradually with the increment of forest biomass from 0.235 for 20-year-old forest to 0.07 for 60-year-old forests. A similar change in the photosynthesizing element ratio is also expected during the growth period of annual crops.

The microenvironment and local microclimate, in which plants live from germination to maturity, are closely related to the physical processes of radiation absorption and heat and mass transfer within and above crop canopies. In connection with the activities of IBP, interest in the micrometeorology of cultivated fields appears to have shifted from the general aspects of surface energy balance to the exchange of heat, water vapor, and CO_2 between the vegetated surfaces and the atmosphere, particularly in terms of plant growth and evaporation (16–18, 120–122, 155).

In relation to the partition of radiant energy into sensible and latent heat within foliage canopies, profiles of net radiation flux were measured within crown canopies (16, 33, 34, 77, 78, 122) and crop canopies (175). The data obtained for forests show a very sharply defined region of radiation absorption in the mid-crown region, with little variation below this level. The radiation data within crop canopies suggest that the net radiation decrease could be expressed in terms of the exponential law.

Since the transfer processes within foliage canopies depend primarily upon the efficiency of air mixing among leaves, considerable attention has been directed toward the wind regime within plant canopies (16, 33, 34, 77, 78, 121, 155). Experimental and theoretical results indicate that the attenuation of wind velocity and of transfer coefficient within the upper half of the canopy are also approximated by an exponential law. The value of the extinction coefficient seemed to be larger in the forb-type canopy than in the grass-type canopy.

More recently the CO_2 environment within plant canopies has received great attention in connection with canopy photosynthesis (4, 116, 117, 158, 173). CO_2 profiles within plant canopies for daylight hours were characterized by a minimum in the middle of the canopies (157, 158). CO_2 concentration in the mid-foliage region of a maize canopy dropped below about 250 ppm when wind was low and solar radiation was intense, implying that CO_2 might limit the photosynthesis of a good maize crop. Research work on the transfer processes of heat and mass within plant canopies has greatly stimulated the development of biophysical models for describing plant behavior in relation to microenvironment (25–30, 36–40, 100, 157). This problem is fully described in a later section.

Canopy Structure

MEASUREMENT OF CANOPY STRUCTURE   As is well known, the canopy structure of plant communities can be specified collectively by the vertical and horizontal distribution of leaf area and by its spatial orientation function. Several methods have been devised and used for determining the canopy structure in plant ecology.

The stratified clipping method of Monsi & Saeki (100) was devised to determine the vertical profiles of each plant element within the canopies, that is, the productive structure of the plant canopies. Goodall (45) and Warren Wilson (164–167) revised the point-quadrat method for convenient investigation of the inclination of leaves and leaf arrangement in the canopy. The two-dimensional and the combined inclined point quadrats based on the use of two point quadrats determine easily the mean inclination angle of leaves, β_L, and the leaf area density f_L, by counting the scores of contacts of the point quadrats with leaf surfaces:

$$\tan \bar{\beta}_L = \frac{\pi}{2}(0.1\, f_{L.0}/f_{L.90})$$

$$f_L = f_{L.90}\, \sec \bar{\beta}_L$$

1.

where $f_{L.0}$ and $f_{L.90}$ are the scores ot contacts with leaves in the horizontal and vertical directions, respectively. Using the point quadrat method, Warren Wilson (164) studied the canopy structure of Italian ryegrass stands.

Nichiporovich (109) devised a leaf protractor for determining the foliage angle density function and showed that the canopy structure of wheat and maize crops tends to become a spherical distribution, that is, the leaf area distribution is the same as the relative frequency of the inclinations of the surface elements of a sphere. The leaf protractor has been frequently used in research on the phytometrics of crop plants (64, 85, 128, 131, 134). Although this procedure may be more precise than the next method to be described, it is difficult to apply this method to the investigation of the spatial leaf arrangement.

Another procedure for studying the leaf area index, the inclination of leaves, and the leaf arrangement in plant canopies is the so-called silhouette method devised independently by Uchijima et al (160, 161) and Loomis et al (91, 92). The silhouette of crop plants restored on chart paper on the basis of the data obtained from phytometric measurements under field or laboratory conditions is used to determine the canopy structure. This method seems to have an advantage over other procedures, because it gives adequate information on the foliage angle density function and the spatial leaf arrangement. The canopy structure of rice, barley, and maize crops has been investigated by using the silhouette method (63–65, 159–161).

VERTICAL PROFILES OF LEAF AREA DENSITY   Examples of the vertical distribution of leaf area within crop and forest stands are shown in Figure 1. In spite of wide differences in plant species, it is possible to recognize two main types of vertical profiles, such as grass and forb types. The canopies of *Fagus crenata* and rice belonging to the grass type can be characterized by a leaf area density profile with its maximum in the middle layer of the canopy. The maximum leaf area density within soybean and *Castanopsis cuspiata* canopies was observed in the upper 8th and 9th tenths of the canopy. The difference in the leaf area density profile between the two types mentioned here is closely associated with the difference in canopy structure, particularly in the foliage angle density distribution function.

At the Institute of Physics and Astronomy (ESSR), the group under the leadership of Ju. Ross has made extensive investigations on the canopy structure of several crops (e.g. maize, wheat, barley, sunflower, and sorghum) in relation to research on radiation regimes and on growth dynamics of crops (131, 134). Ross & Ross (131, 134) have observed a new type of vertical profile of leaf area density in a well grown

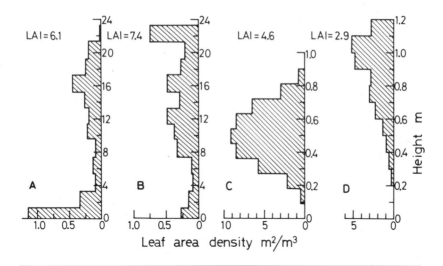

Figure 1 Vertical profiles of leaf area density within plant canopies. *A: Fagus crenata* stand (74). *B: Castanopis cuspidata* stand (74). *C:* Rice stand (62). *D:* Soybean stand (63).

sorghum canopy. It has two peaks of leaf area density, at the upper (7th and 8th tenths) and lower (2nd and 3rd tenths) levels of the canopy. Ross's group intends through such researches to establish a new discipline of science: phytometrics, the investigation of plant and canopy characteristics related to photosynthesis and dry-matter production by means of physics and mathematics.

LEAF DISTRIBUTION WITH RESPECT TO AZIMUTH ANGLE   Although crop plants seem to intend to display leaf area equally with respect to azimuth angle, the management of plants in rows or other configurations and the manipulation of planting rates bring changes upon the canopy structure of crop plants. Experimental results show that the azimuth distribution of leaf area of maize crop is somewhat elongated in the direction perpendicular to the row. A well grown pea canopy was nonpreferential with respect to azimuth angle (131). Nonpreferential distribution of leaf area with respect to azimuth angle was also observed with rice and soybean crops (63, 64).

More recently, Ross & Ross (134) have done detailed measurements of leaf area distribution in sunflower plants in order to make clear the heliotropic phenomenon of leaves. A comparison of leaf distribution functions with respect to the angle of the leaf normal and of the sun, obtained hourly, indicated that the upper leaves followed the movement of the sun well with a phase lag of about 30° when the sun altitude was higher than 25°. The movement of leaves with sun altitude was also observed with a *Platanus* (140).

LEAF DISTRIBUTION WITH RESPECT TO INCLINATION ANGLE   Since Nichiporovich (109) measured the leaf angle density function (LADF) of maize crops by the leaf protractor, much study has been done on the LADF of various kinds of crop plants. de Wit (37) has made extensive measurements of the LADF of several well developed field crops and classified these into the four types, the planophile, erectophile, plagiophile, and extremophile. The planophile type has horizontal leaves dominant (white clover, potato); the erectophile type exhibits dominance of vertical leaves (rye); the plagiophile type, of oblique leaves (maize); in the extremophile type horizontal and vertical leaves occur with equal frequency. He found that the LADF of perennial ryegrass changed from a more erectophile type in early spring to an obviously planophile type in late summer.

Ross & Nilson (131) studied the geometry of leaf orientation function within crop canopies and derived a spatial orientation function to relate the leaf angle density function to the distribution function of leaf normals. Their method has been applied to the quantitative analysis of the canopy structure of crop plants. LADF data obtained by several researchers are summarized in Table 1. Sunflower, pea, and soybean are "planophile," i.e. they display a high percentage of leaf area at low angles of inclination. Although the LADF of a maize crop is nearly equal to a spherical distribution, rice canopy is more nearly vertical than erectophile.

In order to characterize canopy structure, Ross & Ross (134) derived the "leaf area inclination index , x_L" as follows:

$$x_L = \frac{1}{2}\left(\left|\,0.13 - g_1^*\,\right| + \left|\,0.37 - g_2^*\,\right| + \left|\,0.50 - g_3^*\,\right|\right) \qquad\qquad 2.$$

where g_1^*, g_2^*, and g_3^* are the leaf area fractions in each of the inclination angle intervals 0°–30°, 30°–60°, and 60°–90°, respectively. The values of x_L calculated from the above relation are presented in Table 1. Theoretically, the leaf area inclination index is expected to vary between 1.0 for a horizontal-leaved canopy and –1.0 for a vertical-leaved canopy. Phytometrical measurements indicate that the LADF changes with increasing leaf area depth (92, 139, 159, 160). Canopy structure has

Table 1 Leaf angle density function of crop plants

Inclination angle	0-15°	15-30°	30-45°	45-60°	60-75°	75-90°	x_L
Sunflower	0.115	0.260	0.318	0.220	0.071	0.016	0.41
Pea	0.142	0.271	0.243	0.209	0.086	0.049	0.32
Soybean	0.100	0.270	0.305	0.174	0.096	0.027	0.20
Maize	0.065	0.100	0.215	0.300	0.220	0.100	0.18
Barley	0.149	0.095	0.148	0.210	0.259	0.139	0.11
Rice	0.000	0.002	0.029	0.126	0.457	0.387	−0.34
Spherical	0.034	0.100	0.159	0.207	0.241	0.259	0.00
Vertical	0.000	0.000	0.000	0.000	0.000	1.000	−1.00
Horizontal	1.000	0.000	0.000	0.000	0.000	0.000	1.00

been recently analyzed more accurately in rice crops (63, 65). Stratum by stratum treatments have revealed that the canopy structure of rice changes from a nearly ideal type before the heading stage to a quite inverse structure after the heading stage.

Because of the difficulty in making phytometric measurements, comparatively little work has been done on the canopy structure of forests. It is reported that the erectophile, or optimum, canopy structure observed in tropical forests may result from evolutionary adaptation to the very strong radiation intensity in that zone. The canopy structure of deciduous trees (poplar, white birch, and oak) is of the spherical type (121). Rauner (121) also concluded that a pine canopy has relatively vertical needles on the top layer and horizontal needles at the bottom, so that good exposure of photosynthesizing plant elements to radiation beams is assured at any sun altitude. Shiman (140) has devised a goniometer for measuring the angle $\angle \Upsilon_L \ \Upsilon_0$ between the leaf normals Υ_L and the direct sun beam Υ_0. The diurnal change in the distribution function of leaf area with respect to the angle $\angle \Upsilon_L \ \Upsilon_0$ was measured in a *Platanus*. The leaf area distribution function of the angle $\angle \Upsilon_L \ \Upsilon_0$ varied drastically with sun altitude.

SPATIAL ARRANGEMENT OF PLANT LEAVES   Leaf arrangement as illustrated by the spatial distribution of leaf area greatly influences not only the canopy microenvironment but also canopy photosynthesis (42–44, 47, 68, 69, 106, 163, 171). Comparing the morphological characteristics of high vs low yielding cultivars of rice, sweet potato, and soybean, Tsunoda (154) has concluded that the high yielding cultivars tend to have erect leaves arranged in a "gathering type," while the low yielding cultivars tend to have lax leaves arranged in a "dispersing type." The spatial leaf arrangement was measured with soybeans planted in north-south rows (79). Kumura's results (79) indicated that the leaf arrangement was far from a horizontally random one, even when the leaf area index reached the maximum. Ito et al (65) measured the spatial arrangement of leaves of two rice cultivars (Manryo and IR-8) and showed that almost all of the leaves are confined within a narrow space of 20 cm radius, indicating the rice canopies are intensive "gathering type." Kira et al (75) reported that forest crowns are contagious with clumped or aggregated rather than random patterns of leaves, permitting more deep penetration of light beams.

It is not very realistic to assume that foliage in plant canopies is randomly dispersed. As Warren Wilson (165) indicated, the dispersion of foliage depends on the spacing, areas, shapes, and inclinations of leaves and stems. Warren Wilson's method of assessing the dispersion of foliage is to calculate the "relative variance," that is the ratio of the variance to the mean number of contacts per quadrat. The relative variance is equal to 1.0 if the dispersion is random, exceeds 1.0 where there is clumping, and is less than 1.0 where there is regularity.

LIGHT ENVIRONMENT WITHIN PLANT CANOPIES

Energy and mass exchanges such as transpiration, sensible heat transfer, and photosynthesis on foliage surfaces are greatly affected by the radiant energy flux impinging upon the surfaces. The radiation environment within plant canopies is composed of the four kinds of radiant flux, i.e. the direct and diffuse solar radiation fluxes penetrating the canopy and the upward and downward fluxes of complementary diffuse radiation due mainly to transmission and reflection by plant elements.

Light Penetration

Since Monsi & Saeki (100) treated the geometry of light penetration into the canopy in connection with photosynthesis, much theoretical and experimental research has been done on the radiation environment within plant canopies (8–11, 20, 23, 29, 41, 46, 57, 59,61, 62, 67, 70, 78, 82–84, 102, 103, 105, 107, 111, 113, 114, 128, 132, 133, 136, 139, 141, 145, 146, 152, 153, 170, 172–175).

DIRECT SOLAR RADIATION   For measuring the sunlit area within plant canopies Lopukhin's method has been revised and used by a few workers (64, 65). Experimental results indicated that the attenuation of sunlit area with leaf area depth was more rapid for lower than for higher sun altitudes. Consistent with these results, Warren Wilson (165, 166) found in a lucerne canopy that the direct solar radiation Q penetrated farther into the canopy with increased sun altitude, h_o. He also found that the penetration of Q is very strongly affected by the leaf inclination angle β_L.

The extinction of the spatial average intensity of Q at a leaf area depth L at h_o is expressed as follows (132):

$$Q(L, h_o) = Q(h_o) \exp\left\{-\operatorname{cosec} h_o \int_0^L G_L(L, h_o) f_L(L)\, \mathrm{d}L\right\} \qquad 3.$$

where $Q(h_o)$ is the direct solar radiation intensity on the horizontal plane at the canopy top, $Q(L,h_o)$ the spatial average intensity of Q at the canopy depth of L, and $G_L(L,h_o)$ the effective leaf area projection. When $G_L(L,h_o)$ approximates $G_L(h_o)$, the extinction coefficient, k_d [$= G_L(h_o) \operatorname{cosec} h_o$] for direct solar radiation is used for sake of simplicity. Building on Monsi & Saeki's work and assuming horizontally randomly dispersed leaves, Isobe (61, 62), Kuroiwa & Monsi (83) and Anderson (9) have derived the relation between k_d and the two factors h_o and β_L. They showed (a) that k_d is equal to 1.0 irrespective of h_o if the canopy consists solely of horizontal leaves, (b) that k_d for a canopy with leaves inclined at β_L is only constant when $\beta_L \leq h_o$, (c) that when $\beta_L > h_o$, k_d increases more rapidly with lowering h_o, and (d) when $\beta_L = 90°$ k_d decreases monotonically with increasing h_o.

Ross & Nilson (132, 133) have made a more detailed theoretical investigation of the effective leaf area projection. It was found that G_L is 0.5 when the canopy has a spherical leaf distribution, $G_L = \sin h_o$ when the leaves are completely distributed horizontally, and $G_L = 2/\pi \cdot \cos h_o$ when the canopy consists solely of vertical leaves. The G_L data obtained by processing the phytometric data of several crops are presented in Table 2 (63, 64, 133). Rice with very erect leaves behaves in direct solar radiation like vertical foliage, whereas maize and sunflower behave respectively like spherical and horizontal foliages. Another interesting point in Table 2 is that every canopy behaves in direct solar radiation like a spherical canopy when h_o is

Table 2 G_L function of crop canopies

Sun altitude	0°	15°	30°	45°	60°	75°	90°
Sunflower	0.37	0.39	0.47	0.56	0.65	0.72	0.74
Soybean	0.28	0.35	0.47	0.61	0.77	0.81	0.85
Maize	0.46	0.47	0.48	0.52	0.54	0.59	0.60
Rice	0.54	0.53	0.51	0.47	0.44	0.41	0.37
Horizontal	0.00	0.25	0.48	0.70	0.85	0.96	1.00
Vertical	0.63	0.61	0.55	0.45	0.31	0.16	0.00
Spherical	0.50	0.50	0.50	0.50	0.50	0.50	0.50

about 33°. More recently, Ross & Ross (134) proposed a very simple relation for calculating the G_L function from the data of the leaf area inclination index x_L.

The extinction coefficient calculated from the data by the G_L function agreed well with the extinction coefficient determined by the revised Lopukhin's method (64, 132). Ross (128) has improved the penetration theory of direct solar radiation for the case of horizontally heterogeneous plant canopies by introducing a new parameter, the "overlapping index of leaves." Calculations show that $\ln(Q_L/Q_0)$ vs the leaf area depth curve is concave upwards when the leaves are very concentrated. This implies that more direct solar radiation can penetrate into plant canopies than a usual logarithmic law predicts.

SUNLIT LEAF AREA   The sunlit leaf area plays a very important role in canopy photosynthesis (32, 51, 52, 58, 95). de Wit (37) and Nichiporovich (109) studied the sunlit leaf area of a plant canopy with "spherical leaf distribution" and showed that it is given by 2 sin h_0. Warren Wilson (165, 166) investigated the sunlit leaf area index F_d as influenced by leaf inclination angle β_L, sun altitude h_0, and leaf area index F. He derived the following relation

$$F_d = F [1 - \exp(-\phi_h)]/\phi_h$$

$$F_s = F \left\{1 - [1 - \exp(-\phi_h)]/\phi_h\right\}$$

4.

where F_s is the shaded leaf area index and ϕ_h the total area of the projections of the leaves at h_0 onto a horizontal plane per unit area of ground. Using the above relations, he suggested that in temperate latitudes, the maximum sunlit leaf area index is usually between 1 and 2 for "erect-leaved canopies" and near 1.0 for "horizontal-leaved canopies." When a foliage canopy is "under-dispersed" (regular), the value of F_d increases compared to that for a random dispersion canopy. On the other hand, over-dispersion (clumping) decreases the value of F_d.

Consistent with Warren Wilson's conclusions, several researchers (63, 64, 159–161) found in dense canopies a relatively constant percentage of F_d to F during the period of sun altitude higher than 40°, with a sharp drop both in early morning and in late afternoon. The sunlit leaf area index of soybean crop with horizontal leaves was about 1.0 and that of rice crop with very erect leaves about 2.0.

DIFFUSE SOLAR RADIATION   The problem of penetration of diffuse solar radiation into plant canopies was first formulated and solved by Monsi & Saeki (100) in order to reveal the radiation regime within plant communities. They showed that the curves of $\ln(q_L/q_0)$ plotted against leaf area depth are concave upwards except for horizontal-leaved canopies. This is due to the increase of deviation from isotropic distribution of the angle distribution of diffuse solar radiation with increasing leaf area depth. They concluded, however, that when plant canopy is not very dense (F $\leq$ 5.0) the $\ln(q_L/q_0)$ vs leaf area depth curves can be approximated by a linear relation (where q_0 and q_L are diffuse solar radiation intensities at the canopy top and the leaf area depth of L, respectively). The value of the extinction coefficient k_s of diffuse solar radiation was found to decrease monotonically with the leaf inclination angle from 1.0 for a horizontal-leaved canopy to 0.45 for a vertical-leaved canopy, consistent with experimental results obtained with pigweed, green-gland spurge, girasol, miyako bamboo grass, and reed, under overcast sky conditions.

The problem of penetration of diffuse solar radiation into plant canopies was also studied by Isobe (61), Tooming & Ross (152), and Anderson (9).

COMPLEMENTARY DIFFUSE RADIATION (CDR)   Plant elements scatter light to a certain extent through reflection and transmission. The effect of scattering upon the light profile within plant canopies becomes increasingly large with increasing F (61). Saeki (135) introduced an empirical parameter (leaf transmissibility m) to the exponential light profile for considering the scattering effect of plant elements.

Assuming perfectly diffuse reflection and transmission of leaves and the equality of these two quantities, Ross & Nilson (132, 133) and Nilson (111) have formulated the problem of the transfer of radiation flux within plant canopies and solved the related differential equations by using Shwartzshild's method. They have treated this problem fairly rigorously. They showed that the upward flux of CDR decreases exponentially with L, and the downward flux of CDR is zero at the canopy top, increases with L to reach its maximum at a leaf area depth of around 1.0 and decreases again. With increased h_o, the leaf area depth at which the downward flux becomes maximal increased somewhat, particularly in erect-leaved canopies.

Nilson (111) showed that the role of CDR in the radiation regime within plant canopies is larger in the near infra-red region (NIR) than in the photosynthetically active radiation region (PAR). This means that PAR diminishes more rapidly with L than does NIR. This is mainly due to the difference in the scattering coefficient of leaves between PAR ($\omega = 0.03$–0.12) and NIR ($\omega = 0.75$–0.90). Anderson (10) confirmed that the results predicted from Ross & Nilson's model are in good agreement with the transmission measurements of PAR and NIR in sunflower and wheat canopies.

Since the upward flux of CDR at the canopy top characterizes the radiation flux reflected from plant canopies, much attention has been paid to this problem (11, 54, 61, 123, 132, 133). The reflectivity (albedo) of the vertical-leaved canopy decreased drastically with sun altitude, whereas the albedo of the horizontal-leaved canopy is retained at a constant value independently of sun altitude.

Simulation Models of the Canopy in Relation to Light Factors

RANDOM DISTRIBUTION (RD) MODELS   The investigation of light relations in plant communities has progressed mainly by using various simulation models, and this is one of the greatest features distinct from the investigations in other biological and agronomical fields, as Monteith (103) has pointed out. This modeling approach was initiated by the classic work of Monsi & Saeki (100) in 1953, when they derived a theoretical extinction equation of diffuse radiation in a random distribution (RD) foliage. The equation indicated that the attenuation of incident light is approximately exponential with increasing leaf area index (LAI = F) and that the intrinsic extinction gradient (extinction coefficient) is determined by leaf inclination angles in the range between 1.0 for the horizontal and 0.45 for the vertical foliage. Almost all mathematical models thereafter are virtually identical to theirs conceptually. Although these models are usually concerned with the distribution of radiation in relation to canopy photosynthesis, we will temporarily confine the present discussion to the relation between canopy models and the light environment, and canopy photosynthesis will be fully discussed in a later section.

After Monsi & Saeki's model, different forms of the distribution function have been proposed by several workers, e.g. de Wit (37), Horie & Udagawa (54), Kuroiwa (82), and Ross & Ross (134). de Wit (37) identified four types of canopies according to their inclination distribution of leaves, as planophile, erectophile, plagiophile, and extremophile types, as defined earlier. He showed that the distribution function in many crops and pastures can be classified into any one of the above types. At first, a digital computer was intensively employed in his work for the investigation of the light regime in a canopy. Horie & Udagawa (54) expressed the distribution function of leaf inclination angles as a Beta function having two parameters, λ and ν which are, respectively, the mean value of leaf inclination angle and the standard deviation of inclination angle. By a combination of these parameters they obtained foliage models similar to those of de Wit. Kuroiwa (82) studied foliage models having a vertically changing inclination function.

Ross & Ross (134) introduced a new leaf distribution concept—leaf area inclination index—for characterizing the canopy structure of plants. The leaf area inclination index indicates the deviation of the leaf area distribution function from a spherical leaf area distribution. The values of the leaf area inclination index are 1.0 for a horizontal-leaved canopy and −1.0 for a vertical-leaved canopy, whereas the value of the leaf area inclination index for a spherical distribution canopy is equal to zero.

Since the sky luminance was based on a standard overcast sky (SOC) rather than an isotropic or uniform overcast sky (UOC), the assumption of UOC for diffuse radiation may not be practicable. Anderson (9) calculated that the light penetration under the SOC is less than that under the UOC, but that the maximum deviation is less than 10% even for a canopy with vertical leaves; in this canopy the greatest difference between them may occur because much more light comes from higher altitudes in the SOC than in the UOC.

NONRANDOM DISTRIBUTION (NRD) MODELS   The mathematical description of, and solution for, nonrandom distribution (NRD) foliage are both far more difficult than ·for RD foliage. This is the very reason that the mathematical approach has mainly been concentrated on RD foliage. Natural and crop communities may, however, be different—and sometimes widely different—from RD foliage. Lower extinction coefficients than those of the RD model have often been obtained in grass canopies (75, 100), in maize canopies (5), and in some broadleaf forests (75) under diffuse light conditions. Conversely, slightly higher coefficients are also known in forb type communities with a mosaic leaf array (100).

Warren Wilson (165) predicted these variations from the wide range (0.5 to 1.6) of relative variances yielded by vertical point quadrat analysis of real stands. As discussed previously, the relative variance is a convenient measure of foliage dispersion. Acock et al (1) summarized the important attributes of three leaf arrangements, the random, regular, and clumped leaf arrangements, in a table. They can be represented by a Poisson, a binomial (102, 105), and a negative binomial (105) distribution, respectively. Even in a horizontal-leaved canopy, the mean light flux density falling on leaves is not equal to that falling on a horizontal plane except for the case of the random dispersion; the former is greater than the latter for the regular, and the reverse is the case for the clumped arrangement. These conditions must be taken into consideration for more precise calculation of total canopy photosynthesis.

A kind of nonrandom foliage model was first presented by Kasanaga & Monsi (71); they assumed that the model foliage consists of many sublayers with no overlapping within each layer and a fixed fraction s of horizontal leaves. Moreover, these sublayers were assumed to be randomly arranged. Light flux density in their model is discretely reduced according to the values of s. Their measurement showed that the s values are about 0.7 in a horizontally stretched branch of several deciduous trees. A similar parameter to s has been used for the further investigation of light relations in a plant community by Monteith (102), de Wit (37), and Duncan et al (39).

General nonrandom distributions are quite difficult for determining the light environment in relation to the canopy structure. Jackson & Palmer (67) assumed hedgerow orchards as an assembly of continuous solid black blocks. Hedgerow models are so simple that light absorption and related canopy photosynthesis of real orchards cannot be estimated from the model outputs. Mathematical treatments have little success for general NRD foliages.

An interesting approach was attempted by Williams & Soong (169); they built up a foliage model consisting of stiff paper squares which were stuck to the inside walls of a series of tubes lined with reflective aluminum foil, and measured the penetration of light thrown from a slide projector upon this model foliage. Although their study could clarify the effect of leaf area density (LAD) upon light penetration in the foliage, the distribution of incident angles from the light source was utterly different from that under natural light conditions, and it may be technically impossible to correct this serious defect. A substitute model such as theirs will inevitably have great restrictions on model operation.

When a Monte Carlo technique is applied to this problem with the substantial help of an updated electronic computer, logically similar treatments are feasible in a more advanced fashion. As one of the most marked characteristics, this technique is based on a stochastic world so that it can easily offer not only mean values of the system behavior concerned but also their deviations if necessary. This technique was first used by Tanaka (145, 146) to estimate the sunlit leaf area of a tobacco plant under direct solar radiation. Oikawa & Saeki (115) minutely examined the reliability of the Monte Carlo experiments in a random distribution (RD) foliage by comparison with analytical solutions; they constructed a unit foliage with $F = 5$ by arranging randomly 500 leaf laminae of elliptical shape at a specified inclination angle in a unit land area, and carried out an emission experiment of light beams under the UOC in the computer. They found satisfactory agreement between the experimental results and mathematical solutions over all the leaf inclination angles.

They also conducted the same emission experiments for a patch-like stand, where each unit foliage is placed far apart from others without interaction with neighboring foliages; the light enrichment due to lateral light was remarkable, especially in or below the middle stratum of the patch-like stand. We should pay close attention to this point in the measurement of light conditions in natural communities because they tend to be, in most cases, patch-like rather than continuous. Such information can never be deduced by traditional mathematical treatment.

Once the Monte Carlo technique is warranted in its usefulness for the study of light environment, it is applicable to any types of foliage since it need not usually invoke ad hoc hypotheses and absurd assumptions. Oikawa & Saeki (115) then investigated the light conditions in a square-planted (SP) population as one of regular plantings (Figure 2A) by exposing it to the same UOC as used for the RD foliages mentioned earlier (Figure 2B). The population generator of an SP foliage can designate such structural parameters as planting density, leaf divergence angle, leaf size, and phyllotaxis; all of these parameters have been beyond the scope of the usual analytical approaches. According to their results, many new facts were revealed. For instance, it became evident that planting density conspicuously affects the light environment in the SP populations though they have the same LAI, LAD, and leaf inclination angle. Generally, light penetration was largely promoted in the SP populations compared to that in the corresponding RD foliage, except in the SP population with the highest planting density (ND = 6) (Figure 2C).

The Monte Carlo technique will be employed for further investigation of the light environment in various types of populations and great advance may be anticipated in this field hereafter. A lengthy machine running time, however, is one of the greatest difficulties to be faced when we adopt this technique. Next, we must make much effort to reduce experimental errors attendant on such a stochastic technique with every programming manipulation available.

An unexplored problem is the sunfleck theory developed by Miller & Norman concerning light distribution in foliage (98, 99). At first, they determined a mathe-

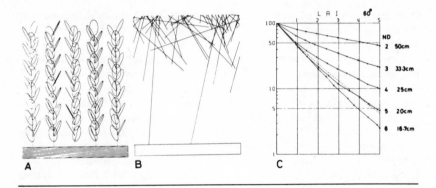

Figure 2 A square-planted (SP) population and light penetration in SP populations (115). *A*: A vertical figure of an SP population with planting density of 5 (*ND* = 5). *B*: Tracks of the first fifty beams emitted to the SP population. *C*: Light penetration in SP populations with different planting densities (solid lines) and in the corresponding random population (dashed line).

matical equation of sunfleck segment length probability under the assumption of zero sun diameter (98). Their theory can predict that sunfleck segment lengths under direct solar radiation may differ greatly among canopies different in a leaf size even though the canopies have the same LAI. Subsequently, the penumbral fuzziness of leaf-edge shadows was mathematically assessed by taking the real angular diameter (~ 0.5) of the sun into consideration (99). Norman et al (114) obtained fairly good agreement of segment distributions between predictions and measurements in a few canopies, where penumbral effects certainly occur in canopies only 2 m in height. The equalizing effect on horizontal light variation due to penumbra, therefore, will be more remarkable in forests occupying a wider foliage space (cf. Figure 1), and leaf layer thickness or leaf size may assume an important role in this point. Application of their theory, however, is still restricted to populations with horizontal leaves.

PHOTOSYNTHESIS IN RELATION TO CANOPY STRUCTURE

As already described, the relationship between canopy photosynthesis and canopy structure has been one of the important subjects in plant and crop ecology (3, 12–15, 49, 50, 56, 66, 100, 104, 137), and many studies have been devoted to the elucidation of this relationship. In the past, a number of excellent reviews have attempted to summarize or synthesize the results of research in various aspects of canopy photosynthesis (Loomis et al 94; Monsi 101; Nichiporovich 108, 110; Yoshida 177).

Experimental Approaches to Canopy Photosynthesis

In their article about the comparative study of photosynthesis of beet plants, Watson & Witis (168) showed that a wild beet with more prostrate leaves was less productive in crowded field conditions than its cultivar with upright leaves. The difference was thought to be due to the increase in mutual shading among leaves within the wild beet canopy. Comparing the leaf arrangements of high and low yielding rice cultivars, Tsunoda (154) also revealed the close correlation between the erectness of leaves and yield. He concluded that high crop yield is generally observed for rice cultivars with erect leaves. The research results obtained in a series of his investigations of high yielding rice cultivars led him to the "plant-type concept" as a guide for breeding high yielding cultivars. Hayashi (48) has made comprehensive studies of the canopy structure of rice cultivars in relation to the search for a plant type maximizing photosynthetic energy utilization. Yoshida (177) summarized desirable characters for high yielding crops, referring to results obtained by many crop scientists.

Critical comparisons of photosynthesis in a canopy with more erect leaves and a plant canopy of more horizontal habit were made by Nichiporovich (110) using the data obtained with beet and soybean crops. Evidently the attenuation of photosynthetic rate with L was less rapid in the beet canopy in which about 50% of the leaf area is arranged in the inclination range between 60° and 90° than in the soybean canopy in which about 80% of the leaf area is distributed in the range between 0° and 30°. Similar results are also reported by Leach & Watson (90) using phytometers. Direct measurements of CO_2 absorption by individual leaves within a sunflower canopy have been made by Horie & Udagawa (54). They showed that the saturation value of leaf photosynthesis rate under strong radiation intensity changes with the canopy depth.

Evidence indicating how the leaf inclination angle influences canopy photosynthesis has been recently reported with rice by Tanaka (147). Tanaka demonstrated clearly by mechanically manipulating the leaf arrangement that a horizontal-leaved canopy shows a plateau type response of photosynthetic rate to radiation, with low photosynthesis, while an erect-leaved rice canopy shows a higher photosynthetic rate (see Figure 3). The rice yield of the horizontal-leaved rice canopy was about 70% that of the vertical-leaved rice canopy. More recently, Angus & Wilson (12) investigated the vertical profiles of net photosynthesis in two wheat cultivars, one an erect leaf type and the other a lax leaf type, using techniques involving $^{14}CO_2$

The patterns of net photosynthesis indicate the localization of carbon dioxide uptake near the surface of the horizontal-leaved canopy (with an extinction coefficient of 0.6–1.0) and at the middle of the vertical-leaved canopy (with an extinction coefficient of 0.2).

Simultaneous measurements of the CO_2 gradient and the turbulent transfer coefficient were made within plant canopies in order to deduce rates of CO_2 exchange at different positions in the canopy (17, 158, 173). The level at which the layer's net photosynthesis reaches the maximum varied between plants, chiefly because of the difference in the canopy structure. Hozumi & Kirita (55) and Yoda (176) have applied the techniques of Saeki (135) to determining the vertical profiles of CO_2 sink and source intensity in a beech forest. They showed that on a sunny day the active layer of photosynthesis expands gradually downward with increasing sun altitude.

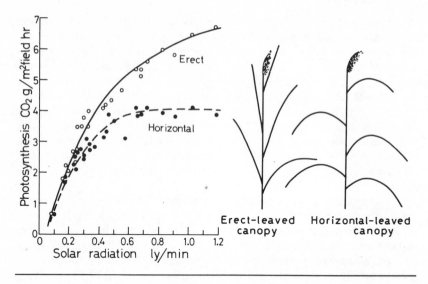

Figure 3 Rice photosynthesis as influenced by leaf arrangement (cultivar Nihonbare) (Tanaka 147).

However, the active layer of photosynthesis was limited to the upper half of the canopy when it was overcast.

Theoretical Approaches to Canopy Photosynthesis

The problems of canopy photosynthesis have been theoretically approached by two different methods. One is based upon the light interception theory and the other is based on the CO_2 transfer theory. Because of the development in the understanding of the physical and biological processes relating to canopy photosynthesis, however, the two methods mentioned here are apparently being gradually synthesized into a more sophisticated model.

CANOPY PHOTOSYNTHESIS MODELS BASED ON LIGHT INTERCEPTION THEORY   The first and most famous model for canopy photosynthesis is the model of Monsi & Saeki (100). Their model was constructed on the basis of the light attenuation law in plant canopies and of light-photosynthesis curves of leaves. This model first enabled us to discuss quantitatively canopy photosynthesis in relation to leaf arrangement and environmental factors.

Much effort devoted to the improvement of canopy photosynthesis models has been mainly directed toward the rigorous treatment of radiation fields within plant canopies and more realistic consideration of photosynthetic processes in single leaves. Although the latter problem has been studied by many researchers (21, 22, 31, 88, 89, 143, 144) and is well reviewed by Acock et al (2) and Laisk (87), it is outside the scope of this review.

For considering the effect of leaf transmissibility, Saeki (135) has used the intensity of radiation E entering a leaf surface in place of radiation flux Q. In this case the gross canopy photosynthesis P_g is given by

$$P_g = \frac{b}{ak_s} \ln \frac{(1-m) + akI_0}{(1-m) + ak_s I_0 \exp(-k_s F)} \qquad 5.$$

where I_0 is the light flux density at the canopy top, k_s the extinction coefficient of diffuse solar radiation, and a and b empirical constants characterizing the shape of the light-photosynthesis curve of leaves. Assuming that the canopy respiration is in proportion to F, the above relation was used to obtain an optimum LAI, F_{opt}, at which the net photosynthesis of the plant canopy reaches its maximum (Saeki 135). In order to present a visual image of the general relationships between structural, physiological, and light characteristics of plant canopies, Saeki (135) has made numerical experiments. The values of F_{opt} increased curvilinearly with increasing solar radiation, agreeing with results obtained experimentally.

Approximating the daily course of solar light flux density by $I_0(t) = I_0^m \sin^2 (\pi t/D)$, where I_0^m is the daily amplitude of light flux density and D the day length, Kuroiwa (81) derived the relation for determining the daily gross photosynthesis $P_{g,d}$ of plant canopies. His relation has been used in modeling the growth dynamics of plants.

Verhagen et al (162) have attempted to extend Saeki's model to plant canopies in which the value of k_s changes with the canopy depth. The plant canopies have been classified into the four hypothetical types: exponential, standard exponential, best exponential, and ideal foliages. The ideal foliages, in which k_s values change with depth so as to bring the even distribution of light among leaves, may not be achieved in practice but it provides an interesting upper limit to the canopy photosynthesis under given light conditions. The photosynthesis of the ideal plant canopy has been investigated by several researchers including Kuroiwa (82), Nilson (112), Ross (124), and Tooming (149).

The nonlinearity of the relationship of leaf photosynthesis and radiation energy requires a more rigorous treatment of the radiation fields within plant canopies in calculating the canopy photosynthesis. The influence of the structure of the radiation field upon the canopy photosynthesis was first formulated by Ross (124, 125). In his model, leaves were divided into sunlit leaves on which direct and diffuse radiation fluxes act and shaded leaves on which only diffuse radiation flux acts. His model revealed afresh the importance of leaf inclination in canopy photosynthesis, as Monsi & Saeki (100) have emphasized. Tooming (149) compared the amounts of photosynthesis among the four kinds of canopy structure (vertical, horizontal, random, and optimum) and concluded that the difference in daily net photosynthesis among these canopies does not exceed 30%. Laisk (86) showed that averaging the radiation field for both space and direction may cause the overestimation of canopy photosynthesis (by about 20%).

The use of high speed digital electronic computers has greatly stimulated the development of procedures for calculating and simulating canopy photosynthesis in relation to canopy structure, optical and physiological properties of leaves, radiation fields, and diffusion conditions. The first step towards the new approach was made by de Wit (37). His numerical experiments showed that the high scattering coefficient of leaves and the high fraction of diffuse radiation increase canopy photosynthesis equally, mainly because of more uniform distribution of light among the leaves. Approaches similar to that attempted by de Wit have been successfully attempted by several workers (39, 54, 62, 80–82). Kuroiwa (82) accounted for contributions of sunlit and shaded leaves to canopy photosynthesis. By taking into account the height dependence of photosynthetic activity of leaves, Horie & Udagawa (54) simulated the CO_2 uptake intensity by layers in a sunflower canopy. The results were in good agreement with those measured by leaf chambers.

Ross & Bikhele (129, 130) have proposed a more improved model. In their model, the height and temperature dependence of photosynthesis and respiration of leaves are considered. A part of the results obtained on an electronic computer is reproduced in Figure 4. The photosynthesis of sunlit leaves decreased drastically with canopy depth. This was attributed to the diminution of the fraction of sunlit area with the canopy depth. In a horizontal-leaved canopy, the contribution of sunlit leaves was approximately independent of sun altitude and about 60%, whereas in a vertical-leaved canopy it increased with sun altitude, that is, was 50% at $h_o = 25°$ and 68% at $h_o = 50°$. When $F > 2.3$, the photosynthesis due to shaded leaves always exceeded that of sunlit leaves.

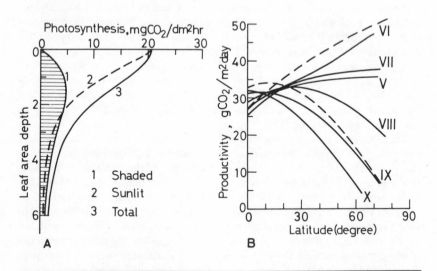

Figure 4 *A*: Profiles of gross photosynthesis of sunlit and shaded leaves (130). *B*: Daily totals of net canopy photosynthesis as a function of latitude and season (151). $F = 6.0$; solid lines denote P for random foliage and dotted lines denote P for spherical foliage.

Using this model, Ross & Bikhele (130) also studied the ecological implication of PAR absorption coefficient κ of leaves. An optimum value of κ at which P_g attains its maximum was found to decrease gradually with sun altitude, namely 0.85, 0.70, and 0.65 at the sun altitude of 25°, 50°, and 90°, respectively. When solar radiation is weak, P_g attains its maximum at κ of 1.0. They concluded from the above results that values of κ in the range between 0.6 and 0.9 as observed with natural plant leaves may result from the evolutionary adaptation of plants to changing radiation regime.

Somewhat different models for canopy photosynthesis have been studied by Tooming's group (150, 151). The basic assumption in these models is that adaptation processes in plants or plant canopies are made so as to maximize the photosynthetic productivity. Tooming pointed out the importance of irradiation intensity of adaptation (IDA) characterizing the radiation niches of plant species.

Simulations by Saeki (135), de Wit (37,) Ross & Bikhele (126, 127, 129, 130) and similar models (24, 39, 54, 62) lead to the following general conclusions: When F is less than about 3.0, the canopy productivity is closely related with percent cover, and a horizontal foliage has an advantage over inclined or vertical leaves, without regard to latitude and sun altitude. With further increase in F, canopies with strongly erect leaves have a considerable advantage over other plant canopies, especially in lower latitudes. The photosynthetic productivity of terrestrial plants varies largely with latitude and season, depending upon the amount of PAR reaching the earth surface (see Figure 4B).

CANOPY PHOTOSYNTHESIS MODELS BASED ON CO_2 TRANSFER THEORY   An alternate approach for modeling canopy photosynthesis is the CO_2 transfer model based on the differential equation describing CO_2 exchange between plant elements and air. Under steady state conditions, CO_2 exchange within canopies can be expressed by

$$-\frac{d}{dz}\left(K\,\frac{dC}{dz}\right) = -f_L(z)\,p(z) + f_L(z)\,r(z) \qquad\qquad 6.$$

where K and C are respectively the turbulent transfer coefficient and the CO_2 concentration in the bulk air among leaves, $f_L(z)$ the height-LAD function, and $p(z)$ and $r(z)$ the height functions of photosynthesis and respiration rates, respectively. The CO_2 transfer model is suitable for assessing the influence of micrometeorological factors, especially air mixing above and within canopies, on the canopy photosynthesis.

Budyko and his co-workers (25–30) have used the CO_2 transfer model in order to clarify the geographic variation of potential productivity of terrestrial plants. Simulation results are in good agreement with those obtained by the light interception model. Using results obtained from numerical experiments, they studied the spatial distribution of coefficients of solar energy utilitation by plants in the USSR. The coefficient was found to vary from 0.1% in arid districts to 1.0% in districts with normal climatic conditions. They also studied the ecological implications of accumulated air temperature using the CO_2 transfer model. The results showed a curvilinear increase of the potential productivity with accumulated effective air temperature.

Even when other parameters were kept constant, canopy photosynthesis increased curvilinearly with increments in wind velocity or the turbulent transfer coefficient at the canopy top, indicating that wind is an important factor affecting canopy photosynthesis (25–30, 156, 157). The effect of the wind factor upon canopy photosynthesis was found to be more considerable in the case with intensive solar radiation and weak wind than with the inverse conditions (Uchijima & Inoue 157).

CO_2 transfer models are also utilized in the study of the CO_2 environment within plant canopies (Menzhulin 97, Paltridge 116, 117, Uchijima & Inoue 157, and Uchijima et al 158). Uchijima & Inoue revealed that the CO_2 profile within plant canopies is related not only to wind velocity and solar radiation intensity but also to soil CO_2 flux. This result seems to describe the difference between the CO_2 profiles observed within maize and rice canopies. CO_2 profiles in canopies were also influenced by the shape of the vertical distribution of LAD (97, 117, 158).

CO_2 transfer models have been used by several workers in order to simulate the influence of artificial CO_2 enrichment on canopy photosynthesis (6, 157, 158). Simulations indicated that artificial CO_2 enrichment does not affect canopy photosynthesis very much when air mixing is strong, because almost all of enriched CO_2 is lost by turbulent transfer into the surface air layer.

CONCLUSIONS

It is apparent from this review of the vast amount of research work on the problems of canopy structure and photosynthesis that our understanding of canopy photosynthesis has been considerably deepened by the activities of IBP during the recent decade. The results obtained have been successfully used in breeding programs towards developing high yielding crops with erect leaf plant types. The results are also interesting and valuable for assessing the photosynthetic productivity of plants—upon which virtually all life on earth depends.

A salient result from recent studies on canopy structure and photosynthesis has been the development of production models based on the theoretical and empirical laws of the major exchange processes of energy and materials between plants and their environment. Special attention has been given to models dealing with light interception and photosynthetic relationships. Such light interception models have provided the mathematical procedure for the assessment of productivity not only of agricultural plant ecosystems but also of natural plant ecosystems, revealing that leaf inclination plays an important role in the photosynthetic productivity of plant canopies. These models have also been remarkably successful in clarifying plant behavior at the ecological level.

In relation to the interception of light by plant canopies, many attempts have been made to measure or formulate canopy structure. These studies made it possible to distinguish the canopies observed in natural and agricultural ecosystems into four types. Canopy structure data from field plantings of crops have been used to simulate the radiation environment and the photosynthesis of plant canopies. However, our information on the phytometric features of plant canopies, including the allocation of photosynthate, has not been as comprehensive as might be desired.

The development of computer technology has greatly facilitated the study of the interactions of plant communities with environmental factors. Although the computer simulation of plant growth seems to have success in integrating knowledge obtained in related disciplines, our efforts in exploring the biological and physical processes involved in plant photosynthesis have not yet necessarily been profound enough. The available information and working models of the relationships of photosynthesis and the environment are still meager and fragmentary. Particularly, the deficiency is greater for forests including fruit tree orchards than for crops with uniform plant stands of relatively low height. Many problems thus remain to be studied.

Attempts to make clear the phytometric characteristics and biological behavior of plant canopies under various environmental conditions and to refine the working models appear to be equally important for improving biophysical simulation of canopy photosynthesis and dry-matter production of plants. Such efforts should necessarily facilitate the increase in man's knowledge on the processes related to forecasting and controlling plant production.

Literature Cited

1. Acock, B., Thornley, J. H. M., Warren Wilson, J. 1970. Spatial variation of light in the canopy. *Proc. IBP/PP Tech. Meet., Třeboň,* 91–102
2. Acock, B., Thornley, J. H. M., Warren Wilson, J. 1971. Photosynthesis and energy conversion. *Potential Crop Production,* ed. P. F. Wareing, J. P. Cooper, 43–75. London: Heinemann. 387 pp.
3. Alessi, J., Power, J. F. 1965. Influence of moisture, plant population, and nitrogen on dryland corn in the northern plains. *Agron. J.* 57:611–12
4. Allen, L. Jr. 1969. Variations in carbon dioxide concentration over an agricultural field. *Agr. Meteorol.* 8:5–24
5. Allen, L. H., Brown, K. W. 1965. Shortwave radiation in a corn crop. *Agron. J.* 57:575–80
6. Allen, L. H. Jr., Jensen, S. E., Lemon, E. R. 1971. Plant response to carbon dioxide enrichment under field conditions: A simulation. *Science* 173:256–58
7. Anderson, M. C. 1964. Light relations of terrestrial plant communities and their measurement. *Biol. Rev.* 39:425–86
8. Anderson, M. C. 1964. Studies of the woodland light climate. I. The Photographic computation of light conditions. *J. Ecol.* 52:27–41
9. Anderson, M. C. 1966. Stand structure and light penetration. II. A theoretical analysis. *J. Appl. Ecol.* 3:41–54
10. Anderson, M. C. 1969. A comparison of two theories of scattering of radiation in crops. *Agr. Meteorol.* 6:399–405
11. Anderson, M. C. 1971. Radiation and crop structure. *Plant Photosynthetic Production: Manual of Methods,* ed. Z. Šesták, J. Čatský, P. G. Jarvis, 412–66. The Hague: Junk. 818 pp.
12. Angus, R., Wilson, J. H. 1972. A comparison of barley cultivars with different leaf inclination. *Aust. J. Agr. Res.* 23:945–57
13. Baker, D. N., Musgrave, R. B. 1964. Photosynthesis under field conditions. V. Further plant chamber studies of the effects of light on corn (*Zea mays L.*). *Crop Sci.* 4:127–31
14. Baker, D. N., Raymond, E. M. 1966. Influence of stand geometry on light interception and net photosynthesis in cotton. *Crop Sci.* 6:15–19
15. Baker, D. N., Myhre, D. L. 1969. Effects of leaf shape and boundary layer thickness on photosynthesis in cotton (*Gossypium hirsutum*). *Physiol. Plant.* 22:1043–49
16. Baumgartner, A. 1956. Untersuchungen über den Wärme and Wasserhaushalt eines jungen Waldes. *Ber. Deut. Wetterdienstes* 5(28):1–53

17. Baumgartner, A. 1967. Ecological significance of the vertical energy distribution in plant stands. *Resources Naturelles,* 5:367–74. New York: UNESCO

18. Baumgartner, A. 1969. Meteorological approach to the exchange of CO_2 between the atmosphere and vegetation, particularly forest stands. *Photosynthetica* 3:127–49

19. Black, J. N. 1960. The significance of petiole length, leaf area, and light interception in competition between strains of subterranean clover (*Trifolium subterraneum L.*) grown in swards. *Aust. J. Agr. Res.* 11:277–91

20. Black, J. N. 1963. The interrelationship of solar radiation and leaf area index in determining the rate of dry matter production of subterranean clover (*Trifolium subterraneum L.*). *Aust. J. Agr. Res.* 14:20–38

21. Bonner, J. 1962. The upper limit of crop yield. This classical problem may be analyzed as one of the photosynthetic efficiency of plants in arrays. *Science* 137:11–15

22. Bray, J. R. 1961. An estimate of minimum quantum yield of photosynthesis based on ecologic data. *Plant Physiol.* 36:371–73

23. Brougham, R. W. 1958. Interception of light by the foliage of pure and mixed stands of pasture plants. *Aust. J. Agr. Res.* 9:39–52

24. Budagovsky, A. M., Ross, Ju. 1966. Fundamental of quantitative theory of community photosynthesis. *Photosynthetic Systems with High Productivity,* ed. A. A. Nichiporovich, 51–58. Moscow: Nauka. 224 pp.*

25. Budyko, M. I. 1964. On theory of influence of climatic factors upon photosynthesis. *Dokl. Akad. Nauk SSSR* 158, No. 2*

26. Budyko, M. I. 1971. *Climate and Life.* Leningrad: Gidrometeorologicheskoe. 472 pp.*

27. Budyko, M. I., Gandin, L. S. 1964. Consideration of physical laws in agricultural meteorology. *Met. Gidrologija* No. 11*

28. Budyko, M. I., Gandin, L. S. 1965. Theory of canopy photosynthesis. *Dokl. Akad. Nauk SSSR* 164, No. 2*

29. Budyko, M. I., Gandin, L. S. 1966. Influence of climatic factors upon plant canopy. *Izv. Akad. Nauk SSSR, Ser. Geogr.* No. 1*

30. Budyko, M. I., Gandin, L. S., Efimova, N. A. 1966. Application of physical methods into study of agroclimatic indexes. *Met. Gidrologija* No. 5*

31. Chartier, P., Chartier, M., Čatský, J. 1970. Resistances for carbon dioxide diffusion and for carboxylation as factors in bean leaf photosynthesis. *Photosynthetica* 4:48–57

32. Closer, J. 1967. The dependence of CO_2 exchange on density of irradiation, temperature and water saturation deficit in *Stipa* and *Bromus. Photosynthetica* 1: 171–78

33. Cowan, I. R. 1968. The interception and absorption of radiation in plant stands. *J. Appl. Ecol.* 5:367–80

34. Cowan, I. R. 1968. Mass, heat and momentum exchange between stands of plants and their atmospheric environment. *Quart. J. Roy. Meteorol. Soc.* 94:523–44

35. Davidson, J. L., Donald, C. M. 1958. The growth of swards of subterranean clover with particular reference to leaf area. *Aust. J. Agr. Res.* 9:53–72

36. Davidson, J. L., Philip, J. R. 1958. Light and pasture growth. *Proceedings Climatology and Microclimatology,* 181–87. New York: UNESCO

37. de Wit, C. T. 1965. Photosynthesis of leaf canopies. *Agr. Res. Rep.* 663:1–57

38. de Wit, C. T., Brower, R., Penning de Vries, F. W. T. 1970. The simulation of photosynthetic systems. *Proc. IBP/PP Tech. Meet., Třeboň,* 47–70

39. Duncan, W. G., Loomis, R. S., Williams, W. A., Hanau, R. 1967. A model for simulating photosynthesis in plant communities. *Hilgardia* 38:181–205

40. Duncan, W. G. 1971. Leaf angles, leaf area, and canopy photosynthesis. *Crop Sci.* 11:482–85

41. Eber, W. 1971. The characterization of the woodland light climate. *Ecol. Stud.* 2:143–52

42. Eagles, C. F. 1972. Competition for light and nutrients between natural populations of *Dactylis glomerata. J. Appl. Ecol.* 9:141–52

43. El-Sharkawy, M. A., Loomis, R. S., Williams, W. A. 1968. Photosynthetic and respiratory exchanges of carbon dioxide by leaves of the grain amaranth. *J. Appl. Ecol.* 5:243–52

44. England, F. 1968. Competition in mixtures of herbage grasses. *J. Appl. Ecol.* 5:227–43

45. Goodall, W. 1956. A new point-quadrat method. *Ecology* 37:627–28

46. Grace, J. 1971. The directional distribution of light in natural and controlled environment conditions. *J. Appl. Ecol.* 8:155–64

47. Haizel, K. A. 1972. The canopy relationship of pure and mixed populations of barley (*Hordeum vulgare L.*), with

mustard (*Sinapis alba L.*) and wild oats (*Avena fatua L.*). *J. Appl. Ecol.* 9:589–600

48. Hayashi, K. 1972. Efficiencies of solar energy conversion in rice varieties. *Bull. Nat. Inst. Agr. Sci.* D 23:1–67**

49. Hesketh, J. D. 1963. Limitations to photosynthesis responsible for differences among species. *Crop Sci.* 3:493–96

50. Hesketh, J. D. 1968. Effects of light and temperature during plant growth on subsequent leaf CO_2 assimilation rates under standard conditions. *Aust. J. Biol. Sci.* 21:235–41

51. Hesketh, J. D., Moss, D. N. 1963. Variation in the response of photosynthesis to light. *Crop Sci.* 2:107–10

52. Hesketh, J. D., Musgrave, R. B. 1962. Photosynthesis under field conditions. IV. Light studies with individual corn leaves. *Crop Sci.* 2:311–15

53. Hadáňová, D. 1967. Development and structure of foliage in wheat stands of different density. *Biol. Plant.* 9:424–38

54. Horie, T., Udagawa, T. 1971. Canopy photosynthesis of sunflower plants—Its measurements and modeling. *Bull. Nat. Inst. Agr. Sci. A* 18:1–56

55. Hozumi, K., Kirita, H. 1970. Estimation of the rate of total photosynthesis in forest canopies. *Bot. Mag.* 83:144–51

56. Hunt, L. A., Cooper, J. P. 1967. Productivity and canopy structure in seven temperate forage grasses. *J. Appl. Ecol.* 4:437–58

57. Hurd, R. G., Rees, A. R. 1966. Transmission error in the photometric estimation of leaf area. *Plant Physiol.* 41:905–6

58. Huxley, P. A. 1969. The effect of fluctuating light intensity on plant growth. *J. Appl. Ecol.* 6:273–76

59. Idso, S. B., Baker, D. G. 1967. Method for calculating the photosynthetic response of a crop to light intensity and leaf temperature by an energy flow analysis of the meteorological parameters. *Agron. J.* 59:13–21

60. Ikusima, I. 1970. Ecological studies on the productivity of aquatic plant communities. IV. Light condition and community photosynthetic production. *Bot. Mag.* 83(987–988):330–341

61. Isobe, S. 1962. An analytical approach to the expression of light intensity in plant communities. *J. Agr. Meteorol.* 17:143–50**

62. Isobe, S. 1969. Theory of the light distribution and photosynthesis in canopies of randomly dispersed foliage area. *Bull. Nat. Inst. Agr. Sci. A* 16:1–25

63. Ito, A. 1969. Geometrical structure of rice canopy and penetration of direct solar radiation. *Proc. Crop Sci. Soc. Jap.* 38:355–63**

64. Ito, A., Udagawa, T. 1971. Phytometrical studies of crop canopies. I. Geometrical structure of soybean canopy and sun light penetration. *J. Agr. Meteorol.* 26:187–95**

65. Ito, A., Udagawa, T., Uchijima, Z. 1973. Phytometrical studies of crop canopies. 2. Canopy structure of rice crop in relation to growth of plant and varieties. *Proc. Crop Sci. Soc. Jap.* In press**

66. Ito, K. 1965. Studies on photosynthesis in sugar beet. III. Photosynthesis in the plant community. *Proc. Crop Sci. Soc. Jap.* 33:492–98**

67. Jackson, J. E., Palmer, J. W. 1972. Interception of light by model hedgerow orchards in relation to latitude, time of year and hedgerow configuration and orientation. *J. Appl. Ecol.* 9:359–76

68. Jeach, G. J. 1969. The relation of photosynthesis by phytometers in the profiles of kale crops to leaf area index above them. *J. Appl. Ecol.* 6:499–505

69. Jennings, P. R. 1964. Plant type as a rice breeding objective. *Crop Sci.* 4:13–15

70. Kallis, A. 1969. Absorption coefficients of PAR of plant communities at different latitudes. *Problems of the Efficiency of Photosynthesis,* ed. Ju. Ross, 44–63. Tartu: Akad Nauk SSR Inst. Phys. Astron. 176 pp.*

71. Kasanaga, H., Monsi, M. 1954. On the light-transmission of leaves, and its meaning for the production of matter in plant communities. *Jap. J. Bot.* 14:304–24

72. Kawashima, R. 1969. Studies on the leaf orientation-adjusting movement and light intensity on leaf surface. *Proc. Crop Sci. Soc. Jap.* 38:718–29**

73. Kawashima, R. 1969. Studies on the leaf orientation-adjusting movement in soybean plants. II. Fundamental pattern of the leaf orientation-adjusting movement and its significance for the dry matter production. *Proc. Crop Sci. Soc. Jap.* 38:730–42**

74. Kira, T., Ogawa, H., Yoda, K., Ogino, K. 1964. Primary production by a tropical rain forest of Southern Thailand. *Bot. Mag.* 77:428–29

75. Kira, T., Shinozaki, K., Hozumi, K. 1969. Structure of forest canopies as related to their primary productivity. *Plant Cell Physiol.* 10:129–42

76. Kishitani, S., Takano, Y., Tsunoda, S. 1972. Optimum leaf-areal nitrogen con-

tent of single leaves for maximizing the photosynthesis rate of leaf canopies: A simulation in rice. *Jap. J. Breed.* 22: 1–10

77. Kondo, J. 1971. Relationship between the roughness coefficient and other aerodynamic parameters. *J. Meteorol. Soc. Jap.* 49:121–24

78. Kondo, J. 1971. Vertical profiles of wind speed and solar radiation above and within forests. *Rep. Res. Group IHD, Jap.* 15–30**

79. Kumura, A. 1965. Studies on dry matter production of soy-bean plant. 2–1. Relation between photosynthesis and light receiving aspect of the population. *Proc. Crop Sci. Soc. Jap.* 33:455–63**

80. Kuroiwa, S. 1968. Theoretical analysis of light factor and photosynthesis in plant communities. III. Total photosynthesis of a foliage under parallel light in comparison with that under isotropic light conditions. *J. Agr. Meteorol.* 24: 75–90**

81. Kuroiwa, S. 1968. Theoretical evaluation of dry-matter production of a crop canopy under insolation- and temperature-climate: summary. *Agroclimatological methods. Proc. Reading Symp.,* 331–32. Paris: UNESCO

82. Kuroiwa, S. 1970. Total photosynthesis of a foliage in relation to inclination of leaves. *Proc. IBP/PP Tech. Meet., Třeboň,* 79–90

83. Kuroiwa, S., Monsi, M. 1963. Theoretical analysis of light factor and photosynthesis in plant communities. I. Relationships between foliage structure and direct, diffuse and total solar radiations. *J. Agr. Meteorol.* 18:143–52**

84. Kuroiwa, S., Monsi, M. 1963. Theoretical analysis of light factor and photosynthesis in plant communities. II. Diurnal changes of extinction coefficient and photosynthesis. *J. Agr. Meteorol.* 19:15–21**

85. Květ, J., Svoboda, J., Fiala, K. 1967. A simple device for measuring leaf inclinations. *Photosynthetica* 1:127–28

86. Laisk, A. 1965. Influence of structure of radiation field on canopy photosynthesis. *Problems of Radiation Regime in Plant Canopies,* ed. Ju. Ross, 73–88. Tallin: Valgus. 127 pp.*

87. Laisk, A. 1968. Prospects of mathematical modelling of leaf photosynthesis function. *Photosynthesis and productivity of plant canopies,* ed. Ju. Ross, 5–45. Tartu: Akad Nauk SSR Inst. Phys. Astron. 199 pp.*

88. Larcher, W. 1969. Physiological approaches to the measurement of photosynthesis in relation to dry matter production by trees. *Photosynthetica* 2:150–66

89. Larcher, W. 1969. The effect of environmental and physiological variables of the carbon dioxide gas exchange of trees. *Photosynthetica* 2:167–98

90. Leach, G. J., Watson, D. J. 1968. Photosynthesis in crop profiles, measured by phytometers. *J. Appl. Ecol.* 5:381–408

91. Loomis, R. S., Williams, W. A., Duncan, W. G. 1967. Community architecture and the productivity of terrestrial plant communities. *Harvesting the Sun,* ed. A. S. Pietro, F. A. Greer, T. J. Army, 291–308. New York: Academic. 342 pp.

92. Loomis, R. S., Williams, W. A., Duncan, W. G., Dovrat, A., Nunez, F. 1968. Quantitative descriptions of foliage display and light absorption in field communities of corn plants. *Crop Sci.* 8:352–56

93. Loomis, R. S., Williams, W. A. 1969. Productivity and the morphology of crop stands: Patterns with leaves. *Physiological Aspect of Crop Yield,* ed. J. D. Eastin, F. A. Haskins, C. Y. Sullivan, C. H. M. van Bavel, 27–51. Madison: Am. Soc. Agron. Crop Sci. Am. 396 pp.

94. Loomis, R. S., Williams, W. A., Hall, A. E. 1971. Agricultural productivity. *Ann. Rev. Plant Physiol.* 22:431–68

95. McCree, K. J., Loomis, R. S. 1969. Photosynthesis in fluctuating light. *Ecology* 50:422–28

96. McCree, K. J., Troughton, J. H. 1966. Non-existence of an optimum leaf area index for the production rate of white clover grown under constant conditions. *Plant Physiol.* 41:1615–22

97. Menzhulin, G. 1971. Gaseous regime within plant canopy inhomogeneous in height. *Tr. Gl. Geofiz. Observ.* 287: 107–14*

98. Miller, E. E., Norman, J. M. 1971. A sunfleck theory of plant canopies. I. Lengths of sunlit segments along a transect. *Agron. J.* 63:735–38

99. Miller, E. E., Norman, J. M. 1971. A sunfleck theory for plant canopies. II. Penumbra effect: Intensity distribution along sunfleck segments. *Agron. J.* 63: 739–43

100. Monsi, M., Saeki, T. 1953. Über den Lichtfaktor in den Pflanzengesellschaften und seine Bedeutung für die Stoffproduktion. *Jap. J. Bot.* 14:22–52

101. Monsi, M. 1968. Mathematical models of plant communities. *Proc. Funct.*

Terr. Ecosystems Primary Prod. Level,
Copenhagen, 131–49. Paris: UNESCO
102. Monteith, J. L. 1965. Light distribution
and photosynthesis in field crops. *Ann.*
Bot. 29:17–37
103. Monteith, J. L. 1969. Light interception
and radiative exchange in crop stands.
See Ref. 93, 99–113
104. Moss, D. N., Musgrave, R. B., Lemon,
E. R. 1961. Photosynthesis under field
conditions. III. Some effects of light,
carbon dioxide, temperature, and soil
moisture on photosynthesis, respiration, and transpiration of corn. *Crop*
Sci. 1:83–87
105. Mototani, I. 1968. Horizontal distribution of light intensity in plant communities. *Rep. 1967 JIBP/PP Photo-*
synthesis Level III Group, Tokyo,
25–28
106. Murata, Y. 1961. Studies on the photosynthesis of rice plants and its culture
significance. *Bull. Nat. Inst. Agr. Sci. D*
9:1–169**
107. Newton, J. E., Blackman, G. E. 1970.
The penetration of solar radiation
through leaf canopies of different structure. *Ann. Bot.* 34:329–49
108. Nichiporovich, A. A. 1956. Photosynthesis and theory for obtaining high
yield. *XV Timirjazevskoe Chtenie.*
Moscow, Akad. Nauk SSR*
109. Nichiporovich, A. A. 1961. Characteristics of plant canopies as optical systems. *Fiz. Rast.* 8:536–46*
110. Nichiporovich, A. A. 1963. On ways for
improving photosynthetic productivity
of plants under field conditions. *Photo-*
synthesis and Problems of Plant Pro-
ductivity, ed. A. A. Nichiporovich,
5–36. Moscow: Akad. Nauk SSR. 158
pp.*
111. Nilson, T. 1968. Calculation of spectral
flux of short-wave radiation within a
plant canopy. *Solar Radiation Regime*
within Plant Canopies, ed. Ju. Ross,
55–80. Tartu: Akad Nauk SSR Inst.
Phys. Astron. 147 pp.*
112. Nilson, T. 1968. On the optimum geometrical arrangement of foliage in the
plant cover. See Ref. 111, 112–46
113. Norman, J. M., Tanner, C. B. 1969.
Transient light measurements in plant
canopies. *Agron. J.* 61:847–49
114. Norman, J. M., Miller, E. E., Tanner,
C. B. 1971. Light intensity and sunfleck-
size distributions in plant canopies.
Agron. J. 63:743–48
115. Oikawa, T., Saeki, T. 1972. Light
regime in relation to population structure—An experimental approach based
on the Monte Carlo simulation model.

Rep. 1971 JIBP/PP-Photosynthesis
Level III Group, Tokyo, 107–16
116. Paltridge, G. W. 1970. A model of a
growing pasture. *Agr. Meteorol.* 7:93–
130
117. Paltridge, G. W. 1972. Experiments on
a mathematical model of a pasture. *Agr.*
Meteorol. 10:39–54
118. Pearce, R. B., Brown, R. H., Blaser, R.
H. 1967. Photosynthesis in plant communities as influenced by leaf angle.
Crop Sci. 7:321–24
119. Ramkrishnan, P. S., Kumar, S. 1971.
Productivity and plasticity of wheat and
Cynodon dactylon (L) Pers, in pure and
mixed stands. *J. Appl. Ecol.* 8:85–98
120. Rauner, Ju. L. 1965. *Heat Balance of*
Natural Plant Stands and Crop Stands.
Moscow: Nauka. 157 pp.*
121. Rauner, Ju. L. 1972. *Heat Balance of*
Plant Canopies. Leningrad: Gidro-
meteoizdat. 210 pp.*
122. Rauner, Ju. L., Rudnev, N. I. 1965.
Vertical profiles of radiation balance
components within a broad leaves tree
forest. *Heat and Radiation Balance,* ed.
Ju. L. Rauner, 91–105. Moscow:
Nauka. 158 pp.*
123. Ross, Ju. 1962. On theory of albedo of
plant canopies. *Nauch. Soobshch., Inst.*
Geologii Geogr. Akad. Nauk SSR 13:
151–64*
124. Ross, Ju. 1964. On mathematical theory
of photosynthesis of plant canopy.
Dokl. Akad. Nauk SSSR 157:1239–42*
125. Ross, Ju. 1965. On the theory of photosynthesis of plant community. *Prob-*
lems of Radiation Regime in Plant
Community, ed. Ju. Ross, 5–24. Tartu:
Akad. Nauk SSR Inst. Phys. Astron.
127 pp.*
126. Ross, Ju. 1966. Role of solar radiation
in photosynthetic activity of plant communities. See Ref. 24, 59–69
127. Ross, Ju. 1970. Mathematical models of
photosynthesis in a plant stand. *Proc.*
IBP/PP Tech. Meet., Třeboň, 29–46
128. Ross, Ju. 1972. A theory of penetration
of direct solar radiation into plant
canopy. *Solar Radiation and Productiv-*
ity of Plant Canopy, ed. Ju. Ross, 122–
47. Tartu: Akad. Nauk SSR Inst. Phys.
Astron. 148 pp.*
129. Ross, Ju., Bikhele, Z. 1968. Calculation
of the photosynthesis of leaf canopies.
Photosynthesis and Productivity of
Plant Canopy, 75–110. Tartu: Akad.
Nauk SSR Inst. Phys. Astron. 134 pp.*
130. Ross, Ju., Bikhele, Z. 1969. Calculation
of the photosynthesis of leaf canopies
(2). *Photosynthetic Productivity of*
Plant Canopy, 5–43. Tartu: Akad.

Nauk SSR Inst. Phys. and Astron. 199 pp.*

131. Ross, Ju., Nilson, T. 1966. Spatial orientation of leaves in plant canopy and its determination. See Ref. 24, 109–25

132. Ross, Ju., Nilson, T. 1967. Radiation regime within horizontal-leaved canopy. Phytoactinometrical Studies of Plant Canopies, ed. Ju. Ross, 5–34. Tallin: Valgus. 179 pp.*

133. Ross, Ju., Nilson, T. 1968. Calculation of photosynthetically active radiation within a plant canopy. See Ref. 111, 5–54

134. Ross, Ju., Ross, V. 1969. Spatial orientation of leaves in crop stands. Photosynthetic Productivity of Plant Stand, ed. Ju. Ross, 60–82. Tartu: Akad. Nauk SSR Inst. Phys. Astron. 199 pp.*

135. Saeki, T. 1960. Interrelationships between leaf amount, light distribution and total photosynthesis. Bot. Mag. 73:55–63

136. Saeki, T. 1963. Light relations in plant communities. Environmental Control of Plant Growth, ed. L. T. Evans, 79–94. New York: Academic. 449 pp.

137. Saito, H., Shidei, T., Kira, T. 1965. Drymatter production by Camellia japonica stands. Jap. J. Ecol. 15:131–39

138. Sato, H., Kira, T. 1960. Plant production on slopes. I. Plant yield as related to total radiation received by different slopes. Physiol. Ecol. 9:70–78

139. Shaw, R. H., Weber, C. R. 1967. Effects of canopy arrangement on light interception and yield of soybeans. Agron. J. 59:155–59

140. Shiman, L. H. 1967. Determination of the orientation of plant leaves in space. Fiziol. Rast. 14:381–83*

141. Suzuki, T., Satoo, T. 1954. An attempt to measure the daylight-factor under crown canopy with the solid angle projecting camera. Bull. Tokyo Univ. Forests 46:169–79

142. Tadaki, Y., Hatiya, K., Miyauchi, H. 1967. Studies on the production structure of forest. XII. Primary productivity of Abies veitchii in the natural forests at Mt. Fuji. J. Jap. Forest. Soc. 49:421–28

143. Takano, Y., Tsunoda, S. 1970. Light reflection, transmission and absorption rates of rice leaves in relation to their chlorophyll and nitrogen contents. Tohoku J. Agri. Res. 21:111–17

144. Takano, Y., Tsunoda, S. 1971. Curvilinear regression of the leaf photosynthetic rate on leaf nitrogen content among strains of Oryza species. Jap. J. Breed. 21:69–76

145. Tanaka, S. 1968. Estimation of sunlit leaf area in tobacco community by Monte Carlo method. (1) Estimation on direct sunlight of a plant. Environ. Control Biol. 7:12–16**

146. Tanaka, S. 1970. Geometrical distribution of leaves in tobacco plant community. Rep. 1969 JIBP/PP-Photosynthesis Level III Group, Tokyo, 16–19

147. Tanaka, T. 1972. Studies on the light-curves of carbon assimilation of rice plants—The interrelation among the light-curves, the plant type and the maximizing yield of rice. Bull. Nat. Inst. Agr. Sci. A 19:1–100**

148. Tanaka, T., Matsushima, S., Kojyo, S., Nitta, H. 1969. Analysis of yield-determining process and its application to yield-prediction and culture improvement of lowland rice. XV. On the relation between the plant type of rice plant community and the light-curve of carbon assimilation. Proc. Crop Sci. Soc. Jap. 38:287–93**

149. Tooming, Kh. 1967. Relationships between photosynthesis, plant growth and geometrical structure of plant canopies at different latitudes. Bot. Zh. 52:601–16*

150. Tooming, Kh. 1970. Mathematical description and net photosynthesis and adaptation processes in the photosynthetic apparatus of plant communities. Proc. IBP/PP Tech. Meet., Třeboň, 103–13

151. Tooming, Kh., Kallis, A. 1972. Calculation of productivity and growth of plant canopy. See Ref. 128, 5–121

152. Tooming. Kh., Ross, Ju. 1964. Radiation regime within a maize canopy and its expression by an approximate formula. Investigation of Atmospheric Physics, ed. Ju. Ross, 63–80. Tartu: Inst. Atmos. Phys.*

153. Tsel'niker, Ju. L., Ed. 1967. Light Regime, Photosynthesis and Productivity of Forest Stands. Moscow: Nauka. 275 pp.*

154. Tsunoda, S. 1964. A Developmental Analysis of Yielding Ability in Varieties of Field Crops. Tokyo: Maruzen. 135 pp.

155. Uchijima, Z. 1962. On the turbulent transfer coefficient within plant layer. J. Agr. Meteorol. 18:1–9**

156. Uchijima, Z. 1966. An improvement of semiempirical method of evaluating the total photosynthesis of a plant community. J. Agr. Meteorol. 22:15–22**

157. Uchijima, Z., Inoue, K. 1970. Studies of energy and gas exchange within crop

canopies (9) Simulation of CO_2 environment within a canopy. *J. Agr. Meteorol.* 26:5–18**

158. Uchijima, Z., Udagawa, T., Horie, T., Kobayashi, K. 1967. Studies of energy and gas exchange within crop canopies (1) CO_2-environment in a corn plant canopy. *J. Agr. Meteorol.* 23:99–108**

159. Uchijima, Z., Udagawa, T., Horie, T., Kobayashi, K. 1968. Studies of energy and gas exchange within crop canopies (4) The penetration of direct solar radiation into corn canopy and the intensity of direct radiation on the foliage surface. *J. Agr. Meteorol.* 24:141–51**

160. Udagawa, T., Uchijima, Z. 1968. Studies of energy and gas exchange within crop canopies (3) Canopy structure of corn plants. *Proc. Crop Sci. Soc. Jap.* 37:589–96**

161. Udagawa, T., Uchijima, Z. 1969. Studies of energy and gas exchange within crop canopies (5) Geometrical structure of barley canopies and the penetration of direct solar radiation into the canopy. *Proc. Crop Sci. Soc. Jap.* 38: 364–76**

162. Verhagen, A. M. W., Wilson, J. H., Britten, E. J. 1963. Plant production in relation to foliage illumination. *Ann. Bot.* 27:627–40

163. Wang, T. D., Wei, J. 1964. Relationships between canopy photosynthesis and leaf area index of several plants. *Acta Sinica Bot.* 12:154–58

164. Warren Wilson, J. 1959. Analysis of spatial distribution of foliage by two-dimensional point quadrats. *New Phytol.* 58:92–101

165. Warren Wilson, J. 1965. Stand structure and light penetration. I. Analysis by point quadrats. *J. Appl. Ecol.* 2: 383–90

166. Warren Wilson, J. 1967. Stand structure and light penetration. III. Sunlit foliage area. *J. Appl. Ecol.* 4:159–66

167. Watson, D. J., French, S. A. W. 1971. Interference between rows and between plants within rows of a wheat crop, and its effects on growth and yield of differently-spaced rows. *J. Appl. Ecol.* 8: 421–45

168. Watson, D. J., Witis, D. J. 1959. The net assimilation rates of wild and cultivated beets. *Ann. Bot.* 23:431–39

169. Williams, C. N., Soong, N. K. 1967. A simple foliage model for studying light penetration. *Ann. Bot.* 31:783–90

170. Williams, W. A., Loomis, R. S., Lepley, C. R. 1965. Vegetative growth of corn as affected by population density. I. Productivity in relation to interception of solar radiation. *Crop Sci.* 5:211–15

171. Williams, W. A., Loomis, R. S., Lepley, C. R. 1965. Vegetative growth of corn as affected by population density. II. Components of growth, net assimilation rate and leaf area index. *Crop Sci.* 5: 215–19

172. Williams, W. A., Loomis, R. S., Duncan, W. G., Dovrat, A., Nunez, F. 1968. Canopy architecture at various population densities and growth and grain yield of corn. *Crop Sci.* 8:303–8

173. Wright, J. L., Lemon, E. R. 1966. Photosynthesis under field conditions (9). Vertical distribution of photosynthesis within a corn crop. *Agron. J.* 58: 265–68

174. Yim, Y. J., Ogawa, H., Kira, T. 1969. Light interception by stems in plant communities. *Jap. J. Ecol.* 19:233–38

175. Yocum, C. S., Allen, L. H., Lemon, E. R. 1964. Photosynthesis under field conditions (6). Solar radiation balance and photosynthetic efficiency. *Agron. J.* 56:249–53

176. Yoda, K. 1971. *Forest Ecology.* Tokyo: Tsukiji-shokan. 331 pp.**

177. Yoshida, S. 1972. Physiological aspect of grain yield. *Ann. Rev. Plant Physiol.* 23:437–64

*Article is written in Russian.
**Article is written in Japanese.

ON THE PRINCIPLES OF THERMODYNAMICS IN ECOLOGY

❖ 4064

Vincent F. Gallucci

The University of Washington, the College of Fisheries and the Center for Quantitative Science in Forestry, Fisheries, and Wildlife, Seattle, Washington

"... just as the essence of food cannot be conveyed in calories, the essence of life will never be captured by even the greatest formulas."

A. I. Solzhenitsyn, *The First Circle* (115)

INTRODUCTION

From a very large literature on thermodynamics and kinetic theory in biology, this review singles out a narrow subset: applications to ecology. However, within this subset there have emerged applications that on the surface appear to be distinct and even unrelated. The intent here is to demonstrate the common basis upon which the several applications in ecology rest, to review a selected part of the relevant literature, and to point out some of the new and exciting theories being proposed to answer fundamental questions in ecology. The chapter may sometimes appear to be more tutorial than review, and there will clearly be material that many readers need not read. However, because the subareas of ecology which use concepts of energy flow seem to have become very specialized, I thought it best to consider them in moderate depth to make each accessible to nonspecialists. Also, it is necessary to present the thermodynamic theory in a version of its general form (the nonequilibrium theory) and a molecular counterpart, statistical mechanics.

Thermodynamics is, historically, the study of heat energy. Contemporary thermodynamics, however, deals with exchanges of all types of energy (and mass). Thus, a more descriptive word would be energetics, and the interest here, bioenergetics. Biological applications are often biochemical, i.e. studies of the disposition of energy stored in the chemical bonds of molecules. Although the energy released in biochemical reactions underlies much of what is reviewed here, it will not be explicitly discussed.

If we momentarily neglect the biological complications of bioenergetics, the physical theory of thermodynamics is high on the scale of scientific obfuscation, a situation which has been recognized (12, 20). Consideration of the complications

329

biology brings to the subject ranges from discussions which itemize the successful analyses (many of which will be cited) to scientific-philosophical debates (16, 49, 78, 124). It is not possible in this review to do justice to even the relatively few viewpoints represented in the above citations.

The use of thermodynamics in biology is, at one level, questioned because of the apparent contradiction of the second law of thermodynamics. The crux of the argument is that biological (including ecological) systems are, at any given instant, highly structured, e.g. as individual organisms or as communities of organisms. In fact, in an evolutionary sense organisms and communities appear to tend toward greater organization or structure (85). However, the second law requires isolated macroscopic systems to evolve in time to a final state of maximum entropy. Within the context of the microscopic model of the system, the final state of maximum entropy is interpretable as a state of maximum disorder, or minimum structure (26). Indeed, the premise underlying much of the literature cited is that the second law is not violated and yet that biological systems are ordered and may tend toward greater order.

At another level, the utilization of thermodynamics in ecology is unquestionably useful and widely accepted. At this level are trophic and productivity studies, studies of energy partitioning by organisms and communities, and, in fact, any study using the calorie as a measure of exchange. I hope that this review will contribute to perceiving these two levels as an artificial dichotomy, the recognition of which should enrich studies of both.

An element of reductionism is implicit in the gathering together of this literature. Reductionism is usually associated with the belief that biological phenomena are explicable solely through the laws of physics, in this case thermodynamics. A more literal interpretation of reductionism is that phenomena at the ecological level of organization are explicable by events at the biomolecular level of organization. Both interpretations will find support in this paper. I think it will be some time before scientists determine how to combine molecular events in such a way as to predictably specify an ecological event. Therefore, for the immediate future, study of ecological phenomena must of necessity involve techniques adapted to macroscopic occurrences. The question is, is thermodynamics an appropriate formalism to use?

It is relevant to the validity of the application of thermodynamics to ecology that thermodynamics, like much of ecology, is commonly known as a phenomenological formalism. That is, the fundamental relationships or laws are observations of nature, the result of curve-fitting, if you will. Is this apparent similarity in formalism exploitable at least for the benefit of ecology? A further possible relationship may be that, in terms of a physical formalism, the complexity of ecological phenomena may correspond to the microscopic counterpart of the thermodynamics formalism.

PHYSICAL THEORY

The intent of this section is to specify vocabulary, to show how selected aspects of the physical theory of thermodynamics interrelate, and to elucidate some concepts central to biological applications, which are discussed in later sections.

Thermodynamics is a macroscopic theory dealing with phenomena that result from the behavior of microscopic entities such as atoms and molecules. It is standard jargon to talk about a "system" and to identify meaningful "system parameters." Let the system be any object of interest with defined boundaries. Thermodynamic system variables which obviously measure macroscopic properties are temperature (T), pressure (P), and volume (V). Less obvious but equally important system parameters are internal energy (U) and entropy (S). Furthermore, depending upon what is in the system, it may be useful to define a chemical potential (μ_j) for the jth chemical species. These system parameters give a good indication of the macroscopic consequences of microscopic interactions. Since these parameters depend upon the system only, it is possible under appropriate conditions to construct functions of them, called system functions, such as the famous Gibbs free energy (G), the Helmholtz free energy (A), and the enthalpy (H) functions.

In general the system is in an environment, perhaps exchanging mass and energy across its boundaries with the environment as a biological organism would. Such a system is termed an "open system." If only energy is exchanged the system is said to be "closed" and if neither energy nor mass is exchanged it is said to be "isolated." That an isolated system can, at most, be approximated in the real world is obvious.

The theory associated with open systems is termed "open system thermodynamics" and is the most general formulation. Not surprisingly, if mass and energy are entering and leaving the system, one is concerned with the rates of exchange or flow rates. If the flow rates are constant in time then many systems will, after a time, reach a steady-state condition. An intuitively satisfying example is the application of heat to a metal rod at a constant rate and the removal of the heat at an equal rate from the other end. The system is the rod and the macroscopic system variable of interest is temperature. The temperature at each point of the rod approaches a steady value because the heat is applied and removed at the same rate. Clearly, removal at a rate less than or greater than the input rate would imply a temperature increase or decrease, respectively, at each point. A less intuitive example could have several different mass flows and several energy flows. The study of the restricted class of open systems where flows take place at a constant rate is termed "steady-state thermodynamics." If the flows across the boundary are zero, the system is isolated and the thermodynamics is termed "equilibrium" or "classical."

To make the (albeit, tenuous) connection to a living system which is continually exchanging mass and energy with its environment, it is desirable to note that the living system is not passive: flows into the system are generally transformed in some way before flowing out. Indeed there is some controversy as to whether an organism is ever in a steady state.

Consider now the content of a not necessarily living system: particulate, microscopic matter, gaseous or in solution, colliding in a seemingly random manner. The exchange of energy from these collisions is termed kinetic energy and thus the related study is kinetic theory, the appropriate formalism for open systems. For isolated or equilibrium systems the formalism of statistical mechanics is well defined. A less well-defined area related to kinetic theory and open systems is nonequilibrium statistical mechanics (100).

With respect to the content of the system, it is common to classify reactions (or processes) as either reversible or irreversible, the former actually corresponding to an idealized class of reactions. The relation between these two types of reactions and the entropy variable S will be explored below.

Thermodynamics of irreversible processes may be subdivided into a part dealing with linear phenomena and a part dealing with nonlinear phenomena. In a sense the linear region is characterized by linear phenomenological laws, which means constant transport coefficients. For example, there is a "law," known as Fourier's law, that the flow of heat is proportional to the gradient of temperature: in one dimension x, where α is a constant, $\partial T/\partial t = \alpha \ (\partial^2 T/\partial x^2)$.

However, if α is a function of temperature (T) the relation is nonlinear. Of the work available in this area that of Prigogine (100) and Glansdorff & Prigogine (41) is outstanding.

Fortunately, for many real world biological problems it is often reasonable to assume both a steady-state condition (for the system with its environment) and the validity of linear (phenomenological) laws, such as Fourier's law and Fick's law (for diffusion). Much of the work in this review is based on the assumption of irreversible processes occurring in a system which is in a steady-state condition where linear relations are acceptable.

When an open system exchanges mass and/or energy with its environment it does so in a way that must be consistent with the laws of thermodynamics. The mechanisms or processes of exchange, that is, the mechanisms of heat and mass transfer in and out of the system, are expressed by empirical relations which may be termed "transport processes." In ecological applications these are usually approximated as linear processes. Examples are convection, conduction, radiation, absorption, diffusion, and transpiration.

THE LAWS OF THERMODYNAMICS   The laws of thermodynamics are well known for the restricted classical theory. It should be emphasized before discussing the restricted and general versions of these laws that they are phenomenological. They are not derived; they are observations of the system variables of macroscopic systems. However, it is possible to derive the equilibrium version of the laws from the (microscopic) theory of statistical mechanics (thereby helping to confirm the latter theory), but, historically, the laws were recognized and used much earlier.

The word "energy" has been used a number of times in the preceding, both as internal energy, U, of the system and as a flow term, in and out of the system. Just as there is no objection to the use of volume, temperature, and pressure without specific definitions, it is typical to rely upon some intuitive feeling for energy, or to settle for the "ability to do work." Perhaps this is because in some way energy flow is associated with caloric flow, a familiar experience. However, the reader may not be as quiescent in the coming entropy flow discussions. From the point of view of the thermodynamics formalism, entropy and energy are equal in their relation to the real world. They are both system variables which fulfill certain mathematical criteria. It should be noted that energy can never be isolated as a thing in itself but is always encountered in its association with matter, human intuition notwithstanding. The first law of thermodynamics is a conservation of energy relation for the amount of mass and energy flowing across a system's boundary, denoted by ϕ, and the amount of mechanical work done on or by the system, denoted by W. The flow across the boundary is heat and mass transfer from the environment flowing inward, and heat and mass transfer to the environment, perhaps the results of internal chemical reactions. The net change of internal energy, U, in a time $\mathbf{d}t$ is:

$$\mathbf{d}U = \mathbf{d}\phi - \mathbf{d}W \qquad \qquad 1.$$

where $\mathbf{d}\phi$ is positive if received by the system and $\mathbf{d}W$ is negative if performed by the system.

The equation for the first law could be made more specific by summing the differential changes in all the known mass and energy inputs and outputs of a biological system. Then,

$$\mathbf{d}U = \Sigma_i \mathbf{d}\phi_i + \Sigma_i \mathbf{d}W_i \qquad \qquad 2.$$

means that the change in the internal energy of the system in a time $\mathbf{d}t$ is the sum of changes in energy input from solar radiation and output from heat producing chemical reactions (examples of $\mathbf{d}\phi$) and changes in the energy equivalent of work done by the biological system. If the system were a single organism W_1 might be the energy cost of foraging, which could be partitioned into costs of muscle movement, respiration, etc. If the system were a population or a community the approach outlined would be equally valid.

If the system is closed (only energy is exchanged with the environment) then ϕ is thermal energy Q. The first law, Equation 1, is then $\mathbf{d}U = \mathbf{d}Q - \mathbf{d}W$.

The second law of thermodynamics expresses a restriction upon the energy changes in terms of the system variable, entropy. The net entropy change of the system in a time $\mathbf{d}t$ is $\mathbf{d}S$ and

$$\mathbf{d}S = \mathbf{d}_e S + \mathbf{d}_i S \qquad \qquad 3.$$

In Equation 3 $\mathbf{d}_e S$ is the total flow of entropy in $\mathbf{d}t$ due to exchanges with the environment or exterior of the system, e.g. the transport processes of radiation, conduction, and convection between the system and environment, and $\mathbf{d}_i S$ is the entropy production in $\mathbf{d}t$ due to irreversible processes inside the system, e.g. diffusion, chemical reactions, and conduction. In the real world no processes are truly reversible, a fact expressed by

$$\mathbf{d}_i S \geq 0 \qquad \qquad 4.$$

For an isolated system

$$\mathbf{d}_e S = 0 \qquad \qquad 5.$$

but $\mathbf{d}_e S$ is otherwise unrestricted with respect to sign; it may be positive, negative, or zero.

If a reaction could occur and have the property of being reversible then $-t$ could be substituted for $+t$ and the relevant quantitative relation would remain valid. In terms of entropy change for that reaction

$$\mathbf{d}_i S = 0 \qquad \qquad 6.$$

In a sense Equation 6 means that the reaction is frictionless: if it is a chemical reaction and if it goes in one direction, it also goes backward. Friction would imply heat energy, which would imply molecules moving more quickly, which would imply less coherence. Although all reactions are irreversible, some may be approximated by assuming reversibility. An intuitively satisfying example of irreversibility is a drop of ink diffusing in a beaker of water. It is difficult to perceive the ink coming back together in negative time over the same paths.

The second law of thermodynamics simply specifies that no process can occur in which the entropy of the system plus its surroundings (the universe) decreases, in other words

$$dS \geq 0 \qquad\qquad 7.$$

is always true. The condition $dS = 0$, i.e. the rate of change of entropy with respect to time is zero, is the exact criterion for a thermodynamic steady state. From Equations 7, 3, and 4 it follows that $d_e S = -d_i S \leq 0$. That is, if a sufficient amount of entropy per unit time emanating from outside the system is added, such that it balances the positive entropy changes generated inside the system, a thermodynamic steady state results.

By contrast, a thermodynamic equilibrium state corresponds to $d_i S = 0$, i.e. S is a maximum, constant value. Such a condition is one of minimum structural coherence, the antithesis of a functional biological system.

The units of entropy are calories per mole per degree, or simply entropy units. Although all processes will, by the second law, tend to maximize the entropy of the universe, entropy and changes in entropy frequently are not easily measured or calculated. However, because biological processes often occur under conditions of reasonably constant volume (V_0), pressure (P_0), and temperature (T_0), it is possible to define the free energy functions mentioned before, Gibbs free energy (G) and Helmholtz free energy (A). That T_0, P_0, V_0 place restrictions upon the open-system properties of a system is obvious. The Gibbs free energy under these restrictions is

$$G = U + PV - TS \qquad\qquad 8.$$

which is easily solved for the change in the internal energy of the system. Furthermore, G is related to the equilibrium constant K of chemical reactions such as $aA + bB \rightleftharpoons cC + dD$. Thus having K supplies G, which yields information about internal energy U (66). Gibbs free energy would correspond to the maximum possible energy available to do work such as respiration.

The Helmholtz free energy may be shown to be related to the microscopic nature of the system by way of the energy distribution (82), as well as being the following function of system variables

$$A = U - TS \qquad\qquad 9.$$

Both A and G have minimum values at thermodynamic equilibrium.

The relationship between the equilibrium microscopic and macroscopic formalisms is that of statistical mechanics and thermodynamics. Certain thermodynamic system variables, such as T, P, V, etc, were identified which have the outstanding feature of being, quite literally, observable or macroscopic variables. If the thermo-dynamic system is a gas in a container, the macroscopic (system) variables T, V, P are incapable of telling us anything at all about the individual gas molecules. It is reasonable then to seek a new set of variables which do refer to individual particles and which correspond in some sense to the macroscopic variables, and to denote these as microscopic variables. The question is, what properties of the molecules will be singled out? Consider a single particle. Obviously, a fundamental variable is displacement from some reference point, the total number of displacements which can be made independently being the number of degrees of freedom of the particle. Thus a "simple" particle, which can move on a line only, has 1 degree of freedom; a particle in space has 3 degrees of freedom; a system of n particles has $3 \times n$ degrees of freedom.

An example of a simple particle is an argon atom; a more complicated particle would be an oxygen molecule, which has more than 3 degrees of freedom. For most purposes any physical (gas) system subjected to analysis would contain many parti-cles (1 mole contains approximately 6.0×10^{23} particles), and the degrees of freedom of the system would be accordingly large. To be able to characterize the motion of a particle it is necessary to know its velocity as well as its position (displacement). Since velocity is not directly measurable it is customary to substitute the momentum (knowledge of the mass and momentum gives the velocity). There are three momen-tum components associated with motion in space (per particle).

Denote (for a single simple particle) the three position and three associated momentum variables by q_1, q_2, q_3, p_4, p_5, p_6, respectively. For n particles there will be 3 (degrees of freedom) $\times$ 2 $\times$ n (particles) = $6n$ such "state" variables. If the system has one particle ($n = 1$), it is customary to denote the instantaneous state of the system as a point in a 6-dimensional Euclidean space; the instantaneous state of a system of n particles would be a point in $6n$-dimensional space. Appropriately, this $6n$-dimensional space is often called the state space or phase space (Γ space).

Having specified the instantaneous state of the system it is of interest to specify the manner in which the system changes with time, i.e. its dynamical behavior. Dynamical behavior is represented by the state of the system being a function of time t. That is, each state variable is a function of time: $\{q_1(t), q_2(t), \ldots, p_6(t), q_7(t), \ldots, p_{6n}(t)\}$. It follows, then, that by picking t values sufficiently close, a continuous curve in Γ space can be approximated which corresponds to the time development of the system of n particles. This curve is called a trajectory for the system.

Using more or less familiar procedures embodied in Newton's laws it is possible to write equations of motion for the particles, i.e. a set of second order differential equations. These equations may be rewritten as a set of twice as many simultaneous first order, coupled, not necessarily linear, differential equations

$$\frac{dq_i}{dt} = \frac{\partial H(q_i, p_i)}{\partial p_i} \qquad i = 1, 2, \ldots, 3n$$

$$\frac{dp_i}{dt} = \frac{-\partial H(q_i, p_i)}{\partial q_i} \qquad i = 1, 2, \ldots, 3n$$

10.

This set of $6n$ equations is called Hamilton's Equations of Motion. The phase space Γ for a mole of gas would be of the order of 4×10^{24}-dimensional. The solution would be a trajectory, or a curve, in Γ space. The functions $H(q_i, p_i)$ may be thought of as specifying the forces acting on the system of particles and are responsible for the dynamical behavior of the system.

In accordance with the Newtonian formalism, a solution to Equation 10 would require the specification of approximately 4×10^{24} initial conditions, e.g. $\left\{ q_1(0), q_2(0), \ldots, p_{6n}(0) \right\}$. That is, a simultaneous knowledge of the three position and three momentum coordinates of all 6×10^{23} particles!

Statistical mechanics gets its name from the association of a probability function with the likelihood of the initial condition being in different parts of Γ space.

By the use of stochastic processes, expected or average values, and related concepts, it is possible to associate the microscopic model's variables, $q_i(t)$ and $p_i(t)$, with the macroscopic model's variables, V, T, P, U, S, etc, and to eventually derive the thermodynamic laws.

More elaborate descriptions of some of the above may be found by referring to Morowitz (82), Prigogine (100), Glansdorff & Prigogine (41), Denbigh (19), Rosen (104), Zemansky (128), or Lehninger (66).

MECHANISMS OF ENERGY TRANSFER IN THE PHYSICAL ENVIRONMENT

There are many modes of organism-physical environment interaction that continue to occur as long as the organism is alive, often at rates which vary according to time of day, location, and season. I am denoting these as passive interactions. In contrast, it is necessary for some organisms to perform some activity which is temporary and which causes a significant change in the expenditure and/or gain of energy. These are denoted as active interactions.

In what ways do fluctuations in the physical environment (save drastic ones) affect the community over daily, seasonal, and evolutionary time periods? The question assumes vast aspects of complexity when it is recognized that communities of organisms, plant and animal, actually modify their climates (77, 85) as they grow and move through the stages of succession. The operation of a sort of feedback mechanism would seem to be involved. The mechanisms of mass and energy transfer between the individual organism and its surrounding climate are a fundamental part of the answer, but it is not clear how to use this information. Some efforts are reported below.

Passive Interactions

Climate, or physical environment, consists of a bath of radiation and fluid (e.g. air) which surrounds each organism. Every organism is adapted to a set of climatic conditions outside of which it cannot survive. Fundamental mechanisms of adjustment vary, e.g. between homeotherms and poikilotherms, and, furthermore, there exist many finer modes of adjustment, e.g. blubber, the fluffing of feathers, hibernation, and the ability to modify the immediate microclimate. For aquatic animals pressure is often a climatic variable.

The thermodynamics used in the study of organism-environment passive interactions is perhaps the most straightforward application of physical theory to ecology. The essence of the procedure is to conceptually reduce, as far as possible, the biology to where an appropriate physical object can act as a guide for the biological experiment and analysis. The analogy guides what parameters are measured, and to some extent, how the measurements are made. The interpretation of measurements, especially for animals, is highly dependent upon the ecology of the organism (8, 117, 129). I tend to think of much of the work in organism-climate interactions as an experimental biophysics of ecology (see 29, 62, 103).

It is important to note, especially in view of the physical theory section, that most experiments are designed to describe the steady-state behavior of the organism, but it is impossible (certainly in the field) to limit consideration to a strict steady state. Perhaps transient states are described by rapidly changing thermodynamic variables as the organism establishes one steady state after another; e.g., just after a warm sun moves behind a cloud on a cold day plants and animals experience abrupt temperature changes. The transients may be quite important but little is known about their role. Subject to the assumption of a steady state, and disregarding energy stored in chemical bonds and generated by metabolic processes, a balance of energy inputs and outputs can be made, the sum total of which must clearly be zero (it may be useful to look at Equation 15). The sum is over the number of calories contributed to the organism from the environment and the number of calories given off by the organism (a negative contribution) as the continual processes of absorption, reflection, radiation, conduction, convection, and evaporation operate.

These processes, or energy mechanisms are the same as would operate between any object and its environment. For example, a small damp aluminum cylinder in air moved into the sunlight will be subject to all six processes. Corresponding to each process is a coupling factor which is associated with the properties of the organism's surface and its efficiency with respect to that energy exchange mechanism. Coupling factors are also independent of the magnitude of the energy exchanged, at least within the range of magnitudes of ecological interest. The coupling factors are not essentially biological either, since they are basically the same factors that would be used for a physical object but now are evaluated for a biological object. The terms coupling factor (31) and coefficients of exchange (37) are equivalent.

THE EXCHANGE MECHANISMS   Subject to possible restrictions imposed by the first and second laws of thermodynamics, it is possible to write several phenomenological relations to express the amount of energy which crosses the interface between the organism and the environment. Consider an organism, plant or animal, in air. Any object above absolute zero temperature gives off radiant energy from its entire surface area. The energy is in the infrared region and is given off in the amount

$$Q_{rad} = E\sigma T^4 \qquad\qquad 11.$$

where E is the emissivity of the object, σ the Stefan-Boltzmann constant, and T the surface temperature (in degrees Kelvin). Emissivity is a comparison of the radiation characteristics of the object to a black body of similar surface construction and is the coupling factor for radiation. The process is rather well known.

An object also receives solar radiation from the sun and infrared radiation from other objects in the environment. The surface of the organism will both absorb and reflect energy from these sources in varying amounts, according to the intensity and spectral composition or wavelengths. For a given spectrum, the amount of absorption and reflectance depends upon the surface area exposed and upon a coupling factor called the absorptivity of the surface (γ). The absorptivity of the surface is a complicated function of the surface's composition (and hence color) and geometry and varies as the spectral composition varies. The amount of energy absorbed (Q_{abs}) is given by a relation similar to Equation 11 with the function γ replacing E. Energy not absorbed is reflected. The absorbed energy is used in many ways, the most familiar use for green plants being photosynthesis. Studies of absorption and reflection have been done for plants (7, 33, 38, 39, 83) and animals (98, 99). All of these studies deal with the absorption of thermal (heat) energy, which is in a lower energy range than ultraviolet (uv) energy. However, in some environments uv may be a major consideration; e.g. mountain climbers have long taken precautions against uv radiation. Little seems to have been done with respect to the effects of uv on plants (14).

In the process of conduction the energy exchange depends upon mass transport, that is, the molecular collisions within the organism. The coupling factor is the conductivity of the surface of the organism (K). The efficiency of the transport of energy to the surroundings will depend first upon whether or not the surface is a good thermal insulator, like fur or blubber, and second upon the conductivity properties of the surroundings, e.g. air. A simple equation for the amount of energy exchanged by conduction is

$$Q_{cond} = K\Delta T/d \qquad\qquad 12.$$

where ΔT is the temperature gradient between two points separated by a distance d.

In air a thin boundary layer forms and adheres to the surface of any warm object. The boundary layer acts as an imperfect insulator between the surface and the free moving air. The depth of the boundary layer depends upon the size, shape, and orientation of the body as well as on the ambient air temperature, wind speed, and other factors. Energy from the organism's surface is transferred to the boundary layer by the process of conduction.

In the process of convection the energy exchange depends upon the motion of the air across the surface of the body, or more specifically, the exchange depends upon the interaction between the boundary layer of air and the air beyond it. Two modes of convection occur, free and forced; the former occurring in still air, the latter in air driven by a wind, for example. The coupling factor is a property of the organism called the convection coefficient (h). A very simplified equation for the amount of energy transfer by convection is

$$Q_{conv} = h\Delta T \qquad\qquad 13.$$

where h represents the heat lost from an object with certain thermal properties when it has a given orientation in a fluid (e.g. wind) flow of a certain velocity, and ΔT

is the temperature difference between the fluid and the surface of the object. Studies of convection, especially free convection, for plant surfaces (37, 92, 118) have been done using an actual plant surface such as a leaf, as well as models of such surfaces.

The process of moisture exchange between the organism and the environment is called transpiration when plants give up the moisture, evaporation or sweating when animals give up the moisture, and condensation when the environment gives up moisture to the organism's surface. Denote the amount of energy transfer by the appropriate process as Q_{evap}. The coupling factor for the three processes is the diffusion resistance of the surface. The diffusion resistance is a function of the complexity of the boundary layer, the permeability of the surface to moisture, and other structural and morphological features of the surface. For the living animal evaporation from the lung surfaces should be included. In general, the vapor pressure or relative humidity of the surrounding air, the air temperature, and the wind velocity are the important climatic features. For example, the waxy cuticle of some desert plants essentially decouples the plant from ongoing moisture exchange processes, whereas the integument of a salamander is permeable to moisture and the skin temperature is strongly influenced by evaporative cooling. Again, the work on plant surfaces is more extensive (6, 21, 30, 32, 35, 65, 105, 120) than similar work on animal surfaces (5, 99).

To summarize, a plant or animal receives and transmits energy from its environment through its surfaces (excluding uptake of nutrient by way of a root structure or feeding activity). The organism may receive energy by radiation, by convection if the organism is cooler than air, and by condensation if the plant temperature is lower than the dew point. The organism loses energy by radiation, by convection if it is warmer than the air, and by evaporation.

THE ENERGY BALANCE   None of the Equations 11, 12, or 13 for Q_{abs}, Q_{rad}, Q_{cond}, Q_{conv}, or Q_{evap}, included terms for surface area and time factors. The surface area and an appropriate unit of time must be chosen. Assume that these factors are incorporated and denote the corresponding terms by $Q_{abs}^{\bullet}$, $Q_{rad}^{\bullet}$, $Q_{cond}^{\bullet}$, $Q_{conv}^{\bullet}$, and $Q_{evap}^{\bullet}$, each of which is now a rate of energy transfer (calories per unit surface area per unit time, say). To write an equation for the balance of energy transfer, define rates $Q_{stor}^{\bullet}$ and $Q_{met}^{\bullet}$ as the rate energy is added to or taken from storage within the organism and the rate energy is consumed or produced by metabolism, respectively. The energy balance equation is written

$$Q_{abs}^{\bullet} = Q_{rad}^{\bullet} + Q_{cond}^{\bullet} + Q_{conv}^{\bullet} + Q_{evap}^{\bullet} + Q_{stor}^{\bullet} + Q_{met}^{\bullet} \qquad 14.$$

If more energy is coming in than is emitted either the organism is storing it in some manner ($Q_{stor}^{\bullet} > 0$), or the organism is heating up. If less energy is coming in than is emitted, the energy must be coming from storage ($Q_{stor}^{\bullet} < 0$) or the organism is getting colder. The alternatives to $Q_{stor}^{\bullet}$ being nonzero occur in very restricted cases (34). Note that neither $Q_{stor}^{\bullet}$ nor $Q_{met}^{\bullet}$ result from the exchange processes operating between the organism and the environment. A steady state in this case means that $Q_{stor}^{\bullet}$ is zero, which is equivalent to saying that the amount of energy being absorbed equals the amount of energy being given off.

Equation 14 is actually a detailed restatement of the first law of thermodynamics, Equation 1. The Q_x^* (where x may indicate any one of the types of Q^*) are all expressed in terms of energy change per unit time and could be formulated in terms of differentials. It would take considerable effort to partition the terms in Equation 14 into differentials of work and mass-energy flow. However, Equation 14 is a relatively straightforward restatement of the first law. The second law, Equation 7, is easily seen to be satisfied for all the Q_x^* processes.

For the purposes of this section, set $Q_{stor} = Q_{met} = 0$ and rewrite Equation 14 as

$$Q_{abs}^* = Q_{rad}^* + Q_{cond}^* + Q_{conv}^* + Q_{evap}^* \qquad 15.$$

A more precise statement of the conservation equation would try to express each of the terms as a function of several independent variables. A possible source of such variables is the environment. Up to now the characteristics of the environment which are central to the exchange processes were only mentioned in passing. These characteristics are, e.g. environmental temperature, wind velocity, humidity, and the intensity of the radiation. They are often considered to be independent variables in the sense that the processes of radiation, absorption, reflection, conductivity, convection, and evaporation can each be expressed, uniquely, as a function in terms of these four variables. The possibility of the function being a constant with respect to one or more of these variables of course exists. It is only certain idealized weather conditions that can be represented by air temperature, wind, humidity, and radiation, but the restriction often is not severe.

For a specific organism it is now theoretically possible to evaluate each Q_x^*, to verify the conservation law in Equation 15, to attempt to generalize results and measurements within classes of organisms (chosen for the similarity of their thermodynamic properties), and to use this information as a jumping off point to study deeper, more essentially biological problems.

The experiments to evaluate a Q_x^* would involve measuring Q_x^* as the independent variable(s) is changed. If two independent variables are involved one is held constant and the other varied, and vice versa (34, as well as other references in this section). Generalization is, as always, a precarious operation, but it is relative. A generalization within the class of deciduous (broad, flat, thin leaves) trees and a generalization within the class of coniferous (small diameter cylindrical needles) trees might not, at times, be unreasonable. However, generalizing from plants to animals is quite another problem. The formalism developed above has been sufficiently general to encompass what is done in both the plant and animal kingdoms.

A natural application of this work is to studies of competition, adaptation, productivity, succession, and diversity (10, 36). Consider, for example, competition in a stand of trees. The competition between individuals is displayed by shading and water consumption. But the effects, e.g., upon a seedling in the forest can only really be known when evaluated in terms of the wind velocity, air temperature, and relative humidity, as well as the light intensity reaching the seedling. Water usage rates and efficiency of use of water are highly dependent upon the relative humidity of the air near the seedling. And there is a transpiration rate dependence upon wind and light intensity (32, 111). The dependence of the Q_x terms upon the climatic variables can now be used in setting up the experiment and the analysis.

It is well to point out how adaptable this formalism is to model building efforts. The computers can do the large amount of necessary bookkeeping. Since approximating equations are known for the processes, they may be programmed. After collecting the necessary experimental data, it may be possible to use the computer laboratory to ask questions about the consequences of changes in limiting factors, speeding up the processes of interaction with climate and with other organisms (competition, cooperation, etc). That it will be necessary to go back to the natural system to learn more about the complexities of the role of some factor or other is all the better for science. Much of the preceding discussion concerning the use of models operating on a thermodynamic formalism is contained in more detail in Gates (34) and Odum (85). Lemon, Stewart & Shawcroft (70) report on the general system aspects of growth in a cultivated field and present a comprehensive computer program (SPAM) which predicts community behavior. Allen, Jensen & Lemon (2) report on a simulation study (using SPAM) of CO_2 enrichment.

There are certain properties of animals that make their analyses even more complicated. Homeostatic organisms are apparently utilizing a feedback mechanism that a steady-state thermodynamic formulation cannot cope with, at least over short periods of time. The ability of animals to vary their metabolic rates, sweating rates, geometry, color, position, etc causes theoretical and experimental difficulties. Nevertheless, there is an active interest in animal responses in both their natural environments and in temperature stressed environments. Barnett & Mount (4) and Fry (27) deal with the responses of organisms to extremes of high and low temperatures; Brock (11) considers only high temperature systems. Stonehouse (116) considers heat transfer in penguins as one of several parts of the overall thermal balance. Moen (79) extends the discussion above to considerations of body weight and growth; Parks (93) focuses on ambient temperature effects upon thermochemical efficiency of growth; Kavanau & Rischer (55) examine the effects of ambient temperature and high activity upon thermoregulatory processes. Brown & Lasiewski (13) examine the energetics of a particular body shape in terms of heat loss and draw conclusions about natural selection; Hadley (46) examines the microclimate of two desert arthropods in terms of a heat exchange budget and finds that incident solar radiation, convection, and reradiation are the main processes; White & Lasiewski (125) examine den temperatures of "hibernating" rattlesnakes as an indication of body heat transfer. These citations are indicative of the range of applications of energy transfer methods in animal ecology.

Active Interactions—Energetics and Foraging

Although plants do not forage they have an intrinsic effect upon the spatial and temporal foraging patterns that do occur in a community (1, 24). An animal, either directly or indirectly, feeds on the plant community via the food web. The foraging activity can be studied in terms of the first law of thermodynamics (Equation 2) and therefore can also be expressed in terms of Equation 14 for Q_{abs}^*. Consider Equation 14 for nonzero Q_{met} and Q_{stor}. The animal forages to survive, to have a nonzero

Q_{met}, and to drive a positive Q_{stor}^*. The Q_{met}, as well as Q_{abs}, keep the transfer terms, Q_{rad}, Q_{conv}, Q_{cond}, and Q_{evap} nonzero to help cope with the vicissitudes of climate.

The variables which characterize climate, i.e. temperature, wind velocity, humidity, and radiation intensity, were shown to be central to the analysis of the five transfer, Q_x, terms. But, in addition, the same climatic variables, some more so than others, also affect Q_{met} and hence Q_{stor}. Lee (64) relates increased biochemical activity to increased temperature, but points out the coupling and feedback which must prevail in a homeotherm. Fry (27) relates increased biochemical activity to increased temperature in poikilotherms, where the adaptative response is a different one. Moen (79) discusses a "critical thermal environment" to describe the combined effects of the climatic variables. Most of the thermal stress literature cited discusses Q_{met}^* variation with temperature.

Different nutrients, each with a different value to the forager, are usually available. Omitting consideration of protein, vitamin, and mineral values, it is the food with higher caloric, or energy, content that is usually assumed to be of the most value to the organism (23). Paine (91) has reviewed the subject of calorimetry in ecology. Verduin (121) discusses the caloric value of plant matter, as well as some incorrect conclusions of others.

Foraging consumes energy. A particular nutrient is probably selected, subject to the constraint that the energy expenditure to acquire that nutrient is not excessive. At the very least this means that energy costs of locating, pursuing, dispatching, and consuming a nutrient are less than the energy gained from assimilation. Each of the activities, location, pursuit, dispatch, and consumption, are thermodynamically similar. They involve physiological work: muscle extension, lactic acid build up and decay, increased cardiac and respiratory activity, etc. These activities, expressible as the rate of change in work per unit time, or dW, plus the inevitable generation of heat, dQ, are written in terms of the first law of thermodynamics, $dU = dQ - dW$, where dW is a sum of the many kinds of physiological work necessary to complete the action. Then dU is the total amount of energy gained or lost in a time dt. Schmidt-Nielsen (107) has evaluated the energy costs of swimming, flying, and running. On a broader scale, the energy costs of reproductive and other activities could be included. The energy cost of survival of a species could be compared to gains from foraging. Heinrich & Raven (48) have studied the energetics of pollinators, Wolf et al (127) have studied the energetics of foraging by hummingbirds, and Orians & Horn (89) have made feeding studies on blackbird populations. Brett (8) has studied the changes in foraging patterns of sockeye salmon subject to natural and laboratory temperature fluctuation.

Natural selection considerations suggest that an optimality principle is operating (15, 23, 54, 75, 108). Presumably a forager would wish to maximize the total energy gain per unit time and simultaneously minimize the time spent foraging (one good reason for doing so would be to minimize his chance of being preyed upon).

Briefly then, the laws of thermodynamics are seen to be fundamental to an understanding of how an organism functions in its physical environment in terms of energy and mass exchange processes. The same laws are fundamental to an understanding of how an organism interacts in its biological community, as seen through the organism's physiology and behavior. Theoretical problems notwithstanding, the laws are useful for practical applications.

ECOSYSTEM PRODUCTIVITY

No discourse on thermodynamics and general ecology is complete without a section on productivity. Rather than review the extensive literature, I will mention just a few of the outstanding publications, a few of the traditional concepts, and a nontraditional formulation to demonstrate the very close relationship between productivity and thermodynamics. Even this well-established field of ecology is apt to be in for exciting new work.

Some attractive sources follow. Hazen (47) is a book of selected reprints. Four of these, Lindeman (71), Slobodkin (112), Ryther (106), and Golley (43) are landmarks. The book by Elton (22) is astonishingly contemporary in its approach. Odum (86) is thorough, with the advantage that the productivity chapters are well integrated with the whole book and references to the literature are plentiful. Phillipson (96) is a short survey of principles. The series of International Biological Program (IBP) handbooks published by Blackwell Scientific Publications of Oxford contains good sources. A book edited by Winberg (126) which first appeared as part of a Russian Language IBP handbook treats the productivity of marine invertebrates. The IBP-related books tend to be experimental in their orientation and therefore complement rather than replace the other sources listed here.

The quest for generalization and underlying principle in science is demonstrated by the concept of trophic levels. The more details available about a community the more it assumes characteristics unique to itself. On the other hand, a number of properties apparently common to all communities are useful for the comparison of different communities, or the same community at different times. The concept of a trophic level is one such common feature.

The approach is to ignore the specific species in a community and group together organisms with similar food requirements (71). Categorize organisms as follows: all autotrophs as primary producers, all herbivores as primary consumers, and all carnivores as secondary, tertiary, etc consumers. From one level to the next communication is in the form of energy flow. Solar energy flows to the primary producers for photosynthesis and temperature adjustment, and a mass flow supplies nutrients. The other trophic levels receive solar (thermal) energy for warmth and mass flow from lower levels. The efficiency of mass transfer is dependent upon definition (112, 113). The number of organisms in successive levels typically follows a pyramiding relationship: ". . . animals at the base of a food chain are relatively abundant while those at the end are relatively few in number, and there is a progressive decrease in between the two extremes" (22). Elton (22) noted that the pyramid is inverted in the case of parasite-host relations.

In each ecosystem there are two basic food chains for herbivores, a grazing chain and a detritus chain (84). The detritus chain, when defined to include the function of feeding on the carrion of consumers and their feces, is commonly denoted as the decomposer trophic level. In general the major pathway for mass flow in forest ecosystems is through detritus organisms; in marine systems the major pathway is through grazing organisms (96). The structure of primary, secondary, and tertiary consumers is attractive, but it is often unrealistic for organisms that feed at multiple trophic levels (omnivorous). This criticism does not diminish the usefulness of the trophic structure for purposes of generalization and classification.

Ecologists often consider mass and energy as equivalent, citing the laws of thermodynamics for support: all forms of energy can be converted completely to heat energy, but incompletely to other forms of energy. By the use of a bomb-calorimeter, the chemical energy (released in a special type of combustion) of a sample of mass ingested by a consumer can be rated for its energy or caloric value. The unit of measurement is the calorie or kilogram-calorie (91, 121). Thus, following the tradition, the references to mass flow between trophic levels may be thought of as energy flow.

Early in this paper I noted that both thermodynamics and ecology are phenomenological sciences. This may explain why a trophic level formulation of organism-organism interactions in a community is coincident with how one would begin to analyze an open system consisting of, say, n smaller open systems. One may make the association between the ecology and the thermodynamics by identifying the ith trophic level with the ith open system, T_i, where $i = 1, 2, \ldots, n$. Allow an input of Q_i solar energy in the ith trophic level T_i, and for this analysis let $Q_1 = Q_2 = \ldots = Q_n = Q$. Of the incident Q calories, an amount Q_i' is reflected by organisms in T_i. The incident radiation $(Q - Q_i')$ is absorbed by organisms in T_i. Organisms in all T_i use $(Q - Q_i')$ as thermal energy; organisms in the primary producer level also use $(Q - Q_i')$ for photosynthesis. Each T_i contains mass M_i, and of this, M_i' is supplied to T_{i+1} and an amount M_{i-1}' is received from T_{i-1}. It is always true that $M_i' < M_i$. The method of supply could be the grazing or foraging of organisms in T_{i+1} on those in T_i. Organisms can function in a level T_{i+1} because of flows M_i' and Q. When they function they respire and do work such as forage, reproduce, etc in an amount W_i. These activities have been discussed in terms of the two laws of thermodynamics, hence the inevitable generation of an amount of heat (infrared energy) Q_i''. The mechanisms, problems, etc, of the transport of Q_i'' from the organisms have been detailed. In each T_i, an amount, dU_i, of energy change will take place in a time interval dt. Thus

$$dU_i = + dQ - dQ_i' - dQ_i'' - dM_i' + dM_{i-1}' - dW_i \qquad 16.$$

or

$$dU_i = d\phi_i - dW_i \qquad 17.$$

where $d\phi_i = + dQ - dQ_i' - dQ_i'' - dM_i' + dM_{i-1}'$. Equations 16 and 17 are in fact the same as Equation 1, the first law of thermodynamics.

For the whole system (community), $dU = \sum_i dU_i$. That $Q_i'' > 0$ is an expression of the second law. One can proceed to identify the usual parameters, standing crop, and gross and net productivity. This analysis assumes $Q = Q_1 = Q_2 = \ldots = Q_n$, which may be unwarranted if the community contains forest floor dwellers as well as trees, e.g. Furthermore, a realistic model would incorporate M_{i-2}' and M_{i-3}', say, to allow for omnivorous organisms.

The numbers Q_i', W_i, etc will be averages or sums over the organisms in T_i. It might be advisable to partition each T_i according to its thermodynamic properties. Then measurements of efficiency of mass flow from T_i to T_{i+1} will be a "thermodynamic efficiency," in contrast to, or supplementing, ecological efficiency, population efficiency, and individual growth efficiency (112). Numerical estimates of an efficiency for particular communities are not difficult to locate (e.g. 86, 96, 113).

An investigation of efficiency in the trophic structure would be eased if "ecological free energy functions" could be defined in analogy to the functions in Equations 8 and 9. Free energy functions express the amount of energy available (free) to do work. However, even the assumption of steady-state conditions on the T_i would not, strictly, permit the use of Gibbs or Helmholtz functions. Slobodkin (112) discusses the possibility of their use in ecology and Patten (94) evaluates free energy functions in a discussion of "community efficiency." Scott (132) critically examines the use of chemical methods as applied to energy transfer in a food chain.

The point is that the trophic level perception of community is suitable for an open system thermodynamics formulation. This formulation is consistent with the thermodynamics of passive and active interactions of an organism with the physical environment.

Even the kinetics formalism, the microscopic version of thermodynamics, is beginning to find application in trophic dynamics, primarily because kinetic energy is exchanged by collision processes, which may be construed as the physical analogue of some ecosystem interactions (119, 122). In terms of the thermodynamics, a major shortcoming of such studies is that a closed system or similar restriction is necessary; in terms of the ecology, a major shortcoming is that a simple transplant of Newtonian dynamics may do a gross injustice to any organism-organism interaction. These shortcomings may not be as severe when a combined open system-kinetic formulation is used (28).

AVANT-GARDE APPLICATIONS

Energy Flow and Biological Order

The idea that the flow of energy in biological systems has some relationship to the orderliness of these systems has been discussed in the thermodynamics literature for some time (9, 109). Morowitz (81) has popularized that idea. The objective of this section is to make the idea accessible to ecologists by fitting it into the overall theme of this review.

The argument is based on the nonequilibrium theory of open systems. The earth is the open system. The sun is a source of solar energy; the surface of the earth receives this energy and reradiates it, either by reflection or as infrared radiation, to outer space, an energy sink. The energy sink is also necessary because if the same amount of energy were not removed from the earth as is flowing to the earth, the planet would show a continually increasing temperature. That is, the surface of the earth has a relatively constant value for its total energy and this is due to the balance between energy flows in and out. It is the flow process that this section focuses on.

There is the hypothesis that the flow of energy is responsible for some fundamental cyclic phenomena. Morowitz (81) mentions three cycles of interest: geological, meteorological (the water cycle), and ecological. For example, an idealized water cycle driven by solar energy consists of sunlight acting on terrestrial water, which causes evaporation and vaporization, i.e. the rise of water molecules to high altitudes. The molecules condense as clouds, the condensation giving off heat; the molecules fall back to earth as rain, again giving off heat as the potential energy of the drops is changed to kinetic energy.

The ecological cycle, associated with trophic dynamics, is the one proposed by Lindeman (71). Solar energy operates on the primary producers in the primary trophic level. Primary consumers and higher levels of consumers operate off this energy. Each transition to the next level involves the transfer of chemical and thermal energy. Ultimately, the decomposers in the trophic structure make the mass products at every level reaccessible to the primary producers. The thermodynamic aspects of the energy transfer have been discussed. It may not be unreasonable to assume that the flow of solar energy is ultimately responsible for the order, or the organization, of trophic levels, and for the resulting community structure, population dynamics, biological diversity and stability.

The total entropy change of the system (earth) in a time dt is dS which, for an open system, is

$$dS = d_iS + d_eS \qquad\qquad 18.$$

where d_iS is the entropy change corresponding to the irreversible processes of life on earth (diffusion, chemical reactions, etc) or in the system. Therefore, from the second law

$$d_iS > 0 \qquad\qquad 19.$$

It is common to associate ordering processes with decreasing entropy, or equivalently, with processes for which $dS < 0$. The processes in the biosphere can be viewed either as satisfying $(S \neq 0)$, $dS < 0$, or $dS = 0$, depending on the time scale. On an evolutionary scale biological systems appear to tend toward increasing structural complexity with a concomitant increase in orderliness, which is associated with $dS < 0$. On a lifetime scale, a steady-state relation $dS = 0$ may be approximated. That the processes contributing to increased orderliness or synthesis approximately balance those corresponding to degradation would seem acceptable. In brief, the interest is in

$$dS \leq 0 \qquad\qquad 20.$$

From Equations 18, 19, and 20, it is easy to show that

$$d_eS \leq -d_iS < 0 \qquad\qquad 21.$$

Although the irreversible energy processes operating strictly inside the system would not seem to support increasing orderliness, the existence of an energy exchange with the exterior, a corresponding negative d_eS, and the second law of thermodynamics of open systems, suggest that increasing order will follow inside the system.

The relation between plant growth and entropy has been studied by Gorski (44). Animal growth and the theory of irreversible thermodynamics have been studied by Zotin & Zotina (130) and Zotina & Zotin (131).

It would be of interest to many who have heard the word negentropy (109) used in connection with thermodynamics and biology to note that the word refers to the decreasing entropy flow inferred by Equation 21. Without this flow of "negative entropy," the result would be a $dS > 0$, and by hypothesis contradict the "living condition."

The arguments above focus on the flow of energy and the second law of thermodynamics. What is not indicated is how the state of low entropy and high order is maintained. Nonequilibrium thermodynamics is usually concerned with steady states near the equilibrium point of the system: the point of maximum disorder. There is evidence to indicate that states of the system far from the system equilibrium point may actually "create" order within the system (41, 100, 101). It suffices to say that the "created" order is dependent upon the system being far from equilibrium and open to a continuous flow of matter and energy, in and out.

Community Structure, Diversity, and Entropy

Community structure in ecology is the study of the interactions of certain species within a trophic hierarchy. Often the study involves only a part of what would be done in a complete trophic analysis. Thus, subject to certain assumptions such as limiting factors, it may be a good approximation to exclude the primary producers and decomposers in a study of the role of competition and/or predation in a specific community (18, 90). The ecology literature of the last eight years or so is replete with examples.

The paper by Paine (90), for one, is concerned with a part of the overall food web of several similar invertebrate communities existing on certain rocky intertidal beaches. He analyzed three subwebs of 45, 11, and 8 species. The community structure (or food web) in each case was well defined and orderly. The interconnections between species or links in the web were carefully analyzed to show both the direction of flow of material (who eats whom) and the relative amounts of flow. Furthermore, based on interpretation of the above structure, Paine has drawn conclusions about the stability of the structure with regard to various perturbations. In the process he qualitatively discussed the diversity of the three communities: "It is suggested that local animal species diversity is related to the number of predators in the system and their efficiency in preventing single species from monopolizing some important, limiting requisite."

In this section I would like to trace the attempts to define indices of diversity by adopting a common one and showing its relation to the entropy function of thermodynamics.

In a theoretical analysis MacArthur (73) concluded that ". . . a given stability can be achieved either by a large number of species each with a fairly restricted diet, or by a smaller number of species each eating a wide variety of other species. . . . The maximum stability possible for m species would arise when there are m trophic levels with one species on each, eating all species below . . ." This ecological generalization may be partially dependent upon the definition of stability that is used (50).

MacArthur is suggesting that, subject to specific criteria about location in the food web (community structure) and the efficiency or rate of transfer of biomass, increased stability results from increased diversity. For the purposes of this discussion, adopt the intuitive notion that stability is a property of the community structure which expresses its resistance to species extinction when perturbed from without. For the moment consider diversity simply as the number of species. It is possible to specify elaborate theoretical conditions (17, 67) around stability but there is often a gap between what can be measured and the conditions.

Consider a highly structured food web with a near maximum number of possible interconnections and a relatively large number of different species (not simply a large number of organisms). Current ecological dogma says that such a food web is quite stable. That is, it does not naturally transform to a different structure unless subjected to an extreme environmental perturbation. Since the food web is a subset of biological elements drawn from the trophic structure, it is reasonable to infer that it is dependent upon the flow of negative entropy in the community (95). In brief, the stability depends upon negative entropy flow, on the number of different species (the diversity), and on their interconnections. Thus, from an essentially biological argument, negative entropy and diversity are related. All of these dependencies or relations are of positive correlation.

The quantification of stability as a property of an arbitrary community enables one to compare the same community at different times, and communities of different sizes and species. The question of course, is what measure to use. Because of the positive correlation of stability and community diversity, the latter seems like a logical measure. However, the definition as simply the number of species is obviously inadequate. One alternative definition, enthusiastically adopted by many, is to exploit the formalism of information theory. In this theory a measure of "information" is developed which when applied in diversity studies is called the Shannon index of diversity.

The standard sources for the fundamentals of information theory are Khinchin (61), Shannon & Weaver (110), and Kullback (63). The latter emphasizes the statistical aspects. The original application of the theory was in the field of theoretical radio communication where it provided an exceedingly powerful means to quantify the content of a signal.

The mathematical model of information theory (at least the one easiest to translate to a diversity index and to statistical mechanics) is formulated by considering a signal composed of an infinite number of discrete symbols of r different kinds, ξ_1, $\xi_2, \ldots, \xi_r$. Associated with j^{th} type of symbol is a probability of its occurring, p_j, $j = 1, 2, \ldots, r$. Then the information content per symbol in an infinite string of symbols is

$$H' = -\Sigma_{j=1}^{r}\ p_j \cdot \log p_j \qquad\qquad 22.$$

where the log function is usually to the base 2. Although the p_j are probabilities, for practical purposes they are often approximated by the relative frequency N_j/N, where N_j is the number of symbols of the jth type and N is the total number of symbols in a finite signal. Then H is the information content per symbol in a signal of size N

$$H = -\sum_{j=1}^{r} (N_j/N)\log(N_j/N) \qquad 23.$$

It will be shown that Equation 23 is also the expression for the information content per organism, or one possible index of diversity.

MacArthur (73) and Margalef (76) suggested using the information theory formalism to calculate an index of species diversity for an ecological community. The initial suggestion to define a diversity index probably originated with Fisher et al (25). In the intervening years as community ecology adopted a more quantitative posture, a large number of studies using a variety of indices have been reported. A large portion of this work has used H, the information content in Equation 23, as the index.

To be more explicit about what (some) ecologists have meant by diversity (97), consider a highly diverse community as one having two properties: many species and a smooth histogram. Then a community with a lower diversity has fewer species with an uneven histogram. The role of the form of the histogram will become clear as the model upon which Equation 23 is based is given an ecological interpretation. Let the $\xi_1, \xi_2, \ldots, \xi_r$ be r species in the community. Let N_j be the number of organisms in the jth species ($j = 1, 2, \ldots, r$) and $N = \sum_{j=1}^{r} N_j$. Then (97) interpret N_j/N as the "probability" that an arbitrarily selected organism will be a member of the jth species. Thus, H in Equation 23 is the information content per organism, or the index of diversity. The histogram is simply a plot of N_j/N versus j. It is easy to verify that the value of H corresponding to a large r value and smooth histogram will be high compared to the H for a low r value and an uneven histogram.

The Shannon or information index, as well as other diversity indices, are discussed by Pielou (97) in some detail. Hurlbert (51) labels diversity a "nonconcept." Further discussions are found in the Brookhaven Symposium (10) and Hutcheson et al (52).

Equation 22 is formally similar to an equation for the entropy associated with an ensemble or a collection of energy states in the theory of statistical mechanics (82).

If S is entropy, k is the Boltzmann constant, and f_j is the probability that some physical isolated system is in some state j

$$S = -k \sum / f_j \cdot \ln f_j \qquad 24.$$

The above form of S is calculated from the statistical mechanics formalism. It is one of the several functions mathematically relating statistical mechanics and thermodynamics (the Helmholtz free energy, A, is another).

For ease of comparison, rewrite Equation 22

$$H = -0.693 \sum_{j=1}^{r} p_j \cdot \ln p_j \qquad 25.$$

There are three aspects of Equations 24 and 25 to discuss. First, despite the formal similarity they have "opposite" interpretations. When only one type of signal is available ($r = 1$), $N_j = N_1 = N$ so $H = 0$. With one type of symbol I can convey "minimum (zero) information." When the entropy, S, is zero the physical system is said to be in the most well-determined energy state possible. In a sense I have "maximum knowledge" of the energy state (in contrast, at equilibrium S is a maximum and I have minimum knowledge of the energy state of my system).

Second, the association of diversity with information, H, is not based on first principles; it is simply an analogy. However, the acceptance of the possible usefulness of the analogy can have disastrous effects if statements about diversity and entropy are not made carefully.

Third, the connection between diversity and stability is dogma. There may or may not be first principles involved.

I believe that with these points in mind it is possible to quickly understand the relatively few articles that have attempted to use entropy, diversity, thermodynamics, and stability concepts as part of one neat formalism.

Odum & Pinkerton (88) seem to have stimulated the interest of ecologists in the possible implications of entropy flow, efficiency, and open systems for ecology. Other papers (76, 87, 94, 95, 112) using combinations of the "avant-garde" ideas just presented followed, all seeking some essential elusive lesson for ecology. Odum and Margalef seem to have felt that the lesson lies in the information and diversity areas. Patten and Slobodkin appear to have concentrated on the possible implications for the trophic level analyses in terms of the thermodynamics and efficiencies of transport between levels. These papers are not uniformly informative because in some the physical theory is not correctly applied, the biological interpretations are strained, or both.

The most frequently heard criticism of the use of the Shannon index is that the more abundant species make an exaggerated contribution. That is, it is insensitive to the presence of rare species. This is undoubtedly true, although Hurlbert (51) feels that it is an irrelevant criticism.

More fundamental, however, is the criticism that the index lacks a measure of specific biological activity. The discussions of diversity in the context of the community food web indicated that the diversity should be dependent upon the specific role that species fill in the community as well as upon how many species are in the community and relative number of organisms in each species. The papers by Paine (90) and MacArthur (73) would seem to support the need to specify a species place in the food web and how efficiently the species functions in that position before diversity has any meaning beyond being a special kind of relative frequency of occurrence term.

Shortly after Shannon's celebrated paper Quastler (102) edited a book on applications of information theory in biology. Since then much work has been done (e.g. 3, 40, 45, 53, 80), mostly at the molecular level. The theory developed about Shannon's work has not as yet fulfilled its expectations.

Statistical Mechanics and Ecology

This review has concentrated on the use of the thermodynamics formalism in ecology as opposed to the statistical mechanics or kinetic theory formalism.

Consider the dynamics of the populations in a community in which many interactions are involved: interspecies competition, interspecies (and often some intraspecies) predation, and intraspecies competition and cooperation (social). In the frame of reference of ecological phenomena, it is the organism-organism interactions which are the "microscopic" events to be described using a statistical mechanics formalism. This type of argument is the basis of the relationships drawn between diversity, statistical mechanisms, and entropy in the earlier discussion of community ecology.

There are a number of points to make about a statistical mechanics of interacting organisms. Points on the favorable side are that both the genuine microscopic and the organism level "microscopic" interactions have certain aspects of their dynamical behavior which are partially predictable but in large measure are masked by their microscopic natures. For example, the Newtonian dynamics of colliding atoms and molecules would be deterministically predictable if the needed initial conditions for the differential equations were known. In a similar sense trophic-type interactions between organisms in a community are often discussed in terms of Volterra-type differential equations. There are similarities between the inability simultaneously to observe a large enough microscopic sample to make predictions about macroscopic (thermodynamic) variables such as heat, work, entropy, and the difficulties in simultaneously observing a large enough sample of field organisms and measuring (and defining) the appropriate variables to predict community events. In both cases, out of seemingly chaotic interactions order seems to result.

The unfavorable aspects of using a statistical mechanics model of community dynamics focus conceptually about the point that the macroscopic events corresponding to molecular dynamics are well-defined variables, usually "qualitatively different" from the variables in the molecular formalism. However, what "qualitatively different" macroscopic ecological variables are observed? Other than diversity, none seems to have been found. This does not mean that none exists or that a search to identify such variables would not be fruitful. The second unfavorable aspect centers upon the shortcomings associated with population dynamics in general, and especially the Lotka-Volterra type models, for it is from such differential equations models that the statistical mechanics model is constructed.

I refer to the set of dynamical population equations, for two species

$$\dot{N}_1 = \alpha_{11}N_1 - \alpha_{12}N_1N_2 \qquad\qquad 26.$$

$$\dot{N}_2 = \alpha_{22}N_2 - \alpha_{21}N_1N_2$$

as Lotka-Volterra equations. It is standard to associate $N_i = N_i(t)$ with the population size or biomass of species i, the dot meaning a derivative with respect to time, t; the α_{ii} is the coefficient of population growth or decay in the absence of the interactions between species. The species interactions are given by the N_1N_2 terms. Of course, larger sets of differential equations than $n = 2$ can be written. The initial expressions of population change in the form of Equation 26 [Lotka 72, Volterra 123 (originally published in 1924)] have not been significantly improved upon despite their popularity in the current literature (68, 74, 97, 114). The later sources amply document the shortcomings and advantages of Equation 26 as seen in contemporary ecological theory.

It has become clear (56–59) that the system in Equation 26 can serve a related role to that of Hamilton's equations (Equation 10) in the classical development of statistical mechanics. Kerner's four papers have been collected into a monograph (60) which includes an extensive introduction.

Hamilton's equations may be written in a form similar to equation 10

$$\dot{\mathbf{q}} = \frac{\partial H(\mathbf{q},\mathbf{p})}{\partial \mathbf{p}} \qquad\qquad 27.$$

$$\dot{\mathbf{p}} = -\frac{\partial H(\mathbf{q},\mathbf{p})}{\partial \mathbf{q}}$$

where $\mathbf{q} = (q_1, q_2, q_3, \ldots, q_{3n})$ and $\mathbf{p} = (p_1, p_2, \ldots, p_{3n})$. When each of n elements in the system has 3 degrees of freedom the elements of these two vectors are the coordinates of a $6n$-dimensional phase space, Γ. In order to solve these equations of motion for the particles it is necessary to know the initial conditions of the system, $\mathbf{q}_0$ and $\mathbf{p}_0$. These generally cannot be known for kinetics problems.

There is an outstanding characteristic of Equation 27 for conservative systems: $dH/dt = 0$ or

$$H(\mathbf{q},\mathbf{p}) = E \qquad\qquad 28.$$

where E is the total energy of the system and is obviously a constant. That is, H is a constant of the motion, a conserved quantity. The existence of a second conservation law, which is geometrical, is postulated. It guarantees that arbitrary volumes in Γ space are conserved, i.e. Liouville's theorem is satisfied. Some, but not all, of these points are discussed in the physical theory section.

The system of population differential equations in Equation 26 is similar to the system in Equation 27 in that they are both systems of autonomous, first order differential equations and both lead to a conservation law for a constant of the motion. In Equation 26 the constant of motion is G (60)

$$\tau_1 (e^{v_1} - v_1) + \tau_2 (e^{v_2} - v_2) = G \qquad\qquad 29.$$

where $v_i = \log N_i/N_{i0}$, N_{i0} is a steady-state population value (corresponding to $\dot{N}_i = 0$) for the ith species, and $\tau_i = \beta_i N_{i0}$, β_i being related to coefficients α_{ij}.

Furthermore, Equations 28 and 29 both describe closed curves in their respective Γ space. Equation 28 has the elements of $\mathbf{q}$ and $\mathbf{p}$ as axes and the Γ space for Equation 29 has v_i, $i = 1,2$ as axes.

The point to be made is that with a few minor transformations a profound similarity can be seen between the mathematical model for interacting gas atoms and the mathematical model for interacting animal species. Both models admit conservation laws, and Liouville's theorem holds in the Γ spaces constructed for each model. Strictly from the mathematics viewpoint, a statistical mechanics of population dynamics is specified.

From the ecological viewpoint, however, it is not clear what these efforts have gained. Certainly there are some pleasing conceptual similarities between the statistical mechanics model and ecological circumstances: both deal with large numbers of interacting bodies, for which initial conditions simply cannot be specified, and the introduction of random variation is desirable and realistic. The shortcomings of a statistical mechanics model of population dynamics are clearly found to reside in the shortcomings of Lotka-Volterra type models. Major contributions to ecological theory are not visible, which is not to say they are not over the horizon. A claim to having supplied deeper insights into the dynamical theory of population change can be justified.

Besides the Kerner (60) reference there are other sources of information about statistical mechanics and community interactions. Goel, Maitra & Montroll (42) have made a major mathematical contribution, other work by Leigh (68, 69) has a more ecological flavor to it with many references to the underlying ideas of community structure, and Rosen (104) has a well-written exposition of the mathematical approach.

SUMMARY

The earth's position as a receiver, reflector, and degrader of energy dictates, a priori, that assemblages of objects utilizing this energy assume an organizational form. The organization is called a community structure, a concept closely allied to trophic structure. The movement of energy (from without and within the community) and/or mass, vertically and horizontally, through the structure is a necessary prerequisite for the structure to maintain itself. Energy from without the system, such as solar energy, is utilized for primary productivity and for maintenance of the relatively narrow temperature range in which the biosphere functions. Energy from within the system is energy resulting from the metabolism of the mass of organisms of one trophic level by organisms of another level.

The concept of a community follows from an appropriate collection of trophic levels. Stability, or the tendency for the community to sustain physical and biological stress, is related to the community diversity. Diversity is expressed in a number of ways, a popular one being similar to the thermodynamic function, entropy.

The dynamics of the population sizes of species that forage are seen to be a special case of the mathematical formalism of statistical mechanics, a field conceptually related to thermodynamics.

Organisms that forage, as well as organisms which compete for mineral nutrient, sunlight, etc, all seem to be subject to the energy transfers described. Community structures, rates, and mechanisms of transfer will differ, but the underlying principles all seem to be subject to the restrictions of thermodynamics.

Ultimately, the reader must judge for himself if the apparent across-the-board correspondence between ecology and the open system thermodynamics formalism is coincidence or a demonstration of first principles.

ACKNOWLEDGMENTS

Drafts of this review were read by Betty Gallucci, Harvey Gold, William Hatheway, Benjamin Jayne, Harold Morowitz, Gordon Orians, Robert Paine, and H. R. van der Vaart, each of whom made comments and constructive suggestions, some of which I have incorporated. The review is better for their efforts. Responsibility for all shortcomings remains with me.

This work was supported in part by the National Institutes of General Medical Sciences Grant in Biometry, GM-01269-09.

Literature Cited

1. Agnew, A. D. Q., Flux, J. E. C. 1970. Plant dispersal by hares (*Lepus capensis L.*) in Kenya. *Ecology* 51:735–7
2. Allen, L. H., Jensen, S. E., Lemon, E. R. 1971. Plant response to carbon dioxide enrichment under field conditions: A stimulation. *Science* 173:256–8
3. Atlan, H. 1968. Application of information theory to the study of the stimulating effects of ionizing radiation, thermal energy, and other environmental factors. *J. Theor. Biol.* 21:45–70
4. Barnett, S. A., Mount, L. E. 1967. Resistance to cold in mammals. In *Thermobiology*, ed. A. H. Rose, 411–77. New York: Academic. 653 pp.
5. Bartlett, P. N., Gates, D. M. 1967. The energy budget of a lizard on a tree trunk. *Ecology* 48:315–22
6. Bazzaz, F. A., Boyer, J. S. 1972. A compensating method for measuring carbon dioxide exchange, transpiration, and diffuse resistances of plants under controlled environmental conditions. *Ecology* 53:343–9
7. Billings, W. D., Morris, R. J. 1951. Reflection of visible and infrared radiation from leaves of different ecological groups. *Am. J. Bot.* 38:327–31
8. Brett, J. R. 1971. Energetic responses of salmon to temperature. A study of some thermal relations in the physiology and freshwater ecology of sockeye salmon (*Oncorhynchus nerka*). *Am. Zool.* 11:99–113
9. Bridgeman, P. W. 1941. *The Nature of Thermodynamics.* Cambridge, Mass.: Harvard Univ. Press. 239 pp.
10. Brookhaven Symposium. 1969. *Diversity and Stability in Ecological Systems.* Upton, New York: Brookhaven Symp. Biol. Vol. 22
11. Brock, T. D. 1970. High temperature systems. *Ann. Rev. Ecol. Syst.* 1:191–220
12. Brostow, W. 1972. Between laws of thermodynamics and coding of information. *Science* 178:123–6
13. Brown, J. H., Lasiewski, R. C. 1972. Metabolism of weasels: The cost of being long and thin. *Ecology* 53:939–43
14. Caldwell, M. M. 1968. Solar radiation as an ecological factor for alpine plants. *Ecol. Monogr.* 38:243–68
15. Charnov, E. 1973. *Optimal foraging: Some theoretical explorations.* PhD thesis. Univ. Washington, Seattle
16. Ciureş, A., Mărgineanu, D. 1970. Thermodynamics in biology: An intruder? *J. Theor. Biol.* 28:147–50
17. Conrad, M. 1972. Stability of foodwebs and its relation to species diversity. *J. Theor. Biol.* 34:325–35
18. Dayton, P. K. 1971. Competition, disturbance and community organization: The provision and subsequent utilization of space in a rocky intertidal community. *Ecol. Monogr.* 41:351–89
19. Denbigh, K. G. 1951. *The Thermodynamics of the Steady State.* New York: Wiley. 103 pp.
20. Dixon, J. R., Emery, A. H. 1965. Semantics, operationalism, and the molecular-statistical model in thermodynamics. *Am. Sci.* 53:428–35
21. Ehrler, W. L., Van Bavel, C. H. 1968. Leaf diffusion resistance, illuminance, and transpiration. *Plant Physiol.* 43:208–14
22. Elton, C. S. 1927. *Animal Ecology.* London: Sidgwick and Jackson. 209 pp.
23. Emlen, J. M. 1966. The role of time and energy in food preference. *Am. Natur.* 100:611–7
24. Feeny, P. 1970. Season changes in oak leaf tannins and nutrients as a cause of spring feedings by winter moth caterpillars. *Ecology* 51:565–80
25. Fisher, R. A., Corbet, A. S., Williams, C. B. 1943. The relation between the number of species and the number of individuals in a random sample of an animal population. *J. Anim. Ecol.* 12:42–58
26. Fox, R. F. 1971. Entropy reduction in open systems. *J. Theor. Biol.* 31:43–6
27. Fry, F. E. 1967. Responses of vertebrate poikilotherms to temperature. See Ref. 4, 375–409
28. Gallucci, V. F. A thermodynamic view of exploited populations. Unpublished

29. Gates, D. M. 1962. *Energy Exchange in the Biosphere.* New York: Harper and Row. 151 pp.
30. Gates, D. M. 1964. Leaf temperature and transpiration. *Agron. J.* 56:273-7
31. Gates, D. M. 1967. Energy exchange between organism and environment. *Biometeorology. Proc. Ann. Biol. Colloq., 28th,* ed. W. P. Lowry, 1-22. Corvallis, Oregon: Oregon State Univ. Press. 171 pp.
32. Gates, D. M. 1967. Water balance in terrestrial ecosystems. In *Transport Phenomena in Atmospheric and Ecological Systems,* 21-36. New York: ASME. 87 pp.
33. Gates, D. M. 1968. Energy exchange between organisms and environment. *Aust. J. Sci.* 31:67-74
34. Gates, D. M. 1968. Toward understanding ecosystems. *Advan. Ecol. Res.* 5:1-35
35. Gates, D. M. 1968. Transpiration and leaf temperature. *Ann. Rev. Plant Physiol.* 19:211-38
36. Gates, D. M. 1969. Climate and stability. *Brookhaven Symp. Biol.* 22:115-27
37. Gates, D. M., Benedict, C. M. 1963. Convection phenomena from plants in still air. *Am. J. Bot.* 50:563-73
38. Gates, D. M., Keegan, H. J., Schleter, J. C., Weidner, V. R. 1965. Spectral properties of plants. *Appl. Opt.* 4:11-20
39. Gates, D. M., Tantraporn, W. 1952. The reflectivity of deciduous trees and herbaceous plants in the infrared to 25 μ. *Science* 115:613-16
40. Gatlin, L. L. 1966. The information content of DNA. *J. Theor. Biol.* 10: 281-300
41. Glansdorff, P., Prigogine, I. 1971. *Thermodynamic Theory of Structure, Stability and Fluctuations.* New York: Wiley-Interscience. 306 pp.
42. Goel, N. S., Maitra, S. C., Montroll, E. W. 1971. *On the Volterra and other Nonlinear Models of Interacting Populations.* New York: Academic. 145 pp.
43. Golley, F. B. 1960. Energy dynamics of a food chain of an old-field community. *Ecol. Monogr.* 30:187-207
44. Gorski, F. 1966. *Plant Growth and Entropy Production.* Krakow, Poland: Zakland Fizjologii Roslin P. A. N. 102 pp. (In English)
45. Griffith, J. S. 1965. Information theory and memory. In *Molecular Biophysics,* ed. B. Pullman, M. Weissbluth, 411-35. New York: Academic. 452 pp.
46. Hadley, N. F. 1970. Micrometeorology and the energy exchange in two desert arthropods. *Ecology* 51:434-44
47. Hazen, W. E. 1970. *Readings in Population and Community Ecology.* Philadelphia: Saunders. 2nd ed. 421 pp.
48. Heinrich, B., Raven, P. H. 1972. Energetics and pollination ecology. *Science* 176:597-602
49. Hubbell, S. P. 1969. *A systems analysis of the ecological bioenergetics of a terrestrial isopod.* PhD thesis. Univ. California, Berkeley. 126 pp.
50. Hurd, L. E., Mellinger, M. V., Wolf, L. L., McNaughton, S. J. 1971. Stability and diversity at three trophic levels in terrestrial successional ecosystems. *Science* 173:1134-6
51. Hurlbert, S. H. 1971. The nonconcept of species diversity: A critique and alternative parameters. *Ecology* 52: 577-85
52. Hutcheson, K., Shenton, L. R., Bailey, R. C. 1972. *Diversity: What is it?* Presented at Advan. Inst. Stat. Ecol., Pennsylvania State University, University Park, Pennsylvania
53. Johnson, H. A. 1970. Information theory in biology after 18 years. *Science* 168:1545-50
54. Katz, P. L. 1972. *Optimal Control Theory Applied to a Predator with Multiple Prey Species.* Presented at Hawaii Int. Conf. Systems Sci., 6th, Honolulu, Hawaii, 1973
55. Kavanau, J. L., Rischer, C. E. 1972. Influences of ambient temperature on ground squirrel activity. *Ecology* 53: 158-64
56. Kerner, E. H. 1957. A statistical mechanics of interacting biological species. *Bull. Math. Biophys.* 19:121-46
57. Kerner, E. H. 1959. Further considerations on the statistical mechanics of biological associations. *Bull. Math. Biophys.* 21:217-55
58. Kerner, E. H. 1961. On the Volterra-Lotka principle. *Bull. Math. Biophys.* 23:141-57
59. Kerner, E. H. 1964. Dynamical aspects of kinetics. *Bull. Math. Biophys.* 26: 333-49
60. Kerner, E. H. 1972. *Gibbs Ensemble: Biological Ensemble. The Application of Statistical Mechanics to Ecological, Neural and Biological Networks.* New York: Gordon and Breach. 167 pp.
61. Khinchin, A. I. 1957. *Mathematical Foundations of Information Theory.* Transl. R. Silverman, M. Friedman. New York: Dover. 120 pp.
62. Kleiber, M. 1961. *The Fire of Life: An Introduction to Animal Energetics.* New York: Wiley. 454 pp.

63. Kullback, S. 1968. *Information Theory and Statistics*. New York: Dover. 399 pp.

64. Lee, D. H. 1967. Principles of homeothermic adaptation. *Biometeorology. Proc. Ann. Biol. Colloq., 28th*, ed. W. P. Lowry, 113–30. Corvallis, Oregon: Oregon State Univ. Press. 171 pp.

65. Lee, R., Gates, D. M. 1964. Diffusion resistance in leaves as related to their stomatal anatomy and micro-structure. *Am. J. Bot.* 51:963–75

66. Lehninger, A. L. 1971. *Bioenergetics*. Menlo Park, California: Benjamin. 2nd ed. 245 pp.

67. Leigh, E. G. 1965. On the relation between the productivity, biomass, diversity of a community. *Proc. Nat. Acad. Sci. USA* 53:777–83

68. Leigh, E. G. 1968. The ecological role of Volterra's equations. In *Some Mathematical Problems in Biology,* ed. Am. Math. Soc., 1–61. Providence, Rhode Island: Am. Math. Soc. 117 pp.

69. Leigh, E. G. 1971. *Adaptation and Diversity*. San Francisco: Freeman. 288 pp.

70. Lemon, E., Stewart, D. W., Shawcroft, R. W. 1971. The sun's work in a cornfield. *Science* 174:371–8

71. Lindeman, R. L. 1942. The trophic-dynamic aspect of ecology. *Ecology* 23:399–418

72. Lotka, A. J. 1956. *Elements of Mathematical Biology*. New York: Dover. 465 pp.

73. MacArthur, R. 1955. Fluctuations of animal populations and a measure of community stability. *Ecology* 36:533–6

74. MacArthur, R. H., Connell, J. H. 1966. *The Biology of Populations*. New York: Wiley. 200 pp.

75. MacArthur, R. H., Pianka, E. R. 1966. On optimal use of a patchy environment. *Am. Natur.* 100:603–9

76. Margalef, D. R. 1958. Information theory in ecology. *General Systems* 3: 36–71

77. Margalef, R. 1963. Succession in marine populations. *Advan. Front. Plant Sci.* 2:137–88

78. McClare, C. W. F. 1971. Chemical machines, Maxwell's demon and living organisms. *J. Theor. Biol.* 30:1–34

79. Moen, A. N. 1968. The critical thermal environment: A new look at an old concept. *BioScience* 18:1041–3

80. Morowitz, H. J. 1955. Some order-disorder considerations in living systems. *Bull. Math. Biophys.* 17:81–6

81. Morowitz, H. J. 1968. *Energy Flow in Biology: Biological Organization as a Problem in Thermal Physics*. New York: Academic. 179 pp.

82. Morowitz, H. J. 1970. *Entropy for Biologists: An Introduction to Thermodynamics*. New York: Academic. 195 pp.

83. Moss, R. A., Loomis, W. 1952. Absorption spectra of leaves, I. The visible spectrum. *Plant Physiol.* 27:370–91

84. Odum, E. P. 1962. Relationships between structure and function in the ecosystem. *Jap. J. Ecol.* 12:108–18

85. Odum, E. P. 1969. The strategy of ecosystem development. *Science* 164: 262–70

86. Odum, E. P. 1971. *Fundamentals of Ecology*. Philadelphia: Saunders. 3rd ed. 574 pp.

87. Odum, H. T., Cantlon, J. E., Kornicker, L. S. 1960. An organizational hierarchy postulate for the interpretation of species-individual distributions, species entropy, ecosystem evolution, and the meaning of a species-variety index. *Ecology* 41:395–9

88. Odum, H. T., Pinkerton, R. C. 1955. Time's speed regulator: The optimum efficiency for maximum power output in physical and biological systems. *Am. Sci.* 43:331–43

89. Orians, G. H., Horn, H. S. 1969. Overlap in foods and foraging of four species of black birds in the potholes of central Washington. *Ecology* 50:930–8

90. Paine, R. T. 1966. Food web complexity and species diversity. *Am. Natur.* 100: 65–75

91. Paine, R. T. 1971. The measurement and application of the calorie to ecological problems. *Ann. Rev. Ecol. Syst.* 2: 145–64

92. Parkhurst, D. F., Duncan, P. R., Gates, D. M., Kreith, F. 1968. Windtunnel modelling of convection of heat between air and broad leaves of plants. *Agr. Meteorol.* 5:33–47

93. Parks, J. R. 1971. Effect of ambient temperature on the thermochemical efficiency of growth. *Am. J. Physiol.* 220:578–82

94. Patten, B. C. 1959. An introduction to the cybernetics of the ecosystem: The trophic-dynamic aspect. *Ecology* 40: 221–31

95. Patten, B. C. 1961. Negentropy flow in communities of plankton. *Limnol. Oceanogr.* 6:26–30

96. Phillipson, J. 1966. *Ecological Energetics*. London: Arnold. 57 pp.

97. Pielou, E. C. 1969. *An Introduction to Mathematical Ecology*. New York: Wiley-Interscience. 286 pp.

98. Porter, W. P. 1967. Solar radiation through the living body wall of vertebrates with emphasis on desert reptiles. *Ecol. Monogr.* 37:273–96

99. Porter, W. P., Gates, D. M. 1969. Thermodynamic equilibria of animals with environment. *Ecol. Monogr.* 39:245–70

100. Prigogine, I. 1967. *Introduction to Thermodynamics of Irreversible Processes.* New York: Wiley-Interscience. 3rd ed. 147 pp.

101. Prigogine, I., Nicolis, G., Babloyantz, A. 1972. Thermodynamics of evolution. *Phys. Today* 25(11):23–8

102. Quastler, H. 1953. *Essays on the Use of Information Theory in Biology.* Urbana, Illinois: Univ. Illinois Press. 273 pp.

103. Rose, A. H., Ed. 1967. See Ref. 4

104. Rosen, R. 1970. *Dynamical System Theory in Biology.* New York: Wiley-Interscience. 302 pp.

105. Rufelt, H., Jarvis, P. G., Jarvis, M. S. 1963. Some effects of temperature on transpiration. *Physiol. Plant.* 16:177–85

106. Ryther, J. H. 1959. Potential productivity of the sea. *Science* 130:602–8

107. Schmidt-Nielsen, K. 1972. Locomotion: Energy cost of swimming, flying, and running. *Science* 177:222–8

108. Schoener, T. W. 1971. Theory of feeding strategies. *Ann. Rev. Ecol. Syst.* 2:369–404

109. Schrödinger, E. 1969. *What is Life?* and *Mind and Matter.* London: Cambridge Univ. Press. 178 pp.

110. Shannon, C. E., Weaver, W. 1964. *The Mathematical Theory of Communication.* Urbana, Illinois: Univ. Illinois Press. 125 pp.

111. Slatyer, R. O. 1967. *Plant-Water Relationships.* New York: Academic. 366 pp.

112. Slobodkin, L. B. 1960. Ecological energy relationships at the population level. *Am. Natur.* 94:213–36

113. Slobodkin, L. B. 1962. Energy in animal ecology. *Advan. Ecol. Res.* 1:69–101

114. Smith, J. M. 1968. *Mathematical Ideas in Biology.* London: Cambridge Univ. Press. 152 pp.

115. Solzhenitsyn, A. I. 1968. *The First Circle,* 399. Transl. T. P. Whitney. New York: Bantam. 674 pp.

116. Stonehouse, B. 1967. The general biology and thermal balances of penguins. *Advan. Ecol. Res.* 4:131–97

117. Stroganov, N. S. 1962. *Physiological Adaptability of Fish to the Temperature of the Surrounding Medium.* Transl. M. Roublev. Jerusalem: For National Science Foundation by Israel Program for Scientific Translations. 108 pp.

118. Tibbals, E. C., Carr, E. K., Gates, D. M., Kreith, F. 1964. Radiation and convection in conifers. *Am. J. Bot.* 51:529–38

119. Ulanowicz, R. E. 1972. Mass and energy flow in closed ecosystems. *J. Theor. Biol.* 34:239–53

120. Van Bavel, C. H., Nakayama, F. S., Ehrler, W. L. 1965. Measuring transpiration resistance of leaves. *Plant Physiol.* 40:535–40

121. Verduin, J. 1972. Caloric content and available energy in plant matter. *Ecology* 53:982

122. Verhoff, F. H., Smith, F. E. 1971. Theoretical analysis of a conserved nutrient ecosystem. *J. Theor. Biol.* 33:131–47

123. Volterra, V. 1931. *Lecons sur la theorie mathematique de lutte pour la vie.* Paris: Gauthier-Villars. 214 pp.

124. Waddington, C. H., Ed. 1969. *Towards a Theoretical Biology.* Int. Union Bio. Sci. Symp. Chicago: Aldine. 351 pp.

125. White, F. N., Lasiewski, R. C. 1971. Rattlesnake denning: Theoretical considerations on winter temperatures. *J. Theor. Biol.* 30:553–7

126. Winberg, G. G., Ed. 1971. *Methods for the Estimation of Production of Aquatic Animals.* Transl. A. Duncan. New York: Academic. 175 pp.

127. Wolf, L. L., Hainsworth, F. R., Stiles, F. G. 1972. Energetics of foraging: Rate and efficiency of nectar extraction by hummingbirds. *Science* 176:1351–2

128. Zemansky, M. W. 1943. *Heat and Thermodynamics.* New York: McGraw. 2nd ed. 390 pp.

129. Zimmerman, J. L. 1965. Bioenergetics of the Dickcissel, *Spiza americana. Physiol. Zool.* 38:370–89

130. Zotin, A. I., Zotina, R. S. 1967. Thermodynamic aspects of developmental biology. *J. Theor. Biol.* 17:57–75

131. Zotina, R. S., Zotin, A. I. 1972. Towards a phenomenological theory of growth. *J. Theor. Biol.* 35:213–25

132. Scott, D. 1965. The determination and use of thermodynamic data in ecology. *Ecology* 46:673–80

ARCTIC TUNDRA ECOSYSTEMS

♦ 4065

L. C. Bliss[1], G. M. Courtin[2], D. L. Pattie[3], R. R. Riewe[4], D. W. A. Whitfield[1], P. Widden[5]

INTRODUCTION

While the roots of the concept are much older (159), integrated studies of the trophic-dynamic interactions within and between components of an ecosystem have been conducted in detail only recently. This is true for the tundra as well as for other biomes. Much of this research was initiated within the International Biological Program (IBP). While we have drawn upon the data presented at the Kevo, Finland (99) and Leningrad, USSR (266) meetings and from other available reports, a more detailed review of tundra ecosystems must await completion of the IBP studies.

Interest in arctic ecosystems has greatly increased in the past five years with the discovery of oil, gas, and minerals in the North American Arctic. Interest in arctic and subarctic regions of the USSR has existed for a longer period of time, stemming from problems of aforestation, limited crop production, the sustained yield of reindeer, fur bearers, and waterfowl, and basic ecologic and taxonomic surveys (246).

There is value in understanding the world's ecosystems, but of greater importance is their long-range management. Within terrestrial arctic systems there is a need to manage vegetation, both as a thermal barrier to permafrost melt and as animal habitat, as well as management of wildlife populations. Only when portions of a system are stressed, and through field and laboratory studies and computer modeling, will we learn the limits of human and natural perturbations. To date most of the literature deals with information on ecosystem structure and function with only limited data on perturbation.

The objective of this paper is to review our knowledge on how these heat-limited terrestrial systems function. The role and impact of both native and Euro-North American peoples in these northern systems is discussed.

[1]Department of Botany, University of Alberta, Edmonton, Alberta.
[2]Department of Biology, Laurentian University, Sudbury, Ontario.
[3]Biology Section, Northern Alberta Institute of Technology, Edmonton, Alberta.
[4]Department of Zoology, University of Manitoba, Winnipeg.
[5]Department of Biology, Loyola College, Montreal, Quebec.

359

Evolution of Arctic Ecosystems

Arctic ecosystems are young (72, see 109), having developed in the Pleistocene, though they may have first appeared in late Miocene or early Pliocene (68). Tundra floras (273) and faunas (106) probably evolved in the highlands of central Asia and the Rocky Mountains. Within these cold-temperate steppe and taiga environments, species evolved which could occupy lowland arctic tundra regions that developed in the Pleistocene. The fossil record and present day faunistic diversity suggest (106) that the new arctic fauna spread to Europe in mid-Pleistocene and to North America in Wisconsin time. Recent studies from northeastern Siberia (213) and the Seward Peninsula of Alaska (88) indicate that tundra-adapted animals date from early to mid-Pleistocene, an earlier time than previously believed. The assemblage of plants, insects, and mammals at the latter site has been studied in an attempt to understand tundra ecosystem evolution (161). Fossil remains of *Equus* and *Cervus* reveal that the Cape Deceit fauna (pre-Cromerian age) may have lived in a grassy tundra, similar to that near Fairbanks, in the late Pleistocene (87, 160). The rapid evolution of micrograzers (*Dicrostonyx*) and macrograzers (*Mammuthus*) in mid-Pleistocene may have resulted from the development and expansion of steppe-tundra environments in Alaska (88); the paleobotanical record from Eurasia, eastern Siberia, and Alaska indicates (79, 81, 87, 160, 272, 273) that a steppe environment also prevailed in the late Pleistocene. The fossil record of herbivores and carnivores from central Alaska (87) demonstrates an abundance of tundra mammals that does not occur today in polar regions and may indicate that less intercontinental migration via Beringia occurred than the previous literature suggests (88).

While late-Pleistocene tundra systems were rich faunistically, present-day ones are much less so, though diversity is greater in the Alaskan than in the Canadian mainland tundra. According to Kaiser et al (129) the numbers of mammal and bird species in Canada range from 20–21 species in the two (103,684 km^2) quadrats of the northwestern Queen Elizabeth Islands, to 59–75 species in the three quadrats near the Mackenzie River Delta and Yukon. The values for Alaska would be of the latter magnitude. Comparable ranges in floristic diversity occur. The same western Queen Elizabeth Islands have ~50–60 species of flowering plants (180) and the Yukon and Northern Alaskan tundra have about 600 species (190).

Fragility of Arctic Ecosystems

To some, arctic systems appear biologically simple and thus fragile (25) when compared with the complex structure and organismic diversity of temperate or tropical grasslands and forests. To others fragility is equated with stability in terms of time required to return to a steady state following perturbation, or to fluctuations of animal populations (15). Regardless of one's definition of fragility and stability, arctic systems possess features that result in their being considered fragile (15, 25).

Ice-rich permafrost, which melts readily if the vegetation mat, mainly peat (90), is removed or compacted, can result in possible subsidence and thermokarst. Natural erosion along the coast and the shorelines of lakes and rivers is a common geomorphic process (37, 153). Once initiated, especially if water erosion is a part of the perturbation, it is difficult to arrest.

Low temperature and sudden spring and fall climatic oscillations confine much of the biological activity to 1–2 months in the High Arctic (mostly north of 73°) and 3–4 months in the Low Arctic. Biological responses include the flourish of plant and animal growth in summer and oscillating microtine-carnivore populations, at least in the Low Arctic, every 3–5 years (211). A low species diversity adds to potential system fragility, for there are few species that can fill the niche of key species should they be eliminated.

Dunbar (72) states that arctic systems are young, immature, and have not reached an evolutionary climax, in part because of the lack of soil rather than low temperature control. Except for Polar Desert Soils, arctic soils are poorly developed counterparts of soils in temperate regions, and while low temperatures reduce rates of chemical and biological development, these soils form in response to the same factors as elsewhere (232, 234, 237). Arctic ecosystems are youthful in that they have occupied the land for only 2000–10,000 years, but this is also true for much of the boreal forest in Canada, though we recognize that in both biomes the species are much older.

The concept that a high gross production/biomass (P/B) ratio characterizes successional or less stable systems and a low ratio characterizes climax systems (179) applies to forested lands but has little significance in the Arctic. Heath or low shrub dominated communities with low P/B ratios (more climax?) frequent slopes and rocky areas with deeper snow and better soil drainage, yet adjacent upland or lowland sedge or grass dominated communities have higher P/B ratios (more successional?) with no indication of directional change between them. Low rates of production and decomposition add significantly to the fragility of these systems because recovery time following perturbation is so great.

Plant succession, so characteristic of forest and shrublands, is a relatively minor feature of tundras. While some have urged new interpretations of succession and climax (111, 196) in relation to permafrost and soil instability, others (23, 53) have concluded that phasic cycles resulting from frost action can occur in observable directionally changing (successional) communities and in nondirectional (climax) communities.

For much of the Arctic, shifts in the relative dominance of the climax species, rather than a different group of species, characterize succession when it can be found (175). This may reflect the minor role that man has played in soil disturbance until very recently in addition to a small flora with few "weedy species." In temperate regions, the larger flora has had a longer time to evolve weedy species in relation to human activity. These species now play a major role in successional communities that often dominate the landscape. With current and anticipated land modification in the Arctic, future generations of ecologists may come to place a higher significance on succession there as well. Weedy species (271) are recognized as an ecological grouping in the arctic and subarctic regions of the USSR (69), though they generally occupy only disturbed soils.

THE ABIOTIC COMPONENT OF ARCTIC ECOSYSTEMS

Macroclimate

There has always been a paucity of climatological data in arctic regions owing to the difficulty of establishing and maintaining weather stations. Meteorological data have been gathered regularly from the polar regions of Canada for only the past 50 years and several of the stations forming the arctic network were not established until the late 1940s (195). Radiosonde data collected from many stations in the network have facilitated description and quantification of the characteristics of the arctic troposphere (70, 93–95).

In studying the interaction of organisms with their environment one tends to consider only the microclimate. Climate near the ground, however, is largely a function of energy exchange phenomena at the ground-air interface. Temperature variations at this interface can, in turn, affect the pressure distribution of the air aloft because of localized heating and cooling.

The long polar night and the equally long polar day serve to give a very steep temperature gradient from equator to pole during the winter and a much shallower gradient during summer. This semiannual fluctuation serves to strengthen and weaken zonal flow over the Arctic that, in turn, causes fluctuations in the positions of the jet streams and the Arctic Front.

The southern boundary of the arctic tundra was long thought to be adequately described by the 10°C July isotherm (142). Bryson (42) feels rather that the mean position of the Arctic Front in summer delimits the transition from boreal forest to arctic tundra. In turn, the more southerly position of the front in winter, when polar air dominates the northern hemisphere circulation, sets the southern boreal limits, at least for North America, a thesis supported by Hare (94). This approach stresses vegetation zones as a function of the sum of all climatic parameters, thus eliminating the artificiality inherent in using a single (mean) parameter in describing vegetational boundaries.

More recently it has been shown that although the Arctic is a heat sink, in part due to high albedo from spring and early summer snow cover and summer cloud cover, net radiation is positive throughout the year (96). In Canada the southern limit of tundra occurs near the 18–19 kly isoline of annual net radiation and in Alaska at the 16 kly isoline. It drops to zero on the north shore of Ellesmere Island.

At the turn of the century the pressure distribution in polar regions was depicted as a shallow surface, high pressure field resulting from radiative cooling, overtopped by a deep low pressure system (70). It has since been shown that the surface high pressure system is not centered over the Polar Arctic Ocean but is displaced over the continental land masses where radiative processes are more extreme (70). The pattern is well defined in winter but weakens during summer with reduction in the latitudinal thermal gradient. The seasonal relationships of surface pressure fields to those aloft were succinctly described by Hare (94) when he wrote: "The complexity of these sea level (pressure) charts has for long obscured the surprising simplicity and uniformity of the Arctic core above the shallow layers responding directly to thermal contrasts between air-sea, air-ice, and air-land surfaces."

Fluctuations in intensity of the zonal flow with season, as mentioned earlier, are believed to control the frequency of cyclonic activity in arctic regions (70). In winter, when the zonal flow is strong, cyclonic systems are maintained in temperate latitudes and it is only with periodic weakening of the flow that cyclones can penetrate northward. In summer, with weaker zonal flow, the incidence of cyclones in northern latitudes increases. More recent work shows that this mechanism appears to be an oversimplification, for Hare (94) suggests that not all lows affecting the Arctic are migrants from mid-latitudes.

The relative infrequency of cyclonic storms, the shallow pressure gradients in summer, and the lack of a moisture source for most of the year, cause the Arctic to be a cold desert with annual water equivalent precipitation of $\sim$ 25 cm in the Low Arctic and as low as 5 cm in the High Arctic (195). Average surface wind speeds, at least in arctic North America, do not seem to be any higher than in temperate regions (195), but Courtin (59) has noted that in summer on Devon Island periods of high wind are generally associated with cyclonic activity while at other times winds are light ($\sim$0–3 m/sec).

The role that mountains play in controlling weather patterns on a global scale cannot be underestimated. Mountain massifs such as the Tibetan plateau and the Rocky Mountains cause acceleration of jet streams and help establish and maintain the wave patterns aloft that control periodicity of cyclonic storms on a large scale (200). Orientation of mountain ranges is also important in controlling movement of polar and tropical air. Only in North America are mountains oriented north-south, allowing deep penetration of tropical air northward and polar air southward.

Passage of frontal systems across mountain barriers can, under certain circumstances, lead to the formation of Foehn winds. Such winds occur on Devon Island (59, 238) and in Greenland (253) and have a strong influence on biological activity.

Microenvironment

Bliss (22) found that wind and insolation as modified by topography are key factors in the growth of tundra plants and segregation of communities. These findings were supported by Warren Wilson (256, 257), who found plant tissue temperatures to be higher than those of the adjacent air, the difference being a factor of wind speed and leaf orientation. The importance of radiant energy in arctic ecosystems has long been recognized and increasing emphasis is being placed on its measurement.

Dramatic environmental changes occur during the brief spring and autumn as the transition takes place from a snow-covered surface with an albedo of about 85% to a snow-free surface with a very low albedo, and vice versa (261). In the High Arctic more than half of the annual energy is received before snow melt (59). In the District of Keewatin, albedo varied from 15–24% depending upon the site (3). At Point Barrow, Alaska the albedo of wet meadow was 15% but this rose to 20% when these sites dried out (261). At Tuktoyaktuk, N.W.T. albedo averaged 15% over low shrub tundra and dropped to 7–10% with disturbance (90). For mid-July on Devon Island the albedo was 14% for dry sites and 11% for wet sites (2).

At Point Barrow net radiation increased tenfold from 37 cal/cm^2 day before snow melt to 380 cal/cm^2 day following melt. The latter value was partitioned into the following fluxes: latent heat 73%, sensible heat 18%, and soil heat 9% (261). At Tuktoyaktuk, N.W.T. net radiation ($\sim$42 cal/cm^2 day) was partitioned into latent heat 60%, sensible heat 21%, and soil heat 19% (90). For Devon Island the percentage values follow closely those of Point Barrow, but during the equivalent summer period net radiation was much higher ($\sim$470 cal/cm^2 day) (2). In the District of Keewatin, on a dry site receiving a net radiation of 377 cal/cm^2 day, 1.5% was dissipated via latent heat, 84% via sensible heat, and 15% via soil heat (3). In the Taimyr, values were intermediate with 40% of the net radiation (321 cal/cm^2 day) being partitioned in latent heat, 30% in sensible heat, and 30% in soil heat (203). In contrast to the above, the net radiation of 181 cal/cm^2 day for Cornwallis Island was nearly balanced by soil heat flux with sensible heat and latent heat flux accounting for only 10% (254).

These data indicate that the partitioning of net radiation is a function of vegetation type and the nature and moisture content of the substrate. For vascular cushion plants the partitioning of energy, especially with respect to latent heat and sensible heat flux, is of adaptive significance in the harsh environment in which they grow (1, 2, 58).

Permafrost

Permafrost is a characteristic of polar rather than alpine regions although it is not totally absent from the latter (124, 149). In North America permafrost is thickest in the Canadian Arctic Archipelago, exceeding 300 m and with ground temperatures at the depth of zero annual amplitude, below $-7°C$ (39). South of the arctic tundra, permafrost becomes thinner and discontinuous; its persistence at the southern fringe is strongly correlated with drainage and type of vegetation (38).

The active layer is generally 30–50 cm in fine textured mineral and organic soils, reaching a depth of 200 cm in well drained sands and gravels (40). Thaw depth depends upon the interaction of summer air temperatures, degree and orientation of slopes, and vegetation and organic matter content of the ground (40, 90).

Massive ice within permafrost has been reported from the Low Arctic (37, 153, 154) and from the High Arctic (39). In the Low Arctic it occurs as ice wedges associated with depressed center polygons, or ice-core hummocks (36), whereas in the High Arctic not only are the former present but also blocks of massive ice in gravel and lacustrine deposits, ice-cored polygonal peat mounds (39), and ice-rich strata in some fine-textured soils (11).

Hydrology

Hydrology in the Arctic is strongly influenced by the cold climate. Permafrost forms an impervious layer that impedes drainage and enhances surface flow, especially during snowmelt. Areas with little or no relief, such as the Alaska Coastal Plain and the Mackenzie Delta, feature an abundance of small shallow lakes and ice wedge polygons.

Dingman (67), studying a small watershed in central Alaska, found that a thick layer of moss overlying the permafrost gave runoff characteristics very different from those in mid-latitudes. Not only was there a considerable lag in increased stream flow following rain but the stream was sluggish during the entire snow-free season.

In the Canadian Arctic Archipelago the regime of arctic rivers varies according to their size. Large rivers flow all year around with minimal flow in winter and peaks corresponding to snowmelt. Small rivers cease in winter and their flow fluctuates violently during the rest of the year (39, 166). Studies of the Mecham River on Cornwallis Island (57) and a small catchment on Devon Island (167) show that 80–90% of the annual snowmelt discharge occurs during the two to three week snowmelt period and that the stream carried a considerable suspended load at this time, as do other rivers (10).

The role of water movement may play an important role in nutrient cycling (104). The presence of surface water is also essential in maintaining lush meadows, especially in the High Arctic.

Soils

In comparison to other soils of the world, those of the Arctic have received little attention. Information from Alaska (36, 71, 233, 236), the Canadian Arctic Archipelago (61, 170, 231, 235), and the USSR (83, 116, 225) is providing a solid basis for understanding pedogenic processes and classification.

Low temperatures and moist conditions weaken the soil forming processes found in temperate regions. The level and poorly drained topography that predominates over much of the Alaska Coastal Plain, the Yukon, and the Northwest Territories, has given rise to hydromorphic soils that have been classified into Tundra and Bog Soils (236). Tundra Soils are characterized by acid-gley processes, pH increasing with depth, and only 25–50% base saturation (230). The only zonal soil, Arctic Brown, constitutes but 1% of all arctic soils and is found on well drained sites. Horizons are better developed, pH increases with depth, and base saturation is usually below 50% (230, 236).

Over much of the Archipelago uniformity of terrain and lack of precipitation have given rise to Polar Desert Soils (49, 61, 170, 235) and lithosols, with Bog Soils and Tundra gleys occupying wet, poorly drained, and vegetated sites (61). The Polar Desert Soil is almost ahumic, well drained, and alkaline in reaction. Carbonate deposits are found on the underside of stones (61).

In the arctic-alpine conditions of the Brooks Range to the south of the Alaska Coastal Plain, Arctic Brown Soils give way to Polar Desert Soils in a similar manner to that in the Altai of Southern Siberia.

ARCTIC VEGETATION AND PRIMARY PRODUCTION

While the Arctic has a distinctive climate and contains unique species and groupings of animals and plants, it does not possess unique vegetation types (19). Shrublands, wet sedge and grasslands, herbfields, steppes, and cushion plant communities have their counterparts elsewhere. Thus the term tundra should not imply a vegetation structure as does prairie, deciduous forest, and tropical rain forest, though the term tundra in much of the literature refers to sedgelands. Many species are circumpolar and grow in a variety of habitats, making it more difficult to recognize communities (19, 53). From the Low to the High Arctic, plant cover decreases greatly and thus the control that vegetation exercises over soil development, geomorphology, and microclimate also decreases.

General Landscape Units

Generalized classification schemes for the Arctic have included Low, Mid, and High Arctic (189); Tundra, Rock Desert, and Ice Desert (190); Subarctic, True Arctic, and Polar Deserts (5); and Hypoarctic Tundra, True Arctic, and High Arctic (273). The various classification systems used in the USSR are summarized by Aleksandrova (5). These various zones and subzones are discussed along with a subdivision of the High Arctic by Bliss (27). The IBP ecosystem studies have been conducted in the Low Arctic, or in shrubland and sedgeland tundra extensions that reach 72°N (Western Taimyr, USSR) and 74°N (Truelove Lowland, Devon Island, Canada).

Low Arctic ecosystems fit into several broad catagories. Lowlands and coastal plain areas are dominated by *Carex, Eriophorum, Alopecurus, Dupontia, Puccinellia,* and other grasses. Lakes are abundant and so are populations of waterfowl and microtines. The poorly drained organic soils are low in available nutrients, high in ice content, and contain much patterned ground. Rolling uplands are generally dominated by *Eriophorum vaginatum* tussock tundra with associated dwarf heath shrubs and scattered forbs. These areas support ptarmigan, various passerines, microtines, and are important calving and summer grazing range for caribou and reindeer. Muskox also utilize these ranges plus shrub tundra and wet sedgelands. The southern arctic tundra (Subarctic Tundra) or low shrub tundra covers small areas in Alaska, large areas in Canada and vast areas in the USSR. Here shrubs of *Betula* and *Salix* (30–100 cm) develop as a layer with an understory of cottongrass tussocks, other sedges, forbs, and dwarf heath shrubs. Many of these species form an understory in the Forest-Tundra transition and Open Woodland (Lichen Woodland) to the south. Caribou, reindeer, ground squirrels, microtines, and numerous birds utilize these areas. Tall shrub complexes occur along rivers and streams and the drier and deeper thaw slopes along lakes. These habitats are important for moose, and where wet sedgelands occur waterfowl and fur bearers abound.

The Canadian High Arctic contains small areas of sedge or grassland tundra and dwarf shrub heath tundra (<3%) (27), areas important to muskox, lemming, and waterfowl. These areas occur in coastal lowlands throughout the islands and in valleys in the eastern islands. While graminoids and mosses account for most of the 80–100% plant cover, a few forbs and prostrate shrubs may occur. Vast areas of the High Arctic are Polar Semi-Desert with 5–20% vascular plant cover, though mosses and lichens, mostly crustose species, raise the total to >50% cover in many areas. Much of the spot medallion, or spotted tundra (USSR), or frost boil, soil hummock tundra occurs here. These areas generally occur from sea level to 100–200 m. Peary's caribou and lemming are common on these lands along with muskox and arctic hare where willow mats are common.

Above 200 or 300 m, or near sea level where the ice pack is more persistent, there are vast areas with little or no vascular plant cover, the Polar Deserts. Lichens and mosses predominate (2–20%) with rarely more than 1–2% cover of vascular plants. Large areas are devoid of plants and in many uplands and mountain slopes rock desert or fellfield predominates. With such a small food base, mammals and birds are rare. The most extensive summary of Alaskan arctic vegetation is that of Britton (35) and Polunin (188) for the Canadian Arctic. More detail on characteristics of the Low and High Arctic and the role of herbivores is contained in Bliss (27).

Primary Production

Standing crop and net annual production have been studied for some time (4, 5, 7, 22, 24, 66), but only with the IBP have production rates and processes been studied in detail. Wielgolaski (265) has classified tundra vegetation into six types on the basis of percent biomass contributed by different life forms. A simplified scheme was used (5) to summarize standing crop data that ranged from Polar Deserts to southern shrub tundra in the USSR.

Most of the available data are only for standing crops of the various components with little information on net primary production, especially for belowground portions. The data summarized in Table 1 illustrate several important aspects of arctic ecosystems. Standing crop of aboveground vascular plants increases roughly 100-fold from Polar Deserts or Semi-Deserts to Low Arctic shrub tundras. Our own studies are confirming this. Cryptogams comprise the majority of the standing crop in Polar Deserts and Semi-Deserts and their standing crop increases 2–10-fold in the Low Arctic. Thus their relative role in annual production decreases. Even in the High Arctic, mosses can have a large standing crop in wet sedge meadows and a high annual production along streams (181), but these are atypical situations.

As with other vegetation types, accurate estimates of net annual production of shoots is compounded by differing phenological development of species, summer browning of current season production, and difficulties in sampling cushion plants growing in heterogeneous patterns (24, 25, 174, 227, 241). Other problems are associated with estimating total roots, live vs dead roots, root production (66, 174, 228), and the recognition of shoot bases growing in a moss mat where definition of a soil surface becomes difficult (265).

General estimates of net primary production and standing crop (aboveground) of tundra ecosystems (202) are high when compared with current data. Net annual aboveground production of vascular plants in the Low Arctic generally ranges from 40–110 g/m² in wetland sedge-grass, cottongrass-dwarf shrub heath, and low shrub-dwarf shrub heath communities (23, 66, 89, 241). Somewhat comparable sedge communities in the High Arctic produced 40–45 g/m² vascular plants (174) and ∼20–30 g/m² mosses (181) per year. Although few data on net production are available from the USSR, it now appears that much of the Low Arctic and the small comparable areas in the High Arctic have similar aboveground production, vascular plants and cryptogams amounting to 60–150 g/m². The similarity of production regardless of community structure and species diversity is somewhat surprising. Net

Table 1 Above and belowground standing crop (g m^{-2} air dry weight) for USSR polar desert to shrub tundra[a]

Plant Parts	Polar Desert moss-lichen polygons[b]	Arctic Tundra herb-dwarf shrub-moss[c]	Typical Tundra Low shrub-moss[d]	Low Arctic Sedge-moss-dwarf shrub[e]	(Subarctic) Shrub-dwarf shrub[f]
Aboveground					
Live vascular plants	6	71	188	218	817
Live cryptogams	123	114	319	261	1345
Dead plants	9	69	366	521	24
Total	138	254	873	1000	2186
Belowground					
Live plants	29	511	1370	2,307	3,716
Dead plants	152	310	4856	7,788	11,300
Total	181	821	6226	10,095	15,016
Total standing crop	319	1075	7099	11,095	17,202
Aboveground vascular: live cryptogams	1:20	1:1.6	1:1.7	1:1.2	1:1.6
Total aboveground live: total aboveground dead	1:14.3	1:2.7	1:1.4	1:1.1	1:0.1
Vascular plants live aboveground: live belowground	1:4.8	1:7.2	1:7.3	1:10.6	1:4.5
Total aboveground: total belowground	1:1.3	1:3.2	1:7.1	1:10.1	1:6.9

[a] Based upon Aleksandrova (5) and references contained therein.

[b] Mosses, lichens, *Phippsia algida*, *Papaver polare*, *Saxifraga rivularis*; roots sampled to 38 cm- Franz Josef Land (Aleksandrova 1969).

[c] Mosses, lichens, *Alopecurus alpinus*, *Luzula confusa*, *Salix polaris*, many forbs; roots sampled to 36 cm- New Siberian Islands (Aleksandrova 1958).

[d] *Betula nana*, *Salix lanata*, *S. glauca*, *Vaccinium vitis-idaea*, *V. uliginosum*, *Empetrum hermaphroditum*, *Carex hyperborea*; roots sampled to 100 cm- Vorkuta region (Shamurin 1966).

[e] *Carex lugens*, *V. uliginosum*, *Empetrum nigrum*, *Arctous alpina*, *Ledum decumbens*, *Betula middendorfii*; roots sampled to 55 cm- Kouzak region (Vikkiryeva-Vasilkova et al 1964).

[f] *Betula nana* (70–80 cm), *Salix glauca*, *S. phylicifilia*, *Carex globularis*, *V. uliginosum*, *V. vitis-idaea*, *V. myrtillus*, *Empetrum hermaphroditum*, *Arctous alpina*, *Ledum palustre*; roots sampled to 100 cm- Vorkuta region (Rakhmaniana 1966).

annual belowground production in sedge or sedge-grass meadows ranges from 130–360 g/m² (66, 174). Ratios of live-aboveground:live-belowground standing crop are generally 1:5 to 1:11 (66, 174, 265), with live roots accounting for about 50–60% of the total root biomass (700–1800 g/m²) (66, 174). Maximum root development typically occurs in the upper 10 cm (66, 174).

At Barrow, Alaska during periods of high lemming population, vascular plant production is held to 3–48 g/m² (66). By reducing litter and increasing depth of the active layer, lemming can influence plant community composition (66) as do geese in the USSR (245). Lemming grazing also stimulates shoot production per unit area (215, 216).

Only a few comparable data are available for the Polar Semi-Deserts of the High Arctic. On gravelly beach ridges, aboveground production of vascular plants averaged 16–20 g/m² and belowground production 3–5 g/m² (228). Ratios of live-aboveground:live-belowground are generally 1:0.5 to 1:0.7 and most of the attached roots are alive (228). Preliminary data (unpublished) from *Luzula confusa* steppe and herbfields indicate an aboveground net production of 3–8 g/m².

Ratios of total-aboveground:total-belowground biomass are low in the Polar Semi-Deserts (1:1–2) (5, 228), but in the Low Arctic are higher in dwarf shrub and low shrub tundras (1:4–10) (5), and highest in sedge-grasslands (1:10–20) (25, 66, 265). These trends and production rates of shoots and roots suggest that low arctic environments may be more limiting for shoots than roots (25, 66). In wet meadow soils root production is greater and decomposition is slower than in upland sites. In the High Arctic, lack of available nutrients and low soil water availability appear limiting to plant growth, including root production (207). Thus permafrost may not be the primary factor limiting growth (241), although cold soils limit both shoot and root growth (63, 271).

Root turnover time at Barrow, Alaska is estimated to be greater than 10 years (66) and on Devon Island 5–10 years (174). Turnover time for shoot production is 3–5 years in Alaska (66, 125) and 2–3 years on Devon Island (174).

The limited data available on aboveground vascular plant production for graminoid communities, photosynthetically available radiation, length of growing season, and plant caloric data show daily productivity rates of 0.9–1.9 g/m² and the efficiency of net primary production of 0.20–0.45% for the growing season in Alaska and Devon Island (22, 66, 125, 174, 198, 241). These data are comparable with some herbaceous communities of temperate regions, though these tundra systems are limited to 50–75 days for their primary production (21, 24).

Chlorophyll and Leaf Area Index

In some communities there is a high correlation between chlorophyll content and dry matter production, but the correlation is much lower when comparing chlorophyll per unit dry weight (25, 243). Tieszen (241) found no relationship between seasonal maximum chlorophyll concentration and dry matter production. In alpine and low arctic sites, communities dominated by heath species have a lower chlorophyll content (300–600 mg/m²) than do wet sedge communities (460–820 mg/m²) (25, 241, 243). Sedge meadows on Devon Island contained 210–283 mg/m² due to lower plant production, although mean tissue chlorophyll level was higher than at Barrow (174, 241). If an increased concentration of chlorophyll in leaf tissue proves common in High Arctic plants, this could reflect an important photosynthetic mechanism for maintaining higher production than would otherwise occur.

Dry sites, be they sedge or heath dominated, have a low chlorophyll content (130 and 180 mg/m^2) (25, 243) and a lower chlorophyll concentration (2–4 mg/g dry wt) (25, 228, 243). At Barrow, with an understory of early leafing forbs with prostrate leaves and a later leafing overstory of graminoids with inclined leaves, Tieszen (241) found that forbs had high chlorophyll concentrations early in the season while the graminoid canopy reached its peak 25–30 days after start of the growing season. This again would permit maximum photosynthesis in the dicots before the graminoid canopy developed and the sun angle reduced in July. Spring release of CO_2 with snowmelt (133) may provide an enriched CO_2 environment for this early burst of growth.

Tieszen (241) found that dicotyledonous species with the highest chlorophyll concentrations had no rosette leaves while rosette species had high chlorophyll per unit area but not per dry weight. Similar studies in the High Arctic might yield interesting data, for there are many species with the rosette form, rapid growth, and small root systems.

Leaf area index is low in graminoid communities: 0.9–1.2 in Alaska (241) and 0.6–0.7 on Devon Island (174). At Barrow chlorophyll concentration peaked first, followed by maximum standing crop of chlorophyll and then maximum plant dry weight (241). On Devon, leaf area index, standing crop of chlorophyll, and plant dry weight all peaked at the same time (174). This may reflect a physiological response to a shortened growing season. Leaf area index is exceedingly low (0.1) in the cushion plants of the Polar Semi-Deserts (228).

Carbohydrates and Nutrients

Early season growth, especially in sedges, is well correlated with rhizome carbohydrate depletion (174) and its gradual replacement later in the summer. Roots appear to play a minor role in this translocation. Lower levels and less seasonal fluctuation of carbohydrate reserves were found in cushion plant communities (228). In both communities oligosaccharides are 10–50 times greater than monosaccharides, and starches amount to 25–40% of the total carbohydrate content (174, 228). These may be important adaptations to low temperature conditions. Generally carbohydrate levels are low in these arctic plants when the data are expressed on a dry weight basis. Simulated grazing has shown that although carbohydrate reserves were lower in the spring following clipping, subsequent plant production was not significantly reduced (11).

Haag (89) has shown that in low arctic wetland sedge and upland willow-birch-dwarf heath shrub communities, available nitrogen limits protein content and dry matter production while phosphorus does not. Nitrogen is metabolized into organic compounds at low temperatures while phosphorus metabolism is limited by low soil temperature and low available nitrogen, factors which limit nucleoprotein formation (143). The xeromorphic characteristics of ericaceous species are postulated as being adaptations that conserve nutrients by emphasizing structural tissue rather than protein synthesis (89). Heath species have lower growth rates and lower photosynthetic levels than graminoids and forbs (91). Heath, *Salix* and *Betula* species have shallow root systems that typically remain in the surficial peat layer while many of the graminoids have root systems that occupy most of the active layer (22). Studies of Dadykin (63) and Younkin (271) show that grasses grow best in warmer soils. Native grasses that invade disturbed soils may achieve over 250 g/m² aboveground net production with only a 30–40% plant cover (271), demonstrating the role of warmer soils and a larger volume of nutrients. These mechanisms, along with high shoot:root ratios (5, 66), perennial vs annual growth (26), and cushion plant form (228), all aid in the conservation of available nutrients in systems where nutrients, low soil and air temperatures, and low plant and soil water potentials are the key factors in low levels of plant production.

In the Low Arctic symbiotic nitrogen-fixation results from legumes, *Alnus crispa* and *Dryas drummondii* (148), as well as blue-green algae and lichens (6). Species of *Peltigera* fixed nitrogen at comparable rates in Alaska (6) and on Devon Island (224). In the High Arctic anaerobic bacteria play the dominant role with lichens and algae playing a minor one (224). Seasonal fixation averaged 190 mg N/m^2 in sedge meadows and 15 mg N/m^2 on dry beach ridges on Devon Island (224). There is little if any nitrogen input from precipitation, salt spray, and animals and little loss via streams on Devon. Thus the Truelove Lowland seems to be a remarkably closed system in terms of nitrogen (224) and the same probably holds for other nutrients.

Photosynthesis

Maximum rates of photosynthesis are 8–20 mg CO_2/dm^2 hr for three graminoid species (241), 25 mg CO_2/dm^2 hr for *Salix pulchra* (242) at Barrow, and 5–16 mg CO_2/dm^2 hr for 10 herb and shrub species in the western Taimyr (274). Maximum rates for *Dryas integrifolia* were 2.5–4.2 mg CO_2/g hr (163) with the highest rates occurring early in the season when the leaves were red and at "night" when leaf temperatures were lower. These cushion plants appear to develop high leaf resistances which result in decreased CO_2 and water fluxes and high leaf temperatures (>40°C). In contrast, *Carex stans* had much higher assimilation rates (12–13 mg CO_2/g hr) (163), values comparable to the data of Shvetsova & Voznesenskii (214) for grasses and sedges in the western Taimyr. In Alaska (242) and on Devon (163) photosynthesis is very active at midnight when light levels are only one-tenth as great as at noon.

Recent data show that *Alopecurus alpinus*, *Arctagrostis latifolia*, and *Dupontia fischeri* are "cool season" grasses that photosynthesize with ribulose-1,5-diphosphate carboxylase. The enzymes are stable at 40°C but since photosynthesis decreases above 15–25°C, increased respiration or increased resistance to CO_2 diffusion must be responsible for the temperature limitation (244). Tieszen (241) has also shown that these arctic grasses have photorespiration which immediately returns 30–45% of the fixed CO_2, and that CO_2 uptake is more dependent on light than temperature. The use of the aerodynamic method (60, 125), simulation studies (173), plus the field and laboratory gas analysis studies are permitting both cross-checks with dry matter production data and a mechanistic analysis of the dynamics of and factors controlling production. More detailed information on the adaptations of arctic plants is contained in recent reviews (21, 26, 123, 207).

THE ROLE OF PRIMARY AND SECONDARY CONSUMERS

Herbivores

Muskox *(Ovibos moschatus)*, the largest high arctic herbivore, does not restrict its grazing to lowlands but frequently forages on slopes and cliffs that would be the domain of mountain goats and sheep in warmer climes. Winter feeding is predominantly in lowlands where forage is abundant, but summer range includes more marginal lands. Muskox prefer vascular plants which contain 4–25% protein and 1–10% fat (240). Insular animals are 56% efficient in digesting their fodder and have an estimated summer intake of 31.5 and 30–34 g/kg dry body weight on Devon Island (112) and in Alaska (267) respectively. Muskox tend to be more intensive grazers than caribou with average herd movements of less than 1 km/day (113).

Muskox biomass on Truelove Lowland, Devon Island averaged between 3 and 4 kg/km² for summer and fall months (113), similar to an estimate for ungulates in the Nelchina Basin, Alaska (87). During winter muskox reach much higher densities in select lowlands (113).

Like muskox, caribou spend 9–10 months each year foraging in areas covered by snow through which they must dig for a substantial part of their nourishment (193). Lichens form a major portion of the winter diet of barren-ground caribou but are less important to the smaller Peary's caribou. Caribou are extensive rather than intensive feeders, and this, coupled with the southward migration of the mainland herds into the taiga for wintering, tends to prevent overgrazing of the slow-growing tundra plants (27).

Caribou energy requisites are highest in spring and summer when calving, lactation, antler development, and molting occur (138). Nutritional studies have been initiated in Alaska (262) and one can expect they will find that a reduced winter food requirement is as characteristic of caribou as it is of moose (128), deer (105), and probably arctic hare (217, 255) in northern latitudes.

Ungulates provide an average standing crop of 0.17 kg/km² in the Alaskan and Canadian tundra (138) and 3 kg/km² in Alaska's Nelchina Basin (87). While these figures represent a nearly 20-fold difference, they dramatically illustrate the paucity of ungulate biomass when compared with the 170 kg/km² found on African savanna (186).

Arctic hares, intermediate sized herbivores, are represented by *Lepus arcticus* in North America and *L. timidus* in Eurasia. The preferred diet of both species is *Salix* (194, 221) which is higher in nitrogen, potassium, phosphorus, and magnesium as well as caloric value than other winter dietary species; the presence of *Salix* may determine distribution of *Lepus arcticus*. Smith (217) estimated an August intake of 700 kcal/hare day for caged animals on Devon while Wang et al (255) estimated a winter daily resting energy consumption of 262–133 kcal/day, caged in complete darkness for ambient temperatures between –24 and 12.5°C. The same individuals in the former experiments were considerably more active.

During winter, hares tend to disperse widely over the available range but in summer they frequently congregate in herds of as many as 700 individuals. They form a major prey species for the wolf and arctic fox.

The smallest of arctic herbivores, the lemming, has been the subject of intensive study near Barrow, Alaska, where brown lemmings *(Lemmus trimucronatus)* and collared lemming *(Dicrostonyx groenlandicus)* are sympatric, since 1955 (187). Three to five year population oscillations were evident in the brown lemming, which reached densitites of 125–250/ha. Neither comparable fluctuations nor modest densities in collared lemming became evident until 1971 when, after an extended period without a brown lemming high, they became abundant. During the Alaskan lows the collective average was 3–5/ha (211) with *Lemmus* comprising more than 90% of the catch (187). A lemming high of 119/ha was recorded at Tareya in western Taimyr, USSR (162).

Food habit studies revealed that lemmings prefer fresh green material summer and winter (131, 171, 221, 245), with *Dicrostonyx* feeding mainly on dicotyledonous plants and *Lemmus* on monocotyledons. In Alaska, adult brown lemming consumed an average 0.32 g dry material/g body weight daily while juveniles consumed an average 0.53 g daily (171). Adults then must consume 170% and juveniles 200% of their body weight daily to maintain weight during the summer. Results of biomass and energy measurements are summarized in Table 2 and may be supplemented by the recent work of Coady (55).

Table 2 Biomass and energy measurements of brown lemming food utilization (from 171)

Component	N	Juveniles (22–26 g) $\overline{X} \pm$ S.E.	N	$\overline{X} \pm$ S.E.
Dry weight ingested/day (g)	7	13.6 ± 0.52	18	22.9 ± 1.89
Dry weight ejected/day (g)	7	8.6 ± 0.49	14	16.0 ± 1.15
Dry weight efficiency (g retained/g ingested)	7	0.37 ± 0.03	14	0.34 ± 0.01
Digested energy/g body weight/day (cal)	7	913 ± 73	13	604 ± 76
Digestive efficiency (cal retained/cal ingested)	7	0.37 ± 0.03	13	0.35 ± 0.02

Carnivores

Although polar bears *(Thalarctos maritimus)* are the largest potential terrestrial carnivore they are functionally marine carnivores preying principally upon seals. Their most significant contribution to terrestrial ecosystems may be in providing leavings and offal for arctic foxes.

The arctic fox *(Alopex lagopus)* is one of the best adapted arctic mammals, being able to withstand exposures of –80°C for one hour with no drop in rectal temperature (120, 210). Because they possess a highly insulative coat, they seldom experience winter temperatures that necessitate increased chemical heat production. In summer their chemical heat production must be increased slightly for more than half of the time (250).

In a full year study, Underwood (250) found that energy intake of foxes differed significantly between summer and winter, with consumption of 263–447 kcal/kg day in July and 54.5–72.2 kcal/kg day in January. Energy of the feces never exceeded 5% of the total energy ingested. Riewe (201) found that four foxes maintained their weight between October and January on 125 kcal/kg day, and on the basis of Speller's (220) data they estimated growing fox pups ingested an average of 490 kcal/kg day. Between 1 May and 30 September one fox would need energy equivalent to that contained in four hundred 28 g lemming, four hundred 30 g snow buntings, and twenty 3.46 kg arctic hare (201). These data help explain the reproductive failure of a large portion of arctic fox populations in years when lemming populations are low. High reproductive potential also requires both parents to hunt (220). Bannikov (17) reported a pair of foxes in the USSR needed 2–30 km² for hunting during the time whelps were in the den, and this expanded following emergence of the young. Their average density on suitable habitat was 0.7/1000 ha before and 0.4/1000 ha after the annual trapping season with 43,000–108,000 animals being harvested each year.

The wolf *(Canis lupus)* is the major large arctic carnivore but is becoming rare (239). Maximum densities of 1 wolf/18 km² on the Thelon Game Sanctuary were reported by Kuyt (145). Throughout its range the wolf preys on the largest sympatric herbivore and a single wolf has been observed to kill a bull muskox (86), though this is no doubt uncommon. Other prey species include caribou (134) and arctic hare; lemmings, birds, arctic foxes, and possibly seals serve as secondary prey species (54, 169). No detailed estimates of energy requisites are available, although Kuyt (145) estimated each wolf ate about 23 barren-ground caribou annually.

The smallest mammalian carnivores are the least and short-tailed weasels *(Mustela rixosa* and *M. erminea)*. *Mustela erminea* are seldom seen and rarely sought commercially in the eastern Canadian Arctic (201). They feed principally upon lemming, with eggs and nestlings supplementing the summer diet. Daily food consumption of a female was 14.3–17.4 g/day dry weight, an equivalent of slightly more than one 28 g lemming per day. Although standing crop and annual production of weasels is low, they took 11.5% of the *Dicrostonyx* population on Devon Island during a time when lemming were not plentiful (201) and the observation of 20% winter lemming predation on Banks Island (158) indicates they are extremely effective predators. Weasels, mostly *M. rixosa,* may be a driving force in creating periodic lemming cycles in the Barrow area by reducing the lemming population and establishing the low from which the population recovers (158, 187).

Birds

Within high arctic ecosystems the number of breeding bird species is low but the number of trophic levels may approach or equal that of more temperate climes. Generally there are few herbivores, frequently only resident ptarmigan and migratory geese. Migratory secondary consumers include several species of shorebirds and ducks, plus a few sparrows, buntings, and longspurs. Jaegers, peregrine falcon, and snowy owl represent both the secondary and tertiary level (182). As most birds shift their diet in conjunction with food availability, they should be considered facultative in terms of trophic level. Reported densities of arctic passerines range from 68–206/km² at Finse, Norway (151) to 68–320/km² in the Cape Thompson, Alaska region (268). At Barrow, Alaska lapland longspur numbers have unaccountably declined from 400–800/km² to 24–50/km². The lowest densities of arctic passerines were 8.8/km² for three successive summers on Devon Island (183).

Density estimates in other parts of the Arctic include 38/km² (all birds) on Ellef Ringnes Island (206) and 122/km² at Chesterfield Inlet, Keewatin and three other early estimates of 71, 155, and 216/km² for the Canadian tundra (206). Saville's estimates were from especially lush tundra regions and thus represent extremely high figures which can be contrasted with 1971 and 1972 post-breeding estimates of 26/km² (all birds) on Devon Island (183).

Energy expended by various bird species during the tundra breeding season includes estimates of 57–75 kcal/day per pair of lapland longspurs *(Calcarius lapponicus)* functioning as insectivores until nestling feeding commenced, when the rate rose to 25.6 kcal/nestling day by fledging (62). Insectivorous snow bunting *(Plectrophenax nivalis)* nestlings used 24.5 kcal/nestling day from hatching to fledging, and adult buntings took 25.7 kcal/day (182). Digestive efficiency of birds eating an insect diet was 72% for buntings (182), 70% for dickcissels (209), assumed to be 70% for Lapland longspurs (62), and 80% for four species of Alaskan sandpipers (178). On a seed diet the digestive efficiency of snow buntings declined to 40% (182).

One interesting aspect of breeding bird energetics was a reduced energy requirement during incubation (62, 182). Bachmitov (12) found a 1–2°C drop below mean temperature during inactivity and an elevation of temperatures of about 1°C associated with migration. This lowering of body temperature may well be related to the reduced energy requisite during incubation and to Norton's (178) observation that both male and female dunlins have metabolic rates 2–3 times as high at the beginning and end of incubation as during brooding.

Bioenergetics of the snowy owl (80) show that thermal conductance of 0.05 cal/(g hr °C) was lower than any published for an avian species and was equivalent to that of the pelt of the arctic fox. On the basis of oxygen consumption studies, Gessaman estimated a daily winter food requisite of 4–7 lemmings or 200–400 g/day with temperatures of –34°C. He makes the point that unless lemmings are extremely abundant the snowly owl cannot exist on the arctic tundra in winter.

MICROFLORA, MICROFAUNA AND DECOMPOSITION IN TUNDRA SYSTEMS

Microbiology

BACTERIAL POPULATIONS Boyd (32) and others have demonstrated diverse groups of soil bacteria including iron oxidizers, nitrogen fixers, ammonia oxidizers, sulfate reducers, and fermentative bacteria in soils and water from arctic Canada, Alaska, and Antarctica. Mesophilic bacteria ranged in numbers from 0 in antarctic dry valley soils to 84 X 10⁶/g soil in sewage-contaminated soils at McMurdo Sound.

Numbers of bacteria from Devon Island, the Taimyr Peninsula, USSR, and possibly Signy Island appear fairly high when compared with other undisturbed tundra sites at lower latitudes (Table 3). It is possible that high soil pH creates a favorable environment for bacteria on Devon Island, but at Taimyr the soils have a low pH (9). Another major factor enhancing bacterial numbers is probably moisture; thus the dry valley soils tend to have low bacterial counts whereas the wetter soils have higher numbers (Taimyr river bank, Devon Island mesic meadow). The numbers of bacteria found in arctic soils do not appear to be much lower than those found in temperate soils. Thus Timonin (249) reports up to 73 X 10^6 and 158 X 10^6 bacteria/g soil for a Canadian podzol and chernozem respectively, and 264 X 10^6 have been reported for an English podzol (82).

The genera of bacteria appear to be fairly varied. Thus on Devon Island *Coryne-bacterium, Bacillus, Pseudomonas, Flavobacterium, Alcaligenes, and Achromobac-ter* were found along with gram-negative rods, gram-positive cocci, and occasional coliform bacteria (264). Similar lists of genera have been reported from S.W. Alaska (76) and Antarctica (34). One major difference, however, was the high numbers of *Corynebacterium* spp. (22.5% of isolates) found on Devon Island, a genus not reported from other polar sites. These lists of genera are as rich as that reported for an English podzol (82), but richer than the list for a heather moor in England (147), where 80% of the isolates were *Bacillus* spp.

FUNGAL POPULATIONS   Most early studies of soil fungi were taxonomic rather than numerical in their approach. This has resulted in an extensive knowledge of species of higher fungi and plant pathogens (139–141, 152, 223), but little knowledge of fungi from the soil or on plant litter. Studies of fungal biomass in soils and litter have been started under the IBP, but few results are available.

The biomass of fungi on Devon Island has been studied using the direct observation method (127); >2000 m mycelium/g soil have been found in a mesic meadow, compared to 400 m/g soil in a raised beach (264). These data cannot be compared

Table 3 Bacterial numbers from tundra soils

Location	Habitat	pH	Millions of Bacteria per g soil
Devon Island (264)	Mesic Meadow	7.6	11–458
	Raised beach	8.0	7.3–36.7
Inuvik (33)	Sphagnum peat	5.0	0.67–2.10
	Disturbed loam	6.5	3.50–26.0
Taimyr (9)	River bank	–	108
	Polygonal bog	–	15
	Hummock	–	0.2
Signy Island (100)	Grassland	5.6	1–16
Antarctica (48)	Wheeler Dry Valley	–	0.001–0.3
	Matterhorn Dry Valley	7.5–8.9	0.0001–0.17

directly to 200 molds/g soil for an antarctic dry valley (48). Data from Signy Island (100) and Moorhouse, England (147) show 66–400 m/cm³ and 205–780 m/cm³ respectively for grassland and a mixed moor, compared to 468 m/cm³ and 168 m/cm³ for a sedge meadow and beach ridge respectively on Devon Island when converted to a unit-volume basis. At Moorhouse only 14% of the mycelium was stained, indicating that most of it was dead (147). A similar pattern may occur on Devon Island, for no distinction was made between live and dead mycelium. When the data from tundra sites (264) are compared with those from temperate region acid mull and moor soils (176), mycelial mass is often greater in the former.

Published data on the taxonomy of tundra soil fungi show a preponderance of sterile forms (46, 73, 117, 147, 264), as do the unpublished data of Bailey (personal communication) from Signy Island. These publications show also that *Penicillium, Chrysosporium, Cephalosporium,* and *Mortierella* are common genera in tundra soils. The genus *Trichoderma,* important in humus from temperate regions, is very rarely found in the tundra.

MICROBIAL BIOMASS   In biomass, fungi are more abundant than bacteria. Thus, on Devon Island, a mean fungal biomass of 30.4 g/m² and 6.4 g/m² for a meadow and a raised beach soil was obtained, whereas the figures for bacteria were 0.16 g/m² for both sites (264). Using similar methods ratios of bacteria:fungi of 1:300, 1:4000, and 1:13,000 were obtained in a limestone grassland, *Juncus* moor, and mixed moor respectively (147). These data are, however, misleading, for total mycelial counts are being compared to viable counts for bacteria. In the moorland soils described above, when direct bacterial counts were used, the ratio was much lower (1:1, 1:6). In the soils of Taimyr, direct bacterial counts ranged from 20X to 710,000X the dilution plate counts for bacteria (9).

MICROBIAL PRODUCTIVITY   Due to methodological problems, few studies of microbial production have been carried out. Pedoscopic (184) estimates of bacterial production in the Taimyr (162) during the summer months range between 0.05–0.25 g/g soil with a mean generation time of 15.5/month. In the same area Aristovskaya and Parinkina (9) reported generation times of 13 hrs in frost-boils to 108 hrs in polygonal bogs. These generation times are much shorter than the 4–5 generations/ month for rice fields (208).

Using the direct count figures for fungal mycelium, it is possible to estimate minimum fungal production. Thus, in the period of June 28th to July 6th, mycelium in the Devon Island meadow site increased from 912 m/g to 1472 m/g, giving a net increase of 18 g/m² of fungus. This figure does not take into account either grazing activities by soil fauna or the decomposition rate of fungus mycelium.

Invertebrates

Invertebrates play an important role in the ecosystem as consumers, decomposers, and parasites (52, 108, 185, 199), and, due to their complex life cycles, it is often difficult to classify a species as to its ecosystem role. Early instars of chironomids may feed on algae or detritus, later instars are predators, and adults are nectar

Table 4 Major groups of soil invertebrates from Devon Island, from combined wet and dry funnel data, No./m^2. Data from Ryan (204).

Site	Nematoda	Acarina	Collembola	Diptera	Copepoda	Ostracoda
Raised Beach	1548	2373	7320	69	–	–
Mesic Meadow	17,194	1500	2150	1315	6541	1621

feeders (65, 121). In addition, a major component of the invertebrate fauna functions in the decomposition process (nematodes, tardigrades, mites, Collembola).

Much of the literature on arctic invertebrates stresses the low diversity of terrestrial arthropods; this has been demonstrated in Sweden (226), Spitzbergen (130), Norway (218, 219), Canada (164, 165), and the USSR (50, 222). The work of Bohnsack (31), however, added 45 species of terrestrial arthropods to an existing list of 300 from the Barrow area (115).

Data from Antarctica (247) showed that the soil fauna consisted mainly of protozoa, nematodes, tardigrades, rotifers, and arthropods, with collembolans and mites being the major arthropod group. The major protozoan group studied on Signy Island was the testate amoebae with an approximate biomass of 2 g/m^2 (97). At Barrow the major insect group was the Chironomidae (157). Ryan (205) reported a terrestrial chironomid emergence of 34 mg/m^2 on a meadow at Devon Island. This compares with 300 mg/m^2 and 128 mg/m^2 emergence per season from aquatic systems in Alaska and the Canadian High Arctic respectively (20, 260). Preliminary data on soil fauna from Devon Island (204) (Table 4) indicate the importance of nematodes, mites, and collembolans in both sites, with dipteran larvae (including chironomids) and ostracods being important in the meadow site; this again illustrates the importance of water in the tundra. From these and other data Ryan (204) estimated a total annual production of 3.3 g/m^2 for the meadow, though, due to low extraction efficiencies of arthropods and nematodes, this may be an underestimate. Thus, Sharp (personal communication) working solely on collembolans, has counted nearly 20,000/m^2 on a raised beach ridge in August 1972. At Prudhoe Bay, numbers of collembolans up to 34,000/m^2 have been reported (155). Estimates of nematodes on Devon Island by Procter (191) have ranged from a low of 1.0×10^5 in a meadow to a high of 2.75×10^6 in the transition zone of a beach ridge with an expected 50 genera and 100 species represented.

Literature values for feeding rates on fungi of proturan collembolans at 15°C vary by an order of magnitude. Healey (101) gives a rate of 0.11–0.17 μg/μg collembolan day whereas a figure of 0.017 μg/μg day was obtained using a radioisotope (144). It is therefore difficult at present to judge the effects of grazing on soil microbial populations.

Decomposition Rates

As stressed elsewhere in this paper, most of the primary production of tundra ecosystems goes straight to the decomposer component. In spite of this fact, very little published data exist on decomposition rates. Studies on Devon Island (264), using litter bags, indicate a 19% dry weight loss in the first year for *Carex stans* with a second year loss of 12%. Studies on *Dupontia fischeri* at Point Barrow (18) show loss rates of 15% and 18% in year one and year two. For the moss, *Chorisodontium aciphyllum,* on Signy Island, a rate loss of only 1.3% was obtained (13). These rates are very low when compared with temperate or tropical forest litter.

INUIT HARVEST OF WILDLIFE IN ARCTIC ECOSYSTEMS

The top carnivore in arctic ecosystems, the Inuit[6] or their ancestors, the Thule, Dorset, Pre-Dorset and Denbigh peoples, has occupied the region for at least the last 5000 years (146, 229). Present distribution extends from East Cape, Siberia eastward approximately 10,000 km to Scoresby Sound, Greenland and from Thule, Greenland southward approximately 3000 km to southern Labrador.

Traditional Subsistence Economy

The Inuit are highly evolved hunters who have been dependent upon arctic wildlife for most necessities of life. They are unique in this respect compared to all other hunter-gatherer peoples of the world, most of whom rely more on plants than on wild game (212).

The possession of an elaborate hunting technology has enabled them to cope with the harshness of their environment and to harvest the wildlife resources. Many of their implements are unique to their culture. Complementary to this specialized hunting they evolved several other techniques which assured them of a nearly continuous food supply, namely, a nomadic way of life, diverse methods of food storage and preservation, versatile diets, extensive systems of community meat distribution and sharing of material goods, religious measures to replenish wildlife populations, and intrinsic controls which limited their own population size and structure (14, 78, 146).

The subsistence economy of the Inuit evolved primarily in association with the arctic coastal environment (16, 41, 75). However, their highly adaptable culture has enabled them to utilize inland resources as well (47, 135, 136).

The Inuit have been exceptionally versatile hunters, able to harvest and utilize whatever resources were available. While often dependent upon a single food source, alternate foods available for only short periods during the annual cycle were of the utmost importance. It was often these relatively scarce resources which sustained them through periods of hardship.

Economy Changes in the Nineteenth and Twentieth Centuries

Since Euro-North American society made contact with the Inuit, there have been significant changes in their ecologic and social environments. Food and material goods which were introduced during the 18th and 19th centuries were readily absorbed into their adaptable culture. This has irrevocably tied them to the cash

[6]The name Inuit ("The People") will be used here in place of the name Eskimo because it is used and preferred by this group of native peoples.

economy of white society. No longer do the Inuit possess a strictly subsistence economy; they are forced to earn sufficient cash to keep the materials and luxuries of the south streaming north.

The whaling industry of the 19th century initiated the contact (74, 269). Between 1920 and 1950 trapping became the mainstay of the Inuit's cash economy (85, 251, 270). At present, trapping plays a far less important role in most areas, although there are a few smaller settlements which are still highly dependent upon trapping, the prime example being Sachs Harbour, Banks Island where in recent years 15–20 trappers have harvested as much as one-third of the entire Canadian catch of arctic fox *(Alopex lagopus)* (251). A major contributor to the Inuit's present-day cash economy is part-time or full-time wage employment, especially in the larger communities. The sale of furs, skins, and handicrafts provides cash income. In many of the larger centers the Inuit have also been forced into a subsidy economy. Continually escalating aspirations are pressuring them even deeper into the white man's cash society.

Post-Contact Utilization of Wildlife

Despite the shift from a subsistence to a cash economy, the Inuit have remained reliant upon their land. The whaling and trapping eras actually intensified their hunting activities (14, 75), for they provided the whalers with food. In the precontact period dogs were used as pack and hunting animals; the average hunter usually owned only one or two dogs. The trapping era demanded greater mobility to tend trap lines and so the hunters increased their dog holdings to as many as 8–15 animals which were then used for hauling sleds rather than as pack animals; this forced the men to increase their hunting activities in order to provide their dogs with food. Introduction of the rifle enabled them to secure most or all of this additional game.

Meat from wildlife is still an important food source in many villages (Table 5). There has also been a growing demand for southern foods in which the proportion of carbohydrates is high. For example, in Grise Fiord in 1954 the 36 persons imported only sugar, tea, and candy for a total retail value of less than $500. By 1971–72 the community of about 90 people consumed more than 70 types of southern food, mostly carbohydrates, which retailed for $23,000.

Wild meat consumed by humans and dogs represents a major source of in-kind income. Wildlife, directly in the form of furs and handicrafts or indirectly in the form of bounties and bonuses, also provides an important cash income.

The economic return from wildlife vs other sources is illustrated by data from Sachs Harbour (251) and our Grise Fiord study. Usher (251) calculated that in 1966–67 the 15 Banks Island hunters earned $29,625 in kind from wild game and $217,233 from the sale of furs. The same year the community earned only $20,014 from wages, statutory payments, relief, and handicrafts. In contrast, the Grise Fiord village in 1971–72 earned $29,180 in kind from wild game and $19,618 from furs, handicrafts, and by-products from animals. Wages and carvings and handicrafts from store-purchased materials provided $78,000 for a greater financial return. These differences result from economic shifts with time and the much larger animal resource on Banks Island than on southern Ellesmere in the High Arctic.

Table 5 Consumption of fresh meat by Inuit and their dogs in five arctic communities

Location	Year	Reference	Number of people	Number of dogs	Harvest (kg/year)	Consumption	
						kg fresh meat per person/year	kg fresh meat per dog/year
Point Hope	(1960–61)	(75)	287	340	269,000	825-adult 551-children	407
Southampton Island	(1961)	(77)	209	400	187,334	136.6	152
Sachs Harbour	(1966–67)	(251)	73	145		144.8	233
Lake Harbour, Baffin Island	(1967–68)	(136)	28	34		261[a]	218[a]
Grise Fiord	(1971–72)	(201)	96	50	108,870	196	387.6

[a]Originally given in kilocalories.

The current trend to replace dog teams with snowmobiles is having a profound effect upon native life style (92, 201, 252). This switch is accelerating the change from a land-based economy to a wage economy. The overhead of owning and operating a snowmobile often surpasses the monetary income derived from the land, which forces the Inuit to subsidize their hunting activities. Since snowmobiles often reduce the time necessary to secure game, the hunters are available for wage employment or can seek other cash incomes to maintain their snowmobiles. In the Eastern Canadian Arctic the adoption of these machines is turning the once-independent and self-reliant hunters into menial laborers who are free to hunt only on weekends in order to satisfy their basic protein and fat nutritional requirements.

A dog team in Grise Fiord costs about $12 per year in equipment depreciation, and since most hunting was for skins and meat for humans, the dogs were fed surplus and thus free meat. In contrast, snowmobiles cost $1,075 per year in depreciation and operating costs. Snowmobiles have raised the Grise Fiord hunter's average annual hunting expenditures on gear from $572 in 1965–67 when dogs were still solely in use to $1,846 in 1969–72 when snowmobiles were used exclusively. Hume (114) reported that hunting is no longer economically feasible in Pangnirtung, Baffin Island due to the costs of owning and operating snowmobiles in that region. In contrast, Usher (252) pointed out that snowmobiles, if properly used, could make trapping both easier and more profitable for the Bankslanders. This undoubtedly is due to the extremely productive arctic fox population in that region which is the basis of their cash economy, an economy atypical in the Arctic.

Snowmobiles, airplanes, and modern hunting and fishing equipment have enabled northern peoples to harvest a greater number of mammals, birds, and fish. This plus rapidly expanding human populations is placing a greater strain on these terrestrial and aquatic systems (to produce an adequate amount of meat), especially fresh water systems that have a slow production rate.

In conclusion, the Inuit in the precontact period were well adapted to the arctic environment and subsisted on wildlife. The energy they harvested was channeled primarily into people and their dogs; lesser amounts passed directly to local scavengers and decomposers. The role of this top carnivore in arctic ecosystems has changed considerably since the 19th century with the ever-increasing input of energy from the south. In return energy has been exported south in the form of skins and handicrafts. The effects of these activities, particularly the economic impact of snowmobiles and the increased efficiency of animal harvest with their use, have yet to be fully understood and measured. Such research is urgently needed as a basis for sound management policies and rational land-use regulations.

INDUSTRIAL DEVELOPMENT

As discussed earlier, arctic ecosystems provided the Inuit with a reasonable wildlife food base in pre-European times. With increased technology and desire for manufactured goods, the economic and ecosystem energy base of these people has been upset. While northern people become more dependent upon southern goods, people from the south look upon northern lands for their heavy metals, oil, and gas. The basic ecologic question is whether development can occur without reducing marine, freshwater, and terrestrial systems to lower environmental and functional levels such as occurred in other biomes. We seek management strategies (123) based upon sound knowledge of ecosystem structure and function. To this end large sums of federal and industrial monies have been invested to establish the limits of acceptable ecosystem modification (8, 28, 30, 118, 119, 150, 192).

While summer seismic and off-road vehicle activity in the mid-1960s proved harmful to the terrain (102, 107, 197), current winter seismic practices and winter snow and snow-ice road utilization have minimized damage (28, 29, 102, 137) in the exploration phase. Terrain scars, both natural and man-made, are quite rapidly invaded by native species (102). This led to numerous studies of revegetation using native and northern adapted grasses (103) for the proposed Alyeska oil and Alaska-Mackenzie Valley gas pipelines. Additional studies have been conducted on the effects of oil spills on native vegetation (168, 258) and the rate of oil decomposition (84). Results show that lichens and mosses have a low survival, most shrubs resprout from latent buds, *Picea mariana* has a delayed response, and wetland sedge tundra recovers more rapidly than upland cottongrass and shrub tundra. Plants respond in a similar manner to diesel fuel in the High Arctic (11). In the Low Arctic oil spills increased bacterial populations and soil respiration (84). Soil moisture and temperature were important factors influencing degradation of crude oil by microorganisms.

Tundra fires, while not common, do occur, and with northern development will probably increase. Net primary production of burned cottongrass-dwarf heath shrub communities approached that of controls in two years (259). Where the organic mat burned, the active layer was 15–25% thicker by autumn. In tundra and woodland tundra, major environmental concerns are the potential for ice-rich permafrost melt, slumpage and sheet erosion, loss of nutrients from the system, and loss of winter forage for caribou and moose.

In the Polar Semi-Deserts and Polar Deserts, sheet and gully erosion during spring melt are common mass-wasting features because of poor cementation of soil particles and low plant cover. Erosion can therefore be a serious problem in maintaining airstrips and roads (28). Wet silty-clay soils also limit summer activity in many areas.

A detailed energy budget study has shown the importance of a 5–10 cm layer of peat (90). Albedo decreased and net radiation increased on a partially rutted winter road and on oil spill and summer burn plots. While portions of this energy were dissipated as sensible and latent heat, soil heat flux increased, especially with exposed mineral soil on the winter road and least on the oil plot. The thin peat layer delays soil thaw because of its high water content (90) and it contains more of the nutrient reserve than the standing crop of vegetation and the litter (122).

The major wildlife concerns focus on potential disruption to migrating caribou with an elevated oil pipeline (gas pipelines will be buried and the gas refrigerated), though animal response to a simulated bermed pipe at Prudhoe Bay is inconclusive (51). With proper route selection and seasonal timing of construction, many ecological problems can be greatly reduced (caribou migration, fish spawning, bird nesting and moulting, winter range of dall sheep, etc). Increased hunting and loss of critical habitat-denning sites for fox and wolves in gravel eskers, and winter sedge and grass meadows for muskox in the High Arctic are probably the greatest problems confronting wildlife. Large amounts of gravel will be used in pipeline construction, but amounts would be far greater if a road or railway were built. Environmental problems associated with drainage and potential barriers to animal migration with frequent passage of trucks or trains loom large.

Top carnivores deserve special attention because of their weak reproductive potential, low numbers, wide distribution, and indispensible role in the ecosystem. In contrast with herbivores, where management includes habitat as well as animal numbers, protecting individual carnivores from hunting pressure is more important (28).

TUNDRA MODELS

Although systems analysis has been applied to problems of tundra ecology within the IBP, most of the results to date are available only in IBP reports and national publications.

Static Energy and Carbon Flow Analysis

Many IBP Tundra Biome projects are preparing ecosystem energy flux diagrams. Danilov (64) gave an energy flow diagram for arthropods, birds, and mamalian carnivores on a tundra plot at the Harp IBP station in the southern Jamal, USSR. Heal (98) presented a preliminary energy flow diagram for a blanket bog at Moor House (U.K.), reproduced here as Figure 1. This shows equal production above and belowground, and a utilization of approximately 6% of the primary production by herbivores. Most of the energy fixed passes directly from plants to microflora, and a much smaller amount to other decomposers.

Figure 2 shows a preliminary energy flow diagram for the Truelove Lowland, Devon Island ecosystem (263). In terms of energy flow, everything but the primary producers and decomposers is incidental. Herbivores take about 1% of the primary production. Thus, any effects of the herbivores on primary production are related to trampling, selective grazing, and nutrient cycling. The effects of carnivores are entirely different. Even with the assumption that many of the carnivores seen on the lowland are present for only a small part of the time and receive only a fraction of their food there, predator consumption is very close to the total production of the vertebrate herbivores. Import to the lowland by sea and lake feeding birds such as loons, terns, and ducks is considered only peripherally in the analysis. Although much more than half of all bird nests are preyed upon probably no more than half of the actual avian production is lost, for more eggs are taken than large nestlings.

Primary Production Models

More effort has been expended in this aspect of tundra modelling than in any other. We consider here a range of model types, from descriptive to highly mechanistic.

Jones & Gore (126) developed simple constant-coefficient, donor controlled, three and five compartment models of energy flow through primary producers, and more

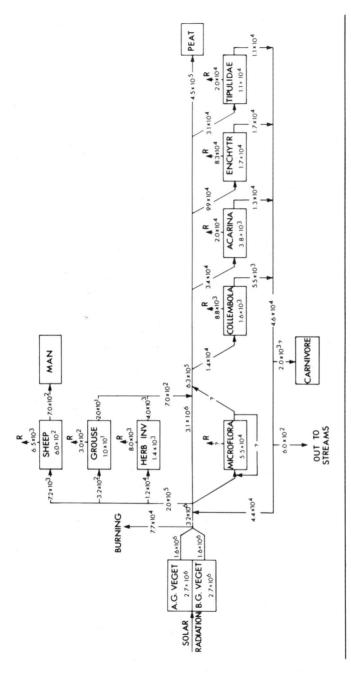

Figure 1 Standing crops and transfers of energy among compartments of a blanket bog at Moor House, England. Standing crops are in cal m^{-2} and transfers in cal m^{-2} yr^{-1}. Arrows labelled R are respiration (modified from Heal 98).

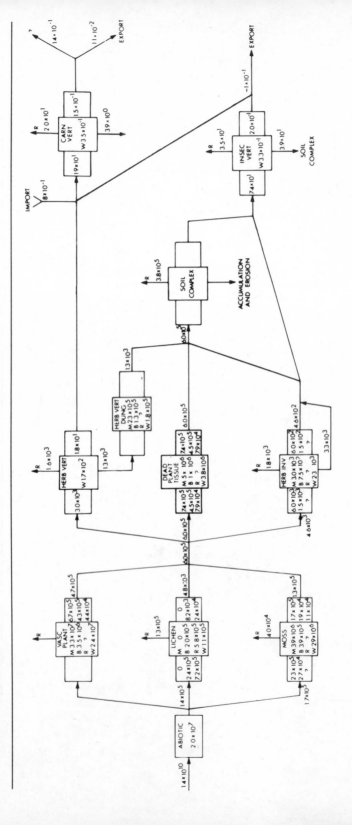

Figure 2 Energy flow diagram for the Truelove Lowland ecosystem, Devon Island, Canada. Standing crops (cal m^{-2}) are shown in the large boxes, for the meadows (*M*), beach ridges (*B*), rock outcrops (*R*), and also weighted averages for the lowland (*W*). Flow rates (cal m^{-2} yr^{-1}) are shown for each habitat division, when known, in the side boxes, and also weighted averages appear outside of the boxes. Arrows

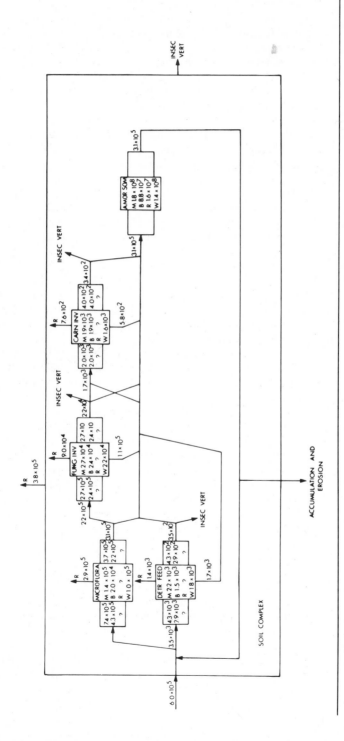

Figure 2 (Continued)

labelled *R* are respiration. The output arrow labelled with a "?" shown on the Carnivorous Vertebrate compartment is weasel production, which has an unknown disposition, perhaps into the larger carnivores (from Whitfield 263).

mechanistic models involving the effects of air temperature and incoming radiation. They expressed cautious optimism that these models might be useful for the purposes of international comparison, and as an example pointed out that their programs indicate ranges of seasonal primary production of 107–313 g/m^2 year for three tundra sites. These results are reasonably close to observed field estimates of net annual shoot production ($\sim$100–450 g/m^2). Nonlinear regression was used to derive empirical models of CO_2 exchange of several species of lichen (132), first finding the dependence on light intensity, temperature, and moisture in the laboratory, and then the dependence of these quantities on solar angle, cloud cover, meteorological station temperature, relative humidity, and time. The third step, computation of productivity from the model and environmental parameters, was not included in the paper.

A detailed process model for *Dupontia fischeri* growing at Barrow, Alaska accounts for radiation and wind profiles in the canopy, individual leaf and stem photosynthetic responses and energy budgets, and leaf water status calculated self-consistently with leaf resistance (173). Use of the model led to the conclusions that: *(a)* because of low solar angles light extinction in the canopy is significantly high despite small leaf area indices; *(b)* primary production peaks at a leaf area index of about 0.5 for horizontal leaves, but at about 1.3 for inclined leaves; *(c)* in view of the individual leaf photosynthesis-temperature curve, the low air temperatures seem to limit the plants to a production lower than their inherent capacity; *(d)* optimal plant shape, when water supply is ample, consists of narrow, steeply inclined leaves of high leaf area index, whereas when water is limiting, the leaves should be broad and horizontal, and the leaf area index low, a scheme which seems to be followed by the plants; *(e)* periodic reduction of standing dead material by lemmings should increase production; and *(f)* in agreement with field observations on grassy tundra species, leaf temperatures are consistently about 0.5°C below air temperature.

The effect of vegetation cover on depth of soil thaw is the subject of a report by Miller et al (172). They drove a physical model of soil heat conduction (177), with a soil surface energy budget model in which the effects of vegetation on turbulent transport, short-wave radiation, and the long-wave radiation regime (both sky radiation and long-wave emission by the vegetation) were incorporated. It proved necessary to add the effect of radiation emitted from the ground and reflected back by the canopy. A four-times increase in leaf area index decreased surface temperature 2°C, and, as even a 0.1° bias in surface temperature noticeably altered thaw depth, this was significant. They suggest that the late season plateau in depth of thaw is due partly to increased leaf area.

Ecosystem Models

A model of lichen-reindeer interaction represented the effects of specified stocking levels of the reindeer on lichen dynamics rather than incorporating reindeer population dynamics (45). At the same time, it provided as output the amount of supplementary feed which would be required by the animals. The lichen dynamics section incorporated the effects of tree cover, precipitation, temperature, canopy density, and competition between horn and reindeer lichens. The model demonstrated that at reasonable animal densities less supplemental feed is needed ultimately if animals are removed from the range when they have reduced lichen standing crop to 75 g/m² instead of waiting until it has reached 25 g/m². Simulating the effects of a burn indicated that reindeer lichen does not attain more than 20 g/m² until 75 years after the burn.

The problem of explaining large-amplitude density fluctuations of lemmings (187, 211) has been tackled through systems analysis. Collier (56) modelled the role of pomarine jaegers in the lemming cycle, showing that the jaeger population can have a significant effect on lemming numbers. A preliminary whole-ecosystem model for the tundra near Barrow, Alaska (248) paid particular attention to lemming interactions with vegetation, jaegers, and weasels. Eliminating the weasel population eliminated the periodic lemming crashes, and resulted in overgrazing and a reduction in plant biomass to a level able to support only a low lemming density. The authors note several significant points at which the model output fails to correspond with reality, including too short a lemming cycle, decline at the wrong time of the year, and incorrect plant biomass behavior.

The most conceptually complete lemming dynamics model to date has coupled a state-oriented compartment model concerned with primary production and decomposition with a lemming population model based upon life table projection (43). After looking at the role of avian predation, weasel predation (including migration of both lemmings and weasels), density dependent migration rates, predator-prey type cycles between the lemmings and their forage, and density dependence of litter size, Bunnell (43) concluded that the "pattern which most closely approximates empirical observations was obtained by maintaining the effects of forage quantity, predation, and density dependence while allowing levels of phosphorus and calcium in the forage to increase the probability that a female would conceive."

The results of a decomposition and nutrient flux model for Barrow, Alaska (44) indicate: *(a)* both by cutting standing vegetation and by gathering and releasing nutrients, lemmings play an important role in decomposition and nutrient flux; *(b)* lemmings modify primary production more by delaying or depressing growth early in the season than by grazing well developed vegetation; and *(c)* the ability of microorganisms to quickly ingest nutrients may partly account for observed slow recovery of plants after lemming highs.

Behavior Models

A preliminary model of habitat selection and temporal organization of shorebird feeding activity accounts for the bird's specific search image, learned preferences for above- and belowground prey and habitats, and for random variation in behavior (156). MacLean (156) pointed out that if the model adequately represents the bird's behavior, it will correctly predict both the frequencies of different habitat use and the temporal pattern of feeding. These two tests together can be used to determine the correctness of the model's conceptualization, as parameters and input cannot likely be juggled to satisfy both tests if the concepts are wrong.

From the material reviewed it is evident that a rapid progression has been made from the early stage of modelling for its own sake to its use as an investigative and · synthetic tool. More progress in this direction can be anticipated in the future along with a greater use of modelling in planning research.

ACKNOWLEDGMENTS

We express our appreciation to our students and colleagues, in Canada and abroad, who have provided data and stimulating discussions that serve as a basis for this review and for their critical review of various sections. Dr. A. F. Mark reviewed the entire manuscript. Much of this review is based upon our Devon Island IBP study which is financed by the National Research Council of Canada (NRC-IBP), the Department of Environment ALUR program within the Department of Indian Affairs and Northern Development, and the PCSP of Energy Mines and Resources. The many (more than twenty) member companies of the Arctic Petroleum Operators Association have provided both financial and logistic support for the Mackenzie Delta and arctic island's research. Use of the Devon Island Camp was provided by the Arctic Institute of North America.

Literature Cited

1. Addison, P. A. 1972. Studies on evapotranspiration and energy budgets on the Truelove Lowland, Devon Island, N.W.T. *Devon Island I.B.P. Project, High Arctic Ecosystem. Project Report 1970 and 1971,* ed. L. C. Bliss, 73–88. Edmonton, Alberta: Dept. Bot., Univ. Alberta. 413 pp.
2. Addison, P. A. 1973. *Studies on evapotranspiration and energy budgets in the high arctic: a comparison of hydric and xeric microenvironments on Devon Island, N.W.T.* M.Sc. thesis. Laurentian Univ. Sudbury, Ontario. 119 pp.
3. Ahrnsbrak, W. F. 1968. *Summertime Radiation Balance and Energy Budget of the Canadian Tundra.* Tech. Rep. No. 37. Madison, Wisc.: Univ. Wisc. Dept. Meteor.
4. Aleksandrova, V. D. 1958. An attempt to measure the over- and underground mass of tundra plants. *Bot. Zh. Leningrad* 43:1748–61
5. Aleksandrova, V. D. 1970. Vegetation and primary productivity in the Soviet Subarctic. *Productivity and Conservation in Northern Circumpolar Lands,* ed. W. A. Fuller, P. G. Kevan, 93–114. Morges, Switzerland: IUCN N. Ser. No. 16. 344 pp.
6. Alexander, V., Schell, D. M. 1972. Seasonal and spatial variation of nitrogen fixation in the Barrow, Alaska tundra. *Arctic Alp. Res.* 5:77–88
7. Andreev, V. N. 1966. Peculiarities of zonal distribution of the aerial and underground phytomass on the Eastern European Far North. *Bot. Zh. Leningrad* 51:1401–11
8. 1972. *Arctic oil pipe line feasibility study.* Calgary, Alberta: Mackenzie Valley Pipe Line Research Ltd. 109 pp.
9. Aristovskaya, A. T. V., Parinkina, O. M. 1972. Preliminary results of the I.B.P studies of soil microbiology in tundra. *Proc. Int. Meet. Biol. Prod. Tundra, 4th,* ed. F. E. Wielgolaski, T. Rosswall, 80–92. Stockholm: Tundra Biome Steering Committee. 320 pp.
10. Arnborg, L., Walker, H. J., Peippo, J. 1967. Suspended load in the Colville River, Alaska. *Geogr. Ann. A* 49:131–44
11. Babb, T. A., Bliss, L. C. 1973. Effects of physical disturbance on arctic vegetation in the Queen Elizabeth Islands. *J. Appl. Ecol.* In press
12. Bachmitov, V. A. 1969. On the question of temperature dependence of bodies on their activity. *Proc. Pop. Ecol. Lab. Vert. 3, Acad. Sci. USSR,* ed. V. N. Bolschakov. Ural Branch, Sverdlovsk
13. Baker, J. H. 1972. The rate of production and decomposition of *Chorisodontium aciphyllum* (Hook, F. & Wils.) Broth. *Brit. Antarctic Surv. Bull.* 27:123–9
14. Balikci, A. 1964. Development of basic socio-economic units in two Eskimo communities. *Nat. Mus. Can. Bull.* 202: *Anthropol. Ser.* No. 69. 114 pp.
15. Banfield, A. W. F. 1972. Are arctic ecosystems really fragile? *Int. Reindeer-Caribou Symp., 1st.* Fairbanks, Alaska: Inst. Arctic Biol. In press
16. Bank, T. P. 1958. The Aleuts. *Sci. Am.* 199:112–20

17. Bannikov, A. G. 1970. Arctic fox in the USSR: Biological premises of productivity. See Ref. 5, 121–32
18. Benoit, R. E., Campbell, W. B., Harris, R. W. 1972. Decomposition of organic matter in the wet meadow tundra. Barrow; a revised word model. *Proc. 1972 Tundra Biome Symp.*, ed. S. Bowen, 111–15. Hanover, New Hampshire: CRREL. 211 pp.
19. Beschel, R. E. 1970. The diversity of tundra vegetation. See Ref. 5, 85–92
20. Bierle, D. A. 1972. Production and energetics of chironomid larvae in ponds of the arctic coastal tundra. See Ref. 18, 182–6
21. Billings, W. D., Mooney, H. A. 1968. The ecology of arctic and alpine plants. *Biol. Rev.* 43:481–529
22. Bliss, L. C. 1956. A comparison of plant development in microenvironments of arctic and alpine tundras. *Ecol. Monogr.* 26:303–37
23. Bliss, L. C. 1962. Adaptations of arctic and alpine plants to environmental conditions. *Arctic* 15:117–44
24. Bliss, L. C. 1966. Plant productivity in alpine microenvironments on Mt. Washington, New Hampshire. *Ecol. Monogr.* 36:125–55
25. Bliss, L. C. 1970. Primary production within arctic tundra ecosystems. See Ref. 5, 77–85
26. Bliss, L. C. 1971. Arctic and alpine plant life cycles. *Ann. Rev. Ecol. Syst.* 2:405–38
27. Bliss, L. C. 1973. Tundra grasslands, herblands, and shrublands and the role of herbivores. *Grassland Ecology: A Symposium*, ed. R. H. Kesel. Baton Rouge: Louisiana State Univ. Press. In press
28. Bliss, L. C., Peterson, E. B. 1973. The ecologial impact of northern petroleum development. *Arctic Oil and Gas: Problems and Possibilities.* Fifth Int. Cong. Paris: Fondation Francaise d'Etudes Nordiques. In press
29. Bliss, L. C., Wein, R. W. 1972. Plant community responses to disturbances in the western Canadian Arctic. *Can. J. Bot.* 50:1097–1109
30. Bliss, L. C., Wein, R. W. 1973. Ecological problems associated with arctic oil and gas development. See Ref. 150, 65–77
31. Bohnsack, K. K. 1968. The distribution and abundance of the tundra arthropods in the vicinity of Pt. Barrow, Alaska. Final report to Arctic Institute of North America; subcontracts: ONR-308, ONR-321. Washington DC
32. Boyd, W. L. 1967. Ecology and physiology of soil microorganisms in polar regions. *Proceedings of the Symposium on Pacific Antarctic Sciences,* 265–275. Tokyo, Japan: JARE Scientific Reports. Special Issue No. 1
33. Boyd, W. L., Boyd, J. W. 1971. Study of soil microorganisms. Inuvik, N.W.T. *Arctic* 24:162–76
34. Boyd, W. L., Staley, J. T., Boyd, J. W. 1966. Ecology of soil microorganisms of Antarctica. *Antarctic Res. Ser.* 8: 125–59
35. Britton, M. 1957. Vegetation of the arctic tundra. *Arctic Biology,* ed. H. P. Hansen, 22–61. Corvallis: Oregon State Univ. Press. 318 pp.
36. Brown, J. 1966. Soils of the Okpilak River region, Alaska. *Res. Rep. U.S. Army Material Command.* Hanover, N.H.: CRREL No. 188. 49 pp.
37. Brown, J. 1970. Environmental setting, Barrow Alaska, section 3. Structure and function of the tundra ecosystem at Barrow, Alaska. See Ref. 5, 50–64
38. Brown, R. J. E. 1966. Influence of vegetation on permafrost. *Proc. Permafrost Int. Conf.* Reprinted. Ottawa: NRC Div. Bldg. Res., Pap. No. 298. 6 pp.
39. Brown, R. J. E. 1972. Permafrost in the Canadian Arctic Archipelago. *Z. Geogmorphol. Suppl.* 13:102–30
40. Brown, R. J. E., Johnston, G. H. 1964. Permafrost and related engineering problems. *Endeavour* 23:66–72
41. Bruemmer, F. 1972. Dovekies. *The Beaver* 303:40–47
42. Bryson, R. A. 1966. Airmasses, streamlines and the boreal forest. *Geogr. Bull.* 8:228–69
43. Bunnell, F. 1972. Lemmings—models and the real world. *Summer Computer Simulation Conf. San Diego, Calif. June 14–16.* p. 1183–93, 1203–4
44. Bunnell, F. 1972. Modeling decomposition and nutrient flux. See Ref. 18, 116–20
45. Bunnell, F., Karenlampi, L., Russell, D. E. 1973. A simulation model of lichen—*Rangifer* interactions in northern Finland. *Rep. Kevo Subarctic Res. Stat.* 10: In press
46. Bunt, J. S. 1965. Observations on the fungi of Macquarie Island. *ANARE, Sci. Reps. Ser. B* 11(78)
47. Burch, E. S. Jr. 1972. The caribou/wild reindeer as a human resource. *Am. Antiquity* 37:339–68
48. Cameron, R. E., King, J., David, C. N. 1970. Microbiology, ecology and microclimatology of soil sites in dry valleys of Southern Victoria Land, Antarctica.

Antarctic Ecology, ed. M. W. Holdgate, 2:702–16. New York: Academic. 391 pp.

49. Charlier, R. H. 1969. The geographical distribution of polar desert soils in the Northern Hemisphere. *Geol. Soc. Am. Bull.* 80:1985–96

50. Chernov, Y. I. 1961. A preliminary study of the faunal population of soils of the arctic tundra of Yakutsk. *Zool. Zh.* 43:326–33

51. Child, K. N. 1972. A study of the reaction of caribou to various types of simulated pipelines at Prudhoe Bay, Alaska. See Ref. 15

52. Chimielewski, W. 1971. The mites (*acarina*) found on bumble bees (*Bombus*) and in their nests. *Ekol. Polska.* 19:58–71

53. Churchill, E. D., Hanson, H. C. 1958. The concept of climax in arctic and alpine vegetation. *Bot. Rev.* 24:127–91

54. Clark, K. R. F. 1968. Food habits and behavior of the tundra wolf on central Baffin Island. PhD thesis. Univ. of Toronto, Toronto. 223 pp.

55. Coady, J. W. 1971. Seasonal and annual energetics of the brown lemming (*Lemmus trimucronatus*) near Barrow, Alaska. *Proc. Alaska Sci. Conf.,* p. 92. College, Alaska: Alaska Div. AAAS

56. Collier, B. D. 1972. A simulation model of the role of pomarine jaegers in Lemming population dynamics. See Ref. 18, 145–8

57. Cook, F. A. 1967. Fluvial processes in the High Arctic. *Geogr. Bull.* 9:262–8

58. Courtin, G. M. 1968. *Evapotranspiration and energy budgets of the two alpine microenvironments, Mt. Washington, New Hampshire.* PhD thesis. Univ. Illinois, Urbana. 172 pp.

59. Courtin, G. M. 1972. Micrometeorological studies of the Truelove Lowland. See Ref. 1, 46–72

60. Coyne, P. I., Kelley, J. J. 1972. CO_2 exchange in the Alaskan tundra: Meteorological assessment by the aerodynamic method. See Ref. 18, 36–9

61. Cruikshank, J. G. 1971. Soils and terrain units around Resolute, Cornwallis Island. *Arctic* 24:195–209

62. Custer, T. W., Pitelka, F. 1972. Time-activity patterns and energy budget of nesting Lapland longspurs (*Calcarius lapponicus*) near Barrow, Alaska. See Ref. 18, 160–4

63. Dadykin, V. P. 1954. Peculiarities of plant behavior in cold soils. *Vop. Bot.* 2:455–72

64. Danilov, N. N. 1972. Birds and arthropods in the tundra biogeocenosis. See Ref. 9, 117–21

65. Davies, I. 1972. A feeding study of some arctic Chironomidae. *Char Lake Project PF-2. Ann. Rep. 1971–72,* ed. F. H. Rigler, 52–9. Toronto: Dept. Zool. Univ. Toronto. 85 pp.

66. Dennis, J., Johnson, P. L. 1970. Shoot and rhizome-root standing crops of tundra vegetation at Barrow, Alaska. *Arctic Alp. Res.* 2:253–66

67. Dingman, S. L. 1966. Characteristics of summer runoff from a small watershed in central Alaska. *Water Resour. Res.* 2:751–4

68. Dorf, E. 1960. Climatic changes of the past and present. *Am. Sci.* 48:341–64

69. Dorogostaiskaya, E. V. 1972. *Weeds of the Far North of the USSR.* Leningrad: Leningrad Sci. Publ. House. 172 pp.

70. Dorsey, H. G. 1951. Arctic meteorology. *Compendium of Meteorology,* ed. T. F. Malone, 942–51. Boston: Am. Meteor. Soc. 1334 pp.

71. Douglas, L. A., Tedrow, J. C. F. 1960. Tundra soils of Arctic Alaska. *Trans. Int. Cong. Soil Sci. 7th, 1960.* 4: 2447–53

72. Dunbar, M. J. 1970. The scientific importance of the circumpolar region and its flora and fauna. See Ref. 5, 71–7

73. Flanagan, P. W. 1971. Decomposition and fungal populations in tundra regions. *The Structure and Function of the Tundra Ecosystem,* ed. S. Bowen, 2:150–5. Hanover, N. H.: CRREL. 282 pp.

74. Foote, D. C. 1964. American whalemen in northwestern Arctic Alaska. *Symp. Arctic Sub-Arctic, Sec. H, AAAS, Montreal.* Montreal, Can.: AAAS 9 pp.

75. Foote, D. C., Williamson, H. A. 1966. A human geographical study. *Environment of the Cape Thompson Region, Alaska,* ed. N. J. Wilimovsky, J. N. Wolfe, 1041–1107. Washington DC: U.S. A.E.C. Div. Tech. Info. PNE–481. 1250 pp.

76. Fournelle, H. J. 1967. Soil and water bacteria in the Alaskan sub-arctic tundra. *Arctic* 20:140–13

77. Freeman, M. M. R. 1969–70. Studies in maritime hunting I. Ecologic and technologic restraints on walrus hunting, Southampton Island, N.W.T. *Folk.* 11–12:155–71

78. Freeman, M. M. R. 1971. A social and ecological analysis of systematic female infanticide among the Netsilik Eskimo. *Am. Anthropol.* 73:1011–18

79. Frenzel, B. 1968. The Pleistocene vegetation of northern Eurasia. *Science* 161:637–49

80. Gessaman, J. A. 1972. Bioenergetics of the snowy owl (*Nyctea scandiaca*). *Arctic Alp. Res.* 4:223–38

81. Giterman, R. E., Golubera, L. V. 1967. Vegetation of eastern Siberia during the Anthropogene Period. *The Bering Land Bridge*, ed. D. M. Hopkins, 232–244. Stanford, Calif: Stanford Univ. Press. 495 pp.

82. Goodfellow, M., Hill, I. R., Gray, T. R. G. 1968. Bacteria in a pine forest soil. *The Ecology of Soil Bacteria*, ed. T. R. G. Gray, D. Parkinson, 500–15. Toronto: Univ. Toronto Press. 681 pp.

83. Gorodkov, B. N. 1939. Peculiarities of the arctic top soil. *Izv. Gosud. Geogr. Abshch.* 71:1516–32

84. Gossen, R., Parkinson, D. 1973. The effect of crude oil spills on the microbial population of soils from the Mackenzie Delta, N.W.T. Unpublished

85. Graburn, N. H. H. 1969. *Eskimos Without Igloos: Social and Economic Development in Sugluk*. Boston: Little, Brown. 244 pp.

86. Gray, D. R. 1970. The killing of a bull muskox by a single wolf. *Arctic* 23:197–8

87. Guthrie, R. D. 1968. Paleoecology of the large-mammal community in interior Alaska during the late Pleistocene. *Am. Midl. Natur.* 79:346–63

88. Guthrie, R. D., Matthews, J. V. Jr. 1971. The Cape Deceit fauna-early Pleistocene mammalian assemblage from the Alaskan arctic. *Quaternary Res.* 1:474–510

89. Haag, R. W. 1973. Mineral nutrition and primary production in native tundra communities of the Mackenzie Delta Region, N.W.T. M.Sc. thesis. Univ. Alberta, Edmonton. 89 pp.

90. Haag, R. W., Bliss, L. C. 1973. Energy budget changes following surface disturbance to upland tundra. *J. Appl. Ecol.* In press

91. Hadley, E. B., Bliss, L. C. 1964. Energy relationships of alpine plants on Mt. Washington, New Hampshire. *Ecol. Monogr.* 34:331–57

92. Hall, E. S. Jr. 1971. The "iron dog" in northern Alaska. *Anthropologica* 13:237–54

93. Hare, F. K. 1951. Some climatological problems for the arctic and subarctic. See Ref. 70, 952–66

94. Hare, F. K. 1968. The Arctic. *Quart. J. Roy. Meteorol. Soc.* 94:439–59

95. Hare, F. K. 1969. The atmospheric circulation and arctic meteorology. *Arctic* 22:185–94

96. Hare, F. K., Ritchie, J. C. 1972. The boreal bioclimates. *Geogr. Rev.* 62:333–65

97. Heal, O. W. 1965. Observations on testate amoebae (Protozoa Rhizopoda) from Signy Island, South Orkney Islands. *Brit. Antarctic Surv. Bull.* 6:43–7

98. Heal, O. W. 1968. The IBP/PT study at Moor House, United Kingdom. *Proceedings Working Meeting on Analysis of Ecosystems: Tundra Zone*, 22–24. Oslo, Norway: Norwegian I. B. C. 87 pp.

99. Heal, O. W., Ed. 1971. *Working Meeting on Analysis of Ecosystems, Kevo, Finland*. London, IBP Cent. Office: Tundra Biome Steering Committee. 297 pp.

100. Heal, O. W., Bailey, A. D., Latter, P. M. 1967. Bacteria, fungi and protozoa in Singy Island soils compared with those from a temperate moorland. *Phil. Trans. Roy. Soc. London B* 252:191–197

101. Healey, I. N. 1967. The energy flow through a population of soil collembola. *Secondary Productivity of Terrestrial Ecosystems*, 695–708. Warsaw: Pantstwowe. 879 pp.

102. Hernandez, H. 1973. Natural plant recolonization of surficial disturbances, Tuktoyaktuk Peninsula Region, N.W.T. *Can. J. Bot.* In press

103. Hernandez, H. 1973. Revegetation studies: Norman Wells, Inuvik and Tuktoyaktuk, N.W.T. and Prudhoe Bay, Alaska. Appendix V. 123 pp. See Ref. 118

104. Hobbie, J. E. et al 1972. Carbon flux through a tundra pond ecosystem at Barrow, Alaska. *U.S. Tundra Biome Rep.* No. 72–1. Hanover, N. H.: CRREL. 29 pp.

105. Hoffman, R. A., Robinson, P. R. 1966. Changes in some endocrine glands of white-tailed deer as affected by season, sex, and age. *J. Mammal.* 47:266–80

106. Hoffmann, R. .S., Taber, R. D. 1967. Origin and history of holarctic tundra ecosystems, with special reference to their vertebrate faunas. *Arctic and Alpine Environments*, ed. W. H. Osburn, H. E. Wright, 143–70. Bloomington: Indiana Univ. Press. 308 pp.

107. Hok, J. R. 1969. *A reconnaissance of tractor trails and related phenomena on the North Slope of Alaska*. U.S. Dept. Interior. B.L.M. 66 pp.

108. Holland, G. P. 1958. Distribution patterns of northern fleas (*Siphonaptera*). *Proc. Int. Congr. Entomol., 10th,* Montreal 1:645–58

109. Hopkins, D. M., Ed. 1967. *The Bering Land Bridge.* Stanford, Calif: Stanford Univ. Press. 495 pp.

110. Hopkins, D. M., Matthews, J. V., Wolfe, J. A., Silberman, M. L. 1971. A Pleistocene flora and insect fauna from the Bering Strait region. *Paleogeogr., Paleoclimatol., Paleoecol.* 9:211–31

111. Hopkins, D. M., Sigafoos, R. S. 1950. Frost action and vegetation patterns on Seward Peninsula, Alaska. *Geol. Surv. Bull.* 974-C:51–101

112. Hubert, B. 1972. Productivity of muskox on Northeastern Devon Island, N.W.T. *Devon Island 1971 Progress Report.* Edmonton, Alberta: Dept. Bot., Univ. Alberta. 8 pp.

113. Hubert, B. 1973. Muskox utilization of the Truelove Lowland, Devon Island. *Devon Island 1972 Progress Report.* Edmonton, Alberta: Dept. Bot., Univ. Alberta. 21 pp.

114. Hume, S. 1972. Cost of gasoline may send arctic hunters back to dogs. *Edmonton Journal,* 28 Feb. p. 51

115. Hurd, P. D., Lindquist, E. E. 1958. Analysis of soil invertebrate samples from Barrow, Alaska. Final Report to Arctic Institute of North America. Washington DC: Projects ONR-173 & ONR-193

116. Ivanova, E. N. 1956. Classification of soils of the northern part of European USSR. *Pochvovedinie* No. 1:70–88

117. Ivarson, K. C. 1965. The microbiology of some permafrost soils in the Mackenzie Valley, N.W.T. *Arctic* 18:256–60

118. Interim Report No. 1. 1971. *Towards an environmental impact assessment of a gas pipeline from Prudhoe Bay, Alaska to Alberta.* Winnipeg: Environ. Protection Board. 28 pp. Also, appendices of research studies

119. Interim Report No. 3. 1973. *Towards on environmental impact assessment of the portion of the Mackenzie Gas Pipeline from Alaska to Alberta.* Winnipeg: Environ. Protection Board. 88 pp. Also, appendices of research studies

120. Irving, L., Krog, J. 1954. Body temperature of arctic and subarctic birds and mammals. *J. Appl. Physiol.* 6:667–80

121. Izvekova, E. I. 1971. On the feeding habits of Chironomid larvae. *Limnologica* 8:201–2

122. Janz, A. 1973. Topographic influence on soil and plant nutrients. *Botanical Studies of Natural and Man Modified Habitats in the Mackenzie Valley, East-ern Mackenzie Delta Region, and the Arctic Islands,* ed. L. C. Bliss, 25–49. ALUR 72–73. Dept. Indian Affairs and Northern Develop. Ottawa. 236 pp.

123. Johnson, P. L. 1969. Arctic plants, ecosystems and strategies. *Arctic* 22: 341–55

124. Johnson, P. L., Billings, W. D. 1962. The alpine vegetation of the Beartooth Plateau in relation to cryopedogenic processes and patterns. *Ecol. Monogr.* 32:105–35

125. Johnson, P. L., Kelley, J. J. 1970. Dynamics of carbon dioxide and productivity in an arctic biosphere. *Ecology* 51:73–80

126. Jones, H. E., Gore, A. J. P. 1972. Descriptive models in comparative ecosystem studies. See Ref. 9, 35–47

127. Jones, P. C. T., Mollison, J. E. 1948. A technique for the quantitative estimation of soil microorganisms. *J. Gen. Microbiol.* 2:54–69

128. Jordon, P. A., Botkin, D. B., Wolfe, M. L. 1971. Biomass dynamics in a moose population. *Ecology* 52:147–52

129. Kaiser, G. W., Lefkovitch, L. P., Howden, H. F. 1972. Faunal provinces in Canada as exemplified by mammals and birds: a mathematical consideration. *Can. J. Zool.* 50:1087–1104

130. Kaisila, J. 1967. Notes on the arthropod fauna of Spitsbergen Island. *Ann. Ent. Fenn.* 33:13–18

131. Kalela, O. 1962. On the fluctuations in the numbers of arctic and boreal small rodents as a problem of production biology. *Ann. Acad. Sci. Fenn. Ser. A4,* No. 66. 38 pp.

132. Karenlampi, L., Tammisola, J. 1970. Preliminary report on models of reindeer lichen CO_2 exchange, environment, and productivity. *IBP I Norden* 5:23–25

133. Kelley, J. J., Weaver, D. F., Smith, B. P. 1968. The variation of carbon dioxide under the snow in the Arctic. *Ecology* 49:358–61

134. Kelsall, J. P. 1953. Predation on barren-ground caribou. *Can. Wildl. Serv. Rep.* #307, Ottawa

135. Kelsall, J. P. 1968. The caribou *Can. Wildl. Ser. Monogr.* #3. Ottawa. 340 pp.

136. Kemp, W. B. 1971. The flow of energy in a hunting society. *Sci. Am.* 225: 105–15

137. Kerfoot, D. E. 1972. *Tundra disturbance studies in the western Canadian Arctic.* Ottawa: ALUR 71-72-11. Dept. Indian and Northern Affairs. 115 pp.

138. Klein, D. R. 1970. Tundra ranges north of the boreal forest. *J. Range Manage.* 23:8–14
139. Kobayasi, Y. et al 1967. Mycological studies of the Alaskan Arctic. *Ann. Rep. Inst. Fermentation,* Osaka. No. 3
140. Kobayasi, Y. et al 1969. The Second Report on the mycological flora of the Alaskan Arctic. *Bull. Nat. Sci. Mus. Tokyo* 12(2)
141. Kobayasi, Y., Turbaki, K., Soneda, M. 1968. Enumeration of the higher fungi, moulds and yeasts of Spitsbergen. *Bull. Nat. Sci. Mus. Tokyo* 11(1)
142. Köppen, W. 1931. *Grundriss der Klimakunde.* Berlin: Walter de Gruyter
143. Korovin, A. E., Sycheva, Z. F., Bystrova, A. Z. 1963. The effect of soil temperature on the amounts of various forms of phosphorus in plants. *Sov. Plant Physiol.* 10:109–12
144. Kowal, N. E., Crossley, D. A. 1971. The ingestion rates of microarthropods in pine mor, estimated with radioactive calcium. *Ecology* 52:444–452
145. Kuyt, E. 1971. Food studies of wolves on barren-ground caribou range in the Northwest territories. *Can. Wildl. Serv. Progr. Notes* #23, Ottawa. 20 pp.
146. Lantis, M. L. 1967. American arctic populations: their survival problems. See Ref. 35, 243–87
147. Latter, P. M., Cragg, J. B., Heal, O. W. 1967. Comparative studies in the microbiology of four moorland soils in the northern Pennines. *J. Ecol.* 55: 445–64
148. Lawrence, D. B., Schoenike, R. E., Quispel, A., Bond, G. 1967. The role of *Dryas drummondii* in vegetation development following ice recession at Glacier Bay, Alaska, with special reference to its nitrogen-fixation by root nodules. *J. Ecol.* 55:793–813
149. Legget, R. F. 1963. Permafrost in North America. *Proc. Permafrost Int. Conf.,* Ottawa: NRCC Tech. Pap. No. 232. 7 pp.
150. Legget, R. F., Macfarlane, I. C., Eds. 1972. *Proceedings of the Canadian Northern Pipeline Research Conference.* Ottawa: NRCC Tech. Memor. 104. 331 pp.
151. Lien, L. E., Ostbye, E., Hagen, A., Klemetsen, A., Skar, H. 1970. Quantitative bird surveys in high mountain habitats, Finse, South Norway, 1967–68. *Nytt Mag. Zool.* 18:245–51
152. Linder, D. H. 1949. Fungi. *Botany of the Canadian Eastern Arctic, Part II. Thallophyta and Bryophyta,* ed. N. Polunin, 234–297. *Nat. Mus. Can. Bull.* 97. 573 pp.

153. Mackay, J. R. 1971. The origin of massive icy beds in permafrost, Western Arctic Coast, Canada. *Can. J. Earth Sci.* 8:397–422
154. Mackay, J. R. 1972. The world of underground ice. *Ann. Assoc. Am. Geogr.* 62:1–22
155. MacLean, S. F. 1971. Population ecology and energetics of tundra soil arthropods. See Ref. 73, 126–30
156. MacLean, S. F. 1972. A simulation model to predict the habitat distribution of shorebird feeding activity. See Ref. 18, 149–55
157. MacLean, S. F., Pitelka, F. A. 1971. Seasonal patterns of abundance of tundra arthropods near Barrow. *Arctic* 24:19–40
158. Maher, W. J. 1967. Predation of weasels on a winter population of lemmings, Banks Island, Northwest Territories. *Can. Field Nat.* 81:248–50
159. Major, J. 1969. Historical development of the ecosystem concept. *The Ecosystem Concept in Natural Resource Management,* ed. G. M. van Dyne, 9–22. New York: Academic. 383 pp.
160. Matthews, J. V. 1970. Quaternary environmental history of interior Alaska: Pollen samples from organic colluvium and peats. *Arctic Alp. Res.* 2:241–51
161. Matthews, J. V. 1973. *Quaternary environments at Cape Deceit (Seward Peninsula, Alaska): A contribution towards an understanding of the evolution of tundra ecosystems.* PhD thesis. Univ. Alberta, Edmonton. 123 pp.
162. Matveyeva, M. V. 1972. The Tareya word model. See Ref. 9, 156–62
163. Mayo, J. M., Despain, D. G., van Zinderen Bakker, E. M. Jr. 1973. CO_2 assimilation by *Dryas integrifolia* on Devon Island, Northwest Territories. *Can. J. Bot.* 51:581–8
164. McAlpine, J. F. 1964. Arthropods of the bleakest barren lands: composition and distribution of the arthropod fauna of the northwestern Queen Elizabeth Island. *Can. Entomol.* 96:127–9
165. McAlpine, J. F. 1965. Insects and related terrestrial invertebrates of Ellef Ringnes Island. *Arctic* 18:73–103
166. McCann, S. B., Cogley, J. G. 1972. Hydrological observations on a small arctic catchment, Devon Island. *Can. J. Earth Sci.* 9:361–5
167. McCann, S. B., Howarth, P. J., Cogley, J. G. 1972. Fluvial processes in a periglacial environment. Queen Elizabeth Islands, N.W.T. Canada. *Inst. Brit. Geogr.* 55:69–82
168. McCown, B. H., Brown, J., Murrmann, R. P. 1971. Effect of oil seepage and

spills on the ecology and biochemistry in cold dominated environments. *Ann. Rept. U.S. Army, Cold Regions Res. Eng. Lab.* Hanover, N.H. 18 pp.

169. McEwen, E. H. 1955. A biological survey of the west coast of Banks Island, 1955. *Can. Wildl. Serv. Rep.* #26, Ottawa. 64 pp.

170. McMillan, N. J. 1960. Soils of the Queen Elizabeth Islands. *J. Soil Sci.* 11:131–9

171. Melchoir, H. R. 1972. Summer herbivory by the brown lemming at Barrow, Alaska. See Ref. 18, 136–8

172. Miller, P. C., Ng, E., Tieszen, L. L., Nakano, Y., Brown, J. 1972. A model of vegetation structure, temperature and depth of thaw of the tundra. Manuscript

173. Miller, P. C., Tieszen, L. L. 1972. A preliminary model of processes affecting primary production in the arctic tundra. *Arctic Alp. Res.* 4:1–18

174. Muc, M. 1973. Primary production of plant communities of the Truelove Lowland, Devon Island, Canada—Sedge meadows. *Proceedings, Conference on Primary Production of Tundra Ecosystems,* ed. F. E. Wielgolaski, L. C. Bliss. Edmonton, Alberta: Tundra Biome Steering Committee. In press

175. Mueller, C. H. 1952. Plant succession in arctic heath and tundra in northern Scandinavia. *Bull. Torrey Bot. Club* 79:296–309

176. Nagel-de-Boois, H. M., Jansen, E. 1966. Hyphal activity in mull and mor of an oak forest. *Progress in Soil Biology.* Braunnschweig: Verlag Friedrich

177. Nakano, Y., Brown, J. 1972. Mathematical modeling and validation of the thermal regimes in tundra soils, Barrow, Alaska. *Arctic Alp. Res.* 4:19–38

178. Norton, D. W. 1973. *Ecological energetics of Calidridine sandpipers breeding in northern Alaska.* PhD thesis. Univ. Alaska, Fairbanks. 163 pp.

179. Odum, E. P. 1969. The strategy of ecosystem development. *Science* 164:262–70

180. Packer, J. G. 1969. Polyploidy in the Canadian Arctic Archipelago. *Arctic Alp. Res.* 1:15–28

181. Pakarinen, P., Vitt, D. H. 1972. The ecology of bryophytes in Truelove Lowland, Devon Island. See Ref. 1, 185–96

182. Pattie, D. L. 1972. Preliminary bioenergetic and population level studies in high arctic birds. See Ref. 1, 281–92

183. Pattie, D. L. 1973. Bioenergetic and population level studies in high arctic birds. See Ref. 113, 15 pp.

184. Perfil'ev, B. V., Gabe, D. R. 1969. *Capillary Methods of Investigating Microorganisms.* Transl. Oliver & Boyd. Toronto: Univ. Toronto Press. 627 pp.

185. Peters, H. S. 1934. Mallophaga from birds of Southampton Island, Hudson Bay. *The Exploration of Southampton Island Hudson Bay,* ed. G. M. Sutton, 35–7. Mem Carnegie Mus. 12(2). Sect. 4, 37 pp.

186. Petrides, G. A., Swank, W. G. 1965. Population densities and range carrying capacity for large mammals in Queen Elizabeth National Park, Uganda. *Zool. Afr.* 1:209–25

187. Pitelka, F. A. 1972. Cycle pattern in lemming populations near Barrow, Alaska. See Ref. 18, 132–5

188. Polunin, N. 1948. *Botany of the Canadian Eastern Arctic, Part III. Vegetation and Ecology.* Nat. Mus. Can. Bull. 104. Ottawa. 304 pp.

189. Polunin, N. 1951. The real arctic: suggestions for its delimitation, subdivision, and characterization. *J. Ecol.* 39:308–15

190. Porsild, A. E. 1951. Plant life in the arctic. *Can. Geogr. J.* 42:120–45

191. Procter, D. 1973. Nematode production. See Ref. 113, 5 pp.

192. *Project description of the Trans Alaska Pipeline System.* 1971. Alyeska Pipeline Service Co. 64 pp.

193. Pruitt, W. O. 1959. Snow as a factor in the winter ecology of barren-ground caribou. *Arctic* 12:158–79

194. Pulliainen, E. 1972. Nutrition of the arctic hare (*Lepus timidus*) in northeastern Lapland. *Ann. Zool. Fenn.* 9:17–22

195. Rae, R. W. 1951. *Climate of the Canadian Arctic Archipelago.* Toronto: Meteorol. Div. Can. Dept. Transp. 89 pp.

196. Raup, H. M. 1951. Vegetation and cryoplanation. *Ohio J. Sci.* 51:105–16

197. Rempel, G. 1970. Arctic terrain and oil field development. See Ref. 5, 243–51

198. Richard, W. H. 1962. Comparison of annual harvest yields in an arctic and a semidesert plant community. *Ecology* 43:770–1

199. Richards, K. 1973. The biology of *Bombus polaris* Curtis and *B. hyperboreus* Schonherr at Lake Hazen, N.W.T. *Hynenoptera: Bombini).* *Quaest. Ent.* 9(2):In press

200. Riehl, H. 1965. *Introduction to the Atmosphere.* New York:McGraw. 365 pp.

201. Riewe, R. R. 1972. Preliminary data on mammalian carnivores including man

in the Jones Sound region N.W.T. See Ref. 1, 315–40
202. Rodin, L. E., Bazilevich, N. I. 1968. *Production and Mineral Cycling in Terrestrial Vegetation,* ed., Transl. G. E. Fogg. Edinburgh: Oliver and Boyd. 288 pp.
203. Romanova, E. N. 1971. Microclimate of tundras in the vicinity of the Taimyr Station. *Biogeocenoses of Taimyr Tundra and Their Productivity,* ed. B. A. Tikhomirov, 35–44. Leningrad: Nauka. 239 pp. Transl. P. Kuchar, Univ. Alberta, Edmonton, Alberta
204. Ryan, J. K. 1972. Devon Island invertebrate research. See Ref. 1, 293–314
205. Ryan, J. K. 1973. Invertebrate production. See Ref. 113, 25 pp.
206. Savile, D. B. O. 1961. Bird and mammal observations on Ellef Ringnes Island in 1960. *Natur. Hist. Pap. Natur. Mus. Can.* 9:1–6
207. Savile, D. B. O. 1972. Arctic adaptations in plants. *Can. Dept. Agr. Monogr. 6.* Ottawa. 81 pp.
208. Schapova, L. N. 1972. The daily fluctuation dynamics of bacterial number in soils of Far East region (Primorsky Krai). *Problems of Abundance, Biomass and Productivity of Microorganisms in Soil,* ed. T. R. Aristovskaya, 136–45. Leningrad: Nauka
209. Schartz, R. L., Zimmerman, J. L. 1971. The time and energy budget of the male dickcissel *(Spiza americana). Condor* 73:65–76
210. Scholander, P. F., Hock, R., Walters, V., Irving, L. 1950. Adaptation to cold in arctic and tropical mammals and birds in relation to body temperature, insulation and basal metabolic rate. *Biol. Bull.* 99:259–71
211. Schultz, A. M. 1969. A study of an ecosystem: The arctic tundra. *The Ecosystem Concept in Natural Resource Management,* ed. G. M. van Dyne, 77–93. New York: Academic. 383 pp.
212. Service, E. R. 1966. *The Hunters.* Englewood Cliffs, N.J.: Prentice-Hall. 118 pp.
213. Sher, A. V. 1969. Early Pleistocene mammals of extreme northeastern Asia and their environments (Abst.) *Congress INQUA, Resumes des Communications, VII.* 135 pp.
214. Shvetsova, V. M., Voznesenskii, V. L. 1970. Diurnal and seasonal variations in the rate of photosynthesis in some plants of Western Taimyr. *Bot. Zh.* 55:66–76. (Int. Tundra Biome Transl. #2)
215. Smirnov, V. S., Tokmakova, S. G. 1971. Preliminary data on the influence of different numbers of voles upon the forest tundra vegetation. *Ann. Zool. Fenn.* 8:154–6
216. Smirnov, V. S., Tokmakova, S. G. 1972. Influence of consumers on natural phytocenosis production variation. See Ref. 9
217. Smith, R. 1973. Feeding experiments with captive arctic hares *(Lepus arcticus monstrabilis).* See Ref. 113, 7 pp.
218. Solhøy, T. 1969. Finseomradets terrestre molluskfauna. *Fauna* 22:207–14
219. Solhøy, T. 1972. Quantitative invertebrate studies in mountain communities at Hardangervidda, South Norway. I. *Nor. Entomol. Tidsskr.* 19:99–108
220. Speller, S. W. 1972. *Food ecology and hunting behavior of denning arctic foxes at Aberdeen Lake, Northwest Territories.* PhD thesis. Univ. Saskatchewan, Saskatoon. 145 pp.
221. Speller, S. W. 1972. Biology of *Dicrostonyx groenlandicus* on Truelove Lowland, Devon Island. See Ref. 1, 257–71
222. Stebaev, I. V. 1959. Soil invertebrates of Salekhard tundras, and the change in their groups due to agriculture. *Zool. Zh.* 38:1559–72
223. Stepanova, I. V., Tomilin, B. 1972. Fungi of basic plant communities in Taimyr tundras. See Ref. 9, 193–8
224. Stutz, C. 1973. *Nitrogen fixation in a high arctic ecosystem.* PhD thesis. Univ. Alberta, Edmonton, Alberta. 57 pp.
225. Svatkov, N. M. 1958. Soils on Wrangell Island. *Pochvov* 1:91–8
226. Svensson, B. G. 1972. *An inventory of the insect fauna at Stordalen mire 1970. International Biological Programme, Swedish Tundra Biome Project.* Stockholm: Technical Report #3
227. Svoboda, J. 1972. Vascular plant productivity studies of raised beach ridges (semi-polar desert) in the Truelove Lowland. See Ref. 1, 145–84
228. Svoboda, J. 1973. Primary production of plant communities of the Truelove Lowland, Devon Island, Canada— Beach Ridges. See Ref. 174
229. Taylor, W. E. 1968. An archaeological overview of Eskimo economy. *Eskimo of the Canadian Arctic,* ed. V. F. Valentine, F. G. Vallee, 3–17. Toronto: McClelland and Steward Ltd. 241 pp.
230. Tedrow, J. C. F. 1963. Arctic soils. 50–55. *Proc. Permafrost Int. Conf.,* NAS-NRC Pub. 1287:50–55. Ottawa
231. Tedrow, J. C. F. 1966. Polar desert soils. *Soil Sci. Soc. Am. Proc.* 30:381–7

232. Tedrow, J. C. F. 1968. Pedogenic gradients of the polar regions. *J. Soil Sci.* 19:197–204

233. Tedrow, J. C. F., Brown, J. 1967. Soils of Arctic Alaska. See Ref. 106, 283–93

234. Tedrow, J. C. F., Cantlon, J. E. 1958. Concepts of soil formation and classification in arctic regions. *Arctic* 11:166–79

235. Tedrow, J. C. F., Douglas, L. A. 1964. Soil investigations on Banks Island. *Soil Sci.* 98:53–65

236. Tedrow, J. C. F., Drew, J. V., Hill, D. E., Douglas, L. A. 1958. Major genetic soils of the Arctic Slope of Alaska. *J. Soil Sci.* 9:33–45

237. Tedrow, J. C. F., Harries, H. 1960. Tundra soil in relation to vegetation, permafrost, and glaciation. *Oikos* 11:237–49

238. Teeri, J. A. 1973. Polar Desert adaptations of a High Arctic plant species. *Science* 179:496–7

239. Tener, J. S. 1963. Queen Elizabeth Islands game survey, 1961. *Can. Wildl. Serv. Occas. Pap.* 4:1–50

240. Tener, J. S. 1965. Muskoxen in Canada, a biological and taxonomic review. *Can. Wildl. Serv. Monogr. Ser.* No. 2, 166 pp.

241. Tieszen, L. L. 1972. The seasonal course of aboveground production and chlorophyll distribution in a wet arctic tundra at Barrow, Alaska. *Arctic Alp. Res.* 4:307–24.

242. Tieszen, L. L. 1972. Photosynthesis in relation to primary production. See Ref. 9, 52–62

243. Tieszen, L. L., Johnson, P.L. 1968. Pigment structure of some arctic tundra communities. *Ecology* 49:370–73

244. Tieszen, L. L., Sigurdson, D. C. 1973. Effect of temperature on carboxylase activity and stability in some Calvin cycle grasses from the Arctic. *Arctic Alp. Res.* 5:59–66

245. Tikhomirov, B. A. 1959. The interrelationship of animal life and vegetation, cover on the tundra. *Acad. Sci. USSR* Leningrad. 83 pp.

246. Tikhomirov, B. A. 1961. The changes in biogeographical boundries in the north of USSR as related with climatic fluctuations and activity of man. *Bot. Tidsskr.* 56:285–92

247. Tilbrook, P. B. 1970. The terrestrial environment and invertebrate fauna of the Maritime Antarctic. See Ref. 48, 886–96

248. Timin, M. E., Collier, B. D., Zich, J., Walters, V. 1972. A computer simulation of the arctic tundra ecosystem near Barrow, Alaska. See Ref. 18, 71–9

249. Timonin, M. 1935. The microfungi in profiles of certain virgin soils in Manitoba. *Can. J. Res. Sect. C* 13:42–46

250. Underwood, L. S. 1971. *The bioenergetics of the arctic fox.* PhD thesis. Pennsylvania State Univ., State College. 85 pp.

251. Usher, P. J. 1971. *The Bankslanders: Economy of a Frontier Trapping Community. Vol. 2. Economy and Ecology.* Ottawa: Can. Dept. Indian Affairs and Northern Develop., Nor. Sci. Res. Group. 169 pp.

252. Usher, P. J. 1972. The use of snowmobiles for trapping on Banks Island. *Arctic* 25:171–81

253. Vibe, C. 1970. The arctic ecosystem influenced by fluctuations in sun spot and drift ice movement. See Ref. 5, 115–20

254. Vowinckel, E. 1966. The surface heat budgets at Ottawa and Resolute, N.W.T. *Meteorology,* No. 84, McGill Univ. Montreal, Quebec

255. Wang, L. C. H., Jones, D. L., MacArthur, R. A., Fuller, W. A. 1973. Adaptation to cold: energy metabolism in a typical lagomorph the arctic hare *(Lepus arcticus). Can. J. Zool.* In press

256. Warren Wilson, J. 1957. Observations on the temperatures of arctic plants and their environment. *J. Ecol.* 45:491–531

257. Warren Wilson, J. 1959. Notes on wind and its effects in arctic-alpine vegetation. *J. Ecol.* 47:415–27

258. Wein, R. W., Bliss, L. C. 1973. Experimental crude oil spills on arctic plant communities. *J. Appl. Ecol.* In press

259. Wein, R. W., Bliss, L. C. 1973. Changes in arctic *Eriophorum* tussock communities following fire. *Ecology. In press*

260. Welch, H. 1972. Chironomid energetics. See Ref. 65, 41–7

261. Weller, G., Cubley, S. 1972. The microclimates of the arctic tundra. See Ref. 18, 5–12

262. White, R. G., Luick, J. R. 1972. Aspects of feed intake and heat production of *Rangifer* in the arctic tundra. *Tundra Biome Newsletter Fall 1971—Winter 1972.* #7, p. 7

263. Whitfield, D. W. A. 1972. Systems analysis. See Ref. 1, 392–409

264. Widden, P., Newell, T., Parkinson, D. 1972. Decomposition and microbial populations of Truelove Lowland. Devon Island. See Ref. 1, 341–58

265. Wielgolaski, F. E. 1972. Vegetation types and plant biomass in tundra. *Arctic Alp. Res.* 4:291–305

266. Wielgolaski, F. E., Roswall, T., Eds. 1972. *I.B.P. Tundra Biome Proceedings IV. International Meeting on the Biological Productivity of Tundra.* Stockholm, Sweden: Tundra Biome Steering Committee. 320 pp.

267. Wilkenson, P. 1971. The domestication of the muskox. *The Polar Record* 16: 683–90

268. Williamson, F. S. L., Thompson, M. C., Hinds, J. Q. 1966. Avifaunal investigations. See Ref. 75, 437–80

269. Wolforth, J. 1966. *The Mackenzie Delta: Its Economic Base and Development.* Ottawa: Can.ᐧ Dept. Indian Affairs and Northern Develop., Nor. Coordination and Res. Centre. MDRPI. 85 pp.

270. Wolforth, J. 1971. *The Evolution and Economy of the Delta Community,* Ottawa: Can. Dept. Indian Affairs and Northern Develop., Nor. Coordination and Res. Centre. MDRPII. 163 pp.

271. Younkin, W. 1973. Autecological studies of native species potentially useful for revegetation, Tuktoyaktuk Region, N. W. T. See Ref. 122, 50–96

272. Yurtsev, B. A. 1966. American-Asian steppe relations and the question of the ancient continental element in the Arctic and Alpine floras of northeastern Siberia. *Probl. Bot.* 8:60–8

273. Yurtsev, B. A. 1972. Phytogeography of northeastern Asia and the problems of Transberingian floristic interrelations. *Floristics and Paleofloristics of Asia and Eastern North America,* ed. A. Graham, 19–54. Amsterdam: Elsevier. 278 pp.

274. Zalenskij, O. V., Shvetsova, V. M., Voznesenskii, V. L. 1972. Photosynthesis in some plants of Western Taimyr. See Ref. 9, 182–6

REPRINTS

The conspicuous number aligned in the margin with the title of each article in this volume is a key for use in ordering reprints.

Available reprints are priced at the uniform rate of $1 each postpaid. Payment must accompany orders less than $10. A discount of 20% will be given on orders of 20 or more. For orders of 200 or more, any Annual Reviews article will be specially printed.

The sale of reprints of articles published in the Reviews has been expanded in the belief that reprints as individual copies, as sets covering stated topics, and in quantity for classroom use will have a special appeal to students and teachers.

AUTHOR INDEX

403

G

Gabe, D. R., 377
Galbraith, I. C., 202
GALLUCCI, V. F., 329-57;
345
Galun, M., 43
Gandin, L. S., 303, 308, 319,
320
Garattini, S., 118
Gardner, H. R., 36
Gardner, J. M., 281, 287-
89
Gardner, R. H., 125
Gasser, D. L., 94
Gates, D. M., 337-41
Gatlin, L. L., 350
Gatschet, A. S., 263, 264
Gauldie, J., 94
Gentry, J., 232
Gentry, J. B., 76, 77, 80, 82,
169
George, W., 143
Georghiou, G. P., 175
Gerloff, R. K., 287, 289
Gershowitz, H., 79
Gessaman, J. A., 375
Getz, L. L., 124
Gibbons, A. M., 289
Gibbs, D. G., 232
Giblett, E. R., 79, 102
Gibson, J. B., 98
Gibson, K. D., 289
Giesel, J. T., 77
Gilardi, E., 289, 292
Gilbert, N., 17, 20
Gillaspy, J. E., 98
Gillespie, D., 275, 278,
279
Gillespie, J. H., 77, 96, 99-
103
Gillespie, S., 279
Gillis, M., 285
Gillman, J., 156
Giterman, R. E., 360
Glansdorff, P., 332, 336,
347
Glendening, G., 9
Gluck, J. P., 124
Goad, W. B., 94
Goel, N. S., 353
Goldman, B. J., 25
Golley, F. B., 232, 343
Golubera, L. V., 360
Gooch, J. L., 97, 99, 169
Goodall, D. W., 32
Goodall, W., 303
Goodfellow, M., 376
Goodge, W. R., 202
Goodman, H. M., 286
Goodman, M., 86
Goosen-De Loo, L., 251
Gore, A. J. P., 384
Gorman, G. C., 56, 57, 75,
180, 216
Gorodkov, B. N., 365
Gorski, F., 347

Gösswald, K., 232, 238
Gotwald, W. H. Jr., 234,
240
Gotz, W., 168
Gould, S. J., 214, 226
Graburn, N. H. H., 380
Grace, J., 308
Grant, P. R., 181, 190,
194, 204, 214, 217, 219,
221, 223-25
Gray, D. R., 374
Gray, T. R. G., 376
Greaves, T., 245
Green, C. R., 33
Green, R. H., 217
Greene, E. L., 263, 265
Greenslade, P., 233
Greenslade, P. J. M., 233,
245, 247
Greenwald, G. S., 153
Greenway, H., 38
Gregory, K. F., 279,
289
Greville, G. D., 99
Griffin, R. A., 42
Griffith, J. S., 350
Griffiths, K. J., 6, 13
Grinnell, J., 207
Gross, W. M., 289, 294
Guerry, P., 284
Guest, W. C., 170, 172,
173
Guhl, A. M., 126
Guinness, F., 146, 149
Gulamhusein, A. P., 145
Gunsalus, I. C., 289
Guthrie, R. D., 167, 360,
372
Gutterman, Y., 33, 46

H

Haag, R. W., 360, 363, 364,
367, 370, 371, 383
Haapala, D. K., 279, 289
Hadáňová, D., 302
Hadley, E. B., 371
Hadley, N. F., 341
Hagen, A., 375
Hagen, D. W., 206
Haigh, J., 96
Haines, R. W., 47
Hainsworth, F. R., 342
Haizel, K. A., 307
Haldane, J. B. S., 175
Hale, C. S., 126
Hall, A. E., 302, 314
Hall, B. D., 275-77, 279
Hall, B. P., 202
Hall, E. A., 46
Hall, E. S. Jr., 382
Hall, H. H., 94
Hall, W. P., 83, 85
Hallberg, B., 95
Halliday, R. B., 100
Hamby, W., 119
Hamlett, G. W. D., 156

Hanabusa, K., 95
Hanaoko, M., 287, 289
Hanau, R., 303, 312, 318,
319
Hanks, J., 154
Hanks, R. J., 35, 36, 38
Hansen, P. A., 286, 289
Hanson, H. C., 361, 366
Hare, F. K., 362, 363
Harley, J. P., 119, 120
Harlow, H. F., 124
Harmeson, J. C., 214
Harper, J. L., 214
Harries, H., 361
Harrington, J. P., 262
Harris, H., 75, 96, 99,
104
Harris, R. D., 194
Harris, R. W., 379
Harris, S. K., 265
Haskins, C. P., 239, 247,
249
Haskins, E. F., 239, 247,
249
Hasler, A. D., 7
Hatiya, K., 302
Hay, M. F., 146
Hayashi, K., 315
Hayes, W. P., 235
Hazen, W. E., 343
Hazlett, B. A., 119
Heal, O. W., 359, 376-78,
384, 385
Healey, I. N., 378
Healey, M. C., 120, 122,
128, 130, 131
Heatwole, H., 56, 58, 214
Heberlein, G. T., 287, 289,
291, 294
Hedrick, P., 106
Heed, W. B., 97, 98, 168
Heinrich, B., 342
Heiple, K. G., 87
Herbert, J., 146
Herman, S. S., 16
Hermann, H. R. Jr., 240
Hernandez, H., 383
Herndon, S. E., 287, 289
Hesketh, J. D., 309, 314
HESPENHEIDE, H. A., 213-
29; 189, 214-18, 220, 221,
223, 225, 226
Heyman, T., 279
Hibble, J., 232, 247, 250
Hiebert, M., 94
Hill, D. E., 365
Hill, I. R., 376
Hill, J. L., 128, 131
Hill, L. R., 289, 292
Hillcoat, B. L., 94
Hillel, D., 36
Hills, E. S., 25
Hinde, R. A., 117
Hinds, J. Q., 375
Hirsch, P., 289
Hitchon, D. E., 241
Hoar, W. S., 152

SUBJECT INDEX

414

CUMULATIVE INDEXES

CONTRIBUTING AUTHORS VOLUMES 1-4

CHAPTER TITLES VOLUMES 1-4

423